The Cereal Rusts

VOLUME II

DISEASES, DISTRIBUTION, EPIDEMIOLOGY, AND CONTROL

CONTRIBUTORS TO THIS VOLUME

J. J. Bouwman
Kira L. Bowen
J. A. Browning
B. C. Clifford
P. L. Dyck
Arthur L. Hooker
L. M. Joshi
E. R. Kerber
N. H. Luig
J. W. Martens
C. C. Mundt
S. Nagarajan
J. E. Parlevliet
J. M. Prescott
L. H. Purdy
A. P. Roelfs
J. B. Rowell
Eugene E. Saari
D. J. Samborski
Marr D. Simons
R. W. Stubbs
P. S. Teng
J. C. Zadoks

The Cereal Rusts

Volume II

Diseases, Distribution, Epidemiology, and Control

Edited by
Alan P. Roelfs
and
William R. Bushnell
Cereal Rust Laboratory
Agricultural Research Service
U.S. Department of Agriculture
University of Minnesota
St. Paul, Minnesota

1985

ACADEMIC PRESS, INC.
(Harcourt Brace Jovanovich, Publishers)
Orlando San Diego New York London
Toronto Montreal Sydney Tokyo

ACADEMIC PRESS, INC.
Orlando, Florida 32887

United Kingdom Edition published by
ACADEMIC PRESS INC. (LONDON) LTD.
24–28 Oval Road, London NW1 7DX

Library of Congress Cataloging in Publication Data
(Revised for vol. 2)
Main entry under title:

The Cereal rusts.

Includes bibliographies and index.
Contents: v. 1. Origins, specificity, structure, and physiology – v. 2. Diseases, distribution, epidemiology, and control.
1. Cereal rusts–Collected works. 2. Cereal rusts–Control–Collected works. I. Bushnell, William R. (William Rodgers) II. Roelfs, Alan P.
SB741.R8C47 1985 633.1'049425 84-15035
ISBN 0–12–148402–5 (alk. paper)

PRINTED IN THE UNITED STATES OF AMERICA

85 86 87 88 9 8 7 6 5 4 3 2 1

To the memory of the pioneers who developed the techniques and concepts that have made economical control of cereal rusts possible

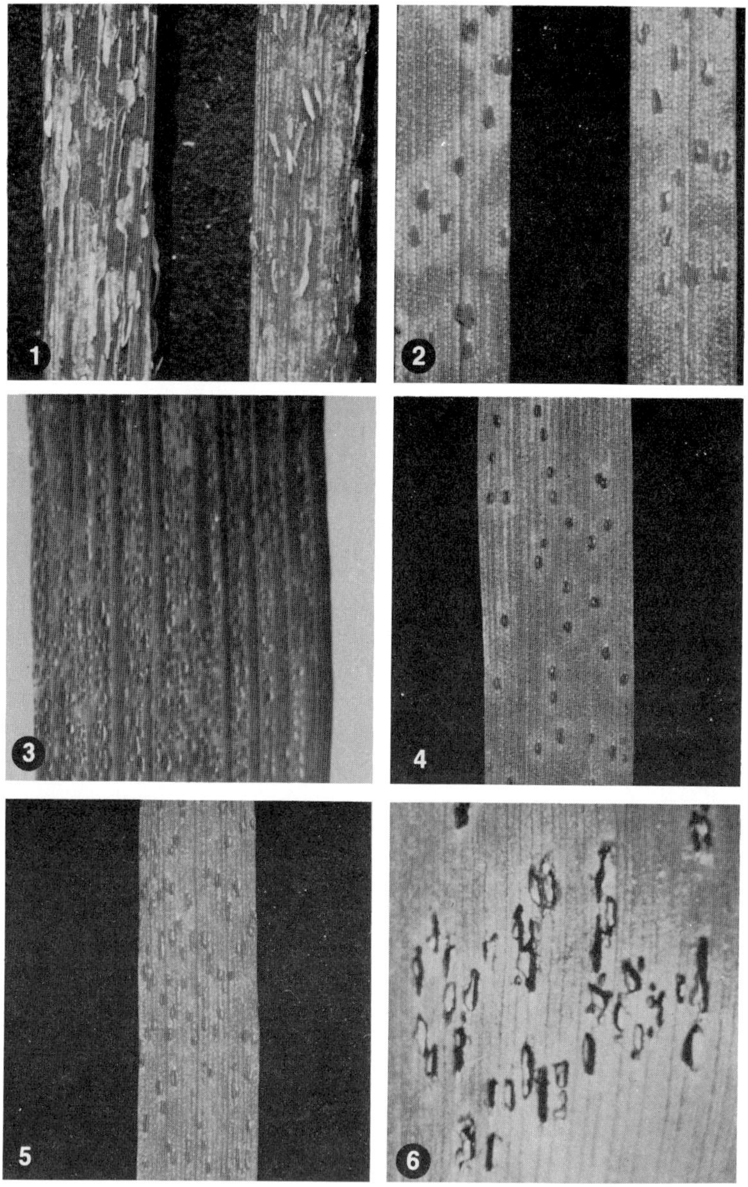

Selected examples of the cereal rusts. (1) Wheat stem rust (*Puccinia graminis* f. sp. *tritici;* Chapter 1), (2) Wheat leaf rust (*Puccinia recondita* f. sp. *tritici;* Chapter 2), (3) Stripe rust (*Puccinia striiformis* f. sp. *tritici;* Chapter 3), (4) Barley leaf rust (*Puccinia hordei;* Chapter 6), (5) Crown rust (*Puccinia coronata;* Chapter 5), (6) Common corn rust (*Puccinia sorghi;* Chapter 7). Photo by D. L. Long, Cereal Rust Laboratory.

Contents

5. Crown Rust

Marr D. Simons

6. Barley Leaf Rust

B. C. Clifford

7. Corn and Sorghum Rusts

Arthur L. Hooker

8.
Sugarcane Rusts

L. H. Purdy

Part II. Disease Distribution

9.
World Distribution in Relation to Economic Losses

Eugene E. Saari and J. M. Prescott

Part III. Epidemiology

10.
Epidemiology in Australia and New Zealand

N. H. Luig

11.
Epidemiology in Europe

J. C. Zadoks and J. J. Bouwman

12.
Epidemiology in the Indian Subcontinent

S. Nagarajan and L. M. Joshi

13.
Epidemiology in North America

A. P. Roelfs

14.
Disease Modeling and Simulation

P. S. Teng and Kira L. Bowen

Part IV. Control

A. Strategies Using Resistance

15. Resistance of the Race-Specific Type

P. L. Dyck and E. R. Kerber

16. Resistance of the Non-Race-Specific Type

J. E. Parlevliet

17. Genetic Diversity and Cereal Rust Management

C. C. Mundt and J. A. Browning

B. Strategies Using Chemicals

18. Evaluation of Chemicals for Rust Control

J. B. Rowell

Contributors

Numbers in parentheses indicate the pages on which the authors' contributions begin.

J. J. Bouwman[1] (329), Department of Phytopathology, Agricultural University, 6709 PD Wageningen, The Netherlands
Kira L. Bowen[2] (435), Department of Plant Pathology, University of Minnesota, St. Paul, Minnesota 55108
J. A. Browning[3] (527), Department of Plant Pathology and Microbiology, Texas A&M University, College Station, Texas 77843
B. C. Clifford (173), Welsh Plant Breeding Station, Aberystwyth SY23 3EB, Wales, United Kingdom
P. L. Dyck (469), Agriculture Canada Research Station, Winnipeg, Manitoba, Canada R3T 2M9
Arthur L. Hooker[4] (207), DeKalb-Pfizer Genetics, St. Louis, Missouri 63141
L. M. Joshi (371), Division of Mycology and Plant Pathology, Indian Agricultural Research Institute, New Delhi-12, India
E. R. Kerber (469), Agriculture Canada Research Station, Winnipeg, Manitoba, Canada R3T 2M9
N. H. Luig (301), Department of Agricultural Genetics and Biometry, The University of Sydney, Sydney, New South Wales 2006, Australia

[1]Present address: Agricultural College, Groenezoom 400, 3315 LA Dordrecht, The Netherlands.
[2]Present address: Department of Plant Pathology, University of Illinois, Urbana, Illinois 61801.
[3]Present address: Department of Plant Pathology, Seed and Weed Sciences, Iowa State University, Ames, Iowa 50011.
[4]Present address: DeKalb-Pfizer Genetics, 3100 Sycamore Road, DeKalb, Illinois 60115.

J. W. Martens (103), Agriculture Canada Research Station, Winnipeg, Manitoba, Canada R3T 2M9

C. C. Mundt (527), Department of Plant Pathology, North Carolina State University, Raleigh, North Carolina 27695-7616

S. Nagarajan (371), Regional Station, Indian Agricultural Research Institute, Simla-171002, India

J. E. Parlevliet (501), Department of Plant Breeding (I.v.P.), Agricultural University, 6709 PD Wageningen, The Netherlands

J. M. Prescott (259), Centro Internacional de Mejoramiento de Maíz y Trigo (CIMMYT), 06600 Mexico City, Mexico

L. H. Purdy (237), Department of Plant Pathology, University of Florida, Gainesville, Florida 32611

A. P. Roelfs (3, 403), Cereal Rust Laboratory, Agricultural Research Service, U.S. Department of Agriculture, University of Minnesota, St. Paul, Minnesota 55108

J. B. Rowell (561), Cereal Rust Laboratory, Agricultural Research Service, U.S. Department of Agriculture, Department of Plant Pathology, University of Minnesota, St. Paul, Minnesota 55108

Eugene E. Saari (259), Centro Internacional de Mejoramiento de Maíz y Trigo (CIMMYT), 06600 Mexico City, Mexico

D. J. Samborski (39), Agriculture Canada Research Station, Winnipeg, Manitoba, Canada R3T 2M9

Marr D. Simons (131), Agricultural Research Service, U.S. Department of Agriculture, Department of Plant Pathology, Iowa State University, Ames, Iowa 50011

R. W. Stubbs (61), Research Institute for Plant Protection (IPO), 6700 GW Wageningen, The Netherlands

P. S. Teng (435), Department of Plant Pathology, University of Minnesota, St. Paul, Minnesota 55108

J. C. Zadoks (329), Department of Phytopathology, Agricultural University, 6709 PD Wageningen, The Netherlands

Preface

The aim of this two-volume treatise is to assist in the worldwide effort to control cereal rusts by bringing together in a single reference source the accumulated knowledge of these important diseases. Not since K. Starr Chester's "The Cereal Rusts," published in 1946, have these diseases been treated comprehensively in a single work. In the interval since then, research on these potentially devastating diseases has proliferated, leading to new principles concerning their nature and new strategies for their control. Contributing to this new knowledge have been biochemists, cytologists, geneticists, physiologists, taxonomists, and epidemiologists, as well as cereal plant pathologists. The work of these diverse specialists as applied to cereal rusts forms the basis of these volumes.

The two volumes will serve the needs not only of cereal rust investigators who have found it increasingly difficult to assimilate the world's cereal rust literature, but also of plant pathologists generally, as a reference source for teaching, extension, and research. Many of the principles of plant pathology have been developed from studies of cereal rusts. Agronomists and other agriculturists concerned with cereal crop production or world food supplies will also find these volumes useful.

The cereal rusts represent the major disease threat to cereal crops, which, in turn, supply two-thirds of the world's edible food. Periodic cereal rust epidemics have plagued mankind since the dawn of agriculture. Consequently, cereal rust diseases were among the first to receive intensive investigation as the science of plant pathology emerged in the 1800s and early 1900s. The rust fungi were soon found to be "shifty enemies" (as E. C. Stakman put it) with a persistent ability to evolve new virulences that could overcome newly introduced resistant cereal cultivars. Periodic rust epidemics persisted into the 1950s and continue to be a threat today.

However, a new science of disease stabilization and management is now emerging that utilizes improved understanding of the complexities of rust disease to slow the evolution of dangerous new virulences, to retard epidemics, and to minimize losses. The knowledge on which these new strategies are based is presented in these volumes. Their contents reflect the great diversity and extent of cereal rust knowledge, including studies at the molecular level, and studies with cells, leaves, and whole plants, with plots and fields, and with epidemics sweeping across continents. In total, the cereal rusts have received more investigation than any other like-sized group of plant diseases.

Contributors to these volumes were asked to provide historical perspectives, give current trends, and project future problems and needs. They were encouraged to emphasize areas of special personal interest and to present their own unique perspectives to their assigned topics. The resulting varied treatments provide a rich compilation of the complex, challenging science of cereal rusts.

Volume I is devoted to fundamental aspects of the cereal rusts. A section on origins treats the contributions of early scientists to knowledge of cereal rusts, the evolution of cereal rusts, and the taxonomy of cereal rust fungi. A section on specificity includes *formae speciales,* race specificity, pathogen–host genetics, histology and molecular biology of host–parasite specificity, and the genetics of rust fungus populations as reflected by virulence frequency. A section on structure and physiology includes germination of urediospores and differentiation of infection structures, infection under artificial conditions, ultrastructure of hyphae and urediospores, development and physiology of teliospores, obligate parasitism and axenic culture of rust fungi, structure and physiology of haustoria, structural and physiological alteration in susceptible hosts, and effects of rust on plant development in relation to translocation. In sum, Volume I presents the historical, evolutionary, taxonomic, structural, genetic, and physiological characteristics of cereal rust fungi and the diseases they cause in cereal crops.

Volume II is devoted to individual cereal rust diseases and their distribution, epidemiology, and control. The major cereal rust diseases are wheat and rye stem rusts, wheat leaf rust, stripe rust, oat stem rust, crown rust, barley leaf rust, corn and sorghum rusts, and sugarcane rusts. Coverage of each is presented in an individual chapter. The distribution and economic importance of the cereal rust diseases on a worldwide basis are presented in a separate chapter.

Wheat and rye stem rusts were placed in a single chapter due to the apparent close relationship of the causal organisms. Leaf rust of rye was omitted due to minimal worldwide importance. The causal organisms of rye and wheat leaf rusts, *Puccinia recondita* f. sp. *secalis* and

P. recondita f. sp. *tritici*, are morphologically similar, but may be genetically quite different. The rusts of millets were omitted because they are generally minor diseases and because millet has limited worldwide use as a cereal (Saari and Prescott, Chapter 9, Volume II). The forage grasses with their numerous rust diseases are beyond the scope of these volumes on the cereal rusts. Although sugarcane is not a cereal, the sugarcane rusts were included because of the importance of sugar as a food crop and the similarity of its agronomic position to cereals. Rice is the only important cereal crop that is not affected by a rust disease.

The chapters on epidemiology of Australia and New Zealand, Europe, the Indian Subcontinent, and North America demonstrate long-distance spore dispersal, genetic epidemiology, overwintering foci, weather trajectory studies, continental epidemics, and disease modeling and simulation.

Disease control is approached through host resistance of race-specific and non-race-specific types, strategies of using and deploying resistances, and methods for evaluating fungicides.

The terminology chosen by the editors is the widely accepted usage in North America. However, in using these volumes, the reader will find several inconsistencies in terminology and concepts reflecting differences in viewpoint among authors. Thus, plant pathologists usually have used the term "rust" for disease, whereas others with a more mycological orientation have used "rust" to designate the fungus. Likewise, the taxonomist, using mainly morphological characters, defines subspecies and varieties; the plant pathologist, using host range, defines *formae speciales* as discussed in Volume I.

Given the large number of cereal rust workers, the selection of authors for these volumes involved difficult choices. We thank the authors, who willingly and capably contributed chapters, and extend thanks to our many other colleagues who reviewed chapters and provided advice and encouragement during this project. Special thanks are given to Gail Bullis and Brenda Anderson, who provided excellent secretarial assistance, to Colleen Curran, who patiently proofread most manuscripts and provided essential logistical support, to Connie Mann, who verified many references with unflagging diligence, and to Dave Casper, Bruce Hitman, and Mark Hughes, who carried on the research work while this volume was being prepared.

Alan P. Roelfs
William R. Bushnell

1984

Contents of Volume I

PART I

Diseases

1

Wheat and Rye Stem Rust

A. P. Roelfs

Cereal Rust Laboratory, Agricultural Research Service,
U.S. Department of Agriculture, University of Minnesota,
St. Paul, Minnesota

THE CEREAL RUSTS,
Volume II

ISBN 0-12-148402-5

I. Introduction

Puccinia graminis Pers. is the cause of several rusts of important cereal crops. Although in antiquity *P. graminis* may have been a single population, the cultivation of groups of dissimilar crops (wheat, oats, barley, and rye) resulted in the development and increase of different pathogen genotypes capable of attacking these crops. One may assume specialization occurred through selection of particular crops. Although crossing between cultures adapted to the same host could result in improved adaptation in the progeny, crossing between *formae speciales* would result in many progeny avirulent on both crops because of the recessive nature of virulence. Thus, the *formae speciales* have developed and been maintained as nearly separate populations of *P. graminis.*

A genetic relationship still exists, however, between many of the *formae specialis.* A diallel series of crosses was made by Johnson (1949) between pairs of f. sp. *tritici, secalis, avenae, agrostidis,* and *poae.* Crosses between f. sp. *tritici* and *secalis* were relatively fertile. Progeny from such crosses often are virulent on a limited number of host genotypes of both rye and wheat. Crosses within cultures of either f. sp. *tritici* or *secalis* can occasionally result in an F_1 culture that can be classified in the other *forma specialis.*

A. The Diseases

1. *Wheat Stem Rust*

Stem rust of wheat is caused by *Puccinia graminis* Pers. f. sp. *tritici.* This disease is also known as black stem rust or summer rust. Early references to stem rust or a disease now thought to be stem rust were reviewed by Chester (1946). Recently, urediospores were found in Israel that have been dated as 3300 years old (Kislev, 1982).

The damage caused by wheat stem rust can be more spectacular than any other cereal disease. Millions of hectares of a seemingly healthy crop with a high yield potential can be totally destroyed in less than a month. These types of epidemics on a smaller scale have been observed in nearly every country where wheat is grown. However, it was the continental epidemics that occurred in Australia and North America that showed the potential destructiveness of wheat stem rust. Stem rust of wheat is probably the most thoroughly studied plant disease. Certainly, more is known of the genetics of host resistance, frequency and distribution of pathogen virulence phenotypes, disease epidemiology and the physiology, biochemistry, and histology of the host–pathogen interaction than for any other plant disease. Although this chapter does not cover all that is known about stem rust of wheat, this and other chapters in this two-volume treatise will review much of the available information.

2. *Rye Stem Rust*

Stem rust of rye caused by *P. graminis* f. sp. *secalis* has been a minor disease worldwide. However, in Brazil in 1982, it totally destroyed the scattered fields of rye throughout the southern half of the country. Stem rust of rye has caused little damage in the United States (Roelfs, 1978), and historically, it probably has been the most serious as a disease in northern Europe. The main reasons generally given for its minor importance worldwide are the limited areas of rye grown in comparison with other cereals and the genotypic variation in rye as a cross-pollinated crop. However, in Brazil where only a few isolated rye fields existed, a severe epidemic was not averted. Rye stem rust was recently introduced into Brazil, and the ryes are old land cultivars. Stem rust has also become more severe on rye in Australia (Tan *et al.*, 1975). Perhaps the rye cultivars have not accumulated or maintained factors for resistance, or perhaps the introduced pathogen is particularly virulent or aggressive. Kingsolver *et al.* (1959) were able to initiate severe epidemics of stem rust on rye in the eastern United States by artificial inoculation. Most of the interest in *P. graminis* f. sp. *secalis* has been due to its close relationship to f. sp. *tritici* and to their putative hybrids (see Chapter 10, this volume) and its evolutionary significance (Green, 1971). However, the transfer of rye chromosomes sections to wheat and the possibility of triticale becoming a crop has also renewed our interest in rye stem rust.

Stakman *et al.* (1930) found that when wheat stem rust race 36 was crossed with rye stem rust race 11, a large range of pathogenic types was obtained including some common races of wheat stem rust. Johnson (1949) crossed races 1 and 30 of wheat stem rust with rye stem rust and recovered race 111 of wheat stem rust among others. In Australia, Watson and Luig (1962) selfed rye stem rust and obtained cultures virulent on Little Club (*SrLC*), Eureka (*Sr6*), and Yalta (*Sr11*). Parasexual recombinants also occur between rye and wheat stem rust (Watson and Luig, 1959). Thus, members of these two *formae speciales* seem to have many genes in common.

3. *Barley Stem Rust*

Stem rust of barley is caused by *P. graminis* f. sp. *tritici* and in North America also by f. sp. *secalis*. Epidemics of barley stem rust were rare historically except when wheat was severely rusted (Roelfs, 1978). Seedlings of most, if not all, barley cultivars have a moderate level of resistance at 18°C (Steffenson *et al.*, 1982b). This resistance is expressed in the seedling stage by a mesothetic response. In adult plants, the resistance is expressed by a reduction in lesion size and number (Steffenson, 1983). This level of resistance in barley, when accompanied by its relatively early maturity and ability to grow at low temperatures, generally provides protection from endogenous inoculum of f. sp. *tritici*.

A single gene with a major effect and one or more genes with lesser effects have been identified in barley. The gene with a major effect was designated as the *T* gene because it provided resistance to *forma specialis tritici* (Powers and Hines, 1933). The *T* gene occurs in the cultivars Chevron and Peatland, selections from a bulk seed lot from the Swiss Seed Experiment Station. Another source of this resistance is the cultivar Kinred, which was selected as an offtype plant from a field of Wisconsin Pedigree 37. The *T* gene resistance has remained effective, even though widely used in the northern United States since Peatland was released in 1926. The *T* gene resistance against some races of *P. graminis* f. sp. *tritici* is less effective in some cultivars than others (Steffenson *et al.*, 1982a,b).

As the frequency of wheat stem rust has decreased in North America, the proportion of isolates of *P. graminis* f. sp. *secalis* isolated from barley has increased (Green, 1971). The level of resistance in barley against rye stem rust seems to be less than against wheat stem rust in North America (Green, 1971; Steffenson, 1983). The *T* gene offers no protection against rye stem rust, and most cultivars are damaged in inoculated nurseries (Steffenson *et al.*, 1982a). Currently, inadequate inoculum exists of *P. graminis* f. sp. *secalis* to threaten North American barley production. However, because this *forma specialis* is virulent on many species of native and introduced grasses, inoculum levels may change. Resistance to *P. graminis* f. sp. *secalis* exists in the cultivars Black Hulless, Heitpas-5, and Valkie (Steffenson *et al.*, 1982a).

In Australia, putative hybrids between *P. graminis* f. sp. *tritici* and f. sp. *secalis* seem to have specialized on barley (Chapter 10, this volume).

4. Puccinia graminis *on the Alternate Hosts*

Many species of *Berberis, Mahonia,* and their hybrid (× *Mahoberberis*) are susceptible to *Puccinia graminis* (Table I). Little specialization occurs on the alternate hosts but is known to occur (Waterhouse, 1929b; Green and Johnson, 1958; Johnson and Green, 1954). Even on susceptible bushes only the tissue (2 weeks old or less) is normally susceptible (Melander and Craigie, 1927). The most important of the susceptible species has been *Berberis vulgaris* L., although other susceptible species exist throughout most of the world. The resistant species of *Berberis* (Table II) tend to have a thick cuticle (Melander and Craigie, 1927), although in a few *Berberis* species and some *Mahonia* species attempts to infect the bush resulted in small hypersensitive flecks, which is an indication of a physiological mechanism of resistance.

Table I

Species of *Berberis, Mahonia,* and × *Mahoberberis* Susceptible to *Puccinia graminis*[a]

Berberis
- *acuminata* Franch.
- *aemulans* Schneid.
- *aetnensis* Presl.
- *aggregata* Schneid.
- × *alksuthiensis* Ahrendt
- *amurensis* Rupr.
- *angulosa* Wall. ex Hook f. et Thoms.
- *aristata* DC.
- *asiatica* Roxb. ex DC.
- *atrocarpa* Schneid.
- × *barbarossa* Watson ex Ahrendt
- *bergmanniae* Schneid.
- *boschanii* Schneid.
- *brachypoda* Maxim.
- *bretschneideri* Rehd.
- × *cerasina* Schrad.
- *chinensis* Poir.
- *chitria* Lindl.
- *chrysosphaera* Mulligan
- *consimilis* Schneid.
- *coriaria* Royle ex Lindl. var. *patula* Ahrendt
- *crataegina* DC.
- *cretica* L.
- × *declinata* Schrad.
- × *declinata* var. *oxyphylla* Schneid.
- *delavayi* Schneid.
- *diaphana* Maxim.
- *dictyoneura* Schneid.
- *dictyophylla* Franch.
- *dielsiana* Fedde
- × *durobrivensis* Schneid.
- *edgeworthiana* Schneid.
- × *emarginata* Willdenow
- × *emarginata* var. britzensis Schneid.
- *empetrifolia* Lam.
- *fendleri* Gray
- *francisci-ferdinandi* Schneid.
- *glaucocarpa* Stapf
- *globosa* Benth.
- *henryana* Schneid.
- *heteropoda* Schrenk
- *holstii* Engler
- *hookeri* Lemaire

(continued)

Table I (*Continued*)

Berberis
hookeri var. *viridis* Schneid.
humido-umbrosa Ahrendt
hypokeriana Airy-Shaw
ilicifolia Forst.
jamesiana Forrest et W.W. Sm.
× *laxiflora* Schrad.
× *laxiflora* var. langeana Schneid.
lecomtei Schneid.
lycium Royle
× *macracantha* Schrad.
× *meehanii* Schneid. ex Rehd.
mitifolia Stapf
morrisonensis Hayata
× *notabilis* Schneid.
nummularia Bunge
oblonga (Regel) Schneid.
orthobotrys Bienert ex Aitch.
× *ottawensis* Schneid.
poiretii Schneid.
polyantha Hemsl.
prattii Schneid.
pruinosa Franch.
× *rubrostilla* Chittenden
× *serrata* Koehne
sibirica Pall.
sieboldi Miq.
Berberis
soulieana Schneid
× *spaethii* Schneid.
stiebritziana Schneid.
suberecta Ahrendt
thibetica Schneid.
turcomannica Kar. ex Ledeb.
umbellata Wall. ex G. Don
× *vanfleetii* Schneid.
vernae Schneid.
vulgaris L.
wilsonae Hemsl.
wilsonae var. *parvifolia* (Sprague) Ahrendt
wilsonae var. *stapfiana* (Schneid.) Schneid.
yunnanensis Franch.
Mahonia
fremontii (Torr.) Fedde
haematocarpa (Wooton) Fedde
napaulensis DC.
nevinii (Gray) Fedde
swaseyi (Buckley) Fedde
trifoliolata (Moric.) Fedde
× *Mahoberberis*
neubertii (Baum.) Schneid.

[a]From Ahrendt (1961).

B. EPIDEMIOLOGY

Details on epidemiology in Australia–New Zealand, Europe, the Indian subcontinent, and North America are covered in separate chapters in this volume. The earlier reports of Chester *et al.* (1951), Rajaram and Campos (1974), and Hogg *et al.* (1969) contain additional historical information on these areas. In the rest of Asia, South America, and Africa, continental studies are not available.

II. Life Cycle

The life cycle of *P. graminis* has been widely studied, but unanswered questions still remain. The first drawing of the fungus was

Table II

Species of *Berberis, Mahonia,* and × *Mahoberberis* Resistant to *Puccinia graminis*[a]

Berberis	*Berberis*
arido-calida Ahrendt	*sargentiana* Schneid.
beaniana Schneid.	*sikkimensis* (Schneid.) Ahrendt
buxifolia Lam.	× *stenophylla* Lindl. var. *diversifolia* Ahrendt
buxifolia var. *nana* A. Usteri	× *stenophylla* var. *gracilis* Ahrendt
calliantha Mulligan	× *irwinii* Byhouwer
candidula Schneid.	*taliensis* Schneid.
cavaleriei Léveillé	*temolaica* var. *artisepala* Ahrendt
× *chenaultii* (Hort. ex Cat. L. Chenault) Ahrendt	*thunbergi* DC.
circumserrata Schneid.	*thunbergi* var. *agenteo-marginata* Schneid.
concinna Hook. f. et Thoms.	*thunbergi* var. *atropurpurea* Chenault
coxii Schneid.	*thunbergi* var. *erecta* (Rehd.) Ahrendt
darwinii Hook.	*thunbergi* var. *maximowiczii* (Regal) Regal
dasystachya Maxim.	*thunbergi* var. *minor* Rehd.
dubia Schneid.	*thunbergi* var. *pluriflora* Koehne
franchetiana Schneid.	*triacanthophora* Fedde
gagnepainii Schneid.	*verruculosa* Hemsl. et Wils.
gilgiana Fedde	*virgetorum* Schneid.
gyalaica Ahrendt	*xanthoxylon* Hasskarl ex Schneid.
heterophylla Juss. ex Poir.	*Mahonia*
× *hybrido-gagnepainii* Suringar	*amplectans* Eastwood
insignis Hook. f. et Thoms.	*aquifolium* (Pursh) Nutt.
julianae Schneid.	*bealei* (Fort.) Carr.
kawakamii Hayata var. *formosana* Ahrendt	*dictyota* (Jepson) Fedde
koreana Palib.	*fortunei* (Lindl.) Fedde
lempergiana Ahrendt	*japonica* (Thunb.) DC.
lepidifolia Ahrendt	*lomariifolia* Takeda
linearifolia Phil.	*nervosa* (Pursh) Nutt.
× *lologensis* Sandwith	*pinnata* (Lag.) Fedde
manipurana Ahrendt	*piperiana* Abrams
pallens Franch.	*pumila* (Greene) Fedde
potaninii Maxim.	*repens* G. Don
replicata W. W. Sm.	× *Mahoberberis*
sanguinea Franch.	*aquicandidula* Krüssmann
	aquisargentiae Krüssmann
	miethkeana Melander et Eade

[a]From Ahrendt (1961).

made by Fontana (1932). DeBary (1866) showed that the two fungi, *P. graminis* on cereals and *Aecidium berberidis* on barberry, were different stages of the same organism. A simple life cycle of *P. graminis* is shown in Fig. 1. Readers interested in more structural detail should consult the appropriate chapters in Volume I of this treatise or Lit-

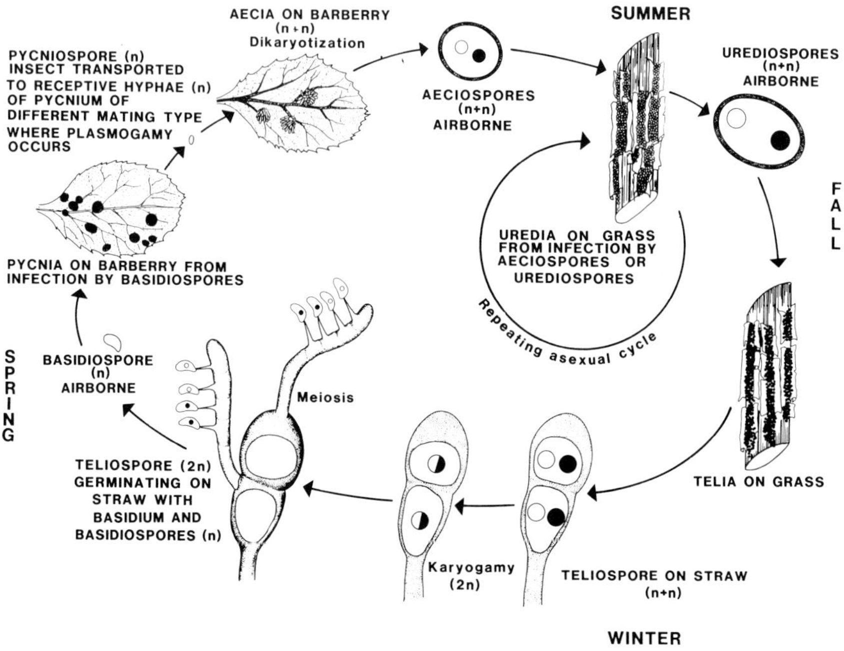

Fig. 1. A simplified life cycle of *Puccinia graminis.* The nuclear condition is shown by n (haploid), n + n (dikaryotic), and 2n (diploid). Spores are all drawn to approximately the same scale. For details and variations in structures and developmental events, see Volume I, Chapters 11 and 12. Germinating teliospore adapted from Buller (1924).

tlefield and Heath (1979) and Littlefield (1981). The life cycle description in this chapter emphasizes the effects of the life cycle on disease development.

A. BASIDIOSPORES

Basidiospores are formed on sterigma on each cell of the basidium. The spores are actively discharged a few centimeters from the sterigma at maturity (Buller, 1958). Basidiospores are small, 7.6 × 6 μm (D. L. Long, personal communication), hyaline, and oval-shaped. They can develop and are released within 20–25 min after the teliospore is moistened (Craigie, 1940); however, some viable teliospores may not germinate for 3–6 or more days (Cotter, 1932). Alternate wetting and drying, freezing and thawing may help to induce teliospore germination. The basidiospores are wind-borne to the alternate host, species of susceptible *Berberis* and *Mahonia,* but seldom cause infection more than 180 to 270 m from the source. Basidiospores are short-lived, surviving only a few hours under ideal conditions. Therefore, basidiospores germinate rapidly by producing a germ tube with a terminal appressorium from which an infection peg develops and directly penetrates the epidermal cell wall. Most of the infections occur on the upper leaf surface, although infections on the berries, stems, and other plant surfaces do occur. Leaves of susceptible cultivars often appear to become resistant to penetration about $2\frac{1}{2}$ weeks after the leaf bud unfolds.

B. PYCNIOSPORES

Approximately 5 days after infection of the barberry by the basidiospore, the initial signs of pycnia development become visible (Cotter, 1932). At 7–14 days after infection, the small, 1.6 × 3.6 μm (D. L. Long, personal communication), haploid, unicellular pycniospores appear in a viscous fluid that forms at the pycnial opening. The pycniospore functions as a gamete and fuses with the receptive hypha that functions as the other gamete. To accomplish fertilization, pycniospores must be transferred to receptive hyphae of a pycnia of a different mating type. This transfer of pycniospores is by insects presumably attracted to the fluid in which the spores are suspended. The function of the pycniospores as sexual gametes was unknown until the work of Craigie (1927).

C. AECIOSPORES

Following fertilization, the nucleus of the pycniospore migrates through the monokaryotic hyphae until it reaches the protoaecium. This process requires 20–25 hr. Following dikaryotization, an aecium develops on the lower surface of the barberry leaf within 7 to 10 days after fertilization of the receptive hyphae (Cotter, 1932). The aeciospores represent the recombination products of the genetic process. The variation in virulence among aeciospores was first shown by Waterhouse (1929a). Aeciospores from an individual aecial horn generally have the same genotype, whereas aeciospores from another horn in the same aecium frequently have different genotypes. Aeciospores are dikaryotic, cylindrical, 16–23 × 15–19 μm, and are generally wind-borne over long distances. Aeciospores are forcibly released following the drying of previously wetted aecia. The optimum temperature for germination is 22°C (Novotelnova, 1935). Large numbers of spores are produced. Stakman (1923) estimated that an average-sized *Berberis vulgaris* bush in southern Minnesota had approximately 35,000 leaves, of which about 28,000 would be infected. On a single multi-infected leaf, 2.3 to 8 million aeciospores were produced. Thus, an average bush could be the source of 64×10^9 spores. Aeciospores have been found at altitudes of 2 km (Stakman and Harrar, 1957). Although the data are incomplete, initial infection from aeciospores is very heavy within a few meters of the source. Isolated infections can occur as far away as 100 m, and there is no reason to suspect that aeciospores are not transported as far as urediospores are (hundreds of kilometers). Aeciospores germinate on the gramineous host upon contact with free water. Germ-tube formation is followed by formation of an appressorium over a host stoma, and a penetration peg enters through the stomatal opening.

D. UREDIOSPORES

A uredium is produced after the successful infection of a grass host by *P. graminis.* Initial signs of the infection usually become visible 5–6 days after inoculation as a light colored spot. Sporulation usually follows infection by 7–14 days, with the shortest period at 30°C. The size of uredia and time required for their development are determined by host resistance–pathogen virulence, pathogen aggressiveness, host maturity, and infection density and environment, particularly temperature and light. Urediospores are dikaryotic, oblong, and about 26–40 × 16–32 μm in size. They can be transported long distances by the wind (Hirst and Hurst, 1967) in a viable condition (Luig, 1977). The terminal velocity (rate at which an object falls) of urediospores in still air ranges from 0.94 to 1.25 cm/sec, increasing with decreasing relative humidity and temperature (Weinhold, 1955; Ukkelberg, 1933). The spores may settle from the air under the influence of gravity, by impaction on objects when striking them, but are most effectively removed from the air by the scrubbing action of raindrops (Rowell and Romig, 1966).

Large numbers of spores are produced over a period of several weeks. A mature uredium produces daily about 23 μg (Katsuya and Green, 1967; Mont, 1970) of urediospores; 1 μg contains about 4.5×10^2 urediospores (Rowell and Olien, 1957). Thus about 10,000 urediospores per day are produced, and with a 5% disease severity (50 pustules per tiller), 5 kg of spores can be produced per hectare (Rowell and Roelfs,

1971). Most of the urediospores produced are deposited within the crop canopy and a large proportion of the remainder within 100 m of the source (Roelfs, 1972). However, due to the large numbers produced, significant numbers of urediospores reach heights of up to 3000 m and have been transported in a viable condition over great distances. The spores are relatively resistant to damage from temperatures 0°–40°C and can withstand greater extremes for short periods. Spores will remain viable at room temperatures for several weeks. Longevity of urediospores rapidly decreases with exposure to high relative humidities (over 80%), and longevity is prolonged at relative humidities of 20–30%.

The optimum temperature for urediospore germination and infection is about 18°C. An appressorium forms over a stoma within 3–6 hr. This part of the process can occur in the dark; however, for further development, approximately 10,000 lux of light is required over about 3 hr for the penetration peg to form and the appressorium to empty. During the light phase, for optimum development, temperatures should gradually rise from 18° to 30°C and a gradual drying should occur. Sporulating uredia usually are present in 10–14 days. Latent periods of 14–20 days are common in the field during early spring and have been observed to be as long as 30 days. The uredial cycle repeats itself every 14–21 days under normal field conditions. Because of varying environmental conditions in the field, some new infections appear almost daily; thus generations overlap and are generally indistinguishable.

E. TELIOSPORES

As host plants mature, uredia gradually develop into telia, producing teliospores. Teliospores are blackish-brown, oblong, diploid, two-celled spores about 40–60 × 16–23 μm. The spores remain attached to the telium by the stalk, and are rather resistant to environmental extremes (Cotter, 1932). Teliospores serve as the overwintering stage of the fungus in climates where freezes are common. Viability of teliospores is reduced with high temperatures, especially if this is accompanied by dry conditions. Teliospore viability varies with the host and pathogen genotypes involved.

Transport of teliospores is generally by man or water when infected wheat or rye straw is moved. The teliospore germinates after a period of weathering, alternating freezing and thawing and drying and wetting. However, under laboratory conditions, teliospores germinate irregularly. Teliospores germinate by producing a basidium, during which time meiosis occurs and the haploid nuclei migrate into the basidium. A basidiospore develops on a sterigma, and a haploid nucleus migrates into the spore.

III. Disease Cycle

The available evidence points to separate populations of *P. graminis* f. sp. *tritici* in the United States, one that undergoes sexual reproduction annually and another that reproduces asexually (Roelfs and Groth, 1980). When the two populations separated is unknown. This may be similar to the development of two forms of *P. graminis* in eastern Europe at a subspecies level (Savile and Urban, 1982). Parasexual mechanisms have been proposed for *P. graminis;* however, if these processes occur in nature, most of the progeny must fail to compete satisfactorily with the preexisting population. The literature on these mechanisms was reviewed by Watson and Luig (1962). An exception may be the *P. graminis* f. sp. *tritici* and *P. graminis* f. sp. *secalis* hybrids (Luig and Watson, 1972) in Australia. In most areas of the world where stem rust is a disease of major importance, the pathogen survives without the sexual cycle. See Chapters 9–12, this volume, for Australia–New Zealand, Europe, the Indian subcontinent, and North America, respectively. Although published information is limited, the sexual cycle is not known to be important in the major wheat- and rye-producing areas of South America, Africa, or eastern or northern Asia.

A. Asexual Cycle

In mild climates the pathogen can reproduce by means of urediospores, and survives noncrop season(s) on volunteer cereal plants or on other gramineous hosts (Fig. 2). Survival is generally difficult for the pathogen during the noncereal growing season, and often the population is reduced to near or below the threshold of detection. However, a few surviving local uredia can produce local inoculum and thereby cause more infections than a severely rusted field 100 km distant from which spores must be transported. In mild climates the asexually reproducing pathogen generally survives the winter on the wheat or rye crop. In tropical climates it often exists as sporulating mycelium in the leaves, whereas in lower latitudes of temperate climates it may be

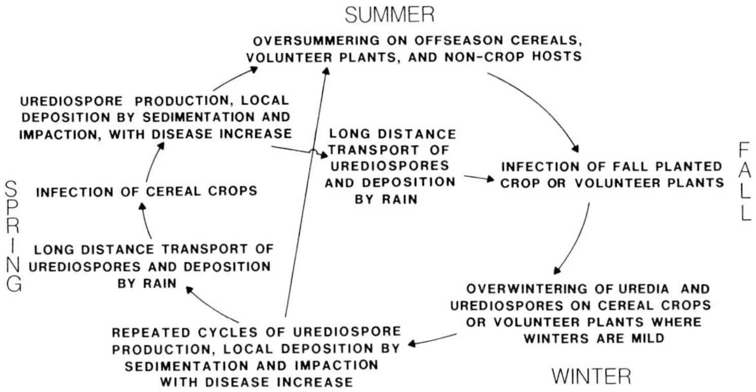

Fig. 2. The asexual disease cycle for stem rust, which is the most common source of disease worldwide. In areas with severe winters, the pathogen may not survive the winter and thus must be reintroduced by long-distance transport of urediospores from areas where winters are milder. Likewise, the pathogen may not survive the summers where it is very hot or dry, and thus must be reintroduced by long-distance transport of urediospores from areas where the summers are milder.

present as nonsporulating mycelium. In the central latitudes of the temperate zone, the pathogen generally fails to survive because the host tissue is winter killed. In these areas epidemics are generated from airborne inoculum from milder climates.

Epidemics seem to be relatively few in areas where stem rust overwinters on the crop itself; the crop is harvested in late spring before conditions are generally favorable for stem rust. Major epidemics are most frequent and severe on spring-planted cereals that grow during the spring and summer, and on fall-planted wheats that mature in the summer.

The major source of variation in the asexual population is by mutation, which is expressed as single gene changes. Although the number of mutations may be great because of the large number of urediospores produced, most mutations must be eliminated through selection (Roelfs and Groth, 1980). Other sources of variation are from exogenous inoculum (Luig, 1977) and parasexualism (Luig and Watson, 1972).

The asexual population is characterized by one or a few predominant phenotypes (Groth and Roelfs, 1982). Related phenotypes often exist (Roelfs and Groth, 1980) that reflect their origin through mutation. This relationship can be seen over the years as a gradual evolution with each new phenotype, clearly related to a previously existing one (Green, 1975) (Chapter 10, this volume).

In North America many of the asexual population phenotypes seem to be losing their ability to produce teliospores, and the teliospores often have a low viability (Roelfs, 1982). These factors could be among the reasons for the separation of the populations into distinct phenotype groups.

B. SEXUAL CYCLE

Where the sexual cycle functions, the pathogen overwinters as teliospores. If straw bearing teliospores is located near a susceptible alternate host (normally *B. vulgaris*), infection will normally occur. Teliospores usually germinate over a period of several days. Infections normally occur on the expanding barberry leaves that are produced in the spring (Fig. 3). The sexual cycle can furnish large amounts of inoculum in the form of aeciospores which initiate local epidemics, and then the resulting production of urediospores can cause regional epidemics (Roelfs, 1982).

The teliospores germinate better when they are produced under cool temperatures (<18°C) than under high temperature. Thus, the sexual populations normally have existed at cooler latitudes—that is, the northern United States, Canada, northern Europe, and at higher elevations farther south. Teliospores also occur on volunteer crop plants or other gramineous hosts late in the fall when temperatures are cool.

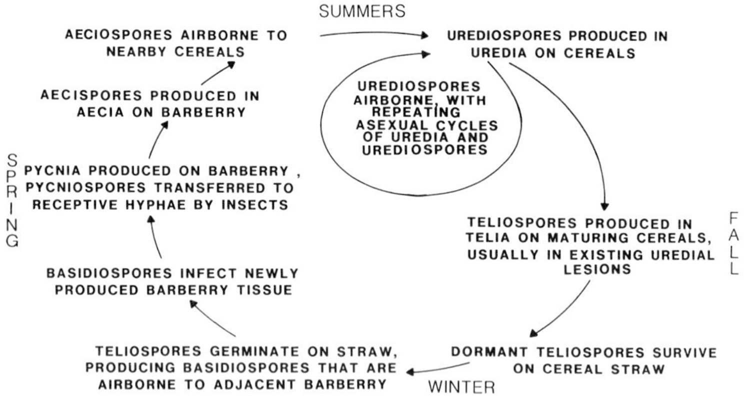

Fig. 3. The complete disease cycle (sexual and asexual) for stem rust, which historically occurred in eastern and northern Europe and in North America where barberry (the alternate host) was present. Viable teliospores are produced in regions without extreme heat as the cereal crop matures. Winters must be severe enough to break teliospore dormancy. For details of the life cycle, see Fig. 1; for details of the asexual disease cycle, see Fig. 2.

The sexually reproducing population is characterized by diversity. The total number of phenotypes is great (usually one phenotype per five or fewer isolates) (Roelfs and Groth, 1980; Groth and Roelfs, 1982). The various phenotypes in the populations tend to occur in a rather even frequency (Groth and Roelfs, 1982). Most cultures have several dominant genes for avirulence, and thus, the heterozygous F_1 cultures often have a narrow host range. However, selfing of heterozygous cultures occassionally results in cultures with a wide host range.

IV. Physiological Specialization

The use of race-specific resistance (see Chapter 13, this volume) to control wheat stem rust requires continued monitoring of the variation in the pathogen population for virulence (Volume I, Chapter 5). This has been and continues to be done annually in many countries. Published information appears regularly in Australia, Brazil, Canada, Egypt, India, Italy, Pakistan, and the United States. Currently, four systems for physiologic race identification in *P. graminis* f. sp. *tritici* are in use.

A. INTERNATIONAL SYSTEM

This system was developed by Stakman and Piemeisel and the most recent key was published by Stakman *et al.* (1962). The set uses 12 differential host cultivars (Table III) representing the three ploidy levels in wheat. These 12 differentials were used worldwide until about 1950, when sources of resistance not represented in them became widely used in commercial wheat cultivars. However, the international differential hosts are still used in many countries and still serve as a basis for communication. Physiologic races are based on combinations of host reaction classes, i.e., the resistant class is indicated by infection types 0, 0;, 1, and 2, mesothetic by infection type X, and susceptible by infection types 3 and 4. To 1983, 344 races have been described worldwide according to this system.

B. MODIFIED POTATO—*PHYTOPHTHORA INFESTANS* SYSTEM

This system was introduced in Australia and New Zealand and uses the international differential set (Table III) and then an ordered set of

Table III

International Differential Cultivars for Wheat Stem Rust and Their Genes for Resistance to *Puccinia graminis*[a]

Cultivar	*Sr* gene(s) known
Little Club	*LC*
Marquis	*7b, 18, 19, 20, X*
Reliance	*5, 16, 18, 20*
Kota	*7b, 18, 28, Kt'2'*
Arnautka	*9d,* plus two genes conditioning infection type X
Mindum	*9d,* plus a gene conditioning infection type X
Spelmar	*9d,* plus a gene conditioning infection type X
Kubanka	*9g,* plus a gene conditioning infection type X
Acme	*9g,* plus a gene conditioning infection type X
Einkorn	*21*
Vernal	*9e*
Khapli	*7a, 13, 14*

[a]Based on Luig *et al.* (1973) and Roelfs and McVey (1979).

Table IV

Additional Wheat Lines Used with the International Series as Differential Hosts for *Puccinia graminis* in Australia and New Zealand

Differential number	Host lines	*Sr* gene(s)
1	McMurachy or Eureka	*6*
2	Yalta	*11*
3	Gamenya	*9b*
4	Mengavi	*36*
5	Gala or Renown	*17*
6	Mentana	*8*
7	Norka	*15*
8	Festiguay or Webster	*30*
9	*Agropyron intermedium* derivative	*Agi*
10	Entrelargo de Montijo	Combination[a]
11	Barleta Benvenuto	8b[b]
12	Coorong Triticale	*27*

[a]Has a gene resulting in a low infection type (;) to pre-1954 strains and a gene resulting in a low infection type (2).

[b]Has a gene resulting in a low infection type (X).

differentials with sequential numbers (Table IV) that is increased as new sources of resistance are found to be effective in the separation of strains. The additional differential hosts were often lines with "single" effective genes for resistance, although the last two cultivars added are known to have at least two effective resistance genes. Nomenclature consists of the international race number (Stakman *et al.*, 1962), followed by ANZ for Australia–New Zealand and then the numbers of the additional differential hosts (Table IV) that are susceptible (Watson and Luig, 1963). Thus, 17 ANZ-2 is a strain of international race 17 that is virulent on *Sr11* and occurs in the Australia–New Zealand geographical area. Watson and Luig (1962) clearly point out that 17 ANZ-2 is *not* a subrace of race 17, as this artificially ranks some host genes as superior to others.

C. FORMULA METHOD

This system was developed and is used in Canada. The formula system is based mainly on differential hosts with a "single" gene for resistance to *P. graminis* (Table V) (Green, 1965). The avirulence/virulence formula is written for each culture, with the effective host genes listed first followed by a slash and then the ineffective host genes. A consecutive C number is assigned each different virulence formula: for example, C38(15B-1L) designates the virulence formula 6, 9a, 9b, 13, 15, 17/5, 7a, 7b, 8, 9d, 9e, 10, 11, 14, 35. The formula number (C38) indicates the 38th virulence formulas assigned a number in Canada. The old race designation was 15B-1L, which included the international race number and B-1L, which was a part of an old supplemental differential nomenclature that is no longer used but is retained to ensure continuity for those combinations for which it existed (Green, 1981).

D. CODED SETS

In the United States, three sets of four host lines, each with a "single" host gene for resistance, are used for physiological race identification (Roelfs *et al.*, 1982) (Table VI). The first set has the resistances present in the international differential set that were important in differentiating races of the asexual population in the United States.

Table V

Lines of Wheat Used as Differential Hosts for *Puccinia graminis* in Canada[a]

Host[b]	*Sr* gene
Prelude*6/Reliance	*5*
Mida-McMurachy-Exchange/6*Prelude	*6*
Na101/6*Marquis	*7a*
Chinese Spring/Hope	*7b*
Chinese Spring/Red Egyptian	*8*
Chinese Spring/Red Egyptian	*9a*
Prelude*4/2/Marquis*6/Kenya 117A	*9b*
H-44-24/6*Marquis	*9d*
Vernstein	*9e*
Marquis*4/Egypt NA95/2/2*W2691	*10*
Chinese Spring/Timstein	*11*
Prelude*4/2/Marquis*6/Khapstein	*13*
W2691*2/Khapstein	*14*
Prelude*2/Norka	*15*
Prelude/8*Marquis*2/2/Esp 518/9	*17*
Marquis*6/2/3*Steward/R.L. 5244	*22*
Agent	*24*
Eagle	*26*
WRT 238-5	*27*
Prelude/8*Marquis/2/Etiole de Choisy	*29*
Webster	*30*
Prelude*4/NHL II.64.62.1	*36*
W3563	*37*

[a]After Green (1981).

[b]The international differentials Marquis, Mindum, and Einkorn and the cultivars Manitou (*Sr5, 6, 7a, 9g, 12, 16,* plus), Selkirk (*Sr6, 7b, 9d, 17, 23, 2*), Sinton (*Sr* genotype unknown), and Neepawa (*Sr5, 7a, 9g, 12, 16,* plus) are tested.

The other two sets have resistance that have become important in more recent years. Races are designated by the international race number followed by a code indicating the virulence/avirulence formula for the culture. Thus, C-33(15B-1L) is similar to 15-TNM. The code for each set is based on a predetermined binary key of the 16 virulence/avirulence formulas possible within the set (listed from the formula of the possible pathogen phenotype avirulent on the four host lines to the last combination that is virulent or all four lines). The combinations of possible avirulent/virulent phenotypes are designated by letters B through T, omitting vowels (Roelfs and McVey, 1972).

Table VI

Lines of Wheat Used as Differential Hosts for *Puccinia graminis* in the United States[a]

Host lines	*Sr* gene
Set I	
ISr5-Ra	*5*
ISr9d-Ra	*9d*
Vernstein	*9e*
ISr7b-Ra	*7b*
Set II	
ISr11-Ra	*11*
ISr6-Ra	*6*
ISr8-Ra	*8*
ISr9a-Ra	*9a*
Set III	
W2691SrTt-1	*36*
W2691Sr9b	*9b*
W2691Sr13	*13*
W2691Sr10	*10*
Testers	
W2691Sr15NK	*15*
ISr16-Ra	*16*
Combination VII	*17 & 13*
Triumph 64	*Tmp*

[a]Bulks of cultures are tested on "universal" resistances *Sr22, 24, 25, 26, 27, 29, 31, 32, 33, 37, Gt,* and *Wld-1,* as well as *Sr30,* which is resistant to most cultures.

E. PHYSIOLOGICAL SPECIALIZATION: RYE STEM RUST

In initial studies by Levine and Stakman (1923), rye cultivars Rosen, Swedish, and Prolific were used. The cultivars Colorless and Dakold were added as differentials by Cotter and Levine (1932). Unfortunately, seed of these cultivars was apparently not maintained. Because rye is cross-pollinated and normally self-sterile, studies of physiological specialization are more difficult than in the *Triticum–P. graminis* system. Thus, a cultivar was considered resistant when 75% or more of the plants responded with infection types 0, 1, or 2. The response was rated mesothetic when 25–75% of infected plants had infection types 3 or 4, or when most of the plants have an infection type X, and susceptible when more than 75% of the plants had infection types 3 or 4 (Cotter and Levine, 1932). A set of self-pollinated rye lines was developed that permits a more detailed investigation of virulence in *P. graminis* f. sp. *secalis* (Tan *et al.*, 1976). These ryes with "single" genes for resistance have been used in North America (Steffenson *et al.*, 1983) and Australia (Tan *et al.*, 1975) to study variation in *P. graminis* f. sp. *secalis*. The Australian work also included hybrids between *P. graminis* f. sp. *secalis* and *tritici* (Tan *et al.*, 1975), as well as *P. graminis* f. sp. *tritici* itself (Luig and Tan, 1978).

Another approach may be to move the rye resistance to a wheat line that is susceptible to *P. graminis* f. sp. *secalis* (Luig and Watson, 1976). Line E, W3498, derived from a cross of Gabo*3/Charter//Little Club. (Luig, 1983), is susceptible to *P. graminis* f. sp. *tritici* and most cultures of *P. graminis* f. sp. *secalis*.

V. Control

A number of methods are available to control wheat and rye stem rust, but none has been totally satisfactory alone. The earliest attempts involved religious practices. These practices varied between areas and existed as early as 1000 B.C. and continued into the first century A.D. (Chester, 1946). In the early 1600s, Worlidge recommended pulling a rope over the grain to knock off the dew. This practice continued until the 1900s in some areas (Chester, 1946). In France in the mid 1600s, laws requiring barberry eradication were passed. In the late 1800s, Farrer in Australia developed early-maturing wheat cultivars to escape the damage of rust (McIntosh, 1976). Resistant cultivars were known as early as 1841 (Chester *et al.*, 1951); however, breeding for resistance did not become common until the early 1900s. The two early successes were the transfers of stem rust resistance from Iumillo durum wheat to bread wheats by Hayes *et al.* (1936), and the transfer of resistance from emmer to bread wheat by McFadden (1930). The former resulted in the cultivar Thatcher, which is still a basis for much of the hard red spring wheats, and the latter resulted in Hope and H-44, which are among the most widely used sources of resistance (*Sr2*). Fungicides have been widely investigated for use in control of stem rust (Rowell, 1968), but only a few examples of commercial control have been documented (Mackie, 1935). *Darluca filum*, a hyperparasite, has been examined as a mechanism for rust control, and Chester *et al.* (1951) summarized this work. The integrated control program for wheat stem rust was reviewed for the United States by Rowell (1973) and for Canada by Green and Campbell (1979).

One of the most spectacular success stories in plant pathology and plant breeding is the complete control of wheat stem rust in the North American Great Plains for over 25 years. This highly destructive pathogen of wheat, which had periodically destroyed millions of hectares of wheat in a single year since the late 1880s, has been successfully overcome. The resistance of the cultivars released since the mid-1950s has endured, even though there has been a continual replacement by improved cultivars. Stem-rust-resistant cultivars have come from many different breeding programs; however, most have historically come from Agriculture Canada at Winnipeg; the United States Department of Agriculture in cooperation with the North Dakota State University at Fargo, with the University of Minnesota at St. Paul, and with the University of Nebraska at Lincoln; CIMMYT (Centro Internacional Mejoramiento de Maíz y Trigo) at Mexico City; and the University of Sydney in New South Wales. These resistant cultivars are generally composed of multigenic resistances that were selected in the field by using a very heavy inoculum density of pathogen cultures differing widely in their phenotypes for virulence. These resistant cultivars have reduced the pathogen population, so that in the past 5 years few samples of stem rust have been found in commercial wheat fields in North America. Nearly all the collections for the annual race surveys in Canada, the United States, and Mexico are made from trap plots, susceptible cultivars in nurseries, or from non-*Triticum* hosts. This reduction in inoculum in turn makes the host resistances used more effective and also reduces the total numbers of pathogen mutations that are likely to occur.

A. CULTURAL METHODS

Currently, cultural methods largely depend on the use of early maturing cultivars and early planting of spring wheats. However, early planting of fall-sown grains may actually increase the chance for fall infection and overwintering of the rust in milder climates. The environmental conditions that favor wheat and rust development are similar. Avoiding excess nitrogen applications and frequent light applications of irrigation water are generally helpful in controlling stem rust. In areas where the disease oversummers, destruction of volunteer wheats and other susceptible grasses several weeks before planting also reduces inoculum level and delays initial infection. Where both winter and spring wheat are grown in the same area, separating these crops by space or by another type of crop can delay the spread between fields. Diversity in the cultivars grown on a farm and spacing between fields of wheat can provide substantial benefits (see Chapter 13, this volume).

B. BARBERRY ERADICATION

Eradication of the alternate host was started in Rouen, France, in 1660, over 100 years before science showed a relation between stem rust and barberry. Reviews of the successful programs in Denmark and the United States were reported on by Hermansen (1968) and Roelfs (1982), respectively. These studies revealed four effects of barberry eradication on stem rust epidemics: (1) delayed disease onset, (2) reduction in initial inoculum level, (3) decreased number of pathogen races, and (4) stabilization of pathogen phenotypes. The available evidence indicates that much of this change may result from eliminating the sexual population of *P. graminis* and a corresponding increase in the importance of the asexual population.

1. *Delayed Disease Onset*

Aeciospores are produced locally in large numbers early in the spring. In contrast, exogenous inoculum arrives only after urediospores are produced elsewhere. For effective inputs of exogenous inoculum, weather conditions must be favorable for urediospore production, liberation, and long-distance transport, and the arriving urediospore must be deposited on a susceptible host genotype when conditions are favorable for infection. The difference in date of disease onset before and after barberry eradication on a regional basis is now difficult to document in many cases, as data on date of disease onset were not collected from infection centers near individual barberry bushes with regularity. In Minnesota, Stakman and Harrar (1957) reported the normal severity for stem rust of wheat in commercial fields was trace to 8% on July 8. From 1965 through 1982, traveling the same route, Roelfs (1982) found only traces of rust, and in most recent years no rust was found even on trap plots of susceptible cultivars.

2. *Reduction in Initial Inoculum Level*

A rust-infected barberry bush can produce around 60×10^9 spores daily, thus resulting in heavy infection in the vicinity of the infected barberry bush (see Section II,C). An exogenous inoculum source is unlikely to provide an equal amount of inoculum to areas remote from it. Exogenous inoculum usually is not concentrated at a point but is generally dispersed in transport and then deposited over a large area (Rowell and Roelfs, 1971). Thus, for effective dispersal of large amounts of exogenous inoculum, environmental conditions must be favorable for disease increase in the source area, for spore transport, for spore deposition, and for infection in the target area. The spores must then possess virulence for the cultivars in the target area. Rarely are these conditions all met, and the number of epidemics are few. However, the epidemics of 1935, 1937, 1953, and 1954 in the United States and Canada show the potential threat of exogenous inoculum (Stakman and Harrar, 1957).

3. *Decreased Number of Pathogen Phenotypes*

Sexual recombination as the major source of new combinations of virulence genes is eliminated with eradication of the barberry. As virulence is normally recessive, mutations for virulence are not always expressed. Parasexualism seems to result in few recombinations that exist in the natural population (Luig, 1977). In one sexually reproducing population, 1 phenotype recombinant was found per 4.2 isolates but only 1 per 148 isolates in an asexually reproducing population (Roelfs and Groth, 1980).

4. *Stabilization of Pathogen Phenotypes*

Not only has elimination of the sexual cycle reduced the number of pathogen phenotypes for virulence, it also has stabilized them over time. This can be demonstrated by comparing the years 1918 through 1921, when a different race (pathogen phenotype) predominated in the race survey each year, with the 48 years since 1934, when only four races dominated the annual populations and two races predominated in 46 of the years (Roelfs, 1982). These four races represent more than four pathogen phenotypes, but a single phenotype has predominated for periods greater than 10 years. Early in the barberry eradication program, great differences existed among the races found between years; thus, in 1918, 28 races were identified; six of these races were not found in 1919, but three different races were found (Roelfs, 1982).

Removal of barberry has resulted in reduced yield losses and made it possible to successfully breed for resistance to stem rust. Many of the resistances currently used could not withstand the very high inoculum densities near an infected barberry bush (for example *Sr36*, *Sr2*, Thatcher, and so forth).

C. RESISTANCE IN THE CROPS

Certain combinations of single host genes for resistance have effectively controlled stem rust for many years. Many examples could be given; however, the story starts with the introduction of the cultivar Selkirk (*Sr6, 7b, 9d, 17, 23, 2*) which terminated the 15B epidemics of the early 1950s, in the northern Great Plains of North America. Selkirk was widely grown until the mid 1960s, when it was replaced by higher yielding, leaf rust-resistant cultivars. Selkirk has remained resistant in field plots through 1983. The cultivars Era (*Sr5, 6, 12, 17,* plus) and Waldron (*5, 11, Wld-1*) were introduced over 10 years ago; each has been grown on over 1 million hectares annually for 10 years, and they have remained as resistant to stem rust as they were when released. The current list of designated host genes for resistance, their sources, low-infection types in seedlings, and comments on their response or effectiveness under field conditions are given in Table VII. Resistance of the race-specific type is discussed in detail in Chapter 15, this volume.

Stomatal exclusion and size of collechyma bundles were proposed as types of morphological resistance in wheat to *P. graminis* (Hart, 1931). They now are both thought to be unlikely mechanisms of resistance. *Puccinia graminis* normally penetrates closed stomata (Volume I, Chapter 10). Webster, the original cultivar studied, has small collenchyma bundles, but cultures virulent on it are known. The small uredia on Webster with most avirulent cultures is now known to be due to *Sr30* (Knott and McIntosh, 1978).

The resistance not yet shown to be race-specific (Chapter 16, this volume) has attracted attention in recent years. Whether this resistance is a unique response or is just a result of combinations of factors that are too small to measure with current technology is unknown in most cases.

Whether a resistance is of the specific or nonspecific type, it would appear that it restricts the disease in one of four ways. Several of the single genes listed in Table VII function in at least two ways.

1. *Reduction in Number of Successful Infections*

Host plants often differ in receptivity to the fungus; equal inoculum densities result in different numbers of uredia. This receptivity can be measured in the field by inoculating plants with a single heavy inoculum density and taking notes 14 days later (Rowell and McVey, 1979). The reduction may be total as with cultures avirulent to *Sr5* on hosts

Table VII

Known Host Genes for Resistance and Their Response to *Puccinia graminis* f. sp. *tritici*[a]

Sr gene	Source[b]	Response to an avirulent culture: Seedling infection type[c]	Adult plant response[d]
2	*T. durum*	None	Few large uredia at nodes, and in spike
5	*T. aestivum*	0	Immune
6	*T. aestivum*	;	Highly resistant
7a	*T. aestivum*	23C	Moderately susceptible
7b	*T. aestivum*	2	Moderately susceptible
8	*T. aestivum*	2	Moderately susceptible
9a	*T. aestivum*	2^-, 23	Moderately susceptible
9b	*T. aestivum*	2^+	Moderately susceptible
9d	*T. durum*	;1	Moderately resistant
9e	*T. dicoccum*	;1	Moderately resistant
9f	*T. aestivum*	;2^-	Rare, culture not evaluated
9g	*T. durum*	2^-	Moderately resistant
10	*T. aestivum*	;1^+N	Moderately susceptible
11	*T. aestivum*	;2^-,2^+3^-	Moderately resistant
12	*T. durum*	0;X^-	Moderately resistant
13	*T. durum*	2^+	Moderately susceptible
14	*T. durum*	12C	Moderately susceptible
15	*T. aestivum*	;1^+N	Nearly susceptible
16	*T. aestivum*	2	Moderately susceptible
17	*T. durum*	;1N	Moderately resistant
18	*T. aestivum*	;$2^{=}$	Rare, culture not evaluated
19	*T. aestivum*	1^-	Rare, culture not evaluated
20	*T. aestivum*	$2^{=}$	Rare, culture not evaluated
21	*T. monococcum*	;$1^{=}$	Resistant
22	*T. boeoticum*	0;	Moderately susceptible
23	*T. aestivum*	23C	Nearly susceptible
24	*A. elongatum*	2^-	Moderately susceptible
25	*A. elongatum*	2^-	Moderately susceptible to susceptible
26	*A. elongatum*	;2^-	Resistant
27	*S. cereale*	0;	Highly resistant
28	*T. aestivum*	;	Rare, culture not evaluated
29	*T. aestivum*	2^+2^-	Moderately susceptible
30	*T. aestivum*	2^-2	Moderately susceptible
31	*S. cereale*	0;	Resistant
32	*Ae. squarrosa*	2^-	Moderately susceptible
33	*Ae. squarrosa*	2	Moderately susceptible
34	*T. aestivum*	3C	Moderately susceptible

(*continued*)

Table VII (*Continued*)

Sr gene	Source[b]	Response to an avirulent culture: Seedling infection type[c]	Adult plant response[d]
35	*T. monococcum*	0;	Resistant
36	*T. timopheevii*	0,0;1^{+},40;	Immune, fewer uredia
37	*T. timopheevii*	0;	Highly resistant
LC	*T. aestivum*	2^{-}	Rare, culture not evaluated
Gt	*T. aestivum*	2	Moderately resistant
dp-2	*T. durum*	2	Resistant
X	*T. aestivum*	23C	Moderately susceptible
McN	*T. aestivum*	2^{-};	Rare, culture not evaluated
Kt'2'	*T. aestivum*	2	Rare, culture not evaluated
Wld-1	*T. aestivum*	2$^{=}$C	Resistant to moderately resistant
Tt-3	*T. timopheevii*	0,;1^{+}C	Resistant
U	*T. aestivum*[e]	21CN	Moderately susceptible
H	*T. durum*[f]	2C	Moderately susceptible

[a]Updated from Roelfs and McVey (1979).
[b]*Triticum* = *T.*, *Agropyron* = *A.*, *Aegilops* = *Ae.*, *Secalis* = *S.*
[c]Infection types at 18°C (plants with *Sr6, 10, 12, 15,* and *17* are more susceptible at higher temperatures, whereas plants with *Sr13* are more resistant); variation is encountered with host genetic background and ploidy level also (Luig and Rajaram, 1972).
[d]Many host resistances are less effective at high temperatures, high inoculum densities, and at plant maturity. Variations also occur with different host genetic backgrounds.
[e]Gene from Red Egyptian other than *Sr6, 8,* and *9a* (Loegering, 1968).
[f]Gene from H-44, other than *Sr7b, 9d,* and *17*. *SrH* was the cause of the differences in Canadian and United States race survey data in the 1970s (Green and Dyck, 1979).

possessing *Sr5*, or partial as with *Sr36* (*Tt-1*) (Rowell, 1981) and *Sr2* (Sunderwirth and Roelfs, 1980). Reduction can also be the result of several host factors, as in Thatcher wheat (Nazareno and Roelfs, 1981). Even among susceptible host cultivars, receptivity varies widely under field conditions (Rowell and McVey, 1979).

2. *Lengthened Latent Period*

Although often considered to be due to nonspecific resistance, a lengthened latent period may result from race-specific resistance. The *Sr36* gene derived from *T. timopheevii* conditions an increased latent period for penetrants that result in successful infections (Rowell, 1981). In the wheat–*Puccinia graminis* system, there are fewer known genotypes with long latent periods than in some other host–rust systems. However, there is a trend for longer latent periods with infections on many host genotypes as the plant ages (Sunderwirth and Roelfs, 1980); this effect often declines or even reverses itself with the start of host senescence.

3. *Reduction in Size of Sporulating Area*

Most resistance of the race-specific type listed in Table VII functions by reducing the sporulating area per lesion. Some resistances like *Sr9e* reduce the sporulating area until few spores are produced per lesion, whereas others like *Sr5* result in no sporulation and some like *Sr29* provide little reduction in numbers of spores produced. The amount of reduction in spore production due to a gene(s) for resistance is often affected by temperature, light, ploidy level of host, and host growth stage.

4. *Reduction in Duration of Uredial Sporulation*

Although some resistance genes seem to reduce the duration of uredial sporulation in infections of seedling leaves, I know of no data from adult plants that would indicate that uredia produce urediospores over varying time periods. However, it may be assumed such resistance exists but remains unreported owing to interactions with host age and environment.

D. FUNGICIDES

The use of fungicides for control of stem rust has been studied for many years (Rowell, 1968). However, chemical control has played a very minor role in stem rust control. The reasons are at least threefold: (1) the effectiveness of the host resistance, (2) the very high rate of disease increase for wheat stem rust under ideal conditions, and (3) the relatively low economic return per hectare of wheat in comparison with the cost of fungicide applications.

E. BIOLOGICAL CONTROL

The hyperparasite of rust, *Darluca filum* (Biv.) Cast., has been widely considered (Chester *et al.*, 1951); however, it seems currently to offer little promise as the rust must be present to have a buildup of the hyperparasite. Another hyperparasite, *Aphanocladium album*, is now being evaluated on a field scale, but its potential is currently unknown. Although *Verticillium niveostratosum*, *V. fungicola*, and *Cephalos-*

porium acremonium were found to be greenhouse parasites of rust, they were thought to have little potential as a practical means of controlling stem rust (Chester, 1946).

VI. Losses

Stem rust develops at warmer temperatures (30°C optimum) than the other rust diseases of wheat and rye. Thus, it is most frequently a disease of the reproductive portion of the host life cycle. Occasionally stem rust can become severe on early sown, fall-planted wheat or on irrigated wheat in tropical areas. Infection by stem rust of seedling wheat or rye under favorable environmental conditions can result in death of tillers or entire plants. A tiller of an adult wheat plant has a surface area of approximately 150 cm^2 including leaf and stem tissue. A disease severity of 100% (6.7 infections/cm^2) destroys the tiller (Rowell and Roelfs, 1976). Severe amounts of disease can halt plant growth or even kill the plant by reducing the photosynthetic area, causing a loss of nutrients and water and disrupting the plant transport system (see Volume I, Chapter 16). Restricted growth often results in small shriveled grain, weakened stems that break or lodge, and in severe cases the death of the plant.

A. Reduction in Photosynthetic Area

Lesions of rust can occupy a significant portion of the host plant tissue. The tissues affected are usually the flag leaf, peduncle, glumes, and awns, the very parts that are the source of most of the nutrients that are transported to the developing grain.

B. Loss of Nutrients and Water

The rupture of the plant epidermal cells by the rust fungus results in a loss of water from the plant. Because the pathogen also uses both water and nutrients from the plant to produce the large volume of urediospores daily, the plant suffers added stress. Infection at an early growth stage results in decreased availability and production of nutrients for plant growth. An infected plant also has less root growth, which aggravates the imbalance in normal water requirements. Such plants are more susceptible to winterkill, produce fewer tillers, have smaller heads, and occasionally have decreased spikelet fertility.

C. DISRUPTION OF NUTRIENT TRANSPORT

Stem rust is characterized by development of uredia on leaf sheaths and peduncle tissue. The fungus often penetrates through the tissue of the true stem. The rupture of the plant tissue by fungus can disrupt transport of nutrients to the roots and cause premature death of the roots (Bushnell and Rowell, 1968). Shriveled kernels result from disruption of nutrient transport to the filling grains (Calpouzos *et al.*, 1976).

D. STEM BREAKAGE AND LODGING

When disease is extremely severe on a portion of the stem, the straw may break, causing the plant spike to break over or fall to the ground. With mechanical harvesting, broken and lodged plants often have the spike below the cutter bar level, making the grain impossible to harvest economically.

E. ESTIMATION OF LOSSES

The several models developed to estimate losses due to stem rust were reviewed by Calpouzos *et al.* (1976). A table relating rust severity at different crop stages to loss was developed by Kirby and Archer (1927). Greaney (1935) found the average loss in spring wheat due to stem rust was 5.4% (range 3.1–9.7) for each 10% of terminal rust severity. Kingsolver *et al.* (1959) related loss in yield to the growth stage at which a 1% disease severity (10 pustules/tiller) occurred. The area under the disease progress curve (y = disease severity and x = days from heading) was used by Buchenau (1970) to predict losses due to stem rust. Calpouzos *et al.* (1976) related loss due to stem rust to the host stage when disease started and the rate at which the disease increased. These models all need more evaluation by using current agricultural practices and cultivars. The model by Calpouzos *et al.* (1976), although the most complex model, varied from extremely accurate to very inaccurate in later tests (Rowell, 1982).

F. WORLDWIDE LOSSES

Losses in Europe due to wheat stem rust were summarized by Hogg *et al.* (1969). They recorded only two mild epidemics in the 1960s, one in Czechoslovakia in 1962 and one in Portugal in 1960. At least seven nation-wide epidemics occurred in the 1920s. In the United States, no major national epidemics have occurred since 1954 (Roelfs, 1978).

Although stem rust is an important disease in Asia, Australia, Africa, and South America, summaries of losses from these areas are unavailable from published sources except for a summary of epidemics by Chester *et al.* (1951).

VII. The Future

Wheat stem rust is the most researched host–pathogen system in agriculture. However, much remains to be learned, in the areas of both applied and basic research. Probably the most elusive property has been the physiologic basis of resistance. Many leads have been followed but without success (Volume I, Chapter 7). The effect of many individual host genes has been studied on seedlings, but their effects on epidemics have not been carefully documented. The effects of combinations of these genes are largely unknown. Most cultivars have some level of resistance in comparison with the most susceptible cultivars known, that is, Line E or Morocco. The genetic basis of this resistance, however, is still unknown.

The mechanism by which the pycniospore fuses with the receptive hyphae, the migration of the nuclei following the fusion, the penetration of the barberry leaf by the infection peg of the basidiospore, and the germination of the teliospore all need clarification.

Genetic studies of the pathogen remain in their infancy. The only factors studied to date are spore color and virulence. Mutation rates for important virulence genes are generally unknown. Although other differences exist among pathogen cultures, they are generally ignored. Certainly future studies will need to consider aggressiveness and adaptation of biotypes. The use of adult plant resistance may necessitate evaluation of cultures on other than seedlings.

Studies of pathogen development must be done for large regions where different crops are grown and cultivars with different resistances are grown, that is, the real world of agriculture. As models of the pathogen–host–environment system are built, values for individual host resistance and pathogen virulence genes will have to be altered as the host growth stage and temperature change. Certain host genes in combination as well as certain pathogen genes in combination will not be equal to the best component (epistasis) or the sum of the components (additive) as is currently generally assumed.

Will stem rust again result in serious epidemics? Yes, the pathogen will change through mutation and selection. Yes, favorable environmental conditions will occur. However, genes for resistance exist and can be bred into well-adapted, high-yielding, and high-quality cultivars. Yet when the environment is ideal and inoculum density high, even resistant cultivars can fail (Roelfs *et al.*, 1972). It appears that a continued development of cultivars with a combination of different types of resistance can reduce the inoculum and prevent the pathogen population from increasing and thus can avoid catastrophic epidemics in the near future.

References

Ahrendt, L. W. A. (1961). *Berberis* and *Mahonia*, a taxonomic revision. *J. Linn. Soc. London Bot.* **57,** No. 369, 1–410.

Buchenau, G. W. (1970). Forecasting profits from spraying for wheat rusts. *S. Dak. Farm Home Res.* **21,** 31–34.

Buller, A. H. R. (1924). "Researches on Fungi," Vol. III, pp. 501–508. Longmans, Green, New York.

Buller, A. H. R. (1958). The violent discharge of aecidiospores. *In* "Researches on Fungi," Vol. III, pp. 552–559. Hafner, New York.

Bushnell, W. R., and Rowell, J. B. (1968). Premature death of adult rusted wheat plants in relation to carbon dioxide evolution by root systems. *Phytopathology* **58,** 651–658.

Calpouzos, L., Roelfs, A. P., Madson, M. E., Martin, F. B., Welsh, J. R., and Wilcoxson, R. D. (1976). A new model to measure yield losses caused by stem rust in spring wheat. *Minn., Agric. Exp. Stn., Tech. Bull.* **307,** 1–23.

Chester, K. S. (1946). "The Nature and Prevention of the Cereal Rusts as Exemplified in the Leaf Rust of Wheat." Chronica Botanica, Waltham, Massachusetts.

Chester, K. S., Gilbert, F. A., Hay, R. E., and Newton, N. (1951). "Cereal Rusts: Epidemiology, Losses, and Control." Battelle Memorial Institute, Columbus, Ohio.

Cotter, R. U. (1932). Factors affecting the development of the aecial stage of *Puccinia graminis. U.S., Dep. Agric., Tech. Bull.* **314,** 1–38.

Cotter, R. U., and Levine, M. N. (1932). Physiological specialization in *Puccinia graminis secalis. J. Agric. Res. (Washington, D.C.)* **45,** 297–315.

Craigie, J. H. (1927). Discovery of the function of the pycnia of rust fungi. *Nature (London)* **120,** 765–767.

Craigie, J. H. (1940). Studies in cereal diseases. XII. Stem rust of cereals. *Can., Dep. Agric., Farmers' Bull.* **84,** 1–39.

DeBary, A. (1866). Neue Untersuchungen uber die Uredineen insbesondere die Entwicklung der *Puccinia graminis* und den Zusammenhang derselben mit *Aecidium berberidis. Monatsber. K. Preuss. Akad. Wiss.* pp. 15–50.

Fontana, F. (1932). "Observations on the Rust of Grain" (P. P. Pirone, transl.), Phytopathol. Classics, No. 2. Am. Phytopathol. Soc., Washington, D.C. (originally published, 1767).

Greaney, F. J. (1935). Method of estimating losses from cereal rusts. *Proc. World's Grain Exch. Conf., 1933,* Vol. 2, pp. 224–235.

Green, G. J. (1965). Stem rust of wheat, rye and barley in Canada in 1964. *Can. Plant Dis. Surv.* **45,** 23–29.
Green, G. J. (1971). Hybridization between *Puccinia graminis tritici* and *Puccinia graminis secalis* and its evolutionary implications. *Can. J. Bot.* **49,** 2089–2095.
Green, G. J. (1975). Virulence changes in *Puccinia graminis* f. sp. *tritici* in Canada. *Can. J. Bot.* **53,** 1377–1386.
Green, G. J. (1981). Identification of physiologic races of *Puccinia graminis* f. sp. *tritici* in Canada. *Can. J. Plant Pathol.* **3,** 33–39.
Green, G. J., and Campbell, A. B. (1979). Wheat cultivars resistant to *Puccinia graminis tritici* in western Canada; their development, performance, and economic value. *Can. J. Plant Pathol.* **1,** 3–11.
Green, G. J., and Dyck, P. L. (1979). A gene for resistance to *Puccinia graminis* f. sp. *tritici* that is present in wheat cultivar H-44 but not in cultivar Hope. *Phytopathology* **69,** 672–675.
Green, G. J., and Johnson, T. (1958). Further evidence of resistance in *Berberis vulgaris* to race 15B of *Puccinia graminis* f. sp. *tritici. Can. J. Bot.* **36,** 351–355.
Groth, J. V., and Roelfs, A. P. (1982). The effect of sexual and asexual reproduction on race abundance in cereal rust fungus populations. *Phytopathology* **72,** 1503–1507.
Hart, H. (1931). Morphologic and physiologic studies on stem-rust resistance in cereals. *U.S., Dep. Agric., Tech. Bull.* **266.**
Hayes, H. K., Ausemus, E. R., Stakman, E. C., Bailey, C. H., Wilson, H. K., Bamberg, R. H., Morkley, M. C., Crim, R. F., and Levine, M. N. (1936). Thatcher wheat. *Stn. Bull.—Minn., Agric. Exp. Stn.* **325,** 1–36.
Hermansen, J. E. (1968). "Studies on the Spread and Survival of Cereal Rust and Mildew Diseases in Denmark," Contrib. No. 87. Dep. Plant Pathol., R. Vet. Agric. Coll., Copenhagen.
Hirst, J. M., and Hurst, G. W. (1967). Long-distance spore transport. *In* "Airborne Microbes" (P. H. Gregory and J. L. Monteith, eds.), pp. 307–344. Cambridge Univ. Press, London and New York.
Hogg, W. H., Hounam, C. E., Mallik, A. K., and Zadoks, J. C. (1969). Meteorological factors affecting the epidemiology of wheat rusts. WMO, *Tech. Note* **99,** 1–143.
Johnson, T. (1949). Intervarietal crosses in *Puccinia graminis. Can. J. Res.* **27,** 45–65.
Johnson, T., and Green, G. J. (1954). Resistance of common barberry (*Berberis vulgaris* L.) to race 15B of wheat stem rust. *Can. J. Bot.* **32,** 378–379.
Katsuya, K., and Green, G. J. (1967). Reproductive potentials of races 15B and 56 of wheat stem rust. *Can. J. Bot.* **45,** 1077–1091.
Kingsolver, C. H., Schmitt, C. G., Peet, C. E., and Bromfield, K. R. (1959). Epidemiology of stem rust. II. Relation of quality of inoculum and growth stage of wheat and rye at infection to yield reduction by stem rust. *Plant Dis. Rep.* **43,** 855–862.
Kirby, R. S., and Archer, W. A. (1927). Diseases of cereal and forage crops in the United States in 1926. *Plant Dis. Rep., Suppl.* **53,** 110–208.
Kislev, M. E. (1982). Stem rust of wheat 3300 years old found in Israel. *Science* **216,** 993–994.
Knott, D. R., and McIntosh, R. A. (1978). Inheritance of stem rust resistance in Webster wheat. *Crop Sci.* **18,** 365–369.
Levine, M. N., and Stakman, E. C. (1923). Biologic specialization of *Puccinia graminis secalis. Phytopathology* **13,** 35 (abstr.).
Littlefield, L. J. (1981). "Biology of the Plant Rusts: An Introduction." Iowa State Univ. Press, Ames.
Littlefield, L. J., and Heath, M. C. (1979). "Ultrastructure of Rust Fungi." Academic Press, New York.

Loegering, W. Q. (1968). A second gene for resistance to *Puccinia graminis* f. sp. *tritici* in the Red Egyptian 2D wheat substitution line. *Phytopathology* **58,** 584–586.

Luig, N. H. (1977). The establishment and success of exotic strains of *Puccinia graminis tritici* in Australia. *Proc. Ecol. Soc. Aust.* **10,** 89–96.

Luig, N. H. (1983). "A Survey of Virulence Genes in Wheat Stem Rust, *Puccinia graminis* f. sp. *tritici*." Parey, Berlin.

Luig, N. H., and Rajaram, S. (1972). The effect of temperature and genetic background on host gene expression and interaction to *Puccinia graminis tritici*. *Phytopathology* **62,** 1171–1174.

Luig, N. H., and Tan, B. H. (1978). Physiologic differentiation of wheat stem rust on rye. *Aust. J. Biol. Sci.* **31,** 545–551.

Luig, N. H., and Watson, I. A. (1972). The role of wild and cultivated grasses in the hybridization of *formae speciales* of *Puccinia graminis*. *Aust. J. Biol. Sci.* **25,** 335–342.

Luig, N. H., and Watson, I. A. (1976). Strains of *Puccinia graminis* virulent on wheat plants carrying gene *Sr*27 derived from Imperial Rye. *Phytopathology* **66,** 664–666.

Luig, N. H., McIntosh, R. A., and Watson, I. A. (1973). Genes for resistance to *P. graminis* in the standard wheat stem rust differentials. *Proc. Int. Wheat Genet. Symp., 4th, 1973* pp. 423–424.

McFadden, E. S. (1930). A successful transfer of emmer characters to *vulgare* wheat. *Agron. J.* **22,** 1020–1034.

McIntosh, R. A. (1976). Genetics of wheat and wheat rusts since Farrer: Farrer Memorial Oration 1976. *J. Aust. Inst. Agric. Sci.* **42,** 203–216.

Mackie, W. W. (1935). Aeroplane dusting with sulphur to combat stem rust of wheat. *Phytopathology* **25,** 892–893 (abstr.).

Melander, L. W., and Craigie, J. H. (1927). Nature of resistance of *Berberis* spp. to *Puccinia graminis*. *Phytopathology* **17,** 95–114.

Mont, R. M. (1970). Studies of nonspecific resistance to stem rust in spring wheat. M.S. Thesis, University of Minnesota, Minneapolis.

Nazareno, N. R. X., and Roelfs, A. P. (1981). Adult plant resistance of Thatcher wheat to stem rust. *Phytopathology* **71,** 181–185.

Novotelnova, N. S. (1935). Some observations on the conditions for the germination of teleutospores and basidiospores of *Puccinia graminis* f. sp. *avenae* and uredospores of *P. triticina*. *Zashch. Rast. (Leningrad)* **4,** 98–106.

Powers, L., and Hines, L. (1933). Inheritance of reaction to stem rust and barbing of awns in barley crosses. *J. Agric. Res.* **46,** 1121–1129.

Rajaram, S., and Campos, A. (1974). Epidemiology of wheat rusts in the western hemisphere. *CIMMYT Res. Bull.* **27,** 1–27.

Roelfs, A. P. (1972). Gradients in the horizontal dispersal of cereal rust uredospores. *Phytopathology* **62,** 70–76.

Roelfs, A. P. (1978). Estimated losses caused by rust in small grain cereals in the United States—1918–76. *Misc. Publ.—U.S., Dep. Agric.* **1363,** 1–85.

Roelfs, A. P. (1982). Effects of barberry eradication on stem rust in the United States. *Plant Dis.* **66,** 177–181.

Roelfs, A. P., and Groth, J. V. (1980). A comparison of virulence phenotypes in wheat stem rust populations reproducing sexually and asexually. *Phytopathology* **70,** 855–862.

Roelfs, A. P., and McVey, D. V. (1972). Wheat stem rust races in the Yaqui valley of Mexico during 1972. *Plant Dis. Rep.* **56,** 1038–1039.

Roelfs, A. P., and McVey, D. V. (1979). Low infection types produced by *Puccinia graminis* f. sp. *tritici* and wheat lines with designated genes for resistance. *Phytopathology* **69,** 722–730.

Roelfs, A. P., McVey, D. V., Long, D. L., and Rowell, J. B. (1972). Natural rust epidemics in wheat nurseries as affected by inoculum density. *Plant Dis. Rep.* **56,** 410–414.

Roelfs, A. P., Long, D. L., and Casper, D. H. (1982). Races of *Puccinia graminis* f. sp. *tritici* in the United States and Mexico in 1980. *Plant Dis.* **66,** 205–207.

Rowell, J. B. (1968). Chemical control of the cereal rusts. *Annu. Rev. Phytopathol.* **6,** 243–262.

Rowell, J. B. (1973). Management of integrated control measures for the prevention of epidemics. *Abstrs. Pap., Int. Congr. Plant Pathol., 2nd, 1973,* No. 888.

Rowell, J. B. (1981). Relation of postpenetration events in Idaed 59 wheat seedlings to low receptivity to infection by *Puccinia graminis* f. sp. *tritici. Phytopathology* **71,** 732–736.

Rowell, J. B. (1982). Control of wheat stem rust by low receptivity to infection conditioned by a single dominant gene. *Phytopathology* **72,** 297–299.

Rowell, J. B., and McVey, D. V. (1979). A method for field evaluation of wheats for low receptivity in infection by *Puccinia graminis* f. sp. *tritici. Phytopathology* **69,** 405–409.

Rowell, J. B., and Olien, C. R. (1957). Controlled inoculation of wheat seedlings with urediospores of *Puccinia graminis* var. *tritici. Phytopathology* **47,** 650–655.

Rowell, J. B., and Roelfs, A. P. (1971). Evidence for an unrecognized source of overwintering wheat stem rust in the United States. *Plant Dis. Rep.* **55,** 990–992.

Rowell, J. B., and Roelfs, A. P. (1976). Wheat stem rust. *In* "Modeling for Pest Management, Concepts, Techniques, and Applications U.S.A./U.S.S.R." (R. L. Tummala, D. L. Haynes, and B. A. Croft, eds.), 2nd U.S./U.S.S.R. Symp., pp. 69–79. Michigan State University, East Lansing.

Rowell, J. B., and Romig, R. W. (1966). Detection of urediospores of wheat rusts in spring rains. *Phytopathology* **56,** 807–811.

Savile, D. B. O., and Urban, Z. (1982). Evolution and ecology of *Puccinia graminis. Preslia* **54,** 97–104.

Stakman, E. C. (1923). The wheat rust problem in the United States. *Proc. Pan-Pac. Sci. Congr., 1st, 1923,* Vol. 1, pp. 88–96.

Stakman, E. C., and Harrar, J. G. (1957). "Principles of Plant Pathology." Ronald Press, New York.

Stakman, E. C., Levine, M. N., and Cotter, R. U. (1930). Origin of Physiologic forms of *Puccinia graminis* through hybridization and mutation. *Sci. Agric. (Ottawa)* **10,** 707–720.

Stakman, E. C., Stewart, D. M., and Loegering, W. Q. (1962). Identification of physiological races of *Puccinia graminis* var. *tritici. U.S., Agric. Res. Serv., ARS* **E617,** 1–53.

Steffenson, B. J. (1983). Resistance of *Hordeum vulgare* L. to *Puccinia graminis* Pers. M.S. Thesis, University of Minnesota, Minneapolis.

Steffenson, B. J., Wilcoxson, R. D., and Roelfs, A. P. (1982a). Field reaction of selected barleys to *Puccinia graminis. Phytopathology* **72,** 1002 (abstr.).

Steffenson, B. J., Wilcoxson, R. D., and Roelfs, A. P. (1982b). Reactions of barley seedlings to stem rust, *Puccinia graminis. Phytopathology* **72,** 1140 (abstr.).

Steffenson, B. J., Wilcoxson, R. D., Watson, I. A., and Roelfs, A. P. (1983). Physiologic specialization of *Puccinia graminis* f. sp. *secalis* in North America. *Plant Dis.* **67,** 1262–1264.

Sunderwirth, S. D., and Roelfs, A. P. (1980). Greenhouse evaluation of the adult plant resistance of *Sr2* to wheat stem rust. *Phytopathology* **70,** 634–637.

Tan, B. H., Watson, I. A., and Luig, N. H. (1975). A study of physiologic specialization of rye stem rust in Australia. *Aust. J. Sci.* **28,** 539–543.

Tan, B. H., Luig, N. H., and Watson, I. A. (1976). Genetic analysis of stem rust in *Secale cereale*. I. Genes for resistance to *Puccinia graminis* f. sp. *secalis*. *Z. Pflanzenzuecht.* **76,** 121–132.

Ukkelberg, H. G. (1933). The rate of fall of spores in relation to the epidemiology of black stem rust. *Bull. Torrey Bot. Club.* **60,** 211–228.

Waterhouse, W. L. (1929a). A preliminary account of the origin of two new Australian physiologic forms of *Puccinia graminis tritici*. *Proc. Linn. Soc. N.S.W.* **54,** 96–106.

Waterhouse, W. L. (1929b). Australian rusts studies I. *Proc. Linn. Soc. N.S.W.* **54,** 615–680.

Watson, I. A., and Luig, N. H. (1959). Somatic hybridisation between *Puccinia graminis* var. *tritici* and *Puccinia graminis* var. *secalis*. *Proc. Linn. Soc. N.S.W.* **84,** 207–208.

Watson, I. A., and Luig, N. H. (1962). Selecting for virulence on wheat while inbreeding *Puccinia graminis* var. *secalis*. *Proc. Linn. Soc. N.S.W.* **87,** 39–44.

Watson, I. A., and Luig, N. H. (1963). The classification of *Puccinia graminis* var. *tritici* in relation to breeding resistant varieties. *Proc. Linn. Soc. N.S.W.* **88,** 235–258.

Weinhold, A. R. (1955). Rate of fall of urediospores of *Puccinia graminis tritici* Eriks. & Henn. as affected by humidity and temperature. *Tech. Rep.—Off. Nav. Res. (U.S.)* ONR Contract No. N90nr 82400, pp. 1–104.

2

Wheat Leaf Rust

D. J. Samborski

Agriculture Canada, Research Station,
Winnipeg, Manitoba, Canada

I. Introduction

Wheat leaf rust, *Puccinia recondita* Rob. ex Desm. f. sp. *tritici,* usually does not cause spectacular damage, but on a worldwide basis it probably causes more damage than the other wheat rusts. The devastating losses previously caused by wheat stem rust in North America

THE CEREAL RUSTS,
Volume II

ISBN 0-12-148402-5

obscured the economic significance of leaf rust in this region for many years. In Eastern Europe, leaf rust has long been recognized as the major rust on wheat.

The early history of research on leaf rust of wheat was described by Chester (1946), who also noted the extensive research conducted in the Soviet Union. This research continues but is not well recognized outside of Eastern Europe. The Soviet journal *Genetica* is available in English translation and contains valuable contributions on genetical aspects of the wheat leaf rust system. In addition, contributions by Soviet and eastern European leaf rust workers to the European and Mediterranean Cereal Rust Conferences are usually in English. These conferences attract many rust workers from outside of Europe and the Mediterranean area and are the primary international forum. The North American Wheat Leaf Rust Research Workers Committee consists of workers from Canada, the United States, and Mexico. In the past, the committee has largely dealt with the mechanics of race identification and nomenclature.

It is not possible to list all the workers who have contributed to our present understanding of leaf rust of wheat. However, important contributions were made by Mains, Caldwell, Johnston, and Chester (United States), Newton, Johnson, and Brown (Canada), Vallega (Argentina), da Silva (Brazil), Waterhouse and Watson (Australia), d'Oliveira (Portugal), Sibilia (Italy), Hassebrauk (Germany), and a number of workers in the Soviet Union whose contributions were described by Chester (1946).

Physiologic specialization in leaf rust of wheat was investigated extensively by the early workers, but there was no consideration of the involvement of specific genes in the parasite, although genes for resistance to leaf rust were recognized. Consideration of specific gene interactions had to wait for general acceptance of the gene-for-gene theory proposed by Flor (1956) for flax and flax rust, which was later elaborated by Person (1959) into a general concept of host–parasite interactions. In the last 20 years there has been considerable progress in the genetic aspects of host–parasite interactions, but there has been little progress in elucidating the chemical products of the genes involved in gene-for-gene interactions. The leaf rust organism can be grown in axenic culture, although growth is limited (Katsuya *et al.*, 1978). Therefore, although availability of nutrients in the host probably influences the rate and amount of rust development, specific nutrients are not likely to be responsible for the specificity shown in gene-for-gene interactions.

II. Distribution and Importance of Wheat Leaf Rust

Wheat leaf rust occurs wherever wheat is grown, and it is the commonest, most widely distributed of all cereal rusts (Chester, 1946). Although the parasite has no doubt occurred on wheat through the course of its development, it is probably more damaging now that large areas tend to be sown to single, genetically homogeneous cultivars or to closely related cultivars.

In the eastern prairies of Canada, leaf rust normally reduces yields by 5–15% when widely grown cultivars are susceptible. However, higher losses can occur if the disease becomes severe before flowering (Samborski and Peturson, 1960). In the United States, epidemics of leaf rust have occurred more frequently on winter wheats in the southern half of the country than in areas further north (Roelfs, 1978). Yield losses in winter wheat can be severe. For example, losses attributable to leaf rust in Oklahoma and Kansas from 1973–1975 have been estimated at 4,110,000 tonnes (Roelfs, 1978).

Leaf rust is now the most important wheat disease in Mexico [International Maize and Wheat Improvement Center (CIMMYT), 1977]. An epidemic of leaf rust affected commercial fields of the cultivar Jupateco 73 in Northwest Mexico during 1976–1977. Severe infections on young plants caused yield reductions up to 40% (Dubin and Torres, 1981).

Leaf rust is endemic in the southern cone countries of South America, and its importance to wheat production has long been recognized. Consequently, wheat breeders in Argentina and Brazil have produced highly resistant cultivars; some of these cultivars have been used extensively as sources of resistance to leaf rust by breeders in other countries. The cultivar Frontana, produced by Beckman in Brazil in 1945 (da Silva, 1966), has been used extensively in North America. An Argentine spring wheat, Klein 33, was the leaf-rust-resistant parent in crosses that produced the cultivars Besostaya 1 and Besostaya 4 at Krasnodar in the U.S.S.R. (Prutskova and Ukhanova, 1972).

Leaf rust is not considered to be a serious problem in Western Europe, but is the most damaging wheat disease in Eastern Europe (Dwurazna *et al.*, 1980; Berlyand-Kozhevnikov *et al.*, 1973; Chumakov, 1963), causing an average yield reduction of 3–5%. It is endemic in the dry delta of Egypt, where irrigation provides moisture conditions suitable for rust infection (Saari and Wilcoxson, 1974). The disease is also severe in most years in Ethiopia (Dmitriev and Gorshkov, 1980). In

India, average losses of 3% have been estimated, although higher losses occur in certain areas if the cultivars are susceptible to leaf rust (Saari and Wilcoxson, 1974). It is now the most important disease of wheat in Pakistan, and a severe epidemic in 1978 resulted in an average national loss of 10% or 830,000 tonnes of wheat (Hussain *et al.*, 1980). The importance of leaf rust in Australia has been well documented (Waterhouse, 1952).

III. Taxonomy

Wheat leaf rust was recognized as a species distinct from the other rusts in 1815 by de Candolle, who described it as *Uredo rubigo-vera* (Chester, 1946). Eriksson and Henning in 1894 described *Puccinia dispersa,* which included leaf rusts of wheat and rye (Chester, 1946). In 1899, Eriksson described wheat leaf rust as *Puccinia triticina* (Chester, 1946). In 1932, Mains subdivided *P. rubigo-vera* into 56 *formae speciales,* one of which, f. sp. *tritici,* corresponded to Eriksson's *P. triticina* (Mains, 1932). In 1956, Cummins and Caldwell suggested that *Puccinia recondita* was the valid name for the leaf rusts of grasses (Cummins and Caldwell, 1956). The term *P. recondita* f. sp. *tritici* is now used by most, although not all, leaf rust workers.

The leaf rusts of grasses are mainly distinguishable by reference to haplont or diplont host relations, and it is not always clear whether they should be considered as *formae speciales* or whether some should be classified as separate species. However, the leaf rusts exhibit parasitic specialization on the hosts for both haplont and diplont phases of the fungus, suggesting that it is more realistic to classify them as *formae speciales* (d'Oliveira and Samborski, 1966; Anikster and Wahl, 1979; Wilson and Henderson, 1966).

IV. Physiologic Specialization

A. Differential Hosts

Wheat leaf rust, *P. recondita* Rob. ex Desm. f. sp. *tritici,* can be further subdivided by the reactions of genetically different strains of wheat to pure isolates of the parasite. Such physiologic specialization in wheat leaf rust was first reported by Mains and Jackson (1921).

Mains and Jackson by 1926 could distinguish 12 physiologic races on 11 differential cultivars (Mains and Jackson, 1926). Three of these differentials were subsequently dropped (Johnston and Mains, 1932), but the remaining eight differentials became accepted internationally.

Since the differentials were used widely and the results were compared, it was important to have cultivars in which the rust reactions were not affected by differences in environment. Three of the eight standard differentials were considered labile with respect to environment, and Basile (1957) proposed new race keys based on the remaining five differentials. This change resulted in more consistent rust reactions but also in the loss of genes *Lr2b, LrB,* and Lr11 from the original series. This was of considerable importance, since only seven genes for resistance to leaf rust were present in the original eight differentials. In practice, therefore, most workers used all the differentials and presented their results according to both keys.

Although physiologic race surveys can be very useful in determining the spread of races from one area to another, their main purpose is to determine the range of variability in the parasite. They are indispensable when deciding which cultivars to use as resistant parents in breeding programs, and in monitoring changes in virulence that affect commercial cultivars and breeding programs. Shortly after resistant cultivars were first released in Canada, it was noted that they were attacked by strains or biotypes of leaf rust that could not be differentiated from avirulent strains on the standard differentials (Johnson and Newton, 1946). Therefore, supplementary differentials were introduced and Hope or its derivatives were the first such differentials used in Canada. Cultures virulent on Hope resistance were designated by the letter "a" after the race number. A number of other supplementary differentials were later introduced to meet the changing needs related to disease control.

The genes for resistance that are present in resistant commercial cultivars or in cultivars that are being used as resistant parents should also be represented in the differential host series. This became obvious after Flor published his studies on flax and flax rust and proposed the gene-for-gene theory (Flor, 1956). Subsequently, backcross lines containing single genes for resistance to leaf rust were developed and were soon adopted as differential hosts in race surveys. These single-gene lines provided an effective means of characterizing the parasite populations in terms of specific gene interactions. A list of the named genes conditioning resistance to leaf rust has been published by Browder (1980).

The single-gene lines developed at Winnipeg with the cultivar

Thatcher as the backcross parent are now used throughout the world. In 1979, 19 single-gene lines were used in Canada, and new lines are added when they become available (Samborski, 1980). North American isolates of leaf rust can apparently detect only genes *Lr1, Lr2a, Lr2b, Lr2c, LrB, Lr3,* and *Lr11* in the old standard differentials. If single-gene lines, each with one of these seven genes, are included in race surveys, the results of present and earlier surveys can be compared. Older data, in terms of reactions of the standard differentials, can be interpreted in terms of specific gene interactions. Although single-gene lines are now the most important differentials, cultivars of known or unknown genotype are often included. All of the differentials used in Canada are important for epidemiological purposes or in relation to the breeding program.

B. NOMENCLATURE OF RACES

Race nomenclature was simple as long as the early standard differentials were used and a key was provided for the identification of physiologic races (Johnston, 1961). The introduction of supplementary hosts required further designations. This resulted in several systems of nomenclature, none of which was completely satisfactory (Young and Browder, 1965). At present, the North American leaf rust workers use a formula system like that introduced by Green for wheat stem rust (Green, 1981). However, the leaf rust workers do not designate strains of leaf rust characterized by a particular virulence formula with a formal race designation. Cultures of leaf rust used in further studies and those that appear in publications are identified by their survey number (Samborski, 1981). In North America, the leaf rust workers consider that race designations based on a set of differential hosts are cumbersome since differentials may be dropped or added. This lack of race designations makes it difficult to follow trends involving gene combinations, particularly since most differentials are retained for a considerable period. Race designations are given in Australia, but their system is unique (Watson and Luig, 1961).

V. Evolutionary Trends in Leaf Rust Races

The determination of evolutionary trends in leaf rust races must be based on race surveys, since trends can only be determined from continuous observations carried out for many years. Before dealing with trends that have been revealed, the mechanisms that most likely contribute to the evolution of leaf rust races will be considered.

A. SOURCES OF VARIABILITY

1. *Alternate Hosts*

Jackson and Mains (1921) demonstrated that *Thalictrum* spp. could function as the alternate host of leaf rust of wheat. Several other alternate hosts have been reported to function in restricted areas. These include *Isopyrum fumarioides* L. in Siberia (Chester, 1946), *Anchusa* spp. in Portugal (d'Oliveira and Samborski, 1966), and *Clematis* spp. in Italy (Sibilia, 1960) and in the Soviet Far East (Azbukina, 1980). It is not clear how important the aecial hosts are in the evolution of new races, but they must play a role by reassortment of variability during sexual recombination. Yamada *et al.* (1973) found infected plants of *Thalictrum thunbergii* DC near wheat fields in Japan. They concluded that this species is not important as a source of inoculum of leaf rust of wheat in Japan but is significant in producing new races. In general, where alternate hosts grow in close proximity to wheat fields, they would have a definite effect on the evolution of leaf rust races. This does not appear to operate in North America. Although some infection can be obtained experimentally (Saari *et al.*, 1968), natural infection of *Thalictrum* spp. native to North America occurs rarely, and thus the alternate host does not play an important role in the origin of new races of leaf rust on this continent.

2. *Asexual Recombination*

Anastomosis of germ tubes and hyphae occurs readily in rust fungi, and it is reasonable to expect mitotic recombination to occur. In fact, new strains have been isolated from hosts inoculated with mixtures of urediospores of two races of various rusts (Webster, 1974). Other workers obtained negative results when races were mixed (Barr *et al.*, 1964; Bartoš *et al.*, 1969b). At the present time, it is not possible to assess the importance of asexual recombination of nuclei in the evolution of leaf rust races.

3. *Mutation*

Mutations provide the basic variation that occurs in the rusts. When one considers the tremendous number of spores that are produced by the leaf rust parasite in any one year, it is obvious that mutations can account for most or all of the changes in virulence that are observed. The dikaryotic nature of the parasite when on wheat provides for considerable conservation of the variation caused by mutation. A number of leaf rust isolates have been selfed, and these cultures have been found to be heterozygous at many loci (Samborski and Dyck, 1968, 1976; Statler, 1977, 1979; Statler and Jones, 1981). Since the parasite reproduces asexually on its main host, deleterious mutations are also conserved; these include spore color mutants. Segregants with orange-colored urediospores are often obtained when pure cultures of leaf rust are selfed, but there is only one report of a color mutant collected from natural populations of leaf rust (Johnston, 1930). In many cases, orange-spored isolates obtained from the aecial host do not grow well on wheat, and could not compete in nature with the normal brown-spored members of the population.

Avirulence in leaf rust is usually dominant; a recessive mutation to virulence at one allele of any locus can therefore only be detected by selfing cultures on the aecial host. Thus, when part of a rust population is heterozygous for virulence at a particular locus, a mutation occurring at the other allele would result in a virulent strain in the rust population. In some cases they will be so infrequent that they cannot be detected by normal survey procedures, but if there is widespread cultivation of a cultivar with the corresponding gene for resistance, such strains have a selective advantage and they increase in the population to the point where they are readily detected.

A few loci conditioning avirulence show incomplete dominance, and in such cases, a mutation at one of the alleles can be detected by the higher infection type that is produced (Samborski, 1963).

B. LONG-TERM CHANGES IN LEAF RUST RACES IN NORTH AMERICA

Annual surveys of leaf rust populations have been carried on for many years in North America, and a considerable body of information is available for analyses of evolutionary trends (Johnston *et al.*, 1968; Johnson, 1956). Since standard differentials were used for many years and the genes for resistance in these cultivars are known, the data can be analyzed in terms of specific gene interactions.

As mentioned, the aecial host is of no epidemiological importance in North America and there do not appear to be any susceptible wild-grass hosts. Leaf rust overwinters largely on winter wheat in Texas and

Oklahoma, although it can and does survive on living wheat leaves farther north. Winds carry the inoculum north in the spring and it reaches Manitoba, Canada, in mid June on average. It is generally agreed that urediospores are blown south in the fall to reinfect winter wheat in the overwintering area. This is an oversimplification of the situation, but the epidemiology will be expanded by Roelfs in Chapter 13, this volume.

1. Influence of Resistance Genes in Overwintering Areas

It is obvious that genes for resistance in cultivars grown in the area where leaf rust overwinters must have a marked effect on the rust population of the whole area the following summer. When the cultivar Agent (resistance gene *Lr24*) was released in Oklahoma and Texas, virulence on *Lr24* had not been detected in North America. In 1971, Agent occupied 13,350 hectares in Oklahoma and virulent strains of leaf rust occurred on it in trace amounts, but cultures virulent on gene *Lr24* were not detected in Canada. In 1972, when Agent occupied 50,600 hectares (about 2.2% of the wheat area) in Oklahoma and leaf rust occurred at 1% severity, these cultures were isolated in Canada (Samborski, 1972).

2. Influence of Resistance Genes in Cultivars Grown in the Rust Area of Western Canada

The cultivar Renown, with resistance to leaf rust from Hope (*Lr14a*), was the first resistant cultivar to be widely grown in western Canada. Renown was released in 1937 but became severely affected by leaf rust in 1943. The cultivar Lee (gene *Lr10*) was licensed in 1950, but virulent strains of leaf rust were soon detected. In 1954, Selkirk (genes *Lr10*, *Lr16*) was released. Since gene *Lr10* was present in Lee, part of the rust population was virulent on this gene when Selkirk was released. Gene *Lr16* does not appear to have been in any other cultivar at this time, and all cultures tested were avirulent on cultivars with *Lr16*. Selkirk rapidly occupied most of the rust area in Canada and the adjoining states. Virulence on hosts with *Lr10* increased rapidly in the rust population, since *Lr10* conditions a high level of resistance and *Lr16* a moderate level that permits considerable sporulation. Virulence on cultivars with *Lr10* thus had a distinct advantage where Selkirk was grown. Virulence on hosts with *Lr16* was not detected until 1961 and increased gradually until 1966, when 56% of the isolates in the leaf rust race survey were virulent on *Lr16* (Samborski, 1967). The release of the cultivar Manitou (gene *Lr13*) and subsequently Neepawa (gene *Lr13*) led to a decline of the area sown to Selkirk after 1966, which was followed by a decline of the proportion of strains in the leaf rust population virulent on cultivars with *Lr16* (Fig. 1). Selkirk has occupied about 10% of the wheat area in Manitoba since 1972; however, no cultures virulent on host plants with *Lr16* have been isolated since 1976.

These results show that gene recycling as proposed by Person (1966) would be effective with gene *Lr16*. It should be emphasized that Selkirk was replaced by other cultivars while only about half of the rust population was virulent on *Lr16*. Virulence on *Lr13* appeared a few years after Manitou was released, but it occurred in that portion of the rust population that was avirulent on cultivars with *Lr16*. The cultivars grown in the rust area of Western Canada since 1966 do not have the Hope or H-44 resistance to leaf rust that was present in Renown, but all cultures isolated in this region during the annual race surveys are still virulent on Hope and H-44 resistance.

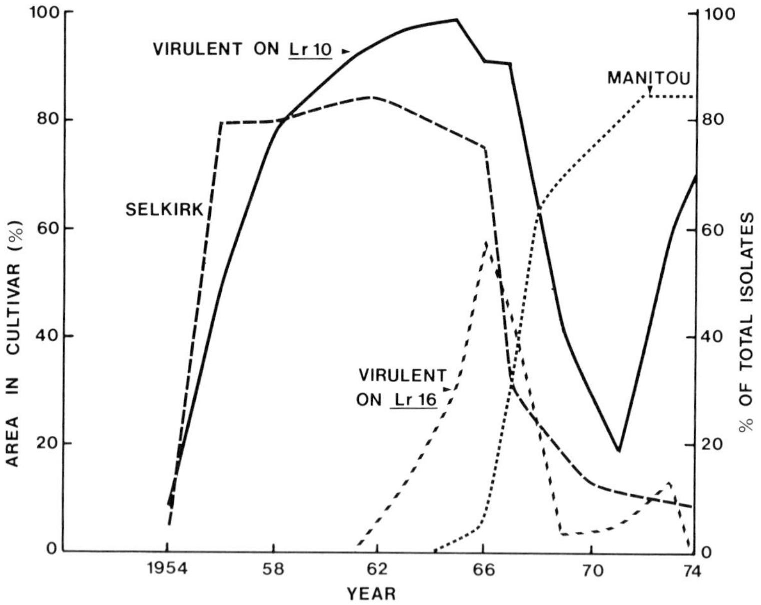

Fig. 1. Percentage of wheat area planted to the wheat cultivars Selkirk and Manitou and the percentage of leaf rust isolates virulent on genes *Lr10* and *Lr16* in Selkirk during the period 1954–1974. Manitou has gene *Lr13* conditioning adult plant resistance. From D. J. Samborski (unpublished results).

3. *Changes in Leaf Rust Races That Are Not Related to Resistance Genes in Commercial Cultivars*

Pawnee, released in 1943–1944, was the first winter wheat cultivar grown in the United States to have a marked influence on leaf rust races (C. O. Johnston, personal communication). Since Pawnee was resistant to race 9, which had been dominant for many years, it was soon replaced by race 5 and race 15. Resistance to race 9 was probably due to the presence of *Lr3,* on which race 9 is avirulent, while races 5, 15, and 126 are virulent. Other cultivars grown in the United States and Canada did not have any resistance corresponding to the genes in the standard differentials, so that changes in prevalence of races 5, 15, and 126 cannot be explained by the action of genes for resistance in the host; the changes in these races are shown in Fig. 2.

The most interesting long-term changes in race prevalence are shown by race 5 and race 15. Race 5 differs from race 15 on the standard differentials only on the cultivar Malakof (gene *Lr1*): race 5 is virulent and race 15 is avirulent on cultivars with gene *Lr1*. Both races appeared to be equally well adapted races, and there is no ready explanation for the dominance of race 15 since the mid 1950s. These changes cannot

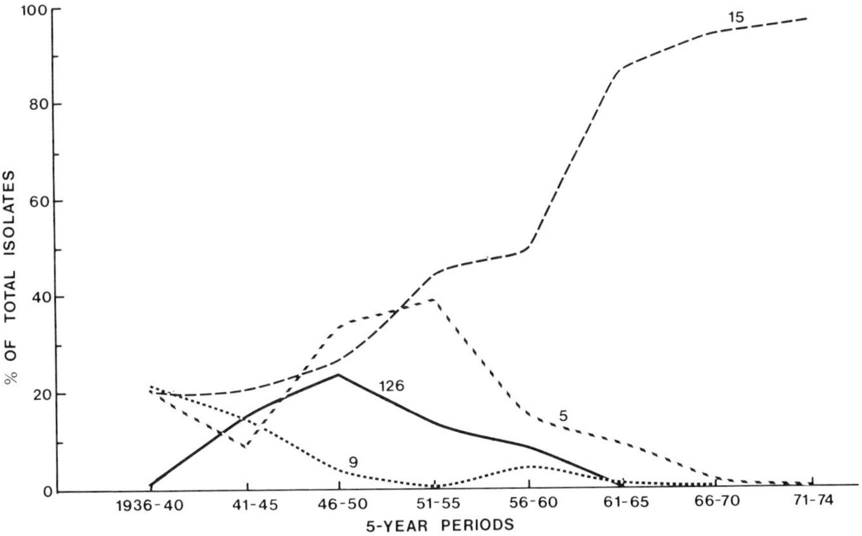

Fig. 2. Changes in prevalence of wheat leaf rust races 5, 9, 15, and 126 during the period 1936–1974. From D. J. Samborski (unpublished results).

be projected further, since cultivars with *Lr1* and *Lr2a* are now widely grown and have influenced subsequent changes in the rust population.

VI. Host–Parasite Genetics in the Wheat Leaf Rust System

Studies on the inheritance of virulence in wheat leaf rust were preceeded by studies on the inheritance of resistance in wheat and by the development of backcross lines containing single genes for resistance. The availability of single-gene lines enabled analyses of avirulence to be carried out with small populations of segregating cultures. The same populations could then be used to study cultivars of unknown genotype, and segregation on these hosts could be compared to that on the single-gene lines. Although it is difficult to produce a large number of cultures on the aecial host, the segregating population equivalent to an F_1 or F_2 can be maintained indefinitely and used repeatedly with new groups of wheats.

A. TELIOSPORE PRODUCTION

Wheat leaf rust does not produce teliospores readily under greenhouse or growth-chamber conditions. Teliospores do form on seedling wheat plants (Jackson and Young, 1967; Takahashi *et al.*, 1965; Freitas, 1972), but this is often very erratic. Teliospores can be most readily obtained by injecting urediospores into the culms of moderately resistant wheat plants in the boot stage. Severe infections are desirable, and teliospores form on the leaf sheath.

B. TELIOSPORE GERMINATION

It is sometimes assumed that teliospores of wheat rusts require a cold treatment in order to germinate. However, with wheat leaf rust a cold treatment is not necessary, and in areas such as Portugal, teliospores germinate in nature during the fall rains after a hot, dry summer. For genetic studies, teliospores will usually germinate if subjected to alternate wet–dry periods without a cold treatment. Teliospores of some cultures germinate readily, while others require a lengthy period of such cycles. Teliospores usually germinate best if they are collected while the plants are still green. Infection of the alternate host is readily carried out by suspending the teliospores over young plants of a susceptible species of *Thalictrum*. At Winnipeg, *Thalictrum speciosissimum* Loefl. is used routinely. Self-fertilization or crosses between rust cultures are carried out by transfer of pycnial nectar in a variety of ways (Samborski and Dyck, 1968; Dyck and Samborski, 1974).

C. GENE-FOR-GENE RELATIONSHIPS IN THE WHEAT–LEAF RUST SYSTEM

Studies on the inheritance of virulence on a number of genes in wheat conditioning resistance to leaf rust have been carried out, and in most cases, classical gene-for-gene interactions were demonstrated (Samborski and Dyck, 1968, 1976; Dyck and Samborski, 1970; Statler, 1977, 1979). However, interactions at two loci, *Lr2* and *Lr3*, each with several alleles, are unusual in that the action of a recessive gene for virulence appears to be modified or inhibited by a second gene (Samborski and Dyck, 1968; Dyck and Samborski, 1974; Haggag *et al.*, 1973). Interactions at the *Lr2* locus are particularly interesting since individual cultures are always least virulent on hosts with *Lr2a*, most virulent on hosts with *Lr2c*, and intermediate on plants with *Lr2b*.

Three types of cultures showing physiologic specialization at the *Lr2* locus are found in the North American leaf rust population. Cultures virulent on plants with *Lr2a* are virulent on the other alleles and are homozygous for gene *p2*. Cultures avirulent on all three alleles are either heterozygous or homozygous for a single gene *P2*, and segregation is identical on the three alleles. Cultures that produce an intermediate infection type on plants with *Lr2a* and that are virulent on plants with *Lr2c* are heterozygous for *P2* and presumably for another gene that affects the expression of *P2*.

D. ANALYSES OF CULTIVARS OF UNKNOWN GENOTYPE

A comparison of the patterns of infection types on backcross lines containing single genes for resistance with the patterns of infection types on cultivars of unknown genotype can often be used to identify genes for resistance in those cultivars (Bartoš *et al.*, 1969a; Browder, 1973; Dyck and Samborski, 1982). Unfortunately, rust cultures with critical combinations of genes for virulence are often not available. For example, culture 10-77 isolated in Manitoba is virulent on seedlings of the cultivar Tobari 66. This seedling resistance appears to be present in many cultivars (Dyck and Samborski, 1982), but culture 10-77 is avirulent on gene *Lr10,* which is present in many Mexican wheats that may also have the type of seedling resistance present in Tobari 66.

A further complication results from the ghost effect reported by Samborski and Dyck (1982). The cultivar Columbus has gene *Lr16* conditioning seedling resistance and gene *Lr13,* which conditions adult plant resistance to wheat leaf rust. Cultures of leaf rust that are virulent on wheat lines containing gene *Lr16* alone are not virulent on seedlings of Columbus. An interaction between genes *Lr16* and *Lr13* results in an incompatible reaction, when compatibility would be expected. In a screening test, it would be concluded that the seedling gene for resistance in Columbus was not gene *Lr16.* It has also been found that genes that singly condition moderate levels of seedling resistance to leaf rust can interact to give much higher levels of resistance (Dyck and Samborski, 1982; Samborski and Dyck, 1982). Therefore, characteristic infection types produced on lines with single genes for resistance cannot always be relied on when interpreting infection types obtained on multigenic cultivars.

Assessment and designation of genotypes becomes difficult when a large number of host cultivars and rust cultures are involved. Computer analysis and machine sorting of data are useful in such cases (Loegering and Burton, 1974; Browder and Eversmeyer, 1976). In many cases, the method proposed by Person (1959) for analyzing host cultivars can be very useful.

VII. Control

A. Resistant Cultivars

Selection of resistant parents is usually done on the basis of their reactions in small plots or rows in experimental fields. When tested in this manner, a cultivar may be highly resistant for many years since selection pressure is negligible under these conditions. However, if a cultivar with this type of resistance is widely grown, selection pressure is severe and races in the population with corresponding virulence may increase rapidly.

It is generally accepted that virulent strains of leaf rust will evolve rapidly on cultivars with single genes for resistance, while cultivars with multigenic resistance may remain resistant much longer. In both cases, the appearance of virulent strains is accelerated if large areas are planted to a single cultivar. Durum wheat (*Triticum durum* Desf.)

cultivars grown in western Canada are, and have always been, resistant to wheat leaf rust. Durums in other parts of the world are not noted for their resistance to leaf rust (diCariello *et al.*, 1977; Dmitriev and Gorshkov, 1980), and the long-lived resistance of durum cultivars in Canada is probably due to the fact that they are grown on a much smaller area than the bread wheats, *Triticum aestivum* L.

The importance of multigenic resistance can be illustrated by the cultivars Manitou and Chris. Both cultivars derived their resistance to leaf rust from Frontana. Manitou and Chris were released at about the same time and were grown commercially in Canada and the United States, respectively. The data in Table I show that the parasite population quickly developed virulence on Manitou, which has gene *Lr13*, while Chris, which also has *Lr13*, was protected by at least one additional gene. Frontana has not been grown commercially in North America and is still highly resistant to leaf rust in this area, although it is now susceptible in Brazil.

The genes for resistance to leaf rust of wheat in Frontana interact to give a higher level of resistance than is conditioned by any gene alone (Dyck and Samborski, 1982). The data in Table I show that when this type of resistance is involved, selection for the highest level of resistance results in selection for the greatest number of genes for resistance.

Table I

Field Reactions of Three Cultivars of Wheat to Infection with a Mixture of Races of *Puccinia recondita*[a]

	Leaf rust reaction[b]		
Year	Frontana	Chris	Manitou
1967	0	0	2R
1968	0	TR	10M
1969	0	TMS	10M
1970	TR	5MS	20M
1971	TR	15MR	50MS
1972	TR	15MR	50MS
1973	TR	20MR	50MS

[a]Leaf rust resistance in Chris and Manitou was derived from Frontana. Manitou has only gene *Lr13*, Chris has *Lr13* plus at least one additional gene, and Frontana has *Lr13* plus at least two additional genes for resistance to leaf rust.

[b]R, resistant; TR, trace resistant; M, mesothetic; TMS, trace moderately susceptible; MS, moderately susceptible; MR, moderately resistant.

B. CHEMICAL CONTROL

Studies on chemical control of cereal rusts began in the last century, but it was soon concluded that chemical control was not economical with the available chemicals (Dickson, 1959). Organic compounds and mixtures of inorganic salts plus dithiocarbamate fungicides were later investigated and showed considerable promise (Rowell, 1968). The recent introduction of systemic fungicides has increased interest in chemical control of wheat leaf rust. Bayleton (Buchenauer, 1976) and Indar (von Meyer *et al.*, 1970) have proven particularly effective in controlling leaf rust of wheat (Rowell, 1972; Abdel Hak *et al.*, 1980; Line and Rakotondradona, 1980) and are of special interest since they can be applied as seed dressings. Indar is highly specific and controls only wheat leaf rust, while Bayleton also controls the other wheat rusts.

Although resistant cultivars will always be the best method of controlling leaf rust, an economical method of chemical control would be valuable for situations when new races of leaf rust develop and new resistant cultivars are not available. This occurred in northwest Mexico in 1976–1977, where a severe attack of leaf rust threatened the prevalent cultivar Jupateco 73. A large area was sprayed with Indar and Bayleton and major yield losses were prevented. New resistant cultivars were released before the next growing season, and there has been no further need for chemical control in this area (Dubin and Torres, 1981).

VIII. Conclusions

If we can learn any useful lessons from past experience, it must surely be clear that constant vigilance and diligent research are necessary if losses from leaf rust of wheat are to be kept to a minimum. It has been demonstrated time and again that cultivars with a single gene for resistance will not remain resistant very long. Yet new cultivars with single genes for resistance are continually being released for commercial use. In many cases, a wheat breeder does not know what genes for resistance are in his parental cultivars, although the information may be available. Close cooperation between breeders and pathologists is obviously essential.

It is probably too late to collect appreciable numbers of land race populations of wheat with the hope of extracting new genes for resistance. The large number of entries in some existing wheat collections does not give much comfort for the future, since most of the entries are usually susceptible to leaf rust and many of the resistant entries have the same genes for resistance. However, genes for resistance to leaf rust of wheat can be transferred from hosts such as *Aegilops* spp. and *Agropyron* spp. and these hosts should be collected wherever possible.

Exotic genes for resistance are no better than any other genes if they are used alone. In North America, cultivars with *Lr9* from *Aegilops umbellulata* Zhuk. and *Lr24* [*Agropyron elongatum* (Host.) Beauv.] were released and virulent strains of leaf rust appeared quickly. In Europe, *Lr26* from *Secale cereale* L. was an important source of resistance, but virulent strains of leaf rust appeared soon after cultivars with this gene were released (Bartoš, 1973).

There is considerable promise for sustained control of leaf rust by gene management. For example, it is obviously good management to have different genes for resistance in spring wheat and winter wheat cultivars in North America. In addition, some genes can be usefully cycled if they are properly managed. However, for the foreseeable future, gene management with respect to wheat leaf rust is likely to be of only theoretical interest, at least in North America. It would entail the control of licensing and withdrawal of cultivars over an entire epidemiological area, as well as removing genes for resistance from particular lines in breeding programs. Although the economic benefits are obvious, the required degree of informal cooperation or governmental action is not likely to be forthcoming. However, gene management can be effectively carried out through the use of multilines and by the cultivation of a number of cultivars with different genes for resistance. If such multilines or cultivars have multigenic resistance, and cooperation between pathologists and breeders is maintained, wheat leaf rust should not be a major threat to wheat production.

References

Abdel Hak, T. M., El-Sherif, N. A., Bassiouny, A. A., Shafik, I. I., and El-Dauadi, Y. (1980). Control of wheat leaf rust by systemic fungicides. *Proc.—Eur. Mediterr. Cereal Rusts Conf., 5th, 1980,* pp. 255–266.

Anikster, Y., and Wahl, I. (1979). Coevolution of the rust fungi on Gramineae and Liliaceae and their hosts. *Annu. Rev. Phytopathol.* **17,** 367–403.

Azbukina, Z. (1980). Economical importance of aecial hosts of rust fungi of cereals in the Soviet Far East. *Proc. Eur. Mediterr. Cereal Rusts Conf., 5th, 1980,* pp. 199–201.

Barr, R., Caldwell, R. M., and Amacher, R. H. (1964). An examination of vegetative recombination of urediospore color and virulence in mixtures of certain races of *Puccinia recondita*. *Phytopathology* **54,** 104–109.

Bartoš, P. (1973). Genetics of stem and leaf rust resistance of the cultivar Kavkaz. *Ved. Pr. Vysk. Ustavu Rastl. Vyroby Praha-Ruzyni,* pp. 63–67.

Bartoš, P., Samborski, D. J., and Dyck, P. L. (1969a). Leaf rust resistance of some European varieties of wheat. *Can. J. Bot.* **47,** 543–546.

Bartoš, P., Fleischmann, G., Samborski, D. J., and Shipton, W. A. (1969b). Studies on asexual variation in the virulence of oat crown rust, *Puccinia coronata* f. sp. *avenae,* and wheat leaf rust, *Puccinia recondita. Can. J. Bot.* **47,** 1383–1387.

Basile, R. (1957). A diagnostic key for the identification of physiologic races of *Puccinia rubigo-vera tritici* grouped according to a unified numeration scheme. *Plant Dis. Rep.* **41,** 508–511.

Berlyand-Kozhevnikov, V. M., Mikhaylova, L. A., and Borodanenko, N. K. (1973). *In* "Catalogue of Wheat Varieties Characterized by Definite Resistant Genes to Brown Rust" (V. F. Dorofeyev, ed.), pp. 3–75. All Union Order of Lenin Sci. Res. Inst. Crop Prod., Leningrad.

Browder, L. E. (1973). Probable genotype of some *Triticum aestivum* 'Agent' derivatives for reaction to *Puccinia recondita* f. sp. *tritici. Crop Sci.* **13,** 203–206.

Browder, L. E. (1980). A compendium of information about named genes for low reaction to *Puccinia recondita* in wheat. *Crop Sci.* **20,** 775–779.

Browder, L. E., and Eversmeyer, M. G. (1976). Recording and machine processing of cereal rust infection-type data. *Plant Dis. Rep.* **60,** 143–147.

Buchenauer, H. (1976). Studies on the systemic activity of Bayleton (triadimefon) and its effect against certain fungal diseases of cereals. *Pflanzenschutz-Nachr. (Bayer)* **29,** 266–280.

Chester, K. S. (1946). "The Nature and Prevention of the Cereal Rusts as Exemplified in the Leaf Rust of Wheat." Chronica Botanica, Waltham, Massachusetts.

Chumakov, A. E. (1963). Principles of wheat rust control in regions with the most frequent epiphytotics. *Trans. All-Union Inst. Prot. Plants* **18,** 280–288.

Cummins, G. B., and Caldwell, R. M. (1956). The validity of binomials in the leaf rust fungus complex of cereals and grasses. *Phytopathology* **46,** 81–82.

da Silva, A. R. (1966). "Melhoramento das variedades de trigo destinadas as diferentes regioes do Brazil," Estud. Tec. No. 33. Minist. Agric., Rio de Janeiro, Brazil.

di Cariello, G., Casulli, F., and Vallega, J. (1977). Razze, efficacia di alcuni geni di resistenza e comportamento di alcune varieta di Grano duro (*Triticum durum* Desf.) e di Grano tenero (*Triticum aestivum* L.) verso *Puccinia recondita* f. sp. *tritici* in Italia. *Phytopathol. Mediterr.* **16,** 51–64.

Dickson, J. G. (1959). Chemical control of cereal rusts. *Bot. Rev.* **25,** 486–513.

Dmitriev, A. P., and Gorshkov, A. K. (1980). The results of some wheat rusts investigation in Ethiopia. *Proc. Eur. Mediterr. Cereal Rusts Conf., 5th, 1980,* pp. 157–159.

d'Oliveira, B., and Samborski, D. J. (1966). Aecial stage of *Puccinia recondita* on Ranunculaceae and Boraginaceae in Portugal. *Proc. Cereal Rusts Conf., 1964,* pp. 133–150.

Dubin, H. J., and Torres, E. (1981). Causes and consequences of the 1976–1977 wheat leaf rust epidemic in Northwest Mexico. *Annu. Rev. Phytopathol.* **19,** 41–49.

Dwurazna, M., Bialota, M., and Gajda, Z. (1980). Resistance of wheat cultivars to rust in Poland. *Proc. Eur. Mediterr. Cereal Rusts Conf., 5th, 1980,* pp. 147–150.

Dyck, P. L., and Samborski, D. J. (1970). The genetics of two alleles for leaf rust resistance at the *Lr14* locus in wheat. *Can. J. Genet. Cytol.* **12,** 689–694.

Dyck, P. L., and Samborski, D. J. (1974). Inheritance of virulence in *Puccinia recondita* on alleles at the *Lr2* locus for resistance in wheat. *Can. J. Genet. Cytol.* **16,** 323–332.

Dyck, P. L., and Samborski, D. J. (1982). The inheritance of resistance to *Puccinia recondita* in a group of common wheat cultivars. *Can. J. Genet. Cytol.* **24,** 273–283.

Flor, H. H. (1956). The complementary genic systems in flax and flax rust. *Adv. Genet.* **8,** 29–54.

Freitas, A. P. C. e (1972). Teliospores of *Puccinia recondita* Rob. on wheat seedlings. *Actas Congr. Uniao Fitopatol. Mediterr., 3rd, 1972,* pp. 127–132.

Green, G. J. (1981). Identification of physiologic races of *Puccinia graminis* f. sp. *tritici* in Canada. *Can. J. Plant Pathol.* **3,** 33–39.

Haggag, M. E. A., Samborski, D. J., and Dyck, P. L. (1973). Genetics of pathogenicity in three races of leaf rust on four wheat varieties. *Can. J. Genet. Cytol.* **15,** 73–82.

Hussain, M., Hassan, S. F., and Kirmani, M. A. S. (1980). Virulences in *Puccinia recondita* Rob. ex. Desm. f. sp. *tritici* in Pakistan during 1978 and 1979. *Proc. Eur. Mediterr. Cereal Rusts Conf., 5th, 1980,* pp. 179–184.

International Maize and Wheat Improvement Center (CIMMYT) (1977). "Report on Wheat Improvement," pp. 81–82. CIMMYT, El Batan, Mexico.

Jackson, H. S., and Mains, E. B. (1921). Aecial stage of the orange leaf rust of wheat, *Puccinia triticina* Eriks. *J. Agric. Res. (Washington, D.C.)* **22,** 151–172.

Jackson, A. O., and Young, H. C., Jr. (1967). Teliospore formation by *Puccinia recondita* f. sp. *tritici* on seedling wheat plants. *Phytopathology* **57,** 793–794.

Johnson, T. (1956). Physiologic races of leaf rust of wheat in Canada 1931 to 1955. *Can. J. Agric. Sci.* **36,** 371–379.

Johnson, T., and Newton, M. (1946). The occurrence of new strains of *Puccinia triticina* in Canada and their bearing on varietal reaction. *Sci. Agric. (Washington, D.C.)* **26,** 468–478.

Johnston, C. O. (1930). An aberrant physiologic form of *Puccinia triticina* Eriks. *Phytopathology* **20,** 609–620.

Johnston, C. O. (1961). Sixth revision of the international register of physiologic races of *Puccinia recondita* Rob. ex Desm. (formerly *P. rubigo-vera tritici*). *U.S., Agric. Res. Serv., ARS* **34-27,** 1–15.

Johnston, C. O., and Mains, E. B. (1932). Studies on physiologic specialization in *Puccinia triticina. U.S., Dep. Agric., Tech. Bull.* **313,** 1–22.

Johnston, C. O., Caldwell, R. M., Compton, L. E., and Browder, L. E. (1968). Physiologic races of *Puccinia recondita* f. sp. *tritici* in the United States from 1926 through 1960. *U.S., Dep. Agric., Tech. Bull.* **1393,** 1–18.

Katsuya, K., Kakishima, M., and Sato, S. (1978). Axenic culture of two rust fungi, *Puccinia coronata* f. sp. *avenae* and *P. recondita* f. sp. *tritici. Ann. Phytopathol. Soc. Jpn.* **44,** 606–611.

Line, R. F., and Rakotondradona, R. (1980). Chemical control of *Puccinia striiformis* and *Puccinia recondita. Proc. Eur. Mediterr. Cereal Rusts Conf., 5th, 1980,* pp. 239–241.

Loegering, W. Q., and Burton, C. H. (1974). Computer-generated hypothetical genotypes for reaction and pathogenicity of wheat cultivars and cultures of *Puccinia graminis tritici. Phythopathology* **64,** 1380–1384.

Mains, E. B. (1932). Host specialization in the leaf rust of grasses, *Puccinia rubigo-vera. Pap. Mich. Acad. Sci., Arts Lett.* **17,** 289–394.

Mains, E. B., and Jackson, H. S. (1921). Two strains of *Puccinia triticina* on wheat in the United States. *Phytopathology* **11,** 40 (abstr.).

Mains, E. B., and Jackson, H. S. (1926). Physiologic specialization in the leaf rust of wheat, *Puccinia triticina* Erikss. *Phytopathology* **16,** 89–120.

Person, C. (1959). Gene-for-gene relationships in host:parasite systems. *Can. J. Bot.* **37,** 1101–1130.

Person, C. O. (1966). Genetic polymorphism in parasitic systems. *Nature (London)* **212,** 266–267.

Prutskova, M. G., and Ukhanova, O. I. (1972). "New Varieties of Winter Wheat." Kolos Publishers, Moscow (transl. by A. K. Dhote, Amerind Publishing Co., New Delhi, 1976).

Roelfs, A. P. (1978). Estimated losses caused by rust in small grain cereals in the United States—1918–76. *Misc. Publ.—U.S., Dep. Agric.* **1363,** 1–85.

Rowell, J. B. (1968). Chemical control of the cereal rusts. *Annu. Rev. Phytopathol.* **6,** 243–262.

Rowell, J. B. (1972). Fungicidal management of pathogen populations. *J. Environ. Qual.* **1,** 216–220.

Saari, E. E., and Wilcoxson, R. D. (1974). Plant disease situation of high-yielding dwarf wheats in Asia and Africa. *Annu. Rev. Phytopathol.* **12,** 49–68.

Saari, E. E., Young, H. C., Jr., and Kernkamp, M. F. (1968). Infection of North American *Thalictrum* spp. with *Puccinia recondita* f. sp. *tritici*. *Phytopathology* **58,** 939–943.

Samborski, D. J. (1963). A mutation in *Puccinia recondita* Rob. ex. Desm. f. sp. *tritici* to virulence on Transfer, Chinese Spring × *Aegilops umbellulata* Zhuk. *Can. J. Bot.* **41,** 475–479.

Samborski, D. J. (1967). Leaf rust of wheat in Canada in 1966. *Can. Plant Dis. Surv.* **47,** 3–4.

Samborski, D. J. (1972). Leaf rust of wheat in Canada in 1972. *Can. Plant Dis. Surv.* **52,** 168–170.

Samborski, D. J. (1980). Occurrence and virulence of *Puccinia recondita* in Canada in 1979. *Can. J. Plant Pathol.* **2,** 246–248.

Samborski, D. J. (1981). Occurrence and virulence of *Puccinia recondita* in Canada in 1980. *Can. J. Plant Pathol.* **3,** 228–230.

Samborski, D. J., and Dyck, P. L. (1968). Inheritance of virulence in wheat leaf rust on the standard differential wheat varieties. *Can. J. Genet. Cytol.* **10,** 24–32.

Samborski, D. J., and Dyck, P. L. (1976). Inheritance of virulence in *Puccinia recondita* on six backcross lines of wheat with single genes for resistance to leaf rust. *Can. J. Bot.* **54,** 1666–1671.

Samborski, D. J., and Dyck, P. L. (1982). Enhancement of resistance to *Puccinia recondita* by interactions of resistance genes in wheat. *Can. J. Plant Pathol.* **4,** 152–156.

Samborski, D. J., and Peturson, B. (1960). Effect of leaf rust on the yield of resistant wheats. *Can. J. Plant Sci.* **40,** 620–622.

Sibilia, C. (1960). La forma ecidica della ruggine bruna delle foglie di grano *Puccinia recondita* Rob. ex. Desm. in Italia. *Boll. Stn. Patol. Veg., Rome* [3] **18,** 1–8.

Statler, G. D. (1977). Inheritance of virulence of culture 73-47 *Puccinia recondita. Phytopathology* **67,** 906–908.

Statler, G. D. (1979). Inheritance of pathogenicity of culture 70-1, Race 1, of *Puccinia recondita tritici. Phytopathology* **69,** 661–663.

Statler, G. D., and Jones, D. A. (1981). Inheritance of virulence and uredial color and size in *Puccinia recondita tritici. Phytopathology* **71,** 652–655.

Takahashi, K., Yamada, M., and Takahashi, H. (1965). Teleutospore formation of leaf rust, *Puccinia recondita* f. sp. *tritici* on young seedling of wheat. I. Isolate-variety relation and some characters of teleutospores. *Ann. Phytopathol. Soc. Jpn.* **30,** 54–61.

von Meyer, W. C., Greenfield, S. A., and Seidel, M. C. (1970). Wheat leaf rust: Control by 4-n-Butyl-1,2,4-triazole, a systemic fungicide. *Science* **169,** 997–998.

Waterhouse, W. L. (1952). Australian rust studies. IX. Physiologic race determinations and surveys of cereal rusts. *Proc. Linn. Soc. N.S.W.* **77,** 209–258.

Watson, I. A., and Luig, N. H. (1961). Leaf rust on wheat in Australia: a systematic scheme for the classification of strains. *Proc. Linn. Soc. N.S.W.* **86,** 241–250.
Webster, R. K. (1974). Recent advances in the genetics of plant pathogenic fungi. *Annu. Rev. Phytopathol.* **12,** 331–353.
Wilson, M., and Henderson, D. M. (1966). "British Rust Fungi." Cambridge Univ. Press, London and New York.
Yamada, M., Takahashi, K., Takahashi, H., and Tanaka, T. (1973). Studies on alternate host, *Thalictrum thunbergii* DC., as an origin of physiologic races of wheat leaf rust, *Puccinia recondita* Roberge ex. Desm. f. sp. *tritici* in Japan. *Rep. Tottori Mycol. Inst.* **10,** 283–302.
Young, H. C., and Browder, L. E. (1965). The North American 1965 set of supplemental differential wheat varieties for identification of races of *Puccinia recondita tritici. Plant Dis. Rep.* **49,** 308–311.

3

Stripe Rust

R. W. Stubbs

Research Institute for Plant Protection (IPO), Wageningen, The Netherlands

I. Introduction

Stripe rust was first described in 1777. The work on stripe rust has been reviewed in a treatise by Prof. Dr. Kurt Hassebrauk, the late Director of the Institute of Botany of the Biologische Bundesanstalt für Land- und Forstwirtschaft in Braunschweig, Federal Republic of Germany. The treatise written in German was published in four parts (Hassebrauk, 1965, 1970; Hassebrauk and Röbbelen, 1974, 1975). The

THE CEREAL RUSTS,
Volume II

ISBN 0-12-148402-5

sections dealing with the genetics of host–pathogen relationships and with resistance breeding were revised in English by Röbbelen and Sharp (1978). This chapter is based on Hassebrauk's writings on *Puccinia striiformis* and includes important recent information and my own research.

II. Nomenclature

Stripe rust was first described by Gadd in 1777 (Eriksson and Henning, 1896). It was considered the causal pathogen of the disease epidemic in rye in Sweden in 1794 by Bjerkander (Eriksson and Henning, 1896). Schmidt (1827) described stripe rust as the third cereal rust under the name *Uredo glumarum.* Westendorp (1854) introduced stripe rust, collected from rye (Hassebrauk, 1965), under the name *Puccinia striaeformis.* Fuckel (1860) named the rust he studied *Puccinia straminis,* but whether it was leaf or stripe rust is in doubt (Hassebrauk, 1965). Finally, Eriksson and Henning (1894) showed that stripe rust was a separate rust of grasses and named it *Puccinia glumarum.* This name remained valid until Hylander *et al.* (1953), followed by Cummins and Stevenson (1956), revived the name *Puccinia striiformis* West. (*P. striiformis* Westend.) (Manners, 1960). The common names stripe rust (roya estriada, ruggine striata, etc.) and yellow rust (roya amarilla, Gelbrost, rouille jaune, gele roest, etc.) were given by Humphrey *et al.* (1924) and by Eriksson and Henning (1894), respectively. A detailed description of the nomenclature and history of stripe rust is given in the monograph on cereal rusts by Eriksson and Henning (1896).

III. Life Cycle and Phylogeny

As far as is known, *Puccinia striiformis* Westend. has a hemiform life cycle comprising only the uredial and telial states. So far, searches for the alternate host have been unsuccessful. Eriksson and Henning (1894) looked among species of Boraginaceae, which followed Ranunculaceae as the alternate host family of wheat leaf rust. Tranzschel (1934) studied species of *Valerianella,* among others, assuming that *Acidium valerianella* could belong to *P. striiformis.* Straib (1937a), following Tranzschel's speculation, inoculated other species of *Val-*

erianella with basidiospores of stripe rust, but neither the pycnial nor the aecidial state was produced. He also failed to isolate stripe rust from aecidiospores collected from species of *Berberis* and *Mahonia,* which according to Mains (1933) could be the alternate host, as *P. striiformis* is closely related to *P. koeleriae* Arth., *P. arrhenatheri* Eriks., and *P. montanensis* Ellis, which have their aecidial state on *Berberis* spp. and *Mahonia* spp. The attempts of Hart and Becker (1939) with *Valerianella* spp., *Berberis* spp., and *Mahonia* spp. were also without success. Due to the failure to find the aecidial host of stripe rust, its existence was doubted by Hassebrauk (1970). According to Rapilly (1979), the alternate host, if ever found, will escape infection because of the short dormancy of teliospores and readily produced basidiospores (Wright and Lennard, 1978).

So far, there is no proof whether stripe rust is heterothallic or homothallic. In studies on somatic recombination in stripe rust, heterothallism is presently assumed (Little and Manners, 1969a; Wright and Lennard, 1980). Goddard (1976a) suggests that *P. striiformis* is heterothallic because it has a haploid chromosome number of six, like other heterothallic *Puccinia* spp. According to Wright and Lennard (1978), *P. striiformis* has a haploid chromosome number of three, which is also a basic number for heterothallic *Puccinia* spp. (McGinnis, 1956).

On the whole, there is little known about the phylogenetic history of *P. striiformis,* and a discovery of its alternate host would certainly contribute to an understanding of it. Maybe the alternate host of stripe rust should be looked for in the Ranunculaceae family, which, as far as is known, has not been fully explored. This suggestion (Viennot-Bourgin, 1934) is based on the fact that *Clematis vitalba* of that family is the alternate host of *P. agropyri* Ell. and Ev., which very much resembles *P. striiformis* in the morphology of its urediospores and teliospores and in its host range and appearance (Guyot, 1966). It may be well that the alternate host of *P. striiformis* only exists or existed in the center of origin. Stripe, leaf, and stem rust of wheat may be phylogenitically related to each other because of their size and number of chromosomes (Goddard, 1976a). Similarity between stripe and leaf rust was found by Meuzelaar (1979) in a pyrolysis mass spectrometric analysis of urediospores of stripe rust of wheat, leaf rust of wheat, crown rust of oats, and coffee rust. Subsequent fundamental research on stripe rust is expected to provide an insight into the phylogeny of stripe rust. Anikster and Wahl (1979) presented an excellent review on the co-evolution of rust fungi and their hosts. Their quotation of Gäumann (1952), namely, "The forms persisting to the present day might well be regarded as living fossils" is indeed pertinent.

IV. Host Range

Puccinia striiformis has its hosts only in the Gramineae family. In the cereal group, wheat and barley are the principal hosts. Although stripe rust on rye was frequently mentioned in early literature, rye is now seldom seen infected by this pathogen. It is likely that in the early days rust infections on rye were mistakenly ascribed to stripe rust (Hassebrauk, 1965). As far as is known, stripe rust has not been observed on oats, corn, or rice.

Hassebrauk (1965) listed about 320 grass species of about 50 genera that were naturally or artificially infected by stripe rust. Based on the division made by Gould (1968), these genera are grouped as follows:

Subfamily: Festucoideae
Tribe: Festuceae. Genus: *Brachypodium, Briza, Bromus, Dactylis, Festuca, Hesperochloa, Lamarckia, Lolium, Poa, Puccinellia, Vulpia*
Tribe: Aveneae. Genus: *Agrostis, Aira, Alopecurus, Arrhenatherum, Avena, Beckmannia, Calamagrostis, Holcus, Koeleria, Lagurus, Milium, Phalaris, Phleum, Trisetum*
Tribe: Triticeae (Hordeae) (Poaceae). Genus: *Aegilops, Agropyron, Asperella, Elymus, Eremopyrum (Agropyron), Haynaldia, Hordeum, Hystrix, Lagurus, Secale, Sitanion, Taeniatherum (Elymus), Triticum*
Tribe: Meliceae. Genus: *Catabrosa, Glyceria*
Tribe: Stipeae. Genus: *Stipa*
Tribe: Monermeae. Genus: *Monerma, Parapholis*
Subfamily: Eragrostoideae
Tribe: Eragrosteae. Genus: *Muhlenbergia*
Tribe: Chlorideae. Genus: *Chloris*

The most susceptible genera are *Aegilops, Agropyron, Bromus, Elymus, Hordeum, Secale,* and *Triticum.*

Several wild grasses can harbor stripe rust pathogenic to wheat and barley, but their role in the epidemiology of stripe rust differs from area to area. In the United States in the mountains in the Pacific Northwest and California, wild grasses harboring stripe rust during the summer are considered to be important sources for infecting wheat in the lower elevations (Hendrix *et al.*, 1965; Tollenaar and Houston, 1967). However, these sources did not contribute to the epidemic of stripe rust in the Sacramento Valley in California in 1975: stripe rust from volunteer wheat initiated that epidemic (Line, 1976). In Oregon, also, more importance is given to oversummering of stripe rust in residual green wheat than that on grasses (Shaner and Powelson, 1973). In Montana, Sharp and Hehn (1963) consider grasses of minor importance in the epidemiology of stripe rust in the wheat area they studied. In Europe, stripe rust on grasses was not important in the northwest (Zadoks, 1961), whereas it is important in the Harz in Germany (Becker and

Hart, 1939). In central Asia, Abiev *et al.* (1982) consider stripe rust on wild grasses to be an important source for infecting wheat in southeastern Kazakstan, but in northern China it is of little importance (Chen *et al.*, 1957).

V. Continental Dispersal

A. PATHWAY PATTERNS

The center of origin of *P. striiformis* is assumed to be Transcaucasia, where wild grasses were the primary host (Hassebrauk, 1965; Leppik, 1970). In view of our present knowledge, it may also be assumed that this is the lone center of origin of the fungus. From this center, stripe rust dispersed in all directions and developed genetic diversity in the absence of its alternate host. The most obvious eastern route for the fungus is along the mountain ranges to China and then to Japan. Mongolia is the northern range of stripe rust (Schmiedeknecht and Puncag, 1967). Via the Aleutians and Alaska, stripe rust entered the American continent, where it first occurred on wild grasses and then on cereals (Humphrey *et al.*, 1924). The mountain ranges of western part of the American continents made up the pathway to southern Chile. The Andean–Patogonian valleys are considered to be the entrance path to Argentina (Vallega, 1955). Stripe rust from Transcaucasia moved southward to Saudi Arabia and Yemen along the mountains and then into East Africa, i.e., Ethiopia and Kenya. So far, stripe rust has not been found in South Africa and in the countries of west Africa.

Stripe rust moved into Europe from the Near East. It migrated to Australia in 1979 (O'Brien *et al.*, 1980) and then to New Zealand in 1982 (Beresford, 1982). As the race first detected on wheat in Australia was similar to the one in Europe, it may be assumed that intercontinental air transportation was the means of introduction (also see Chapter 10, this volume). Presumably, this is the same way that European barley stripe rust reached Colombia in 1975. Within 7 years, this race moved southward to Ecuador, Peru, Bolivia, Chile, and then to Argentina (Dubin and Stubbs, 1984), covering a distance of about 6500 km and indicating the possible route followed by stripe rust when it initially entered South America.

According to Rapilly (1979), efficient dispersal of urediospores of stripe rust by wind is limited to short distances, because of the susceptibility of the spores to ultraviolet (UV) light; their susceptibility is three times higher than that of stem rust spores (Maddison and Manners, 1972). This supports the experience of H. J. Dubin (personal communication) working with stripe rust in the Andes region at 3000 m, where the intensity of UV light is high. Spores collected in the field early in the morning had a higher viability than those collected at noon, which had been exposed to UV light for a longer period. However, in spite of environmental factors, stripe rust urediospores can cover long distance within a short time, as reported by Zadoks (1961). Zadoks estimated that during the stripe rust epidemic in the United States in 1958, this rust had covered, step by step, a distance of about 2400 km—i.e., from northern Mexico, the source of inoculum, to North Dakota—within 6 months. In Europe, stripe rust urediospores have arrived in viable state after being wind transported 800 km (Zadoks, 1961), or over 1200 km from northern France to Algeria (R. W. Stubbs, unpublished).

When *P. striiformis* evolved and when it was introduced to the different areas in the world, except for Australia, is unknown. The answer may possibly be found by combining all evidence concerning the evolution of this rust, as well as other cereal rusts, on a global map and subsequently by shifting the continents into positions according to the evolution of the earth. Anikster and Wahl (1979) stated that "during 250–300 million years of phylogenetic history (Gäumann, 1952; Savile, 1954), the Uredinales have undergone profound changes in morphologic and biologic traits." This implies that cereal rusts evolved and dispersed during a period 225 million years ago, when our earth consisted of two continents, Laurasia (North America, Europe, and Asia) and Gondwana (South America, Africa, the Indian subcontinent, Antarctica, Australia, and New Zealand). Stripe rust is absent in western and southern Africa and until recently in Australia and New Zealand. Stripe rust may have not occurred in South America until this continent was joined to the North American continent. The absence of stripe rust in parts of Gondwana may have been caused by an unsuitably hot climate, with such exceptions as the mountains in Ethiopia and Kenya. Stripe rust occurs in the isolated Pulney and Nilgiri hills (about 2500 m) in southern India (see Section V,B), which once was connected with eastern Africa. In the Nilgiri hills, stripe rust forms still prevail with relative low pathogenicity, and primitive wheat such as *Triticum dicoccum* is still cultivated. The stripe rust in these hills may be as old as that in Ethiopia. In Laurasia there was a larger climatic area for stripe rust to develop, with the mountainous region of the Middle East, Central and East Asia, and further to east in North

America. West Europe may have been invaded by stripe rust after the separation from North America and as the climate then became influenced by the sea.

Stripe rust may also be much younger than stem or leaf rust and therefore with a less developed means of dispersal.

In a simple, abstract, and philosophical manner, this approximates the origin of the distribution of stripe rust in relation to the Earth's history. Reshaping this sketch will require a team of plant pathologists with "living fossils" at their hand and geologists with moving tectonic plates and the accompaning changes in climate.

B. REGIONAL PERFORMANCE AND CENTERS OF RESEARCH

1. North America

In Canada, stripe rust was first detected in 1918 on *Hordeum jubatum* L.; a few years later it was found in wheat and barley (Fraser and Conners, 1925). It is common on graminaceous hosts in southwestern Sasketchewan, Alberta, and British Columbia (Brown, 1956) and can be serious on wheat, as in southwestern Alberta in 1963 (Creelman, 1964).

In the United States, stripe rust was first recognized as a serious new wheat rust by Carleton (1915). Apparently, stripe rust had already been observed on wild grasses in 1892, described under the names *Puccinia agropyri* and *P. rubigo-vera* (Humphrey, 1917). Stripe rust has also been observed on barley and rye (Hungerford and Owens, 1923) but has been of economic importance only on wheat. Initially, stripe rust was found only in the western states, until 1941 when it was also observed in the central plains in Texas (Humphrey, 1941). Epidemics on wheat have been reported in Texas in 1957 (Futrell, 1957) and 1958 (Futrell *et al.*, 1959) and in Kansas in 1958 (Pady and Johnston, 1959). In the Pacific Northwest in 1960 and 1961, serious stripe rust epidemics occurred in wheat in Washington and Oregon (Hendrix, 1964; Shaner and Powelson, 1971); in 1961 they also occurred in California for the first time (Tollenaar and Houston, 1967). In the Pacific Northwest for the period 1935–1979, the frequency and severity of stripe rust epidemics was related to climatic variations, particularly temperatures in winter and spring, and also to a certain extent to the type of resistance of the wheat cultivars (Coakley, 1978; Coakley and Line, 1981).

So far, barley and rye have remained free from stripe rust epidemics. In the United States, stripe rust research is concentrated in Montana,

Oregon, and Washington State Universities. A race survey is being performed by the United States Department of Agriculture, Cereal Disease Laboratory, Washington State University.

2. *Central America*

In Mexico, stripe rust was detected for the first time by Holway on *Hordeum jubatum* in 1896 (Hassebrauk, 1965). In Mexico and Guatemala, stripe rust is a feared wheat disease at high altitudes. In Guatemala severe epiphytotics occurred in 1955 and 1956, with losses in yield up to 40% (Le Beau *et al.*, 1956).

The Centro Internacional de Mejoramiento de Maiz y Trigo (CIMMYT) is located in Mexico and has a major program for the development of stripe-rust-resistant wheats worldwide. A worldwide wheat screening and evaluation program has been established (see Chapter 9, this volume).

3. *South America*

Stripe rust was first discovered by Holway on *Hordeum chilense* in Chile in 1919 and 1 year later on *Agropyron altenuatum* in Ecuador (Arthur, 1925). Its first appearance on wheat was observed in Argentina in 1929 (Rudorf and Job, 1931), although it may have already been present on wheat in western South America without being recorded. In 1930, stripe rust was also observed on barley for the first time (Hirschhorn, 1933), but the stripe rust infections on barley as mentioned in literature may have been caused by the form attacking wheat until 1975, when the form specifically infecting barley was found in Colombia. Since 1976, stripe rust has been the most feared disease in barley in the Andean region (Colombia, Ecuador, Peru, and Bolivia). Barley, along with the wild barleys that serve as collateral hosts of stripe rust, is present throughout the year in the Andean zone (Dubin and Stubbs, 1984). In this area and in central Chile, stripe rust on wheat is a feared disease as well. In Argentina, Uruguay, and, to a much lesser extent, in Brazil and Paraguay, stripe rust appears regularly in wheat but seldom reaches epidemic levels (Grillo, 1937; Marchionatto, 1931; Stubbs *et al.*, 1974; R. W. Stubbs, unpublished). In Venezuela, stripe rust was observed on wheat (Standen, 1952).

As far as is known, no detailed studies of the epidemiology of stripe rust in Central and South America exist. We assume that in the tropical and subtropical zone of the Andean region with their small differences in temperature, the rainfall distribution and the growth of both the sown and volunteer wheat and barley are the main epidemiological parameters effecting epidemic development. The epidemiological cycle in North America (Line, 1976; Shaner and Powelson, 1973) and in Europe (Zadoks, 1961) may be applicable to the temperate zone of South America. Studies on physiologic specialization of stripe rust in South America have mainly been performed in Braunschweig, Federal Republic of Germany, and in Wageningen, the Netherlands.

4. Europe

Europe has one of the longest histories of stripe rust development and is considered the birthplace of studies on the pathogen. Epidemics have been reported in varying frequencies in all wheat-growing countries (Hassebrauk, 1965). Europe can be divided into some 20 agroclimatic subareas (Thran and Broekhuizen, 1965), and for some subareas the risk of epidemics by stripe rust has been calculated (Rijsdijk and Zadoks, 1976, 1979). A well-known subarea with a high risk of epidemics comprises the countries of northwest Europe, namely, England, the Netherlands, Belgium, northern France, and northern Germany. In this area, the frequency of stripe rust epidemics is relatively high (particularly in wheat), the pathogen exhibits some of the most highly evolved physiological races, and a long series of initially resistant cultivars has been used (see Section VII). In Europe, stripe rust is more common on wheat than barley, as the wheat area is predominantly winter wheat, enabling the fungus to survive in its uredial stage during the winter. In the area with high risk of epidemics, barley is mainly cultivated as a spring crop and is consequently seldom severely attacked. Winter barley is in the low-risk area of France and Germany. The European barley cultivars are generally highly resistant to the stripe rust form infecting wheat, and the reverse applies—i.e., the wheat cultivars are highly resistant to the form infecting barley. A serious stripe rust epidemic in northwest and central Europe in wheat and barley occurred in 1961 due to very favorable weather conditions for the oversummering and overwintering of both stripe rust forms and for their development in spring (Hassebrauk, 1962). Dispersal patterns of the pathogen are described by Zadoks in Chapter 11, this volume, and the corresponding race virulence pattern are presented in Section VII.

The study on the physiologic specialization of *Puccinia striiformis* was interrupted by World War II and resumed in 1956 when a severe stripe rust epidemic occurred on wheat in the Netherlands in 1955 (Zadoks, 1961). Since then, the occurence of stripe rust in Europe is annually recorded by means of trap nurseries sent to about 150 locations in Europe by the Research Institute of Plant Protection in Wageningen, the Netherlands. This institute, serving many countries in the world in the identification of stripe rust races, maintains a unique collection of stripe rust specimens collected in various countries and preserved as urediospores in liquid nitrogen (−196°C). Physiologic race surveys are also performed in Cambridge (United Kingdom), Braunschweig (Federal Republic of Germany), Praha (Czechoslovakia), and Fundulea (Romania).

5. *Middle East*

In the Middle East, with Transcaucasia as the presumed center of origin of stripe rust, the pathogen occurs in all countries. In Afghanistan, it is the most important rust disease in wheat (Ghaffor, 1970) and barley (Saari and Prescott, 1978), with losses up to 90% in epidemic years. In Iran, with about 6 million hectares of wheat (Anonymous, 1983), stripe rust is important in all wheat-growing regions, particularly along the Caspian Sea where epidemics occur once every 3–4 years (Niemann *et al.*, 1968; Khazra and Bamdadian, 1974). On barley, severe infections have only been observed in specific cultivars (Niemann *et al.*, 1968); it is likely that these cultivars are susceptible to the stripe rust form infecting wheat, as exhibited by the local barleys grown in Saudi Arabia. In Turkey, with about 9 million hectares of wheat (Anonymous, 1983), stripe rust is most prevalent in the cooler plateaus in central and eastern Anatolia, Thrace, and mountainous areas (Iren, 1981); in the past 50 years epidemics have been recorded once every 3–5 years (Hassebrauk, 1965).

In Iraq, stripe rust is a serious wheat disease in the mountainous area in Kurdistan. In Syria, Lebanon, Jordan, Israel, Saudi Arabia, Yemen Arab Republic, and Yemen Peoples Democratic Republic, stripe rust regularly appears on wheat and/or barley and can become serious (Anonymous, 1969; R. W. Stubbs, unpublished).

In the Middle East, little is known about the epidemiology of stripe rust and particularly the role of wild grasses, including the indigenous *Triticum* spp. and *Hordeum* spp. Wheat is primarly sown in the autumn, and early infections result from inoculum produced on the oversummering hosts of stripe rust. In the semiarid areas that make up the greater part of the wheat area, weather conditions in the spring determine if further development of stripe rust will occur, as in 1968–1969 when this rust showed high disease levels throughout the Middle East due to a prolonged wet and cool spring (Anonymous, 1968; Özkan and Prescott, 1972; R. W. Stubbs, unpublished).

National institutes performing race surveys are located in Izmir, Turkey, and in Tehran, Iran.

6. *Central Asia*

In the Asian part of the Soviet Union, stripe rust is the most serious rust disease in the foothills and hills in the mountainous part in Azerbaijan, Kirgizia, and Kazakhstan (Cumakov, 1963). In southeastern Kazakhstan, epidemics of stripe rust occur every 3–4 years in wheat (Abiev *et al.*, 1977). In this region, losses in yield can reach 50–60% (Urazaliev and Zeinalova, 1979).

7. *East Asia*

In China, with a wheat area of about 27 million hectares (Anonymous, 1983), stripe rust is considered to be the most important rust disease (S. M. Chen, personal communication; Li and Liu, 1957). In contrast to Japan, the Asian part of the Soviet Union, and the countries in southern Asia, stripe rust has not been observed on barley (Hassebrauk, 1965; S. M. Chen, personal communication). This phenomenon could result either from the Chinese barley cultivars being highly resistant to the stripe rust races infecting barley in the bordering countries or from the lack of spore migration to China due to physical or climatic barriers. Studies on the epidemiology of stripe rust since 1949 have indicated that, in northern China, early fall infections are caused by stripe rust oversummering on residual green wheat and/or on late-maturing spring wheats in the mountainous areas (Chen *et al.*, 1957). Studies on physiologic specialization were initiated in 1944 (Fang, 1944), and race surveys are now annually being performed in the Institute of Plant Protection in Beijing.

In Japan stripe rust was already known in the 1890s and exists from Kyushu in the south to Hokkaido in the north (Hemmi, 1934; Ito, 1909). It is an important disease of both wheat and barley, and epidemics in both crops were recorded in 1950–1956 (Kajiwara *et al.*, 1964a). Oversummering and overwintering on residual wheat and barley were shown by Narita and Mano (1962).

8. *South Asia*

In India, with about 22 million hectares of wheat (Anonymous, 1983), serious outbreaks of wheat rusts were recorded as early as 1786 (Nagarajan and Joshi, 1975). Stripe rust was first described in India by Butler (1903). The same form of stripe rust infects both wheat and barley, but it occurs in epidemic form only in the northwestern regions, northern foothills, and adjacent plains, and in the Nilgiri and Pulney hills in the south (Joshi, 1976). The pathogen is practically absent in central and southern India, where higher temperatures inhibit its development in the crop as well as its survival during the summer (Joshi *et al.*, 1974). In the Nilgiri and Pulney hills, two wheat crops are grown annually (Joshi, 1976), which may explain the survival of stripe rust in the isolated hills of southern India. Mehta (1940, 1952) indicated that in the plains in northern India, wheat rusts cannot oversummer due to extremely high temperatures, and that the Himalayas, particularly in central Nepal, were the primary sources of infection for the new crop. Later studies (Joshi *et al.*, 1977) showed that the hills in central Nepal are the only area where significant oversummering occurs. In the Simla hills in northwest India, stripe rust was found to survive on off-season plants (Joshi *et al.*, 1976). Weather analysis showed that in north India epidemic years are associated with a significantly greater number of rainy days during January to mid-April (Nagarajan and Joshi, 1978). Since 1967, a cereal disease surveillance has been conducted by the Indian Agricultural Research Institute in New Delhi (Joshi *et al.*, 1974), using mobile and trap nurseries. The identification of stripe rust races is done by this institute in Flowerdale, Simla.

In Pakistan, where wheat is cultivated on about 7 million hectares (Anonymous, 1983), stripe rust is important in the foothills in the north and is also present in the central region and western upland areas (Hassan, 1968). Serious damage to wheat occurred in 1954, 1958 (Hassan, 1968), and 1978 (Kidwai, 1979). The latter epidemic was associated with cool and wet winter conditions as influenced by certain weather systems called "western disturbances" originating around the Caspian Sea (Nagarajan *et al.*, 1982; also Chapter 2, this volume). The identification of stripe rust races since 1983 has been done by the Cereal Disease Laboratory in Murree.

In Nepal, stripe rust on wheat and barley is a dominant disease in the hilly areas, sometimes with losses of 100% (Karki, 1980).

In Bangladesh, stripe rust occurs on barley. In Burma, severe stripe rust infections were reported in 1938 (Seth, 1939). So far, stripe rust has not been reported or observed in Sri Lanka, Vietnam, Malaysia, the Philippines, and Indonesia (E. E. Saari, personal communication).

9. *Africa*

In the northern Africa, stripe rust occurs in all countries from Morocco to Egypt and infects wheat and barley. The disease has appeared seriously on wheat in Tunisia in 1977 (Ghodbane, 1977) and in Egypt in 1961 (Mohamed, 1963), 1967, and 1968 (Abdel-Hak *et al.*, 1972). In Egypt, where stripe rust was first reported in 1904 (Sydow and Sydow, 1904), neither the fungus nor its hosts oversummers. Initial infections are ascribed to sources in the north (Abdel-Hak *et al.*, 1966), in contrast to a general northward and eastward epidemiological pattern in the adjacent countries (Hogg *et al.*, 1969).

In East Africa, stripe rust infecting wheat and barley in the highlands is a serious problem, particularly in Ethiopia (Wodageneh, 1974; Ciccarone, 1947) and in Kenya (Martens and Oggema, 1972). In Ethiopia, a stripe rust epidemic in wheat occurred in 1983 (J. C. Zadoks, personal communication). Occasionally, stripe rust has also appeared in Uganda, Tanzania, Zambia, Malawi, and Madagascar (H. Bonthuis, personal communication).

Stripe rust has not been reported in western and southern Africa, although in South Africa wheat is being grown where the environmental conditions are suitable for stripe rust.

10. Australia and New Zealand

In Australia, stripe rust was widespread in New South Wales, Victoria, South Australia, and Tasmania (Anonymous, 1980) when it was detected for the first time on wheat in 1979 (O'Brien *et al.*, 1980).

Stripe rust entered New Zealand in 1982, probably as the result of spread from Australia by airborne urediospores (Beresford, 1982), covering a distance of about 2000 km (Chapter 10, this volume). In Australia, wheat stripe rust is able to infect a wide range of grass genera, i.e., *Bromus* spp. and *Phalaris* spp. (Wellings, 1982). *Puccinia striiformis* f. sp. *dactylidis* on *Dactylis glomarata* L. was also recorded for the first time in Australia in 1979, but it had been observed in New Zealand in 1975 (Wellings, 1982).

VI. Physiologic Specialization

A. Mechanism of Variability

As the sexual state has not been recorded for *Puccinia striiformis*, the development of genetic variation must be due to other mechanisms, such as mutation, somatic recombination, and parasexuality. So far, only the first two mechanisms have been specifically associated with *P. striiformis*.

Gassner and Straib (1932a) were the first to describe the role of mutation in the formation of new races of *P. striiformis* and estimated a mutation frequency of 1.6 per 100,000–200,000 urediospores. Straib (1937b) adopted the view that new races were formed only by way of progressive mutation. Oort (1955) shared this view and grouped the races known at that time into related series, each having a parent race from which they supposedly arose. Races with increased virulence have been artificially produced by Stubbs (1968a) by irradiating urediospores or infected wheat seedlings with X-rays. The mutants differed from the parent race by only one virulence factor, and one mutant was virulent to *Triticum spelta album* (*Yr5*), generally resistant at that time. Johnson *et al.* (1978) obtained a similar result with UV rays and indicated the high mutability of the genetic locus of virulence to Compair (*Yr8*). A decrease in virulence has only been reported by Kajiwara *et al.* (1968), who also reported the change to avirulence to Chinese 166 (*Yr1*).

The creation of new races by the fusion of germ tubes of urediospores of stripe rust was first shown by Litle and Manners (1969a,b), who ascribed the origin of the two new races they found to the reassortment of whole heterokaryotic nuclei. Somatic recombinants were also produced by Taylor (1976) using an albino race of stripe rust, by Goddard (1976b), and by Wright and Lennard (1980). In these three experiments there were always two races differing in virulence on the differential Chinese 166. It seems remarkable that no other recombinants were detected. One recombinant exhibited a decrease in virulence on Chinese 166 as compared to the parent (Taylor, 1976). The predominant appearance of recombinants with virulence on Chinese 166 corresponds with the ease with which stripe rust in nature has produced virulent races to the resistance of Chinese 166. Evidently, there are differences between virulence factors with regard to their exchange in the process of somatic recombination of the corresponding resistance genes.

The reciprocal exchange of whole chromosomes between nuclei as a mechanism of variation appears not to be applicable to stripe rust (Wright, 1976).

B. *FORMAE SPECIALES*

Eriksson (1894) distinguished five specialized varieties (*formae speciales*) of *Puccinia glumarum* Ericks. et Henn.—namely, f. sp. *tritici, hordei, secalis, elymi,* and *agropyri*—and recognized that some *formae speciales* were more highly specialized than others. Criticism on subdividing *P. glumarum* into *formae speciales* came from Sydow and Sydow (1904), Newton and Johnson (1936), and Straib (1935b, 1936), but the subdivision was supported by Zadoks (1961), among others, who stated that "physiologic specialization on the generic level is as real as that on the varietal level." His statement, particularly regarding f. sp. *tritici* and f. sp. *hordei,* is supported not only by recent greenhouse and field data, but by the results of an enzyme analysis of barley stripe rust and wheat stripe rust urediospores (A. C. Newton and C. E. Caten, personal communication). In this analysis, in which 10 enzymes were examined, clear differences separated the 13 cultures of stripe rust into two distinct groups, namely, the wheat form and the barley form. No differences were detected between races within each group. The results of Newton and Caten are in fact similar to those obtained by Meuzelaar (1979), also indicating host specifity on the generic level. It is therefore justified to refer wheat stripe rust and barley stripe rust as *P. striiformis* Westend. f. sp. *tritici* Eriks. and *P. striiformis* Westend. f. sp. *hordei* Eriks. As for rye stripe rust, there are no data to support the use of f. sp. *secalis.* Rye can be infected by both the wheat and barley form. The stripe rust form infecting *Dactylis glomerata,* previously described as race G (Manners, 1950), was given varietal rank: *P. striiformis* Westend. var. *dactylidis* (Manners, 1960). Consequently, the wheat, barley and rye form of stripe rust are referred to as *P. striiformis* Westend. var. *striiformis* by Manners (1960). The nomenclature for stripe rust on *Poa pratensis* L. is *Puccinia striiformis* Westend. f. sp. *poae* (Tollenaar and Houston, 1967).

C. PHYSIOLOGIC RACES

After the discovery of "biological forms" by Stakman and Piemeisel (1917), Hungerford and Owens (1923) were the first to report that there were also "strains" in *Puccinia glumarum tritici.* Differences between stripe rust forms were also recognized by Gassner and Straib (1929, 1930), but Allison and Isenbeck (1930) established the existence of races in *P. glumarum tritici.*

In Europe, extensive studies were made in Braunschweig, Germany, by Gassner and Straib (1932b, 1934) and Straib (1935a, 1937a), introducing a differential set of cultivars of wheat, barley, and rye, which was widely used (Table I). After World War II a new era on physiologic race studies began in 1955 when Dr. Eva Fuchs took over the stripe rust work in Braunschweig. In cooperation with Dr. J. C. Zadoks, Wageningen, the Netherlands, an international race survey was initiated by means of yellow rust trials sown at many locations in Europe (Zadoks, 1961). A worldwide race survey was conducted by the author.

Difficulties in reproducing the race classification of Gassner and

Table I

Races of Stripe Rust Described on Selected Cultivars of Wheat, Barley, and Rye[a]

	Wheat												Barley		Rye
Race	Michigan Amber	Blé Rouge d'Ecosse	Strubes Dickkopf	Webster	Holzapfels Früh	Vilmorin 23	Heines Kolben	Carstens V	Spaldings Prolific	Chinese 166	Rouge Prolifique Barbu	*Triticum dicoccum* var. *tricoccum*	Fong Tien	Heils Franken	Petkuser Roggen
1	9	9	9	8	9	9	9	6	0	0	0	9	8	2	1
2	9	9	9	8	9	9	2	6	9	0	9	9	8	2	1
3	9	9	9	8	9	9	2	6	3	3	0	9	8	2	1
4	9	9	9	8	9	8	2	6	0	0	3	9	8	2	1
5	9	9	9	8	4	2	2	9	3	0	3	9	8	2	1
6	9	9	9	8	4	2	2	4	3	0	3	9	8	2	1
7	9	9	9	3	4	2	2	9	3	0	3	9	8	2	1
8	9	9	9	4	4	2	2	4	0	0	3	9	8	2	1
9	9	9	2	8	9	3	9	2	0	0	0	9	2	2	1
10	9	9	2	3	2	2	9	2	0	0	0				
11	9	9	2	8	2	2	2	2	0	0	0				
12	9	9	3	2	2	2	2	4	0	0	0	9	8	2	1
13	9	2	2	7	2	2	2	2	0	9	0	9	6	2	1
14	9	3	2	2	2	2	2	2	0	0	0	9	8	2	1
15	9	9	2	2	2	2	2	2	9	0	3	9	8	2	1
16	9	9	2	8	8	3	3	2	0	0	0	9	2	2	1
17	9	9	9	8	8	2	2	4	8	0	3	9	8	2	1
18	9	3	0	2	2	0	3	0	0	0	0	9	8	2	1
19	9	2	2	2	2	2	8	2	0	0	0	9	8	2	1
20	9	2	2	9	9	2	8	2	0	0	0	9	8	2	1
21	9	9	0	1	1	2	9	0	0	0	0	9	8	2	1
22	9	9	9	8	8	2	2	4	8	0	8	9	8	2	1
23	2	2	2	0	0	0	2	0	0	6	2	9	9	2	1
24	2	2	0	0	0	0	2	0	0	0	0	9	9	9	1
25	9	2	0	8	8	2	2	0	0	0	0	9	2	2	1
26	9	9	9	4	4	2	2	9	9	0	3	9	8	2	1
27	9	9	9	4	4	2	2	9	2	6	2	9	8	2	1
28	1	0	0	0	0	0	1	0	0	0	0	9	8	2	1
29	9	9	2	4	4	2	2	4	3	0	2	9	8	2	1
30	9	6	0	8	8	2	3	1	3	0	2	9	8	2	1

31	9	2	2	9	9	2	8	2	0	9	0	9	8	2	1
32	9	9	2	8	8	3	2	2	0	0	0	9	8	2	1
33	7	0	0	2	2	0	6	0	0	2	2	9	9	2	1
34	7	0	0	0	0	0	3	0	0	0	0	9	6	2	9
35	9	9	9	8	8	9	9	2	9	0	9	9	8	2	1
36	2	0	0	2	2	0	1	0	0	0	0	9	9	2	1
37	9	6	7	8	8	2	9	6	3	0	2	9	8	2	1
38	9	6	6	8	8	2	3	6	3	0	2	9	8	2	1
39	9	6	1	8	8	2	3	6	9	0	7	8	2	1	
40	9	2	2	8	8	2	8	2	0	0	0	9			
41	9	9	9	7	7	2	2	9	3	0	2	9	8	2	
42	9	3	3	7	7	2	6	2	0	9	0	9			
43	9	9	2	9	9	2	2	2	9	0	9	9	8	2	
44	9	7	2	2	2	2	9	2	0	0	0	9	8	2	
45	2	2	2	0	0	0	2	0	0	7	2	9	9	2	
46	2	2	2	0	0	0	2	0	0	7	2	9	9	2	
47	2	2	0	0	0	0	2	0	0	0	0	9	9	1	
48	6											9	9	9	
49	9	9	3	8	9	3	2	2	0	0	0				
50	9	9	9	3	9	9	2	2	3		3				
51	9	9	8	8	2	2	2	2	3	0	3				
52	9	8	2	9	9	9	2	2	0	0	0				
53	9	9	9	6	8	2	2	9	7	8	2	9	8	2	
54	9	9	9	4	7	2	9	8	2	0	4		8		
55	9	9	9	8	9	8	2	9	1	1	1		9		
56	2	2	2	2	1	0	2	1	0	2	2	9	9	2	
57	2	2	2	2	2	2	2	2	2	2	2	3	7	2	0
58	9	9	9			9	2		3	8	0	9			
59	9		1			6	9	0	1	1		9			
60	9	9	9			2	9	2	2	9	9	9	8		
A[b]	9	2	2	7	2	2	8	2	2	2	2	9	9	2	1
D	9	2	2	7	2	2	8	2	7	2	2	9	9	2	1
E	9	8	7	7	2	2	8	2	2	2	2	9	9	2	1
F	9	8	7	8	2	2	8	8	2	2	2	9	9	2	1
G	2	2	2	2	2	2	9	2	2	2	2	9	9	2	1
H	9	8	7	8	2	7	9	2	2	2	2	9	9	2	1

[a]Modified and based on data of Fuchs (1956, 1965). Entries are infection types in the scale 0–9 (McNeal *et al.*, 1971).
[b]Recorded in India (Mehta, 1933; Prasada and Lele, 1952).

Straib forced Fuchs (1960) to combine races not sharply differing from each other into race groups, e.g., 27/53, 5/6, and 7/8. The addition of supplemental cultivars led to a change in nomenclature. This was done by adding a letter to the race number—e.g., race 32A referred to a susceptible reaction on Lee (*Yr7*) (Fuchs, 1965). In India the letter A was used for virulence to Kalyansona (Singh *et al.*, 1978). The letter B indicated virulence to Cappelle Deprez (*Yr3a*) (Batts, 1957), as well as to Heine's VII (*Yr2*). In the Netherlands, races were named according to the cultivars on which they were found or that they would specifically attack (Ubels *et al.*, 1965; Zadoks, 1961). A reevaluation of race identification and race nomenclature of wheat stripe rust was made by Johnson *et al.* (1972), introducing the binary notation, which is also being used in India (Nagarajan, 1983; Nagarajan *et al.*, 1984). The differential cultivars currently used in Europe are presented in Table II. It was agreed that for seedling tests in growth rooms, a light period of 16 hr of 10,000 lux minimal at 18°C and a dark period at 11°C should be maintained. The division in virulence and avirulence would correspond with the infection types 7–9 and 0–6, respectively, according the scale proposed by McNeal *et al.* (1971).

A detailed study on physiologic specialization of wheat stripe rust on adult plants has been made by Zadoks (1961) utilizing the technique of race nurseries. In England a similar technique is employed using polyethylene tunnels (Priestley and Doodson, 1976). According to the conditions, namely, environment and plant stage, under which stripe rust exhibits physiologic specialization, Zadoks (1961) has distinguished "greenhouse" and "field" races corresponding with resistance of the overall type and mature type, respectively. One greenhouse race may comprise several field races (Ubels *et al.*, 1965; Zadoks, 1961).

In North America, Newton *et al.* (1933) isolated physiologic races in Canada. In the United States, Bever (1934) was the first to establish the presence of two physiologic races. Extensive race studies began in the 1960s when stripe rust appeared seriously on wheat and races were found infecting wheat cultivars previously resistant to stripe rust, such as Suwon 92 (Purdy and Allan, 1966) and Moro (*Yr10*) (Beaver and Powelson, 1969). A system for differentiating races was introduced using the avirulence and virulence formula corresponding with infection types 0–4 and 5–9, respectively (Line *et al.*, 1970). A coded virulence formula has been used since 1972 (Line, 1972, 1980). In contrast to Europe, the conditions for seedling tests in growth rooms are 12 hr of light (10,000 lux or higher) and 12 hr of dark at temperatures gradually changing from a minimum of 2°C (dark period) to a maximum of 18°C during the light period. The differential cultivars were Lemhi,

Table II

A Comparison of the Various Sets of Differential Hosts Currently Used for *Puccinia striiformis* f. sp. *tritici*

World set[a]		European supplemental[a]		Indian supplemental[b]		North American[c]		Chinese[d]	
Yr gene	Host	*Yr* gene	Host	*Yr* gene	Host	*Yr* gene	Host	*Yr* gene	Host
1	Chinese 166	*4b*	Hybrid 46	*4b*	Hybrid 46		Lemhi		Trigo Eureka
7	Lee	*7* & ?	Reichersberg 42	*2*	Heines VII	*1*	Chinese 166		Fulhard
6	Heines Kolben	*6* & ?	Heines Peko	*8*	Compair	*2*	Heines VII		Bima 1
3	Vilmorin 23[e]		Nord Desprez	*5*	*Triticum spelta*	*10*	Moro		Lutescens 128
10	Moro	*8*	Compair		*album*		Paha		Xibei Fenshou
	Strubes Dickkopf		Carstens V		Sonalika		Druchamp		Xibei 54
	Suwon 92/Omar		Spaldings Prolific		Kalyansona	*9*	Riebesel 47/51		Quality
9	Riebesel 47/51 or	*2*	Heines VII				Produra		Mentana
	Clement						Yamhill		Kansu 96
					Kathia		Stephens		Virgilio
					Barley local	*7*	Lee		Abbondanza
					WL 711		Tadorna		Beijing 8
					Arjun				Early Premium
					WL 410				Funo
					WH 147				Danish 1
					Webster				Jubilejna 2
					Triticum dicoc-				Feng Chan 3
					cum tricoccum				Strubes Dickkopf
					Himani				Lovrin 13
									Kangyin 655
									Tanshan 1
									Shuiyun 11
									Zun 4

[a]See Johnson *et al.* (1972).
[b]See Nagarajan (1983), Nagarajan *et al.* (1984). The world set is also used.
[c]See Line (1980).
[d]According to S. M. Chen, K. N. Wang and X. S. Xie (personal communication).
[e]Provisionally assigned the *Yr3* gene.

Chinese 166, Heines VII, Moro, Paha, Druchamp, Riebesel 47-51, Produra, Yamhill, Lee, Stephens, and Tadorna (Table II).

In South America, Rudorf and Job (1934), using the differentials of Gassner and Straib, recognized that in Argentina the races differed from those in Europe, and in Chile with the same differentials Volosky de Hernandez (1953) established the presence of four groups of races.

In Asia, physiologic race studies were initiated by Mehta (1933) in India, and studies were done annually since with periodical reviews. Races were identified according the scheme developed by Gassner and Straib, and new races were denoted with an alphabetical letter (Table I). In India there was little or no change in the race spectrum till the late 1960s, possibly due to the low selection pressure exerted on the pathogen and to limited possibilities for pathogen survival during the summer. A new virulence factor rendering ineffective the resistance of Kalyansona was found to be present in three races, presumably introduced from the Near East (Sharma *et al.*, 1972). Virulence to Kalyansona has also been found in races prevailing in China, Europe, Africa, and South America.

In China, Fang (1944) established the presence of nine races, C1–C9, with eight differential cultivars. The differential set was revised by Wang *et al.* (1963) and extended by S. M. Chen, K. N. Wang and X. S. Xie (personal communication) (Table II). In Japan, Kajiwara *et al.* (1964b), using the differentials of Gassner and Straib supplemented with Lee (*Yr7*) and Reichersberg 42, found two races only differing from those in China by being avirulent on Spaldings Profilic.

D. Effects of Environment

Environmental conditions are more critical for stripe rust than for other cereal rusts. Hassebrauk (1970) has reviewed in detail the various aspects of the pathogen and the disease.

In studies of physiologic specialization, either in a greenhouse or growth room, the most critical aspect is maintaining low night temperature, preferable 15°C or lower, for the host, particularly prior to inoculation. Daytime temperatures are less critical. A detailed study of the effect of temperatures for the host, prior to and after inoculation, has been made by Brown and Sharp (1969).

The effect of light on disease was shown by Gassner and Straib (1934), Manners (1950), and Stubbs (1967). In these studies, Carstens V, Holzapfels Früh, and Chinese Spring were highly sensitive to changes in light intensity. Low light intensities gave susceptible reactions, whereas high light intensities yielded resistance reactions. Seedling plant tests must therefore be done at 10,000 lux or more to assure consistent reaction types.

Relative humidity (RH) is another important factor. The general tendency is an increase in infection type with increasing RH (Zadoks, 1961). Stripe rust uredia on the leaves can be parasited by *Verticillium lecani* at an RH of higher than 90% (Mendgen, 1981). For long-term storage of urediospores either under vacuum or liquid nitrogen, predrying of moist spores at an RH of approximately 40% is essential.

Stripe rust is more sensitive to air pollution than other cereal rusts, as demonstrated by Sharp (1967). A high concentration of large ions in the air due to air pollution decreased the germination rate of urediospores. The effects of tobacco smoke on germination and infectivity of spores was shown by Melching *et al.* (1974). My spore germination studies indicated differences in sensitivity to air pollution between stripe rust cultures from different areas in the world. The northwest European cultures were much better adapted to air pollution than those from the Andes, East Africa, or Asia. A few northwest European cultures were even highly resistant. It was also noticed that inoculations made during days with high air pollution resulted in an increase in the length of the latent period and a shift of infection types toward resistance. This may explain why infection types are difficult to reproduce even working under strictly controlled conditions.

VII. Host Resistance

Röbbelen and Sharp (1978) have reviewed the earlier work on host–pathogen relations, selection of resistant cultivars, and breeding for relative resistance and tolerance. Stripe rust was the first disease in which host resistance was shown to be inherited in a Mendelian manner (Biffen, 1906). Lupton and Macer (1962) studied the seedling expressed resistance in seven cultivars and assigned *Yr* designations. Current listings of *Yr* genes are maintained in an International Register (McIntosh, 1973). The *Yr* genes most important in physiologic race studies are given in Table II. Resistance other than that of *Yr* genes is common. Some resistances are durable (Johnson, 1981), and some resistance genes seem to have little if any effect independently but are very useful in combination (Sharp and Volin, 1970; Wallwork and Johnson, 1984; Gramma *et al.*, 1984). Resistance genes with minor effects and only detectable at certain temperature regimes are also known (Lewellen *et al.*, 1967; Lewellen and Sharp, 1968). Resistance that functions only in the adult plant stage is also common and widely used. It is beyond the scope of this chapter to discuss all types of resistance and to list all known resistant cultivars. However, it should be noted that almost all wheat cultivars exhibit some undefined type of resistance, either to a low or high degree as compared to the most susceptible known wheats, such as Michigan Amber, *Triticum spelta saharense,* and Taichung 29. This type of resistance was named *rest resistance* and is assumed to be race nonspecific (Stubbs, 1977; Zadoks, 1961).

A. EVOLUTION OF RESISTANCE AND VIRULENCE

The effects of resistance on virulence in stripe rust could be studied in several areas of the world. However, data on host cultivars grown, their resistance to stripe rust, and disease epidemiology (see Chapter 1, this volume) for Europe are well documented and easily available.

In northwest Europe, wheat is primarily a winter crop, although some spring wheat is also grown. Stripe rust epidemics have been frequently reported. Epidemics are initiated from inoculum from volunteer plants from the previous crop that infects the crop in the fall (Chapter 1, this volume). Wheat cultivars are developed by various private companies in the area and introduced in several countries. Wheat culture is intensive, with large imputs of fertilizers and pesticides. The resulting wheat crop is among the highest yielding in the world. The important wheat cultivars of German, French, or Dutch origin used in the Netherlands are shown in Fig. 1 and their resistance genes are given in Table III. Generally, spring wheat cultivars had a different basis of resistance. Resistance gene *Yr6* is most common in them. During 1946–1983, 29 important cultivars were grown, and most possessed resistance to stripe rust when they were introduced to the farmers. The pathogen developed virulence to all but Juliana, Carstens VI, and Arminda. Juliana and Arminda have the adult-plant type of resistance, while Carstens VI resistance is also expressed in the seedling stage. Several cultivars—Alba with an adult-type resistance, Felix with two-gene seedling type resistance, Staring with *Yr3* and an adult plant type resistance, and Manella with *Yr2* and an adult plant type resistance—all lasted over 10 years.

The failures were just as dramatic. Heines VII with *Yr2* failed in 5 years. The resistance of Falco failed in the second year; however, it should be noted that Alba, with the same resistance, had already been grown for 11 years. The adult plant resistance of Dippes Triumph

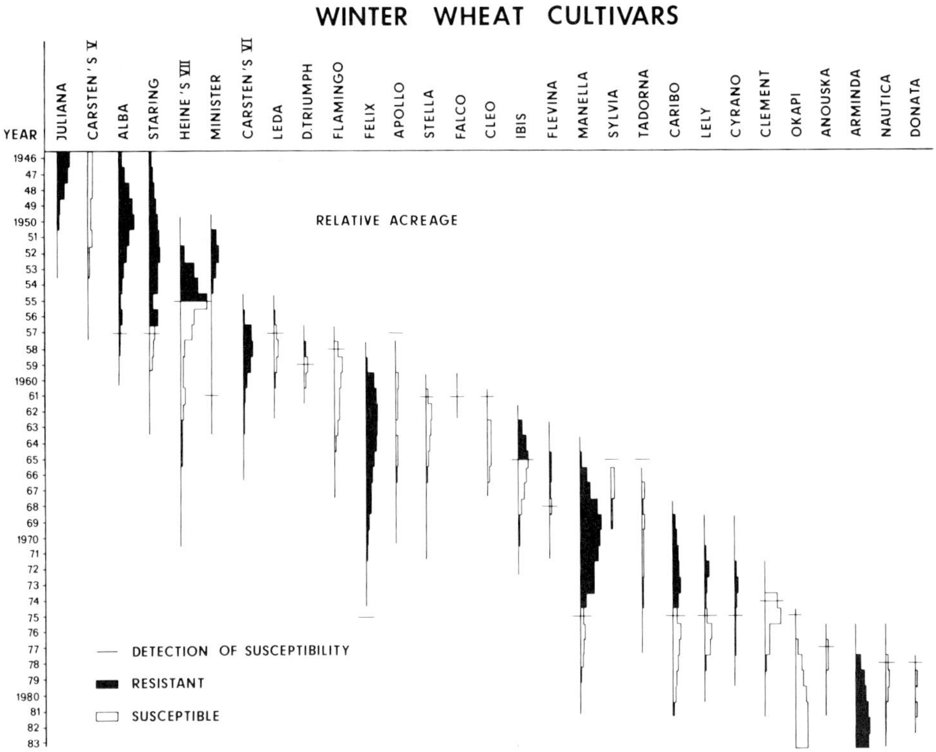

Fig. 1. The importance and resistance of selected winter wheat cultivars in the Netherlands from 1946 through 1983. The width of the band is proportional to the area planted to the cultivars. The dark and light areas of the bands indicate those years in which the cultivars were resistant and susceptible to *Puccinia striiformis* f. sp. *tritici*, respectively. Horizontal lines indicate the first year virulence was observed on the cultivar.

failed in the third year of usage. Clement was introduced with *Yr9* resistance in 1972, and the resistance was overcome in 1975. Because of the use of Ibis (*Yr1* and *Yr2*) and development of virulence on this combination, virulence existed to the cultivars Sylvia and Tadorna (both *Yr1* and *Yr2*) before they were released. The *Yr7* gene in Thatcher was first introduced with Flevina but was overcome in the fifth year. Lely with *Yr7* and Alba adult plant resistance failed in 7 years.

Carstens V was susceptible to stripe rust prior to 1946 and was out of production by 1957. Virulence to this cultivar either disappeared or decreased to an undetectable level. In 1958, Felix was released with the Carstens V resistance and *Yr3*. This combination was adequate for Felix to remain resistant until 1974. Then in 1968, Caribo was released

Table III

Resistance of Important Wheat Cultivars of German, French, and Dutch Origin Grown in The Netherlands[a]

		Genotype[c]	
Cultivar	Origin[b]	Overall resistance	Adult resistance
Alba	Trésor/Jacob Cats	?	14
Anouska	Prof. Marchal/Carstens VIII	12 & 2 or 3	
Apollo	Alba/Criewener 192	?	14
Arminda	Carstens 854/Ibis	?	13 & ?
Caribo	Cappelle Desprez/Carstens VIII	12	
Carstens V	Carstens III/Dickkops//Dickkopfs/Criewener	12	
Carstens VI	Carstens Dickkopf/Kladener Altmarkischer Braun	12 & ?	?
Clement	Hope/Timstein//3 × Heines VII/3/Riebesel 47-51/2 × Heines VII/4/Cleo	9 & 2?	
Cleo	Heine VII/Minister	2 & 3	
Cyrano	Witiko/Merlin	12 & ?	
Dippes Triumph	Tassilo/Carstens V//Dehrenburger Silber	?	15
Donata	Cebeco 150/2/Cleo//Norin 10/Flevina	7 & 9 & 2?	
Falco	Minister/Lovink		14
Felix	Tassilo/Carsten V//Marquillo/leaf rust resistance line	3 12 & 3?	
Flamingo	Heines IV//Tassilo/Svalöfs Kronen	2 & 6	13
Flevina	Heines VII × 3//Hope/Timstein	2 & 7	
Heines VII	Hybride à courte paille/Svalöfs Kronen	2	
Ibis	as Tadorna	1 & 2	13
Juliana	Essex gladkaf/Wilhelmina	?	14 & ?
Leda	Jubilé/Zanda	2	13 & 14
Lely	Cebeco 30/Flevina	2 & 7	14
Manella	Alba/Heines VII	2	14
Minister	Benoist 40/Prof. Deloz	2 & 3	
Nautica	Mildress/Manella	2 & 9	14
Okapi	Cappelle Desprez/Carstens VIII	12	
Staring	Vilmorin 23/Juliana	3	14
Stella	Rimpaus Bastard II/Prof. Delos//Prof. Delos/Hybride du Joncquois	2 & 3	
Sylvia	as Tadorna	1 & 2	13
Tadorna	Merlin/5/Heines VII/4/Teutonen/Hindukush 516/3/Chinese 165/Panzer III//Heines IV	1 & 2	13

[a]Based on seedling and adult plant tests.
[b]From Zeven and Zeven-Hissink (1976).
[c]Resistance genes correspond with virulence genes in Table IV.

Table IV

Evolution of *Puccinia striiformis* f. sp. *tritici* in Northwest Europe

Number of genes	Gene combinations[a]
1	*0*
2	*0.2 0.3 0.12 0.14 0.15*
3	*0.1.2 0.1.3 0.1.12 0.2.3 0.2.6 0.2.13 0.3.14*
4	*0.1.2.3 0.1.2.13 0.2.3.6 0.2.6.13 0.2.13.14 0.3.4.11*
5	*0.1.2.3.6 0.1.2.3.7 0.1.2.3.9 0.1.2.3.12 0.1.2.6.13*
6	*0.1.2.3.4.11 0.1.2.6.7.13 0.2.3.4.6.11 0.2.3.4.7.11 0.2.3.4.9.11*
7	*0.1.2.3.4.9.11 0.2.3.4.7.9.11 0.2.3.4.7.11.14*
8	*0.1.2.3.4.7.9.11*
9	*0.1.2.3.4.7.9.11.14*

[a]Test hosts were 0 = Strubes Dickkopf, 1 = Chinese 166, 2 = Heines VII, 3 = Vilmorin 23, 4 = Hybrid 46, 6 = Heines Kolben, 7 = Lee, 9 = Riebesel 47/51 (Clement), 11 = Suwon 92/Omar, 12 = Carstens V, 13 = Heines IV (adult plant resistance), 14 = Alba (adult plant resistance) and 15 = Dippes Thriumph (adult plant resistance).

with only the Carstens V resistance, and the similar Okapi was released in 1975. Virulence to Carstens V reoccurred 7 years after the release of Caribo. However, the culture combined virulence on Carstens V with virulence on Felix, which was no longer important but had grown as a resistant cultivar for 16 years.

It is generally known that northwest European wheat cultivars have a small genetic base of resistance, as they are closely related (Stubbs, 1968b; Wienhues and Giessen, 1957). Many cultivars possess the same resistance derived from one common ancestor. *Yr* genes were used either singly or in combinations, but stripe rust developed virulence to all of them. The evolution of stripe rust virulence corresponding with the breakdown of resistance of the cultivars in Table III is presented in Table IV. It illustrates the step-by-step increase of virulence in races, either due to mutation or somatic recombination. It also demonstrates the unintentional inputs of humans in the evolution of stripe rust.

B. Effects of Mutation Rates

Irradiation with X-rays has been used to produce mutations for virulence in *P. striiformis* (Stubbs, 1968a). Mutants virulent on *Yr1*, *Yr2*, *Yr3* commonly occurred, while a mutant with virulence to *Yr5* was only found once. This agrees well with the study of worldwide virulence frequencies (Table V), which shows worldwide distribution of

Table V

Zonal Distribution and Frequency[a] of Wheat Yellow Rust Virulence Factors in Populations of *Puccinia striiformis* f. sp. *tritici*

Zone	Wheat yellow rust virulence factors (WYV)[b]												Number of races
	1	2	3	4	5	6	7	8	9	10	11	12	
West and central Europe	3	4	4	3	0	3	2	1	1	0	2	3	39
West Mediterranean	2	2	3	2	0	3	2	3	0	0	2	0	17
East Europe	1	3	3	1	0	4	3	3	0	1	2	2	25[c]
East Mediterranean (including east Africa)	1	1	3	1	0	3	3	3	1	2	4	1	29
Southeast Asia	2	1	1	0	1	2	4	4	0	0	4	0	26
East Asia	4	2	3	0	0	3	3	2	1	0	0	1	[d]
North America	2	4	3	0	0	2	1	0	0	1	2	2	21[e]
South America (northern Andes)	0	1	3	2	0	3	1	0	1	0	3	2	20
South America (southern Chile and Argentina)	1	4	4	4	0	2	3	0	1	0	3	3	22
Australia and New Zealand	0	4	4	4	0	3	2	0	0	0	4	0	5[f]

[a]Frequency in percentage virulence in the race population: 0 = not known, 1 = less than 10%, 2 = 11–25%, 3 = 26–50%, and 4 = over 50%.

[b]Test hosts were 1 = Chinese 166, 2 = Heines VII, 3 = Vilmorin 23, 4 = Hybrid 46, 5 = *Triticum spelta album*, 6 = Heines Kolben, 7 = Lee, 8 = Compair, 9 = Riebesel 47/51 or Clement, 10 = Moro, 11 = Suwon 92/Omar, and 12 = Carstens V.

[c]Supplemented with data from Negulescu and Ionescu-Cojocaru (1976).

[d]Based on Fang (1944) and R. W. Stubbs, unpublished data.

[e]Based on Line (1980) and R. W. Stubbs, unpublished data.

[f]Based on McIntosh (1983) and Wellings (1983).

virulence on *Yr1*, *Yr2*, and *Yr3*, while virulence on *Yr5* is rare. In New Zealand the original race introduced was avirulent on *Yr7*, but virulence was found by 1982 (Wellings, 1983). Similarly, in northwest Europe, virulence on *Yr7* readily appeared when this resistance gene was used in breeding programs. The same phenomenon took place in southern South America. Evidently the mutation rate to virulence on *Yr7* is high, which may also apply to virulence on *Yr9*. The latter virulence was found in four zones (Table V) as soon as the host with *Yr9* was set out to select it. It is anticipated that this virulence will become widespread if the resistance of *Yr9* becomes widespread in host cultivars.

Studies on mutation rates could be valuable in determining the expected useful life of resistance genes.

C. EFFECTS OF FUNGICIDES

In recent years, with the use of fungicides, wheat growers have been able to grow cultivars that are stripe-rust-susceptible but otherwise desirable. For example, Okapi was susceptible at the time of its release (Fig. 1) but the area planted with this cultivar increased and continues at high level due to availability of fungicides for stripe rust control, such as triadimefon (Bayleton) propiconazole (Tilt, Desmel, Radar) and fenpropimorph (Corbel, Mistral).

The effect of continued use of fungicides on the disease is unknown. Proper use of fungicides should keep disease severities low, thus reducing the numbers of urediospores produced. This should result in few mutants. Mutants produced will have little selective advantage on a susceptible host, and many should be lost from the population by chance due to low disease levels. However, the potential exists for mutants to arise that are resistant to the fungicide used. These would have a great selective advantage. Fungicide-resistant mutants may be expected in stripe rust, because current strains differ greatly in their sensitivity to pollution levels (see Section VI,D).

VIII. Stripe Rust on Wheat

A. DISTRIBUTION OF VIRULENCE GENES

The coevolution of stripe rust and wheat has gone in different directions, with different degrees of divergence in different regions. When a new gene for virulence arises through mutation in one region, it can be spread to another region by wind dispersal of urediospores or by unintentional human activity. Additionally, the same mutation can occur independently in more than one region. However, it may not be detected unless the corresponding host resistance gene is present as a selecting agent.

The world distribution of virulence factors among races is presented in Table V. The division into zones has been made on the basis of race patterns and their virulence composition. Australia is only represented by five races that have evolved since the stripe rust introduction from Europe in 1979 (see Chapter 10, this volume).

Virulence for Chinese 166 (*Yr1*) is worldwide, and its frequency is especially high in east Asia where the cultivar originated. All races in China and Japan attack this cultivar (Fang, 1944; Kajiwara *et al.*, 1964b). Virulence on Chinese 166 was present in the early stripe rust collections from North America (Bever, 1934), which is in line with its supposed Asian source (Hungerford and Owens, 1923). The high frequency of wheat yellow rust virulence factor (WYV1) in the west and central Europe zone is related to the introduction of Chinese 166 in wheat breeding programs.

Virulence to Heines VII (*Yr2*) and Vilmorin 23 (*Yr3*) also occur worldwide. Most European cultivars possess *Yr2* and/or *Yr3*, and the cultivation of similar cultivars in southern South America explains the high frequency of WYV2 and WYV3. Similarly, virulence to Hybrid 46 (*Yr4b*) is common in both zones.

Virulence to *Triticum spelta album* (*Yr5*) has not yet been detected, except for a race in India (Nagarajan, 1983).

Virulence to Heines Kolben (*Yr6*) and Lee (*Yr7*), both spring wheat cultivars, has worldwide distribution. Resistance of Heines Kolben (*Yr6*) is derived from an old French cultivar Blé de Saumur and, according to McIntosh *et al.* (1981), *Yr7* from *Iumillo durum*, which is widely used, through Thatcher (*Yr7*). Possibly, both resistant genes are also present in other old cultivars or land races.

The frequency for virulence on *Yr8* derived from *Aegilops comosa* L. (Riley *et al.*, 1968) is high in the area where this grass is common. It is supposed that races with that virulence have spread to the east historically over the pathway along which the pathogen still migrates (Section V,A). Similarly, this virulence has moved into the east European and west Mediterranean zones. Its occurrence in the west and central Europe zone has resulted from a separate pathogen mutation, but *A. comosa* itself is still resistant to the races found (Johnson *et al.*, 1978). The *Yr8* resistance was not introduced into the western Hemisphere breeding programs until recently. Neither *A. comosa* nor virulence for it is currently present.

Virulence for *Yr9* occurs in the west and central Europe zone, where it was first detected in the mid-1970s (Stubbs *et al.*, 1977). In the southern American zone, virulence for *Yr9* was found in a trap nursery and before any known use of the resistance in commercial cultivars. In China, the breakdown of resistance of Lovrin 10 and Lovrin 13 (*Yr9*) (Li *et al.*, 1984) indicated the presence of virulence for *Yr9*. Virulence for *Yr9* was detected in East Africa by H. Bonthuis (personal communication). The *Yr9* resistance was obtained from a wheat–rye cross (Zeller and Fuchs, 1983) and is widely used in breeding programs internationally. The differential Riebesel 47/51 possessing *Yr9* was widely used in disease resistance breeding but is still highly resistant in the adult plant stage.

Yr10 was derived from PI 178383 of Turkish origin (Harlan, 1976). This may explain why virulence for it exists in the east Mediterranean zone and may explain its movement to the east Europe zone. Virulence in North America is probably the result of a separate mutation. *Yr10* was introduced in the United States and commercially used. Virulence to it was detected by Beaver and Powelson (1969).

Virulence for Suwon 92/Omar (WYV 11), detected for the first time in the United States (Purdy and Allan, 1966), is nearly worldwide, but nothing is known about the distribution of hosts possessing the corresponding resistance gene. WYV 11 has not been detected in East Asia, but only limited samples have been available for study.

Virulence for Carstens V (WYV 12) is common in the west and central Europe zone and southern South America zone, possibly as the result of cultivating the similar cultivars in both zones. No easy explanation can be given for the absence of WYV 12 in the west Mediterranean and southeast Asia zones when it is otherwise so widely distributed.

It is evident that parallel evolution of virulence of stripe rust has occurred frequently in the past two decades due to the international exchange of host resistances. Microevolution of stripe rust gradually changed into macroevolution. The data presented also indicate that stripe rust in all parts of the world possesses the same genetic background of pathogenicity, possibly due to its monotopic origin. Evaluation of host resistance should therefore be made with all available virulences of stripe rust and in all available combinations.

IX. Stripe Rust on Barley

Stripe rust on barley can be either of the form which infects wheat (*P. striiformis* f. sp. *tritici*) or of the form that infects barley (*P. striiformis* f. sp. *hordei*).

The importance and geographical distribution of stripe rust on barley are discussed in Section V.

The early studies of Gassner and Straib (1932b) made no distinction between the diseases. Barley cultivars were added as differentials by Straib (1935a). Studies on physiologic specialization have not gained the importance of those on wheat. So far, the nomenclature of races introduced by Gassner and Straib (see Table I) has not been changed by the Europeans or until recently by the Indian stripe rust workers (Nagarajan, 1983; Nagarajan *et al.*, 1984). In Japan, the differential cultivars of Gassner and Straib (see Table I) were used, and were supplemented with Yechao 35368 (Chinese wheat cultivar) and Sagigake (Japanese barley cultivar) to distinguish four races, A, B, C, and D, respectively (Kajiwara *et al.*, 1964a). Working with 10 European and seven Japanese cultures of barley stripe rust, Kajiwara (1964) introduced a new classification of barley stripe rust races. Seven races were distinguished with the differential hosts Fong Tien (barley), Bavaria (barley), Sakigake (barley), Yechao 35368 (wheat), and Hokkaido Chevalier (barley). These cultivars were included in my studies, but the reaction types on the differentials were variable. The current differential barley cultivars used in Europe are Cambrinus [=Heils Franken (Table I) = Astrix (Bayles and Priestley (1983)], which is only susceptible to race 24 = 48 (Table I), Mazurka, Varunda, Emir (widely used in breeding programs), Abessinian 14, Bigo, I 5, and Topper (susceptible check). In India, Barley Local and Himani (Table II) are used to differentiate the barley from the wheat stripe rust. In the nomenclature of races no distinction has been made between the wheat and barley form of stripe rust (Nagarajan, 1983).

The virulence spectrum of barley stripe rust in Europe and South America since race 24 was introduced in Colombia in 1975 are similar. The presence of race 24 has also been reported in the Nilgiri hills in India (Prasada *et al.*, 1967). Although in China stripe rust on barley has not been observed (see Section V), Fang (1944) isolated a culture from barley virulent on Heils Franken. Virulence to Bigo is confined to England, where it was first detected (Priestley and Byford, 1978), and in the Netherlands. According to trap-nursery data, virulence on I 5 occurs in South America (Dubin and Stubbs, 1984). The barley stripe rust cultures from the Middle East and East Africa analyzed in Wageningen have not shown clear differences in virulence on the differential cultivars supplemented with barley cultivars grown in those regions.

Interest in resistance of barley to stripe rust has increased in the past 20 years. Major studies were undertaken in India by Bahl and Bakshi (1963), Bakshi and Bahl (1965), Bakshi *et al.* (1964), Luthra (1966), and

Jain and Agrawal (1964). They designated the resistance loci as *Ps* (*P. striiformis*), *Ps1–Ps5*. In Germany, Nover and Scholz (1969) studied the inheritance of resistance in five barley cultivars and assigned *Yr* symbols to the genes identified. In Bigo, Abessinian 14, and BBA 2890, *Yr* occurs; in Abed Binder 12, *Yr2* occurs; and in I 5, *Yr3* occurs. Johnson (1968) found an additional gene in Europe, Cambrinus, and Deba Abed and designated it *Yr4*. In the European studies, *P. striiformis* f. sp. *hordei* was used, while in the Indian studies, *P. striiformis* f. sp. *tritici* (Macer and van den Driessche, 1966) was used.

In northwest Europe the impression is that barley stripe rust, like wheat stripe rust, is highly sensitive to environmental parameters. The effect of pre- and postinoculation temperature has been shown by Kellock and Lennard (1982). Under certain undefined greenhouse conditions, the hyphal growth may be completely restricted, resulting in a formation of urediosori as for leaf rust. The first-leaf reaction may differ greatly from the second-leaf reaction. In repeated tests, some cultivars, such as Belfort barley, exhibited an immune reaction on the first leaf and a susceptible reaction on the second and following leaves.

So far, stripe rust on barley has not been a major problem in northwest Europe, notwithstanding the fact that some susceptible spring barleys are being grown. However, as the acreage of winter barley is increasing, it is expected that the disease will acquire greater importance. In South America, particularly in those regions where the disease pressure is high, the disease is a problem and the pathogen may develop greater pathogenicity, as it has done on wheat. In other regions of the world the pathogen has not shown much diversity, but it may change with the improvement of barley cultivation.

References

Abdel-Hak, T., Kamel, A. H., Keddis, S., and Shafik, E. (1966). Epidemiology of wheat rusts in U.A.R. (Egypt). *Minist. Agric. Plant Prot. Dep. Cereal Dis. Res. Div., Tech. Bull.* No. 1, pp. 1–45.

Abdel-Hak, T., Stewart, D. M., and Kamel, A. H. (1972). "The Current Stripe Rust Situation in the Near East Region," Proc. Reg. Wheat Workshop, Vol. 1. Ford Found., Beirut, Lebanon.

Abiev, S., Zhakhanow, A., Kenesarina, G., and Esengulova, B. (1977). (Infection of wheat by yellow rust). *Zashch. Rast. (Leningrad)* **8,** 41.

Abiev, S. A., Zhakhanov, A., Kenesarina, G., and Esengulova, B. Zh. (1982). (Specialization of yellow rust of wheat in South-Eastern Kazakhstan). *Bot. Mater. Gerbariya, Inst. Bot. Akad. Nauk Kaz. SSR* No. 12, pp. 96–98.

Allison, C., and Isenbeck, K. (1930). Biologische Specialisierung von *Puccinia glumarum tritici* Erikss. und Henn. *Phytopathol. Z.* **2,** 87–98.

Anikster, Y., and Wahl, I. (1979). Coevolution of the rust fungi on *Gramineae* and *Liliaceae* and their hosts. *Annu. Rev. Phytopathol.* **17,** 367–403.

Anonymous (1968). "Wheat Rust Diseases. Wheat and Barley Improvement Project, Iran," U. N. Dev. Programme. FAO, Rome.

Anonymous (1969). "FAO Information Bulletin, Near East Wheat and Barley Improvement and Production Project," Spec. Rev., Vol. VI, No. 1, pp. 1–93. FAO, Rome.

Anonymous (1980). "Research Report 1978–1980." Biol. Chem. Res. Inst., Dep. Agric. N. S. W. 10.

Anonymous (1983). "1983 World Wheat Facts and Trends. Report Two: An Analysis of Rapidly Rising Third World Consumption and Imports of Wheat." Centro Internacional de Maíz y Trigo (CIMMYT), El Batan, Mexico.

Arthur, J. C. (1925). The grass rusts of South America, based on the Holway collection. *Proc. Am. Philos. Soc.* **64,** 131–223.

Bahl, P. N., and Bakshi, J. S. (1963). Genetic of rust resistance in barley II. The inheritance of seedling resistance to four races of yellow rust. *Indian J. Genet. Plant Breed.* **23,** 150–154.

Bakshi, J. S., and Bahl, P. N. (1965). Inheritance of resistance to four races of yellow rust in two varieties of barley. *Indian J. Genet. Plant Breed.* **25,** 239–242.

Bakshi, J. S., Bahl, P. N., and Kohli, S. P. (1964). Inheritance of seedling resistance to some Indian races of yellow rust in the crosses of rust resistant barley variety E.B. 410. *Indian J. Genet. Plant Breed.* **24,** 72–77.

Batts, C. C. V. (1957). The reaction of wheat varieties to yellow rust, *Puccinia glumarum,* 1951–1956. *J. Natl. Inst. Agric. Bot. (G. B.)* **8,** 7–18.

Bayles, R. A., and Priestley, R. H. (1983). "U.K. Cereal Pathogen Virulence Survey," 1982 Annu. Rep. U.K. Cereal Pathogen Virulence Survey Committee, Cambridge. 60–67.

Beaver, R. G., and Powelson, R. L. (1969). A new race of stripe rust pathogenic on the wheat variety Moro, C.I. 13740. *Plant Dis. Rep.* **53,** 91–93.

Becker, H., and Hart, H. (1939). Das Auftreten und die Verbreitung von Gelbrost im Ostharz und der darn angrenzenden Weizenanbaugebieten. *Z. Pflanzenkr. (Pflanzenpathol.) Pflanzenschutz* **49,** 449–481.

Beresford, R. M. (1982). Stripe rust (*Puccinia striiformis*) a new disease of wheat in New Zealand. *Cereal Rusts Bull.* **10,** 35–41.

Bever, W. M. (1934). Physiologic specialization in *Puccinia glumarum* in the U.S. *Phytopathology* **24,** 686–688.

Biffen, R. H. (1906). Mendels law of inheritance and wheat breeding. *J. Agric. Sci.* **1,** 4–48.

Brown, A. M. (1956). A check list of plant rusts in Canada. *Publ.—Can. Dep. Agric.* **976,** 1–50.

Brown, J. F., and Sharp, E. L. (1969). Interactions of minor host genes for resistance to *Puccinia striiformis* with changing temperatures. *Phytopathology* **59,** 999–1001.

Butler, E. J. (1903). "The Indian Wheat Rust Problem," Part 1, Bull. 1, Dep. Agric., India.

Carleton, M. A. (1915). A serious new wheat rust in this country. *Science* **42,** 58–59.

Chen, S. M., Chou, C. P., Lee, S. P., Wang, K. N., Ou-Yang, Hung, S. W., Lu, S. I., Yang, T. M., and Wu, W. C. (1957). (Studies on the epidemiology of stripe rust of wheat in North China). *Acta Phytopathol. Sin.* **3,** 63–85 (in Chinese with English abstract).

Ciccarone, A. (1947). Il problema della ruggine dei grani in Ethiopia. Tre anni di observazioni (1938, 1939, 1940). *Riv. Agric. Subtrop. Trop.* **41,** 1–42.

Coakley, S. M. (1978). The effect of climate variability on stripe rust of wheat in the Pacific Northwest. *Phytopathology* **68,** 207–212.

Coakley, S. M., and Line, R. F. (1981). Quantitative relationships between climatic variables and stripe rust epidemics on winter wheat. *Phytopathology* **71,** 461–467.
Creelman, D. W. (1964). A summary of the prevalence of plant diseases in Canada in 1963. *Can. Plant Dis. Surv.* **44,** 1–4.
Cumakov, A. E. (1963). Principy bor'by s ržavčinoj pšenicy v rajonach naibolee častych épifitotij. *Tr. Vses. Nauchno-Issled. Inst. Zashcn. Rast.* **18,** 280–288.
Cummins, G. B., and Stevenson, J. A. (1956). A check list of north American rust fungi (*Uredinales*). *Plant. Dis. Rep., Suppl.* **240,** 109–193.
Dubin, H. J., and Stubbs, R. W. (1984). *Puccinia striiformis f. sp. hordei* (PSH): Cause of barley yellow rust epidemic in South America. *Phytopathology* **74** (Abstr.), 820.
Eriksson, J. (1894). Über die Spezialisierung des Parasitsimus bei den Getreiderostpilzen. *Ber. Dtsch. Bot. Ges.* **12,** 292–331.
Eriksson, J., and Henning, E. (1894). Die Hauptresultate einer neuen Untersuchung über die Getreiderostpilze. *Z. Pflanzenkr.* **4,** 197–203.
Eriksson, J., and Henning, E. (1896). "Die Getreideroste." Norstedt & Söner, Stockholm.
Fang, C. T. (1944). Physiologic specialization of *Puccinia glumarum* Erikss. and Henn. in China. *Phytopathology* **34,** 1020–1024.
Fraser, W. P., and Conners, I. L. (1925). The *Uredinales* of the prairie provinces of Western Canada. *Trans. R. Soc. Can.* **19,** 275–308.
Fuchs, E. (1956). Der Stand der Rassenspezialisierung beim Gelbrost *Puccinia glumarum* (Schm.) Erikss. et Henn. in Europe. *Nachr. Dtsch. Pflanzenschutzd. Braunschweig* **8,** 87–93.
Fuchs, E. (1960). Physiologische Rassen bei Gelbrost (*Puccinia glumarum* (Schm.) Erikss. et Henn.) auf weizen. *Nachr. Dtsch. Pflanzenschutzd. Braunschweig* **12,** 49–63.
Fuchs, E. (1965). Untersuchungen über die physiologische Spezializierung des Weizengelbrostes (*Puccinia striiformis* West. f. sp. *tritici* Erikss. et Henn.) in den Jahren 1959–1964 and über das Anfälligkeitserhalten einiger Weizensorten. *Nachr. Dtsch. Pflanzenschutzd. Braunschweig* **17,** 161–176.
Fuckel, L. (1860). Enumeratio fungorum Nassovia. *Jahrb. Ver. Naturkd. Herzogthum Nassau* **15,** 9.
Futrell, M. C. (1957). Wheat stripe rust epiphytotic in Texas in 1957. *Plant Dis. Rep.* **41,** 955–957.
Futrell, M. C., Lahr, K. A., Porter, K. B., and Atkins, I. M. (1959). Second stripe rust epiphytotic in Texas hits wheat crop in 1958. *Plant Dis. Rep.* **43,** 165–167.
Gassner, G., and Straib, W. (1929). Experimentelle Untersuchungen über das Verhalten der Weizensorten gegen *Puccinia glumarum. Phytopathol. Z.* **1,** 215–275.
Gassner, G., and Straib, W. (1930). Über das Auftreten einer neuen Gelbrostform auf Weizen. *Züchter* **2,** 313–317.
Gassner, G., and Straib, W. (1932a). Über Mutationen in einer biologischen Rasse von *Puccinia glumarum tritici* (Schmidt). Erikss. u. Henn. *Z. Indukt. Abstammu- Vererbungsl.* **63,** 154–160.
Gassner, G., and Straib, W. (1932b). Die Bestimmung der Biologischen Rassen des Weizengelbrostes (*Puccinia glumarum* f. sp. *tritici* (Schm.) Erikss. u. Henn.). *Arb. Biol. Reichsanst. Land= Forstwirtsch.* **20,** 141–163.
Gassner, G., and Straib, W. (1934). Weitere Untersuchungen über biologisch Rassen und über die Spezialisierungsverhältnisse des Gelbrostes *Puccinia glumarum* (Schm., Erikss. u. Heen). *Ebenda* **21,** 121–145.
Gäumann, E. A. (1952). "The Fungi." Hafner, New York.
Ghaffor, A. (1970). Wheat Improvement and Production Programme in Afghanistan (1966–69). *Proc. FAO/Rockefeller Found. Wheat Semin., 3rd, 1970,* pp. 48–52.

Ghodbane, A. (1977). Stripe rust on wheat. *FAO Plant Prot. Bull.* **25** (4), 212.
Goddard, M. W. (1976a). Cytological studies of *Puccinia striiformis* (yellow rust of wheat). *Trans. Br. Mycol. Soc.* **66,** 433–437.
Goddard, M. V. (1976b). The production of a new race, 105E137 of *Puccinia striiformis* in glasshouse experiments. *Trans. Br. Mycol. Soc.* **67,** 395–398.
Gould, F. W. (1968). "Grass Systematics." McGraw-Hill, New York.
Gramma, A., Gerechter-Amitai, Z. K., and van Silfhout, C. H. (1984). Additive gene action for resistance to *Puccinia striiformis* f. sp. *tritici* in *Triticum dicoccoides. Euphytica* **33,** 281–287.
Grillo, H. V. S. (1937). Lista preliminar dos fungos assignalados en plantas do Brasil. *Rodriguésia* **2** (Num. esp. 1936), 39–96.
Guyot, L. (1966). Specific *glumarum*-like features of the Mediterranean rust (*Puccinia agropyri* E. et E., forme Europeene) of grasses and cereals. *Proc. Cereal Rusts Conf., 1964,* pp. 68–71.
Harlan, J. R. (1976). Diseases as a factor in plant evolution. *Annu. Rev. Phytopathol.* **14,** 31–51.
Hart, H., and Becker, H. (1939). Beitrage zur Frage des Zwischenwirtes für *Puccinia glumarum. Z. Pflanzenkr. (Pflanzenpathol.) Pflanzenschutz* **49,** 559–566.
Hassan, S. F. (1968). Cereal rusts situation in Pakistan. *Proc. Cereal Rusts Conf., 1968,* pp. 124–125.
Hassebrauk, K. (1962). Die Gelbrostepidemie 1961 in Deutschland. *Nachtr. Dtsch. Pflanzenschutzd.* **14,** 22–26.
Hassebrauk, K. (1965). Nomenklatur, geographische Verbreitung und Wirtsbereich des Gelbrostes, *Puccinia striiformis* West. *Mitt. Biol. Bundesanst. Land= Forstwirtsch., Berlin-Dahlem* **116,** 1–75.
Hassebrauk, K. (1970). Der Gelbrost *Puccinia striiformis* West. 2 Befallsbild. Morphologie und Biologie der Sporen. Infektion und weitere Entwicklung. Wirkungen auf die Wirtspflanze. *Mitt. Biol. Bundesanst. Land= Forstwirtsch., Berlin-Dahlem* **139,** 1–111.
Hassebrauk, K., and Röbbelen, G. (1974). Der Gelbrost *Puccinia striiformis* West. 3. Die Spezialisierung. *Mitt. Biol. Bundesanst. Land= Forstwirtsch., Berlin-Dahlem* **156,** 1–150.
Hassebrauk, K., and Röbbelen, G. (1975). Der Gelbrost, *Puccinia striiformis* West. 4. Epidemiology. - Bekämpfungsmassnahmen. *Mitt. Biol. Bundesanst. Land= Forstwirtsch., Berlin-Dahlem* **164,** 1–183.
Hemmi, T. (1934). On the distribution of cereal rusts in Japan and the relation of humidity to germination of urediniospores of some species of *Puccinia. Proc. Pac. Sci. Congr., 5th, 1933,* pp. 3187–3194.
Hendrix, J. W. (1964). Stripe rust, what it is and what to do about it. *Circ.—Wash., Agric. Exp. Stn.* **424,** 1–6.
Hendrix, J. W., Burleigh, J. R., and Tu, J.-C. (1965). Oversummering of stripe rust at high elevations in the Pacific Northwest-1963. *Plant Dis. Rep.* **49,** 275–278.
Hirschhorn, J. (1933). Dos royas de la cebeda, nuevas para la Argentina. *Rev. Fac. Agron., Univ. Nac. La Plata* **19,** 390–397.
Hogg, W. H., Hounam, C. E., Mallik, A. K., and Zadoks, J. C. (1969). Meteorological factors affecting the epidemiology of wheat rusts. *WMO, Tech. Note* **99,** 1–143.
Humphrey, H. B. (1917). *Puccinia glumarum. Phytopathology* **7,** 142–143.
Humphrey, H. B. (1941). Brief notes on plant diseases. *Plant Dis. Rep.* **25,** 337–338.
Humphrey, H. B., Hungerford, C. W., and Johnson, A. G. (1924). Stripe rust (*Puccinia glumarum*) of cereals and grasses in the United States. *J. Agric. Res. (Washington, D.C.)* **29,** 209–227.

Hungerford, C. W., and Owens, C. E. (1923). Specialized varieties of *Puccinia glumarum* and hosts for variety *tritici*. *J. Agric. Res. (Washington, D.C.)* **25,** 363–401.
Hylander, N., Jørstad, I., and Nannfeldt, J. A. (1953). Enumeratio uredionearum Scandinavicarum. *Opera Bot.* **1,** 1–102.
Iren, S. (1981). Wheat Diseases in Turkey. *Bull. OEPP* **11,** 47–52.
Ito, S. (1909). Contributions to the mycological flora of Japan. II. On the *Uredineae* parasitic on the Japanese *Gramineae*. *J. Coll. Agric., Tokyo Imp. Univ.* **3,** no. 2, 180–262.
Jain, K. B. L., and Agrawal, R. K. (1964). Mature plant resistance of barley varieties to Indian races of stripe rust. *Indian J. Genet. Plant Breed.* **24,** 203–208.
Johnson, R. (1968). Genetics of resistance of barley to yellow rust. *Int. Congress Plant Pathology, 1st, 1968,* Abstracts, p. 99.
Johnson, R. (1981). Durable disease resistance. *In* "Strategies for the Control of Cereal Disease" (J. F. Jenkyn and R. T. Plumb, eds.), pp. 55–64. Blackwell, Oxford.
Johnson, R., Stubbs, R. W., Fuchs, E., and Chamberlain, N. H. (1972). Nomenclature for physiologic races of *Puccinia striiformis* infecting wheat. *Trans. Br. Mycol. Soc.* **58,** 475–480.
Johnson, R., Priestley, R. H., and Taylor, E. C. (1978). Occurrence of virulence in *Puccinia striiformis* for Compair wheat in England. *Cereal Rusts Bull.* **6,** 11–13.
Joshi, L. M. (1976). Recent contributions towards epidemiology of wheat rusts in India. *Indian Phytopathol.* **29,** 1–16.
Joshi, L. M., Saari, E. E., and Gera, S. D. (1974). Survey and epidemiology of wheat rust in India. *In* "Current Trends in Plant Pathology" (S. P. Raychaudhuri and J. P. Verma, eds.), pp. 150–159. Dept. of Botany, Lucknow University, Lucknow.
Joshi, L. M., Goel, L. B., and Sinha, V. C. (1976). Role of the western Himalayas in the annual recurrence of yellow rust in northern India. *Cereal Rusts Bull.* **4,** 27–30.
Joshi, L. M., Nagarajan, S., and Srivastava, K. D. (1977). Epidemiology of brown and yellow rusts of wheat in North India 1. Place and time of appearance and spread. *Phytopathol. Z.* **90,** 116–122.
Kajiwara, T. (1964). Beiträge zur Kenntnis der Physiologischen Spezialisierung des Gerstengelbrostes (*Puccinia striiformis* West. f. sp. *hordei* Erikss.). *Nachr. Dtsch. Pflanzenschutzd. Braunschweig* **4,** 58–59.
Kajiwara, T., Ueda, I., and Iwata, Y. (1964a). Untersuchungen über die Physiologische Spezialisierung des Gerstengebrostes (*Puccinia striiformis* f. sp. *hordei*) in Japan. *Phytopathol. Z.* **50,** 313–328.
Kajiwara, T., Ueda, I., and Iwata, Y. (1964b). Untersuchungen über die Physiologische Spezialisierung des Weizengelbrostes *Puccinia striiformis* West. f. sp. *tritici* Erikss. et Henn. in Japan. *Phytopathol. Z.* **51,** 19–28.
Kajiwara, T., Ueda, I., and Iwata, Y. (1968). (Susceptibility of wheat varieties to the Japanese isolate of wheat yellow rust *Puccinia striiformis* f. sp. *tritici*). *Bull. Natl. Inst. Agric. Sci., Ser. C* **22,** 243–257.
Karki, C. B. (1980). Report on "Evaluation of Nepalese Wheat and Barley Varieties in the Seedling Stage on their Resistance to Yellow Rust." Res. Inst. Plant Prot. (IPO), Wageningen, The Netherlands.
Kellock, L. J., and Lennard, J. H. (1982). The development of barley yellow rust (*Puccinia striiformis*) on different cultivars in relation to pre- and post-inoculation temperatures. *Cereal Rusts Bull.* **10,** 42–53.
Khazra, H., and Bamdadian, A. (1974). The wheat diseases situation in Iran. *Proc. FAO/Rockefeller Found. Wheat Semin., 4th, 1973,* pp. 292–299.
Kidwai, A. (1979). Pakistan reorganise agricultural research after harvest disaster. *Nature (London)* **227,** 169.

Le Beau, F. J., Sosa, O. N., and Fumagalli, A. (1956). Yield reduction in wheat by stripe rust. *Plant Dis. Rep.* **40,** 886.

Leppik, E. E. (1970). Gene centres of plants as sources of disease resistance. *Annu. Rev. Phytopathol.* **8,** 324–344.

Lewellen, R. T., and Sharp, E. L. (1968). Inheritance of minor reaction genes combinations in wheat to *Puccinia striiformis* at two temperature profiles. *Can. J. Bot.* **46,** 21–26.

Lewellen, R. T., Sharp, E. L., and Hehn, E. R. (1967). Major and minor genes in wheat for resistance to *Puccinia striiformis* and their response to temperature changes. *Can. J. Bot.* **45,** 2155–2172.

Li, C. C., and Liu, H. W. (1957). (Preliminary studies on the trend of occurrence and development of the stripe rust of wheat (*Puccinia glumarum* (Sch.) Erikss. and Henn.) in provinces Shensi, Kansu, and Chinghai). *Northwest Agric. Coll. J. Sian* **1,** 33–45.

Li, Z. G., Shang, H. S., Yin, X. L., Qiang, Z. F., Zhao, Y. Q., Lu, H. P., Hong, X. W., Song, W. X., and Lin, S. J. (1984). Studies on the breakdown of Lovrin cultivars to stripe rust (*Puccinia striiformis* West.). *Scientia Agricultura Sinica* **1,** 68–74.

Line, R. F. (1972). Patterns of pathogenicity of *Puccinia striiformis* in the United States. *Proc.—Eur. Mediterr. Cereal Rusts Conf., 3rd, 1972,* Vol. 1, pp. 181–185.

Line, R. F. (1976). Factors contributing to an epidemic of stripe rust on wheat in the Sacramento Valley of California in 1974. *Plant Dis. Rep.* **60,** 312–316.

Line, R. F. (1980). Pathogenicity and evolution of *Puccinia striiformis* in the United States. *Proc.—Eur. Mediterr. Cereal Rusts Conf., 5th, 1980,* pp. 93–96.

Line, R. F., Sharp, E. L., and Powelson, R. L. (1970). A system for differentiating races of *Puccinia striiformis* in the United States. *Plant Dis. Rep.* **54,** 992–993.

Little, R., and Manners, J. G. (1969a). Somatic recombination in yellow rust of wheat (*Puccinia striiformis*). I. The production and possible origin of two new physiologic races. *Trans. Br. Mycol. Soc.* **53,** 251–258.

Little, R., and Manners, J. G. (1969b). Somatic recombination in yellow rust of wheat (*Puccinia striiformis*). II. Germ tube fusions, nuclear number and nuclear size. *Trans. Br. Mycol. Soc.* **53,** 259–267.

Lupton, F. G. H. and Macer, R. C. F. (1962). Inheritance of resistance to yellow rust (*Puccinia glumarum* Erikss. et Henn.) in seven varieties of wheat. *Trans. Br. Mycol. Soc.* **45,** 21–45.

Luthra, J. K. (1966). Inheritance of seedling resistance to races 57 and G of yellow rust in Barley. *Indian J. Genet. Plant Breed.* **26,** 356–359.

Macer, R. C. F., and van den Driessche, M. (1966). Yellow rust (*Puccinia striiformis* Westend.) of barley in England—1960–65. *J. Agric. Sci.* **67,** 255–265.

McGinnis, R. G. (1956). Cytological studies of chromosomes of rust fungi. II. The relationship of chromosome number to sexuality in *Puccinia. J. Hered.* **47,** 255–259.

McIntosh, R. A. (1973). A catalogue of gene symbols for wheat. *Proc. Int. Wheat Genet. Symp., 4th, 1973,* pp. 893–937.

McIntosh, R. A. (1983). "Annual report 1983," pp. 5–6. Plant Breed. Inst., University of Sydney, Sydney, Australia.

McIntosh, R. A., Luig, N. H., Johnson, R., and Hare, R. A. (1981). Cytogenetical studies in wheat XI. Sr 9g for reaction to Puccinia graminis tritici. *Z. Pflanzenzüecht.* **87,** 274–289.

McNeal, F. H., Konzak, C. F., Smith, E. P., Tate, W. S., and Russell, T. S. (1971). A uniform system for recording and processing cereal research data. *U. S., Agric. Res. Serv., ARS* **34-121,** 1–42.

Maddison, A. C., and Manners, J. G. (1972). Sunlight and viability of cereal rust uredospores. *Trans. Br. Mycol. Soc.* **59,** 429–443.
Mains, E. B. (1933). Studies concerning heteroecious rusts. *Mycologia* **25,** 407–417.
Manners, J. G. (1950). Studies on the physiologic specialization of yellow rust (*Puccinia glumarum* (Schm.) Erikss. & Henn.) in Great Britain. *Ann. Appl. Biol.* **37,** 187–214.
Manners, J. G. (1960). *Puccinia striiformis* Westend. var. *dactylidis* var. nov. *Trans. Br. Mycol. Soc.* **43,** 65–68.
Marchionatto, J. B. (1931). La presencia de la roya "amarilla," Secc. Prop., Inf., Circ. No. 836, pp. 3–5. Minist. Agric. Nac., Buenos Aires.
Martens, J. W., and Oggema, M. W. (1972). "The Yellow Rust Situation in Kenya," Proc. Reg. Wheat Workshop, Vol. 1. Ford Found., Beirut, Lebanon.
Mehta, K. C. (1933). Rusts of wheat and barley in India. *Indian J. Agric. Sci.* **3,** 939–962.
Mehta, K. C. (1940). Further studies on cereal rusts in India. Part 1. *Indian Counc. Agric. Res. Sci. Monogr.* **14,** 1–224.
Mehta, K. C. (1952). Further studies on cereal rusts in India. Part 2. *Indian Counc. Agric. Res. Sci. Monogr.* **18,** 1–368.
Melching, J. S., Stanton, J. R., and Koogle, D. L. (1974). Deleterious effects of tobacco smoke on germination and infectivity of spores of *Puccinia graminis tritici* and on germination of spores of *Puccinia striiformis, Pyricularia oryzae* and an *Alternaria* species. *Phytopathology* **64,** 1143–1147.
Mendgen, K. (1981). Growth of *Verticillium lecanii* in pustules of stripe rust (*Puccinia striiformis*). *Phytopathol. Z.* **102,** 301–309.
Meuzelaar, H. L. C. (1979). "Coffee Rust Control," Report on a seminar held in Paipa/Colombia, pp. 181–198. German Agency for Technical Cooperation (GTZ), Eschborn, Federal Republic of Germany.
Mohamed, H. A. (1963). The status of wheat rusts in the U.A.R. (Egypt) in the period 1959–1962. *Robigo* **14,** 13–14.
Nagarajan, S. (1983). "Annual Report 1983." Indian Agric. Res. Inst., Regional Station, Flowerdale, Simla.
Nagarajan, S., and Joshi, L. M. (1975). A historical account of wheat rust epidemics in India, and their significance. *Cereal Rusts Bull.* **3,** 29–33.
Nagarajan, S., and Joshi, L. M. (1978). Epidemiology of brown rust and yellow rust of wheat over North India. II. Associated meteorological conditions. *Plant Dis. Rep.* **62,** 186–188.
Nagarajan, S., Kranz, J., Saari, E. E., and Joshi, L. M. (1982). Analysis of wheat rusts epidemic in the Indo-Gangetic Plains 1976–78. *Indian Phytopathol.* **35,** 473–477.
Nagarajan, S., Bahadur, P., and Nayar, S. K. (1984). Occurrence of a new virulence, 47S102 of Puccinia striiformis West., in India during crop year, 1982. *Cereal Rusts Bull.* **12,** 28–31.
Narita, T., and Mano, Y. (1962). "Researches on the Oversummering, Overwintering, and Source of Primary Infection of Stripe Rust in Wheat and Barley," Spec. Rep., Plant Dis. Insects Forecasting Serv., No. 12, pp. 1–36. Soc. Plant Prot., Min. Agric. For., Japan.
Negulescu, F., and Ionescu-Cojocaru, M. (1976). Aspects of *Puccinia striiformis* f. sp. *tritici* physiologic specialization in Romania. *Proc.—Eur. Mediterr. Cereal Rusts Conf., 4th, 1976,* pp. 81–84.
Newton, M., and Johnson, T. (1936). Stripe rust, *Puccinia glumarum* in Canada. *Can. J. Res., Sect. C* **14,** 89–108.
Newton, M., Johnson, T., and Brown, A. M. (1933). Stripe rust in Canada. *Phytopathology* **23,** 27–28.

Niemann, E., Scharif, G., and Bamdadian, A. (1968). Die Getreideroste in Iran. Wirtsbereich, Unterscheidung, Bedeutung, Bekämpfung. *Entomol. Phytopathol. Appl.* **27,** 25–41.

Nover, I., and Scholz, F. (1969). Genetic studies on the resistance of barley to yellow rust (*Puccinia striiformis* Westend.). *Theor. Appl. Genet.* **34,** 150–155.

O'Brien, L., Brown, J. S., Young, R. M., and Pascoe, T. (1980). Occurrence and distribution of wheat stripe rust in Victoria and susceptibility of commercial wheat cultivars. *Australas. Plant Pathol.* **9,** 14.

Oort, A. J. P. (1955). Verspreiding en verwantschap van physiologische rassen van gele roest (*Puccinia glumarum*) van tarwe in Europa. *Tijdschr. Plantenziekten* **61,** 202–219.

Özkan, M., and Prescott, J. M. (1972). Cereal rusts in Turkey. *Proc.—Eur. Mediterr. Cereal Rusts Conf., 3rd, 1972,* Vol. 2, pp. 183–185.

Pady, S. M., and Johnston, C. O. (1959). Stripe rust of wheat in Kansas in 1958. *Plant Dis. Rep.* **43,** 159.

Prasada, R., and Lele, V. C. (1952). New physiologic races of wheat rusts in India. *Indian Phytopathol.* **5,** 128–129.

Prasada, R., Joshi, L. M., Singh, S. D., Misra, D. F., Goel, L. B., Kumari, K., Sharma, S. K., Joshi, P. C., and Ahmad, S. T. (1967). Occurrence of physiologic races of wheat and barley rusts in India during 1962–64 and their sources of resistance. *Indian J. Agric. Sci.* **37,** 273–281.

Priestley, R. H., and Byford, P. (1978). "U.K. Cereal Pathogen Virulence Survey," 1977 Annu. Rep. U.K. Cereal Pathogen Virulence Survey Committee, Cambridge, 12–16.

Priestley, R. H., and Doodson, J. K. (1976). Physiologic specialization of *Puccinia striiformis* to adult plants of winter wheat cultivars in the United Kingdom. *Proc.—Eur. Mediterr. Cereal Rusts Conf., 4th, 1976,* pp. 87–89.

Purdy, L. H., and Allan, R. E. (1966). A stripe rust race pathogenic to 'Suwon 92' wheat. *Plant Dis. Rep.* **50,** 205–207.

Rapilly, F. (1979). Yellow rust epidemiology. *Annu. Rev. Phytopathol.* **17,** 59–73.

Rijsdijk, F. H., and Zadoks, J. C. (1976). Assessment of risk and losses due to cereal rusts in Europe. *Proc.—Eur. Mediterr. Cereal Rusts Conf., 4th, 1976,* pp. 60–62.

Rijsdijk, F. H., and Zadoks, J. C. (1979). A data bank on crop losses: First experiences. *Bull. OEPP* **9,** 297–303.

Riley, R., Chapman, V., and Johnson, R. (1968). Introduction of yellow rust resistance of *Aegilops comosa* into wheat by genetically induced homoeologous recombination. *Nature (London)* **217,** 383–384.

Röbbelen, G., and Sharp, E. L. (1978). Mode of inheritance, interaction and application of genes conditioning resistance to yellow rust. *Fortschr. Pflanzenzücht., Beih. Z. Pflanzenzücht.* **9,** 1–88.

Rudorf, W., and Job, M. (1931). La existencia de *Puccinia glumarum tritici* (Schmidt) Erikss. et Henn. en los paises del Rio de la Plata. *Arch. Soc. Biol. Montevideo* **5,** Suppl., 1363–1370.

Rudorf, W., and Job, M. (1934). Untersuchungen bezüglich des Spezialisierung von *Puccinia graminis tritici, Puccinia triticina* und *Puccinia glumarum tritici,* sowie über Resistenz und ihre Vererbung in verschiedenen Kreuzungen. *Pflanzenzüchtung* **19,** 333–365.

Saari, E. E., and Prescott, J. M. (1978). Barley diseases and their surveillance in the region. *Proc. Reg. Winter Cereal Workshop—Barley, 4th, 1977,* Vol. 2, pp. 320–330.

Savile, D. B. O. (1954). The fungi as aids in the taxonomy of the flowering plants. *Science* **120,** 583–585.

Schmidt, J. K. (1827). "Allgemeine ökonomisch-technische Flora oder Abbildungen und Beschreibungen aller in bezug auf Ökonomie und Technologie, merkwürdigen Gewächse," Vol. I, p. 27. Jena, Germany.

Schmiedeknecht, M., and Puncag, T. (1967). *Puccinica*-Arten aus der Mongolischen Volksrepublik. *Foddes Repertorium* **74,** 177–199.

Seth, L. N. (1939). Report of the Mycologist, Burma, Mandaley, for the year ended 31st March 1939.

Shaner, G., and Powelson, R. L. (1971). Epidemiology of stripe rust of wheat 1961–1968. *Oreg. State Univ., Agric. Exp. Stn., Techn. Bull.* **117,** 1–31.

Shaner, G., and Powelson, R. L. (1973). The oversummering and dispersal of inoculum of *Puccinia striiformis* in Oregon. *Phytopathology* **63,** 13–17.

Sharma, S. K., Joshi, L. M., Singh, S. D., and Nagarajan, S. (1972). New virulence of yellow rust on Kalyansona variety of wheat. *Proc.—Eur. Mediterr. Cereal Rusts Conf., 3rd, 1972,* Vol. 1, pp. 263–266.

Sharp, E. L. (1967). Atmospheric ions and germination of urediospores of *Puccinia striiformis. Science* **156,** 1359–1360.

Sharp, E. L., and Hehn, E. R. (1963). Overwintering of stripe rust in winter wheat in Montana. *Phytopathology* **53,** 1239–1240.

Sharp, E. L., and Volin, R. B. (1970). Additive genes in wheat conditioning resistance to stripe rust. *Phytopathology* **60,** 1146–1147.

Singh, D., Goel, L. B., Sharma, S. K., Nayar, S. K., Chatterjee, S. C., Sinha, V. C., and Bahadur, P. (1978). Prevalence and distribution of physiologic races of wheat rusts in India during 1969–74. *Indian J. Agric. Sci.* **48,** 1–7.

Stakman, E. C., and Piemeisel, F. J. (1917). Biological forms of *Puccinia graminis* on cereals and grasses. *J. Agric. Res. (Washington, D.C.)* **10,** 429–495.

Standen, J. H. (1952). Host index of plant pathogens of Venezuela. *Plant Dis. Rep., Suppl.* **212,** 59–106.

Straib, W. (1935a). Auftreten und Verbreitung biologischer Rassen des Gelbrostes (*Puccinia glumarum* (Schm.) Erikss. et Henn.) im Jahre 1934. *Arb. Biol. Reichsanst. Land= Forstwirtsch., Berlin-Dahlem* **21,** 455–466.

Straib, W. (1935b). Infektionsversuche mit biologische Rassen des Gelbrostes auf Gräsern. *Arb. Biol. Reichsanst. Land= Forstwirtsch., Berlin-Dahlem* **21,** 483–497.

Straib, W. (1936). Über Gelbrostanfälligkeit und resistenz der Gerstenarten. *Arb. Biol. Reichsanst. Land= Forstwirtsch., Berlin-Dahlem* **21,** 467–473.

Straib, W. (1937a). Untersuchungen über das Vorkommen physiologischer Rassen des Gelbrostes (*Puccinia glumarum*) in den Jahren 1935–1936 und über die Agressivität einiger neuer Formen auf Getreide und Gräsern. *Arb. Biol. Reichsanst. Land= Forstwirtsch., Berlin-Dahlem* **22,** 91–119.

Straib, W. (1937b). Über Resistenz bei Gerste gegenüber Zwergrost und Gelbrost. *Züchter* **9,** 305–311.

Stubbs, R. W. (1967). Influence of light intensity on the reactions of wheat and barley seedlings to *Puccinia striiformis. Phytopathology* **57,** 615–619.

Stubbs, R. W. (1968a). Artificial mutation in the study of the relationship between races of yellow rust of wheat. *Proc.—Eur. Mediterr. Cereal Rusts Conf., 2nd, 1968,* pp. 60–62.

Stubbs, R. W. (1968b). *Puccinia striiformis* Westend. f. sp. *tritici.* The evolution of the genetic relationship of host and parasite. *Int. Congress Plant Pathology, 1st, 1968,* Abstracts, p. 198.

Stubbs, R. W. (1977). Observations on horizontal resistance to yellow rust (*Puccinia striiformis* f. sp. *tritici*). *Cereal Rusts Bull.* **5,** 27–32.

Stubbs, R. W., Fuchs, E., Vecht, H., and Basset, E. J. W. (1974). The international survey of factors of virulence of *Puccinia striiformis* Westend. in 1969, 1970 and 1971. *Ned. Graan-Centrum Tech. Ber.* **21,** 1–88.

Stubbs, R. W., Slovenčikov́, V., and Bartŏs (1977). Yellow rust resistance of some European wheat cultivars derived from rye. *Cereal Rusts Bull.* **5,** 44–47.

Sydow, P., and Sydow, H. (1904). "Monographia Uredinearum," Vol. 1. Leipzig, Germany.

Taylor, B. C. (1976). The production and behaviour of somatic recombinants in *Puccinia striiformis. Proc.—Eur. Mediterr. Cereal Rusts Conf., 4th, 1976,* pp. 36–38.

Thran, P., and Broekhuizen, S. (1965). "Agro-ecological Atlas of Cereal Growing in Europe," Vol. 1. Elsevier, Amsterdam.

Tollenaar, H., and Houston, B. R. (1967). A study on the epidemiology of stripe rust, *Puccinia striiformis* West., in California. *Can. J. Bot.* **45,** 291–307.

Tranzschel, W. (1934). Promežutočnye chozjaeva rzavčiny chlebov i ich der UdSSR. (The alternate hosts of cereal rust fungi and their distribution in the UdSSR). *Bull. Plant Prot., Ser. 2,* pp. 4–40 (in Russian with German summary).

Ubels, E., Stubbs, R. W., and s'Jacob, J. C. (1965). Some new races of *Puccinia striiformis. Neth. J. Plant Pathol.* **71,** 14–19.

Urazaliev, R. A., and Zeinalova, Yu. D. (1979). (Inheritance of resistance to yellow rust in intraspecies winter wheat hybrids). *Sel. Semenovod. (Moscow)* No. 2, pp. 26–27.

Vallega, J. (1955). Wheat rust races in South America. *Phytopathology* **45,** 242–246.

Viennot-Bourgin, G. (1934). La rouille jaune des graminées. *Ann. Ec. Natl. Agric., Grignon, Ser. 3,* **2,** 129–217.

Volosky de Hernandez, D. (1953). Estudios preliminares sobre el *Puccinia glumarum* (Schm., Erikss.) del trigo en Chile. *Agric. Tec. (Santiago)* **13,** 159–165.

Wallwork, H., and Johnson, R. (1984). Transgressive segregation for resistance to yellow rust in wheat. *Euphytica* **33,** 123–132.

Wang, K. N., Hong, S. V., Si, C. M., Wang, C. S., and Shen, C. P. (1963). Studies on the physiologic specialization of stripe rust of wheat in China. *Acta Phytopathol. Sin.* **2,** 23–36.

Wellings, C. R. (1982). "Annual Report 1982," pp. 24–27. Plant Breed. Inst., University of Sydney, Sydney, Australia.

Wellings, C. R. (1983). "Annual Report 1983," pp. 19–20. Plant Breed. Inst., University of Sydney, Sydney, Australia.

Westendorp, G. D. (1854). Quatrième notice sur quelques *Cryptogames* récemment découvertes en Belgique. *Bull. Acad. R. Sci. Belg.* **21,** 229–246.

Wienhues, F., and Giessen, J. A. (1957). Die Abstammung europäischer Weizensorten. *Z. Pflanzenzücht.* **37,** 218–230.

Wodageneh, A. (1974). Country reports and programmes: Ethiopia. *Proc. FAO/Rockefeller Found. Wheat Semin., 4th, 1973,* pp. 64–68.

Wright, R. G. (1976). Variation in *Puccinia striiformis. Proc.—Eur. Mediterr. Cereal Rusts Conf., 4th, 1976,* pp. 42–44.

Wright, R. G., and Lennard, J. H. (1978). Mitosis in *Puccinia striiformis* 1. Light microscopy. *Trans. Br. Mycol. Soc.* **70,** 91–98.

Wright, R. G., and Lennard, J. H. (1980). Origin of a new race of *Puccinia striiformis. Trans. Br. Mycol. Soc.* **74,** 283–287.

Zadoks, J. C. (1961). Yellow rust on wheat studies in epidemiology and physiologic specialization. *Tijdschr. Plantenziekten* **67,** 69–256.

Zeller, F. J., and Fuchs, E. (1983). Cytologie und Krankheitsresistenz einer 1A/1R- und mehrerer 1B/1R-Weizen-Roggen Translokationssorten. *Z. Pflanzenzücht.* **90,** 285–296.

Zeven, A. C., and Zeven-Hissink, N. C. (1976). "Genealogies of 14,000 Wheat Varieties." Netherlands Cereals Centre-NGC, Wageningen, and Int. Maize and Wheat Improvement Center (CIMMYT), Mexico.

4

Oat Stem Rust

J. W. Martens
Agriculture Canada Research Station,
Winnipeg, Manitoba, Canada

I. Introduction

Oats, *Avena* spp. L., and its parasite, *Puccinia graminis* Pers. f. sp. *avenae* Eriks. and E. Henn., are first known to have appeared in the literature almost 2400 years ago. Oats were referred to by the Greek writers Dieuches (400 B.C.) and Theophrastus (371–286 B.C.), and by the Romans Cato the Elder (234–149 B.C.), Cicero (106–43 B.C.), and

THE CEREAL RUSTS,
Volume II

ISBN 0-12-148402-5

Pliny the Elder (23–79 A.D.), among others (Coffman, 1961). Numerous writers have interpreted biblical references to blight, blasting, and mildew of grain to mean the rusts and the smuts, but there is probably merit in Arthur's (1929) suggestion that these pathogens were only a few of many factors that could have caused the problem. Aristotle the Greek (384–322 B.C.) writes of rust being produced by warm vapor, of devastating rust, and of rust years. His pupil Theophrastus noted that cereals are more affected by rust than are legumes (Arthur, 1929). The prayer of the officiating priest at a Robigalia ceremony, as given by the Roman poet, Ovid (43 B.C.–17 A.D.), "Stern Robigo, spare the herbage of the cereals, . . . withold we pray, thy roughening hand," could scarcely be more explicit. It leaves little to the imagination of anyone who has read a rust nursery. Pliny referred to rust as the greatest pest of the crops.

However, a detailed account of the stem rust condition, almost certainly including oat stem rust, had to wait almost 18 centuries, when the Italians Tozzetti and Fontana independently described it in 1767 (see Fontana, 1932; Tozzetti, 1952). The quality of their descriptions is most remarkable.

For almost a century after Persoon first named the organism *P. graminis* in 1797 (Arthur, 1929), it was regarded as a single species capable of attacking all the common cereals and grasses. It was not until 1894 that Eriksson in Sweden recognized it as a distinct biologic race and named it *P. graminis* f. *avenae* (Eriksson, 1894). Thirty years later, Stakman *et al.* (1923) first reported specialization within *P. graminis avenae.*

It is not practical to list all of the investigators who have contributed to our knowledge and understanding of the organism, but some of the pioneers deserve special mention. These include Eriksson and Henning in Sweden, who laid the foundation for pathogenic specialization studies, Stakman and Levine in the United States, Waterhouse and Watson in Australia, and Bailey, Gordon, and Johnson in Canada.

Stakman, Levine, and Bailey (1923) first reported specialization within *P. graminis avenae* describing four avirulence/virulence combinations based on a study of over 100 collections from five countries using two differentials that carried genes *Pg-1* and *-2* and a universally susceptible cultivar, Victory.

II. Distribution and Importance

Stem rust of oats occurs almost everywhere that oats are grown and periodically has caused severe crop losses. The ancients recognized oats and also wrote of "rust years" earlier than 300 B.C. (Arthur, 1929; Chester, 1946). Eriksson and Henning in 1896 (Chester, 1946) described the occurrence of rust years on a worldwide basis during the period 1660–1892. In 1889 oat rust was so destructive in Sweden that the royal government funded research on rust prevention, which led to the historic and fundamental work of Eriksson and Henning.

In North America there were severe epidemics in 1904, 1916, 1923, 1927, 1935, 1938, 1943, 1949, 1953, 1955, and 1977 (Craigie, 1957; Roelfs and Long, 1980). Roelfs (1978) and Roelfs and Long (1980) report that during the 59-year period from 1918 to 1977, oat stem rust epidemics causing yield losses of over 5% occurred in 8 years in Minnesota and North Dakota, 7 years in South Dakota, 5 years in Texas, 4 years in Iowa and Michigan, 3 years in Illinois, Kansas, and Nebraska, 2 years in Wisconsin, and 1 year in Oklahoma and Pennsylvania. The greatest nationwide losses were sustained in 1953, when they were estimated at 25, 10, 7, 5, and 4% in Minnesota, Iowa, Wisconsin, North Dakota, and South Dakota, respectively. Total United States losses for that year were 947,450 tonnes. In western Canada, Greaney (1936) estimated the average annual losses for Manitoba and Saskatchewan from 1929 to 1934 at 128,581 tonnes, with losses of 463,767 tonnes sustained in 1930. There were also important epidemics in Canada in 1944, 1945, 1947, and 1950 (Green *et al.*, 1961) but loss data are not available. The 1970 epidemic caused losses estimated at 92,572 tonnes in Manitoba (Martens, 1971), and the 1977 epidemic, the most severe in decades, caused losses of about 35% or 385,000 tonnes in Manitoba and eastern Saskatchewan (Martens, 1978). There were moderate losses in Manitoba and Saskatchewan again in 1981 (J. W. Martens, unpublished).

III. Pathogenic Specialization and Virulence Dynamics

Fortunately, the results of almost all of the investigations since Stakman, Levine, and Bailey first described specialization in oat stem rust can be interpreted in genetic terms because all of the differential lines used were later shown to have single genes for resistance. Since that time, a number of different systems of nomenclature have been used to describe the virulence characteristics of races of the organism (Stewart and Roberts, 1970). In this chapter, the results obtained since specialization was discovered will be described in terms of contemporary nomenclature (Martens *et al.*, 1979). This method describes the avir-

Table I

Avirulence–Virulence Combinations of *Puccinia graminis* f. sp. *avenae* Known to Occur in North America[a]

North American numbers	Avirulence–virulence formula (*Pg* genes)	North American numbers	Avirulence–virulence formula (*Pg* genes)
NA 1	1,2,3,4,8,9,13,16,a/15	NA 27	9,13,15,16,a/1,2,3,4,8
NA 2	1,2,3,4,8,13,16,a/9,15	NA 28	9,13,15,16/1,2,3,4,8,a
NA 3	1,2,3,4,8,16,a/9,13,15	NA 29	9,13,16,a/1,2,3,4,8,15
NA 4	1,2,3,8,16,a/4,9,13,15	NA 30	13,16,a/1,2,3,4,8,9,15
NA 5	1,2,4,8,9,13,16,a/3,15	NA 31	1,3,8,16,a/2,4,9,13,15
NA 6	1,2,4,8,13,16,a/3,9,15	NA 32	1,8,16,a/2,3,4,9,13,15
NA 7	1,2,4,8,16,a/3,9,13,15	NA 33	1,4,8,13,16,a/2,3,9,15
NA 8	1,2,8,16,a/3,4,9,13,15	NA 34	1,3,4,8,16,a/2,9,13,15
NA 9	1,3,8,13,16,a/2,4,9,15	NA 35	2,4,8,13,16,a/1,3,9,15
NA 10	1,4,8,9,13,16,a/2,3,15	NA 36	2,8,13,16,a/1,3,4,9,15
NA 11	1,8,9,13,16,a/2,3,4,15	NA 37	2,8,16,a/1,3,4,9,13,15
NA 12	1,8,13,16,a/2,3,4,9,15	NA 38	1,2,3,4,8,13,15,16,a/9
NA 13	1,13,16,a/2,3,4,8,9,15	NA 39	1,2,4,8,9,13,15,16,a/3
NA 14	2,3,4,9,13,15,16,a/1,8	NA 40	1,3,8,13,15,16,a/2,4,9
NA 15	2,4,8,9,13,15,16,a/1,3	NA 41	1,4,8,13,15,16,a/2,3,9
NA 16	2,4,9,13,15,16,a/1,3,8	NA 42	1,4,8,16,a/2,3,9,13,15
NA 17	2,4,9,13,15,16/1,3,8,a	NA 43	1,8,9,13,15,16,a/2,3,4
NA 18	2,4,9,13,16,a/1,3,8,15	NA 44	2,3,9,13,15,16,a/1,4,8
NA 19	3,8,9,13,16,a/1,2,4,15	NA 45	2,4,8,13,15,16,a/1,3,9
NA 20	3,8,13,16,a/1,2,4,9,15	NA 46	2,4,8,13,15,16/1,3,9,a
NA 21	3,9,13,15,16,a/1,2,4,8	NA 47	2,4,8,16,a/1,3,9,13,15
NA 22	4,8,9,13,16,a/1,2,3,15	NA 48	2,4,13,15,16,a/1,3,8,9
NA 23	4,9,13,15,16,a/1,2,3,8	NA 49	2,8,13,15,16,a/1,3,4,9
NA 24	8,9,13,16,a/1,2,3,4,15	NA 50	3,4,9,13,15,16,a/1,2,8
NA 25	8,13,16,a/1,2,3,4,9,15	NA 51	3,13,15,16,a/1,2,4,8,9
NA 26	8,16,a/1,2,3,4,9,13,15	NA 52	8,13,15,16,a/1,2,3,4,9

[a]Revised by Roelfs and Martens from Martens *et al.* (1979).

ulence–virulence of any culture in terms of a simple formula (Table I). The formula consists of numbers corresponding to those assigned the specific genes for resistance in the differential lines. Numbers designating the genes that condition resistance to the culture are written first, followed by a slash line, and the numbers designating ineffective genes. Infection types 0 to 2+ are considered to indicate a resistant or incompatible host response; infection types 3 and 4 are considered to indicate a susceptible or compatible host response (Stakman *et al.*, 1962). The designation of the mesothetic reaction, especially from the early published reports, becomes a matter of judgement. If what is expressed is potentially effective resistance in terms of disease control,

X− or better, then the reaction is considered resistant; an X+ is considered susceptible.

The terminology for describing host–parasite interactions presents certain problems since there is a lack of convention or agreement among pathologists and breeders. For the purposes of this chapter, pathogenicity is taken to mean the ability of the pathogen to cause the stem rust condition; avirulence or virulence qualitatively describes the interaction; and aggressiveness refers to competitive ability. While it would be more accurate to refer to avirulence or virulence of a specific pathogen race on plants, cultivars, or lines with specific genes for resistance, for the sake of conciseness "virulence on *Pg-X*" will be permitted. Similarly, avirulence–virulence combination is more accurate and precise than the term "race," but the latter will often be used because it is more concise and very widely understood.

It should also be noted that the progressive increase in detected variability and virulence in the pathogen population is, in part, related to the availability of the corresponding host gene differentials. The specialization studies began with genes *Pg-1, -2,* and *-3; Pg-4* was added in 1950; *Pg-8* in 1961; *Pg-9* in 1964; *Pg-13* in 1969; and *Pg-15, -16,* and *-a* in 1978.

A. EARLY INVESTIGATIONS

Following the pioneering work of Stakman *et al.* described earlier, Bailey (1925) published a race identification key based on three cultivars, later shown to possess single genes *Pg-1, -2,* and *-3,* and on the use of three reaction categories, resistant, mesothetic, and susceptible. In contemporary terms, four races were described: 1,2,3/ and 1,2/3 from North America; 2,3/1 from South Africa and Sweden; and 3/1,2 from Sweden only. By 1933, Gordon and Bailey (1928), and Gordon (1933) reported the occurrence of all of these races plus /1,2,3 and 1/2,3 from Canada. Waterhouse (1929) found all of the described races, except 3/1,2, and a new race 2/1,3 in Australia.

In 1930 Tedon reported five races from Sweden (Gordon, 1933). Thus from early specialization studies, it is evident that many of the possible avirulence–virulence combinations, based on the differentials then available, occurred in widely separated parts of the world.

B. NORTH AMERICA

North America is considered as one epidemiological unit with three subregions: the Great Plains region, where *Berberis* L. is not a significant factor, stretches from the Gulf of Mexico some 4000 km to the northern limits of oat production; the eastern region, with a different pathogen population, where *Berberis* is a factor; and the much smaller western region, where the disease is less important and the population unlike that of the rest of the continent. The southern part of the continent may be further divided into additional ecological subregions (Roelfs *et al.*, 1982). The full range of variability of the North American population can usually be found in Canada, where the last generations of the annual cycles are produced. The graphic presentation of the succession of races in Canada and the performance of the pathogen population since 1921 (Fig. 1) is fairly representative of what is known about this aspect of the pathogen. From 1921 to 1943, the pathogen population in North America was relatively stable (Levine and Smith, 1937; Newton and Johnson, 1944; Green *et al.*, 1961; Stewart and Roberts, 1970). Although most of the possible avirulence–virulence combinations on the differentials then available were observed in nature, the one-gene virulence race, 1,2/3 predominated until 1943 when the first of a series of shifts occurred, largely in response to resistance genes in the host population on the continent. The distribution of resistance genes in the host population on the continent may be considered to have had three main phases: from about 1942 to 1955, when genes *Pg-1* and *-2* were important in breeding programs; from 1956 to 1978, when *Pg-4,* singly and in combination with *Pg-2,* represented the main genes for resistance; and the present phase with the addition gene combinations *Pg-2, -4, -9* (McKenzie *et al.*, 1976), *Pg-1, -9, -13* (McKenzie *et al.*, 1981), and *Pg-2, -9, -13* (cultivar Dumont; McKenzie *et al.*, 1984). These gene combinations have yet to influence the pathogen population in a significant way.

Although the old races were still important in the late 1940s, races with a wider virulence range increased sharply in both the eastern and Great Plains regions, but not the western region of the continent. By 1950, increasing differences between the eastern and Great Plains regions not attributable to host gene patterns became apparent. The one-gene-virulence races were disappearing, and races 3/1,2 and /1,2,3 were becoming important components in the eastern population, an event that did not occur in the Great Plains regions until 1961. As new genes for resistance were discovered, virulence on them, in various combinations, was usually also found. In most cases (e.g., *Pg-4, -9, -13,* and *-a*), virulence was present in the pathogen population before cultivars with these genes were grown over a significant area. Thus, the resistant cultivars simply caused selective advantage shifts in the pathogen population. However, changes in the host population do not account for all

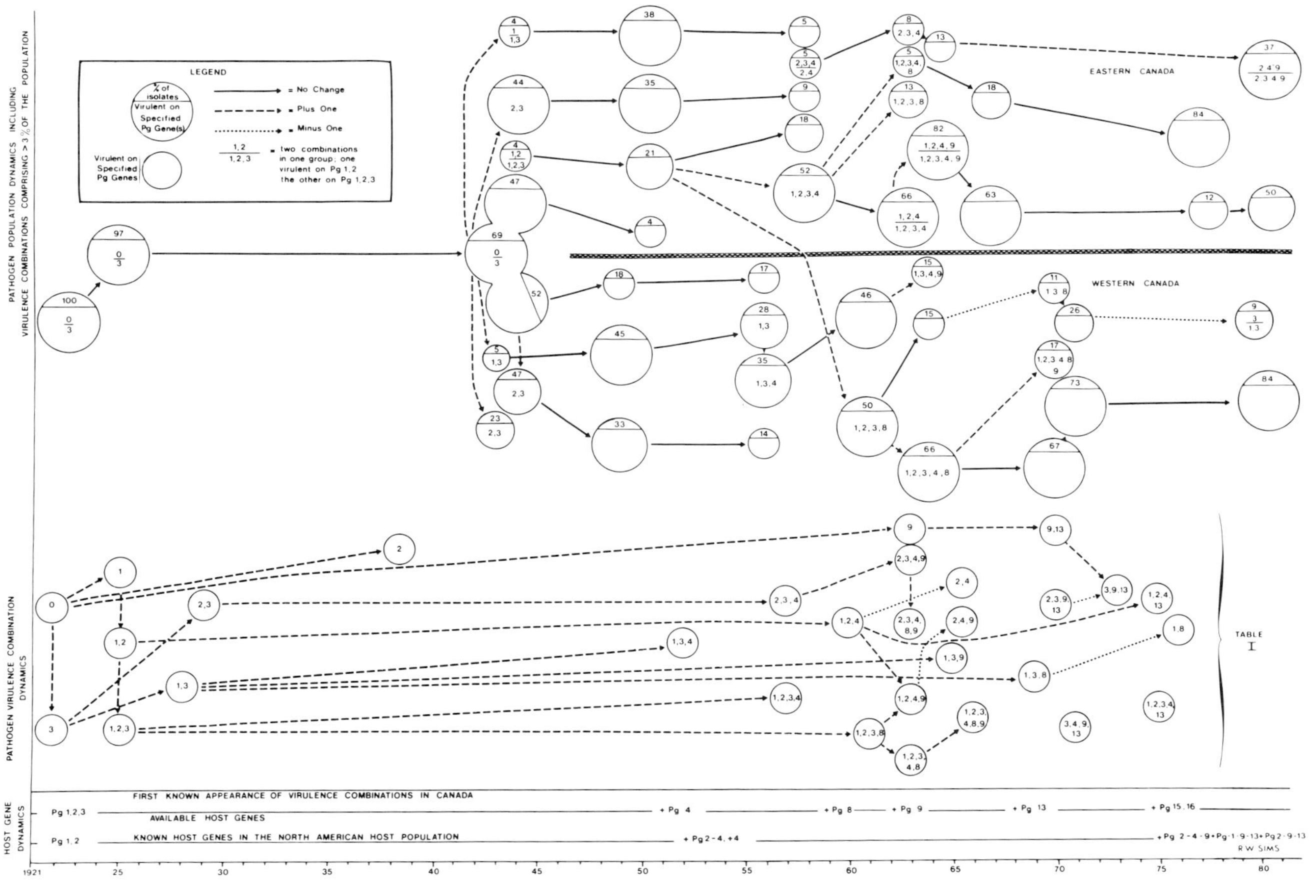

Fig. 1. The dynamics of pathogenicity, the pathogen population composition in *Puccinia graminis* f. sp. *avenae,* and the host population resistance genotype in Canada from 1921 to 1981. Revised from Martens and McKenzie (1979).

the pathogen changes that occurred. Subsequent to 1956–1957, the eastern and Great Plains populations became quite distinct. In the east, virulence on *Pg-2* resistance was very common, but in the Great Plains region it was rare until 1961. By the time *Pg-8* and *-9* became available as differentials in 1960 and 1963, respectively, races 1,8/2,3,4,9 and 2,9/1,3,4,8 were the most common in the east and the Great Plains, respectively. In the western region, races 1,2,3,4,8/9 and 1,2,4,8/3,9 continued to persist. From 1965 to 1981, the eastern population was characterized by combined virulence on all of the available genes except *Pg-8, -13, -16,* and *-a* and in the Great Plains population on all genes except *Pg-9, -13, -16,* and *-a* (Fig. 1; Martens and McKenzie, 1979; Roelfs *et al.*, 1978, 1982; Martens, 1981). Virulence on the differentials with *Pg-13, -16,* and *-a* is very rare or nonexistent on the continent. The appearance and persistence of race 2,4,9,13/1,3,8 at significant levels throughout most of the period since 1969, despite the fact that most of the commercial cultivars in the Great Plains region are resistant to it, clearly indicates that factors other than virulence are determinants of the pathogen population composition. None of the virulence carried by this race, except that on *Pg-1,* is helpful to survival anywhere on the continent. In the western region, Mexico, and the southern United States, races with virulence on only *Pg-15,* or on *Pg-3* and *-15* continue to persist (Roelfs *et al.*, 1980; Martens, 1978, 1981). All of these results are based on field isolates, collected from commercially grown *Avena sativa* L., which may have either no known resistance genes, or genes *Pg-1, -2, -4* singly and *Pg-2* in combination with *Pg-4,* or from wild oats (*A. fatua* L.) that is not known to have any genes for stem rust resistance. However, there is recent evidence to suggest that such survey techniques do not detect all of the variability present in the pathogen population on the continent, even though many of the host plants have no resistance. When trap nurseries with various host genotypes, but especially lines with gene *Pg-15,* were planted and sampled extensively, races not found in the field isolates, however large the field samplings, were detected (Martens and McKenzie, 1979). In 1980, a total of 21 identified races, on 10 differentials, were isolated from field-survey and trap-nursery isolates, thus representing the greatest pathogen variability ever reported for one season in the 60-year history of specialization studies in oat stem rust in Canada (Martens, 1981).

C. EURASIA

The Scandinavian pathogen population is characterized by a great deal of variability with at least 16 races including all possible combinations in the four-gene system used (Leijerstam, 1966; Mac Key and Mattsson, 1972), a phenomenon that has prevailed since studies began. The most virulent race reported by Stakman *et al.* (1923) was an isolate from Uppsala, Sweden. Gordon (1933) also noted the relatively wide virulence range of Swedish isolates. The most common races also include those with the most genes for virulence, even though the genes are not necessary for survival in Scandinavia. *Berberis vulgaris* occurs in Scandinavia and may be a factor in the life cycle of the pathogen (Leijerstam, 1966). From 1961 to 1966, four races, including 4,9/1,2,3,8, were identified in Italy. In 1974–1975, 10 races were identified, including the widely virulent race /1,2,3,4,8,9 (Paradies *et al.*, 1976), indicating a wide range of virulence that is apparently not related to resistance genes in the host.

Kostic (1966) reported that oat stem rust is a serious problem in Yugoslavia in some years. Races 1,2,4/3 and 4/1,2,3 were the most common, but four other races were also isolated.

Sebesta (1973a) isolated six races in Czechoslovakia in 1967–1968, with the most widely virulent race comprising 64% of all isolates identified, but found no virulence on *Pg-4* resistance. For the 11-year period 1965–1975, 10 avirulence–virulence combinations were identified in Czechoslovakia, Austria, and Romania. Race 4,9/1,2,3,8 was most common during this period (Sebesta and Zwatz, 1980). Genes *Pg-13, -4,* and the adult plant gene *Pg-11* were the most effective against this central European pathogen population.

Suzdalskaya *et al.* (1978) have recently published an account of the dynamics of this pathogen in the European and Transcaucasian (Georgian) regions of the U.S.S.R. They report that races 3,4/1,2, 4/1,2,3, 3/1,2,4, and /1,2,3,4 prevailed in all of the regions studied 1959 to 1963. From 1964 to 1968, 2,9/1,3,4,7,Sa and 2,3,9/1,8,Sa (Sa = Saia) were common; and from 1970 to 1975, 1,2,8,9,Sa/3,9 and 3,4,Sa/1,2,8,9 prevailed. In 1971 and 1974, race /1,2,3,4,8,9 was also observed. For the 6-year period ending in 1975, the following average annual frequency of virulence was observed in the isolates studied: *Pg-1* = 59%; *Pg-2* = 62%; *Pg-4* = 21%; *Pg-8* = 56%; *Pg-9* = 75%; and *Saia* = 15%. Virulence on Saia was found most commonly in isolates from the Volga–Vyatka region, only rarely from the central region, Byelorussia and the Ukraine, and never from the north western region, the Baltic Sea area or Georgia. The authors conclude that the high frequency of virulence on the specified types of resistance in the pathogen population of the U.S.S.R. cannot be explained by the selective effect of resistance genes in the host population, since most of the oat cultivars grown are susceptible to stem rust and have no genes that could produce a noticeable selective effect on the pathogen population. Shikina (1974) reported 15 races occurring in the northern Caucasus with six races predominating, suggesting considerable variability in the populations.

Mehta (1940) observed stem rust of oats in the Nilgiri Hills of southern India at altitudes of over 2000 meters. Although the sample size was small, he identified four races, 2,3/1, 2/1,3, 3/1,2, and /1,2,3, indicating considerable variability and a wide range of virulence.

D. THE MIDDLE EAST AND EAST AFRICA

Variability and apparently unnecessary virulence characterize the pathogen populations of these regions. From 1926 to 1938, five races were identified from the Middle East, including 1,2,3/, 1,2/3,½,3,2/1,3, and /1,2,3. The latter was the most widely virulent race, and comprised 30% of all isolates identified during that period (Wahl *et al.*, 1966). This pattern has continued in Israel (Sztejnberg and Wahl, 1967, 1976), with race 4,9/1,2,3,8 being the dominant race for many years.

Limited data from East Africa indicate a similar pattern (Green *et al.*, 1970; Martens *et al.*, 1976). Races 3,8/1,2,4 and 8/1,2,3,4 were prevalent from 1960 to 1964. In 1968, races 9/1,2,3,4,8, 3,9/1,2,4,8, and /1,2,3,4,8,9 were identified with the first comprising 75% of the isolates. In 1971 races 3,9,13/1,2,4,8, 9,13/1,2,3,4,8, 13/1,2,3,4,8,9, and /1,2,3,4,8,9,13 were identified in collections from Ethiopia and Kenya. The genes for resistance in the commercial oat population in this region do not provide an obvious reason for this sustained wide range of virulence.

E. AUSTRALASIA AND SOUTH AMERICA

From 1928 to 1938, races 1,2,3/ and 1,2/3 comprised over 85% of over 700 isolates identified in Australia. Races 2,3/1, 2/1,3, and /1,2,3 were also found (Waterhouse, 1952; Luig and Baker, 1973). The dominant races continued to prevail from 1939 to 1951 with 57% of all isolates identified; with races 2,3/1, 2/1,3, and 1/2,3 comprising the balance. From 1970 to 1972, Luig and Baker (1973) identified 15 races (22 strains) in eastern Australia. There was considerable virulence in the population on *Pg-1, -2, -3, -4,* less on *Pg-9,* and none was observed on *Pg-8* resistance. In contrast to the Australian race spectrum, only three races with relatively narrow virulence range, 1,2,3,4,8,9,13/, 1,2,3,4,8,13/9, and 1,2,4,8,13/3,9 were identified in New Zealand from over 60 isolates in 1975–1976 (Martens *et al.*, 1977).

Although only limited data are available on the South American rust population, it appears genetically variable with many genes for virulence (Orjuela *et al.*, 1962; Martens *et al.*, 1976). In Colombia, stem rust occurs wherever oats are grown and is a limiting factor in its production. Orjuela *et al.* reported 22 different avirulence–virulence combinations, including some not found elsewhere, and some that were virulent on all of the resistance genes used. They concluded that the extreme physiologic variation of *P. graminis avenae* in that country was surprising. *Berberis* spp. are known to occur in Columbia, but it is not known if they are a factor in the life cycle of the rust there. Coelho (1976) reported 10 races from Brazil but noted that none of them were virulent on *Pg-8* resistance.

F. VIRULENCE AND COMPETITIVE ABILITY

Leonard (1969) cultured a heterogeneous population of the pathogen for eight generations in the greenhouse and found that the races with unnecessary virulence genes had survival values of 14–46% lower than races without them. He concluded that multiline cultivars could be used to stabilize pathogen populations. Martens (1973) cultured mixtures of races of differing virulence range on susceptible plants in growth chambers for five uredial generations at various temperatures, and in the field. He found that the races with fewest genes for virulence maintained or increased their levels in the growth chambers in all cases, but were consistently out-performed by races with more genes for virulence under field conditions. He concluded that the number of genes for virulence carried by a given race, other than those required for successful parasitism, was probably not the key determinant affecting the frequency of races in nature. Sebesta (1973b) found that the most widely virulent race was also the most aggressive. Suzdalskaya *et al.* (1978) concluded that the very frequent occurrence of virulent races in the U.S.S.R. cannot be explained by the selective effect of the resistance genes of the host plant, since the majority of oat cultivars grown are susceptible to stem rust and could not produce a noticeable selection effect on the pathogen population.

Sztejnberg and Wahl (1976) noted that race 4,9/1,2,3,8 has been prevalent in Israel for many years. They concluded that its prevalence cannot be ascribed to the preferential selection pressure of the host, and "this shows that a race with a wide range of virulence is not inferior in fitness even when the virulence is not necessary for survival."

Mac Key and Mattsson (1972) found isolates with every possible avirulence–virulence combination for the system they were using in Sweden. They noted that it is disturbing that a pathogen population like that in Scandinavia maintains so many genes for virulence. The population is fairly well isolated geographically under rather extreme ecological conditions and apparently has not been subjected to the selective effects of specific genes for resistance in the host. Similar evidence from Australia (Luig and Baker, 1973) and East Africa (Green *et al.*, 1970; Martens *et al.*, 1976) leaves no doubt that stabilizing selection is not a key determinant affecting pathogen population distribution in those parts of the world where it has been studied.

Why does this organism carry so many genes for virulence, and how do they arise? It is possible that in some cases the genes for resistance are more widely distributed in the host population than is recognized. Genes *Pg-3* and *-9* could be cases in point. However, with the genes from *A. sterilis* L., *Pg-13* and *-15*, it is most unlikely that they ever occurred in North America before they were intentionally introduced and identified. But the corresponding virulence genes were already present before the pathogen had an opportunity to "recognize" these genes for resistance. Perhaps these genes impart selective advantage other than virulence on *Avena* spp. Possibly the pathogen parasitizes hosts other than *Avena* where these "unnecessary" genes are in fact useful, if not necessary.

IV. Environmental Factors Affecting the Host–Parasite Interaction

The presence or absence of the gene(s) for virulence in the pathogen and resistance in the host are only two of the determinants affecting the course of the interaction. Sensitivity to temperature in some interactions has been recognized almost since the beginning of race studies. Gordon (1933) demonstrated the thermostability of host genes *Pg-1* and *-2* and the sensitivity of *Pg-3*. He also noted that telia were formed more rapidly at 24°–28°C than at 12°–16°C. Roberts (1962) examined the phenomenon of temperature sensitivity and found that the expression of gene *Pg-4* was dependent on a certain temperature during a critical period between the inoculation and the fleck stage.

Plants kept at 22°C after inoculation would not express resistance if grown at 30°C for at least 3 days prior to the fleck stage. Plants kept at 30°C after inoculation required at least 4 days at 22°C before flecking to completely express the characteristic resistance. Preinoculation temperatures had no effect on the expression of resistance. The response was shown to be quantitative in the case of partial treatment, and it was localized rather than systemic. Martens *et al.* (1976) showed that the temperature sensitivity of *Pg-4* and other genes was independent of the host genotype, but that there was a race effect. High-temperature breakdown of *Pg-4* resistance occurred at a higher temperature with some races than with others, but at 30°C breakdown was complete for all races. They found the optimal temperature for rust development to be 20°–25°C and confirmed the observations of Gordon regarding telial development.

Genes *Pg-1* and *-2* are temperature-insensitive; genes *Pg-9, -8, -13, -15, -16, -4,* and *-3* are temperature-sensitive in order of increasing sensitivity (Martens *et al.*, 1967, 1979).

The effect of temperature, light, and host genotype on prepenetration development of *P. graminis avenae* was studied by Kochman and Brown (1976a). Germination of uredospores occurred at the relatively wide range of 10°–30°C, but optimum conditions for germ-tube growth and formation of appressoria was 20°C in darkness. Germination was inhibited at 35°C. Under optimum conditions, maximum germination and formation of appressoria was attained within 4 hr after inoculation. No host effect on germination or prepenetration development was detected among the six wild and cultivated lines, including one with genes *Pg-2* and *-4.* This observation suggests the absence of any prepenetration defence mechanisms. However, optimum development conditions were different for the penetration phase (Kochman and Brown, 1976b). Penetration was greatest at 30°–35°C and at light intensities of 5625 lux or above. Maximal penetration was achieved in a dew period of 16 hr.

V. Genetics and Cytology of the Pathogen

A. INHERITANCE OF VIRULENCE AND SPORE COLOR

Gordon and Welsh (1932) appear to be the first investigators to have studied the inheritance of virulence in *P. graminis avenae.* They selfed race 1/2,3 and from 20 aecial cups they isolated races 1/2,3 (15 times), /1,2,3 (seven times), and 2/1,3 (once), demonstrating that the isolate was heterozygous at two of the three loci studied. Johnson and Newton (1940) selfed a series of races and found two different cultures of race 1,2/3 to be homozygous and one culture to be heterozygous for virulence on *Pg-3* resistance; race 2,3/1 was heterozygous for virulence on *Pg-2* resistance; two different cultures of race 1/2,3 were heterozygous for virulence on *Pg-1* resistance; and races 2/1,3 and 1,3/2 were homozygous.

In crosses between races, they generally found avirulence to be dominant. In reciprocal crosses, they noted evidence of cytoplasmic inheritance of virulence on *Pg-3* resistance. In almost all crosses involving virulence on *Pg-3,* the hybrid was similar to the maternal parent. In crosses between races with normal brick-red- and orange-colored uredia, they found red to be dominant. In further studies, with hybrids of races 1,3/2 and 2/1,3, Johnson (1949) showed avirulence on *Pg-1* and *Pg-2* resistance to be dominant, and he concluded that they were governed by two pairs of complementary genes, while he confirmed that avirulence on *Pg-3* resistance was maternally inherited. However, Green (1965), working with selfs and crosses of 10 different cultures, concluded that avirulence on resistance conferred by genes *Pg-1* and *Pg-2* is governed by single dominant genes and not two complementary genes each as previously indicated. Green also confirmed the extrachromosomal nature of the inheritance of avirulence on *Pg-3* resistance. All the cultures used in the study were homozygous for virulence or avirulence on resistance conferred by gene *Pg-4,* and in all crosses between avirulent and virulent races, avirulence was dominant in the F_1. All the progenies of all the selfing and crossings were avirulent on resistance conferred by gene *Pg-8.* In further studies, Green and McKenzie (1967) confirmed that avirulence on *Pg-1, Pg-2,* and *Pg-4* resistance was controlled by single dominant genes. However, in some crosses the action of the dominant avirulence gene on *Pg-2* resistance was modified by a recessive gene and type 2+ rather than type 1 infections were produced. Avirulence on plants with *Pg-9* resistance also appeared to be controlled by a single gene. Although the results were not conclusive, evidence suggested that avirulence on cultivars with *Pg-8* resistance is recessive and that on hosts with *Pg-13* resistance avirulence is dominant (Martens *et al.,* 1970). There is no evidence for linkage between virulence genes, and generally gene-for-gene relationships occur within this host–parasite system (Martens *et al.,* 1970; Mac Key and Mattsson, 1972).

B. MUTABILITY

Puccinia graminis avenae can be readily mutated in the uredial stage by means of chemical treatment (Teo and Baker, 1975). Ethyl methanesulfonate (EMS) induced variability in virulence, uredospore color, and the rate of telial development. Within the ranges of 9.8–19.6 × 10^{-2} *M* EMS, 20°–30°C, and 2.0–5.4 hr treatment duration, increases in any of the treatment variables increased mutation rates. However, with increasing severity of treatment, spore viability decreased from 96 to 19%. Recurrent mutagen treatment of successive uredial generations was shown to produce variability not induced by the first treatment.

C. CROSSES BETWEEN *PUCCINIA GRAMINIS TRITICI* AND *PUCCINIA GRAMINIS AVENAE*

Johnson and Newton (1933) were successful in hybridizing these two *formae specialis* in one of 32 attempts. They crossed a race of *Puccinia graminis* f. sp. *tritici* that was homozygous for avirulence on all the oat hosts used, *Dactylis glomerata* L. and *Phalaris canariensis* L., with race /1,2,3 of *P. graminis* f. sp. *avenae.* Selfed progeny resulted in segregates of races /1,2,3, 2/1,3, and 3/1,2 and segregates virulent on wheat, which indicated a successful hybrid. The hybrids were virulent on oat cultivars with no resistance genes, genes *Pg-1* and *Pg-2,* the wheat cultivars Little Club and Liguleless, *D. glomerata* and *P. canariensis,* and some barley cultivars, but they were avirulent on oats with gene *Pg-3,* and on nine wheat cultivars that the *tritici* parent could attack.

D. CYTOLOGY

McGinnis (1953) examined germinating sporidia of *P. graminis* during nuclear division and found at metaphase a haploid number of six chromosomes. Prophase chromosomes were observed to be united to form a continuous chain. Discrete individual chromosomes could not be seen at this stage. At metaphase, chromosomes appeared loosely paired with residual terminal attractions, suggesting that the basic haploid chromosome number could be three.

Craigie and Burrows (1967) noted departures from the usual binucleate condition in uredial mycelium, urediospores and germ tubes in race /1,2,3 of *P. graminis* f. sp. *avenae.* They observed occasional trinucleate and tetranucleate cells, which reproduced in kind. The trinculeate condition developed when one nucleus of a binucleate cell divided and the other did not, or by the migration of one nucleus from a tetranucleate cell. Trinucleate cells occurred mainly in younger mycelium, and their frequency varied widely from one infection site to another. These cells tended to revert to the binucleate condition either by dissolution of the extra nuclei or by their migration from the cells concerned.

VI. Host Resistance and Control Strategies

A. ORIGIN AND NATURE OF THE *PG* GENES

The *Pg* genes that have been important historically, that have influenced pathogen dynamics, or that are presently potentially important in terms of breeding for resistance are all included in the international differential set (Table I). The genes that are absent from Table I are described by Simons *et al.* (1978). All of these genes are available as single-gene lines in the Rodney O background, which are the basis for the international differential set (Martens *et al.*, 1979).

Gene *Pg-1* was introduced to the United States, presumably from northern or central Russia, in about 1850 in the cultivar White Russian, which became one of the progenitors of some 15 cultivars on the continent (Coffman, 1977). The dominant thermostable gene was used extensively in the north central states of the United States for many years (Stewart and Roberts, 1970), but not in Canada.

Gene *Pg-2,* also dominant and thermostable, appears to have been introduced to the United States from the U.S.S.R. at least three different times. The first introduction involved the cultivar Green Russian, which, according to Coffman (1977), was brought into North Dakota by settlers from the U.S.S.R. in about 1870. The exact date and where the settlers came from is not clear (Coffman, 1977), but other sources (Quisenberry and Reitz, 1974) indicate a migration from the Ukraine to Dakota in the early 1870s, suggesting the Ukraine as a source of Green Russian. The second source of the gene was the cultivar Kherson, introduced by F. W. Taylor of Nebraska. It was named after the area of origin near Odessa on the Black Sea. The third was with the introduction of the cultivar Sixty-Day, believed to be very similar to Kherson, which was sent to the United States Department of Agriculture from Southern Podolia, also in the Ukraine (Coffman, 1977). Thus, *Pg-2* appears to have been fairly common in the host population of the southern Ukraine during the latter part of the nineteenth century. Green Russian and Kherson became progenitors of some 130 cultivars in North America (Coffman, 1977; Coffman *et al.*, 1961), indicating how widespread and important *Pg-2* became.

Gene *Pg-3*, the last of the "pre-1950" genes, is also dominant. It was introduced to Canada from France in 1888–1889 in the cultivar Joanette. Its importance is more historical than practical, since neither the introduction nor gene *Pg-3* were used extensively in oat breeding. *Pg-3* has been in the international differential set almost since the beginning of race studies. In many respects the rust reaction it confers is unlike that of the genes previously described. It confers a mesothetic reaction to some races, a highly resistant one to avirulent races at low temperatures, and it is thermolabile (Roberts, 1962; Martens *et al.*, 1967). It is either closely linked to a gene conferring crown rust resistance, or itself confers resistance to both rusts (McKenzie *et al.*, 1968). Moreover, virulence on this type of resistance is inherited extrachromosomally (Green and McKenzie, 1967).

Gene *Pg-4* is a dominant, thermolabile gene that was introduced to the United States via South Africa in the cultivar Hajira in 1919 (Welsh and Johnson, 1951). Authors agree on a North African origin, but Welsh and Johnson (1951) thought that it originally came from Egypt in 1904, while Coffman (1977) gives Algeria as the probable source. However, Hajira was not important as a cultivar or as germ plasm until the discovery of *Pg-4*, which was present in about 10% of the plants in the cultivar (Welsh and Johnson, 1951). This source of resistance has been used very widely, and together with *Pg-1* and *Pg-2* has in the past been the basis for breeding for stem rust resistance wherever there were breeding programs with that objective.

Gene *Pg-5* is probably the same as *Pg-4* (Welsh and Johnson, 1951).

Genes *Pg-6* and *Pg-7* are dominant genes in the diploid species *A. strigosa* Schreb., C.D. 3820, conditioning resistance to a wide range of races (Murphy *et al.*, 1958; Dyck and Zillinsky, 1962). However, it has not yet been possible to transfer this resistance to the hexaploid level. The two genes may be the same (Dyck, 1966).

The origin of *Pg-8*, a thermolabile, recessive gene, is somewhat difficult to trace. Browning and Frey (1959) identified the source as a hull-less cultivar from Africa (C.I. 3030 and C.I. 3031), but it was also present in C.I. 2710, an unnamed Chinese cultivar. Welsh *et al.* (1961) isolated the gene from R.L. 524.1, which is a Hajira × Banner hybrid, and the resistance most likely came from Hajira. Thus, wherever else it may have occurred, it appears to have been present in North Africa. As far as is known, *Pg-8* has not been used in breeding programs. It appears to be effective against eastern Australian races (Luig and Baker, 1973). It is also effective against races found in the *Berberis* areas of eastern North America, and it is effective in parts of South America

(Coelho, 1976) and parts of Russia (Suzdalskaya *et al.*, 1978). In the Great Plains region of the continent, over 90% of all field isolates are virulent on this type of resistance.

Pg-9, a thermolabile, recessive gene, probably originated from the U.S.S.R. It was introduced to the United States in 1930 (Coffman, 1977) in the cultivar Ukraine. Since its discovery it has been identified in lines from numerous breeding programs (McKenzie and Green, 1965), and in the introduction Kerkhiachskii 41 from the U.S.S.R. and in Sante Fe introduced from Argentina in 1945 (Coffman, 1977). As far as is known it has not been used in breeding programs until recently (McKenzie *et al.*, 1976). Like gene *Pg-3*, *Pg-9* is closely associated with a gene for resistance to *P. coronata* (McKenzie *et al.*, 1965); because of this association, it may be present in some North American cultivars. *Pg-9* is effective against the most common and widely virulent races of the Great Plains region, but not of the eastern region of North America.

Gene *Pg-10* is a dominant or partially dominant gene identified in the hexaploid hull-less lines C.I. 1575, C.I. 2641, and C.I. 2824 and conditions a moderately susceptible infection type to race NA 25 (Pavek and Myers, 1965). Because of the infection type it confers, this gene has not been used in differential sets.

Gene *Pg-11* is unique among the known genes for stem rust resistance in oats in that it is expressed only in the adult plant stage. Adult plants with *Pg-11* are resistant to all races that have been tested. It is an incompletely recessive gene that was isolated from C.I. 3034, a line that also carried gene *Pg-1* (McKenzie and Martens, 1968). C.I. 3034 originated from Rhodesia in about 1926, as a rust-resistant selection from a badly rusted field of the cultivar Burt. Burt is a selection from cultivar Red Rustproof, which is a Mexican oat that was brought back from Mexico to South Carolina, United States, by a soldier in 1848 or 1849 (Coffman, 1977). If the progenitor of C.I. 3034 was really Burt via Red Rustproof, we must try to explain the presence of *Pg-1*, which is not present in Burt and is widely believed to be of U.S.S.R. origin, occurring in White Russian and its derivatives. This cultivar was introduced to the United States in about 1850, the same time as Red Rustproof; since both have been widely used in breeding programs, natural outcrossing and/or admixtures might have occurred. Thus there are at least three choices for the putative origin for gene *Pg-11:* a *de novo* mutant from southern Africa; a western Black Sea origin along with *Pg-1, -2, -9,* and *-15;* and an Iberian origin via Burt, Red Rustproof, and Mexico.

There appears to be an association between the resistance and yellow plant color, weak straw, and somewhat reduced yield in the absence of rust. The seedlings with *Pg-11* are susceptible and have near-normal chlorophyll and carotenoid levels, but with increasing age and the onset of adult plant resistance the pigment content decreases more rapidly than in plants without the gene (Harder *et al.*, 1971). No significant change in the ratios of chlorophyll to carotenoids occurred during the shift from susceptibility to resistance. A gene affecting chlorophyll levels may be tightly linked with *Pg-11*. Alternately, gene *Pg-11* may not be a rust resistance gene in the conventional sense, but rather a progressively effective, sublethal, pigment deficiency gene that incidentally causes rust resistance. This type of resistance has been used successfully in cultivar improvement in Mexico (A. P. Roelfs, personal communication).

The *Pg-a* complex (Martens *et al.*, 1981) consists of gene *Pg-12* and a complementary or interacting gene(s). *Pg-12* is a recessive gene that was isolated from Kyto, a cultivar introduced from Yugoslavia via Finland by the United States Department of Agriculture in 1939 (Martens *et al.*, 1968). The cultivar Osmo expresses a rust reaction similar to that of Kyto (Green and McKenzie, 1964).

The pedigrees of Kyto and Osmo (Baum, 1972) indicate that they have the Swedish cultivar Victory = Milton = Probsteier and unknown local Finnish cultivars, one of them from the Kuopio region of south central Finland, in common. Since Victory is the universally susceptible host for oat stem rust, Kuopio, Finland, becomes the prime candidate for the origin of *Pg-12*, as far as it can be traced. In the seedling stage, Kyto and Osmo are resistant to all races that have been tested. The infection type ranges from (;) to (2) and is associated with severe necrosis but no chlorosis of the surrounding leaf tissue. Resistance diminishes as the plants develop, and while some resistant reactions can still be observed in adult plants, they appear moderately susceptible (Martens *et al.*, 1968). In interaction with another gene(s), *Pg-12* provides highly effective resistance against all but two of the oat stem rust races known to occur in North America (Martens *et al.*, 1981).

Gene *Pg-13* is a recessive gene isolated from *A. sterilis* collected near Tunis, Tunisia (McKenzie *et al.*, 1970). It is the first stem-rust-resistance gene to have been found in this species and is also one of the most effective genes available to breeders (Roelfs *et al.*, 1982; Martens, 1981; Table I). Although races with virulence on *Pg-13* have been observed, it has been found effective in most of the countries where it has been tested (Martens *et al.*, 1976). Fidler and Dumont, cultivars with *Pg-13* in combination with several other genes, have recently been released (McKenzie *et al.*, 1981; McKenzie *et al.*, 1984).

Gene *Pg-14* is a partially dominant gene isolated by Mac Key and Mattsson (1972) from Milford (C.I. 5039), Winter Turf (C.I. 1570) and other lines. Milford is a Welsh cultivar, released in 1947, with both the Swedish Milton = Victory = Probsteier line and Winter Turf = Grey Winter line prominent in its pedigree. Winter Turf is a very old cultivar that was introduced to the United States from England by George Washington in about 1764 (Coffman, 1977). It was apparently not noted for its stem rust resistance. From this evidence it becomes difficult to identify a putative origin for this gene.

Gene *Pg-15* is a partially dominant gene isolated from *A. sterilis* (Martens *et al.*, 1980) collected east of Uskudar on the Black Sea near Istanbul, Turkey. Races avirulent on *Pg-9* are usually also avirulent on *Pg-15* in the Great Plains region (Roelfs *et al.*, 1980; Martens, 1981). Lines with this gene have been shown to be highly effective in trap nurseries for detecting variability in the pathogen population (Martens and McKenzie, 1979). This gene has not yet been used in commercial cultivars.

Resistance to oat stem rust is known to occur in the diploid species *A. strigosa* Schreb., from which Murphy *et al.* (1958) and Dyck (1966) isolated a dominant gene. In the diploid background, this resistance is highly effective in many parts of the world. Stem rust resistance has also been found in collections of other diploid species, including *A. longiglumis* Durieu. from Mamura-Tiflet and Rabat-Meknes regions of Morocco (J. W. Martens, T. Rajhathy, and R. I. H. McKenzie, unpublished). Resistance also occurs in the tetraploid species *A. barbata* Pott *ex* Link. in the Middle East (Dinoor and Wahl, 1963), North Africa (J. W. Martens, unpublished) and Turkey (Martens *et al.*, 1980).

Repeated attempts to transfer resistance from the lower ploidy levels to the hexaploid *A. sativa* have been largely unsuccessful (Mac Key and Mattsson, 1972; Dyck, 1966; Rajhathy and Thomas, 1974). *Pg-16* (Martens *et al.*, 1979) is a highly effective gene from *A. barbata* that may be successfully transferred to the hexaploid level.

Thus, the best available evidence suggests that most of the known genes for stem rust resistance in hexaploid oats originate from two relatively small geographic areas: the region around the western Black Sea (Russian–Turkish) yielded *Pg-1, -2, -9, -15,* and possibly *Pg-11,* and the North African region yielded *Pg-4, -8, -13,* and possibly *Pg-3* via France. Further plant-collecting expeditions in quest of new genes for resistance should include these regions and should sample the indigenous host populations extensively. Finland and England are putative origins for genes *Pg-12* and *Pg-14*, respectively, as far as has been possible to trace them.

B. BREEDING FOR ENDURING STEM RUST RESISTANCE

In view of the variability and the wide virulence range of the pathogen throughout the world, and the limited numbers of available resistance genes, breeding for enduring disease control is a major challenge. How can it best be done?

The best intermediate-term prospect for the efficient control of stem rust in oats lies in the synthesis of cultivars carrying several effective genes for resistance with the germ plasm presently available. The basis for optimism regarding the effectiveness of this approach is the wheat stem rust example in North America (Roelfs, 1978; Green and Campbell, 1979) and Australia (Watson, 1981). On the Great Plains of North America, there have been virtually no losses due to stem rust for over 30 years, despite very large areas of near monoculture wheat, because of the use of complex resistance. The presently available genes for resistance offer considerable scope for complex resistance breeding, despite the problem of allelism. Seven of the seedling resistance genes in the hexaploid host occur in three independent linkage groups: genes *Pg-1, -2,* and *-8* in one group; genes *Pg-4* and *-13* in another; and genes *Pg-3* and *-9* in a third group. Genes *Pg-12* and *-15* occur independently of these groups and of each other (McKenzie *et al.*, 1970; Martens *et al.*, 1980).

Genes such as *Pg-8, -13,* and *-16* if it can be transferred, and the *Pg-a* complex, in combination with *Pg-1, -2,* and *-4,* offer considerable scope for complex resistance breeding. If possible, genes should not be used singly in the production of new cultivars.

The apparent effectiveness of the *Pg-a* complex, which clearly demonstrates transgressive segregation for resistance, indicates that the search for similar additional systems in the existing germ-plasm banks should be pursued.

The search for new sources of resistance in the hexaploid state should be intensified. In view of how little plant collecting has been done in the past, extensive sampling of the populations in regions known to have yielded effective resistance genes in the past should almost certainly result in the discovery of additional genes.

The prospects of success with mutation breeding are not encouraging. Attempts to induce *de novo* resistance by irradiation, chemical mutagen treatment or low-level chronic irradiation of growing plants using large populations have produced no new sources of useful resistance (McKenzie and Martens, 1974; Harder *et al.*, 1977). The greatest potential for mutation breeding may be in the transfer of resistance from lower ploidy levels to the hexaploid level.

VII. Conclusions

Puccinia graminis avenae is a highly variable pathogen that carries genes for virulence in excess of what is required for its survival in most parts of the world where it has been studied. Host genes for resistance to this pathogen are rare relative to those for *P. coronata* Cda. on the same host, even though both host–parasite systems are known to have coevolved in the same region for a very long time.

Most of the known genes for resistance are believed to originate from two relatively small geographic areas: the region around the western Black Sea, and North Africa. The prospects of controlling the disease by the use of the existing genes for resistance in multigene cultivars are good. The North African gene *Pg-13,* in combination with previously discovered genes, and the *Pg-a* complex are particularly promising. The search for new sources of resistance should be continued and intensified. Efforts to transfer resistance from diploid and tetraploid species of the genus to the hexaploid level should also be pursued.

References

Arthur, J. C. (1929). "The Plant Rusts." Wiley, New York.

Bailey, D. L. (1925). Physiologic specialization in *Puccinia graminis avenae* Erikss. and Henn. *Minn., Agric. Exp. Stn., Tech. Bull.* **35,** 1–33.

Baum, B. R. (1972). Material for an international oat register. *Can. Dep. Agric., Plant Res. Inst.* **895,** 1–266.

Browning, J. A., and Frey, K. J. (1959). The inheritance of new sources of oat stem rust resistance. *Plant Dis. Rep.* **43,** 768–771.

Chester, K. S. (1944). Naumov, N.A. Rusts of Cereals in the USSR. *Phytopathology* **34,** 513–514.

Chester, K. S. (1946). "The Nature and Prevention of the Cereal Rusts as Exemplified in the Leaf Rust of Wheat." Chronica Botanica, Waltham, Massachusetts.

Coelho, E. T. (1976). Distribution, prevalence and a new race of *Puccinia graminis avenae* in Rio Grande do Sul. *Pesquis. Agropecu. Bras.* **11,** 19–22; cited in *Rev. Plant Pathol* **59,** 3217.

Coffman, F. A. (1961). Origin and history. *In* "Oats and Oat Improvement" (F. A. Coffman, ed.), pp. 15–40. Am. Soc. Agron., Madison, Wisconsin.

Coffman, F. A. (1977). Oat history, identification and classification. *U.S. Dep. Agric. Tech. Bull.* **1516,** 1–356.

Coffman, F. A., Murphy, H. C., and Chapman, W. H. (1961). Oat breeding. *In* "Oats and Oat Improvement" (F. A. Coffman, ed.), pp. 263–329. Am. Soc. Agron., Madison, Wisconsin.

Craigie, J. H. (1957). Stem rust of cereals. *Publ.—Can. Dep. Agric.* **666,** 1–44.

Craigie, J. H., and Burrows, V. D. (1967). Trinucleate cells and other nuclear abnormalities in race 6 of oat stem rust. *Can. J. Bot.* **45,** 1063–1075.

Dinoor, A., and Wahl, I. (1963). Reaction of non-cultivated oats from Israel to Canadian races of crown rust and stem rust. *Can. J. Plant Sci.* **43,** 263–270.

Dyck, P. L. (1966). Inheritance of stem rust resistance and other characteristics in diploid oats, *Avena strigosa. Can. J. Genet. Cytol.* **8,** 444–450.

Dyck, P. L., and Zillinsky, F. J. (1962). Segregation for crown rust and stem rust resistance in diploid and autotretraploid *Avena strigosa. Can. J. Genet. Cytol.* **4,** 469–474.

Eriksson, J. (1894). Ueber die Specialisirung des Parasitismus bei den Getreiderostpilzen. *Ber. Dtsch. Bot. Ges.* **12,** 292–331.

Fontana, F. (1932). "Observations on the Rust of Grain" (P. P. Pirone, transl.), Classics, No. 2, Am. Phytopathol. Soc., Washington, D.C. (originally published, 1767).

Gordon, W. L. (1933). A study of the relation of environment to the development of the uredinial and telial stages of the physiologic forms of *Puccinia graminis avenae* Erikss. and Henn. *Sci. Agric. (Ottawa)* **14,** 184–237.

Gordon, W. L., and Bailey, D. L. (1928). Physiologic forms of oat stem rust in Canada. *Sci. Agric. (Ottawa)* **9,** 30–38.

Gordon, W. L., and Welsh, J. N. (1932). Oat stem rust investigations in Canada. *Sci. Agric. (Ottawa)* **13,** 228–235.

Greaney, F. J. (1936). Cereal rust losses in western Canada. *Sci. Agric. (Ottawa)* **16,** 608–614.

Green, G. J. (1965). Inheritance of virulence in oat stem rust on the varieties Sevnothree, Richland and White Russian. *Can. J. Genet. Cytol.* **7,** 641–650.

Green, G. J., and Campbell, A. B. (1979). Wheat cultivars resistant to *Puccinia graminis tritici* in western Canada: Their development, performance, and economic value. *Can. J. Plant Pathol.* **1,** 3–11.

Green, G. J., and McKenzie, R. I. H. (1964). Dangerous new races of oat stem rust and sources of resistance to them. *Can. J. Plant Sci.* **44,** 418–426.

Green, G. J., and McKenzie, R. I. H. (1967). Mendelian and extrachromosomal inheritance of virulence in *Puccinia graminis* f. sp. *avenae. Can. J. Genet. Cytol.* **9,** 785–793.

Green, G. J., Johnson, T., and Welsh, J. N. (1961). Physiologic specialization in oat stem rust in Canada from 1944 to 1959. *Can. J. Plant Sci.* **41,** 153–165.

Green, G. J., Martens, J. W., and Ribeiro, O. (1970). Epidemiology and specialization of wheat and oat stem rusts in Kenya in 1968. *Phytopathology* **60,** 309–314.

Harder, D. E., Martens, J. W., and McKenzie, R. I. H. (1971). Changes in chlorophyll and carotenoid content in oats associated with the expression of adult plant resistance to stem rust conferred by gene pg-11. *Can. J. Bot.* **49,** 1783–1785.

Harder, D. E., McKenzie, R. I. H., Martens, J. W., and Brown, P. D. (1977). Strategies for improving rust resistance in oats. *In* "Induced Mutations against Plant Diseases," pp. 495–498. IAEA, Vienna.

Johnson, T. (1949). Inheritance of pathogenicity and urediospore color in crosses between physiologic races of oat stem rust. *Can. J. Res., Sect. C* **27,** 203–217.
Johnson, T., and Newton, M. (1933). Hybridization between *Puccinia graminis tritici* and *Puccinia graminis avenae. Proc. World's Grain Exhibition Conf. Can., 1933,* Vol. 2, pp. 219–223.
Johnson, T., and Newton, M. (1940). Crossing and selfing studies with physiologic races of oat stem rust. *Can. J. Res., Sect. C* **18,** 54–67.
Kochman, J. K., and Brown, J. F. (1976a). Effect of temperature, light and host on prepenetration development of *Puccinia graminis avenae* and *Puccinia coronata avenae. Ann. Appl. Biol.* **82,** 241–249.
Kochman, J. K., and Brown, J. F. (1976b). Host and environmental effects on the penetration of oats by *Puccinia graminis avenae* and *Puccinia coronata avenae. Ann. Appl. Biol.* **82,** 251–258.
Kostic, B. (1966). Physiologic races of *Puccinia graminis* var. *avenae* Erikss. et Henn. and reactions of some oat varieties. *Proc. Cereal Rusts Conf., 1964,* pp. 222–229.
Leijerstam, B. (1966). Physiologic specialization in oat stem rust in Scandinavia. *Proc. Cereal Rusts Conf., 1964,* pp. 230–232.
Leonard, K. J. (1969). Selection in heterogeneous populations of *Puccinia graminis* f. sp. *avenae. Phytopathology* **59,** 1851–1857.
Levine, M. N., and Smith, D. C. (1937). Comparative reaction of oat varieties in the seedling and maturing stages to physiologic races of *Puccinia graminis avenae,* and the distribution of these races in the United States. *J. Agric. Res. (Washington, D.C.)* **55,** 713–729.
Luig, N. H., and Baker, E. P. (1973). Variability in oat stem rust in eastern Australia. *Proc. Linn. Soc. N.S.W.* **98,** 53–61.
McGinnis, R. C. (1953). Cytological studies of chromosomes of rust fungi. I. The mitotic chromosomes of *Puccinia graminis. Can. J. Bot.* **31,** 522–526.
McKenzie, R. I. H., and Green, G. J. (1965). Stem rust resistance in oats. I. The inheritance of resistance to race 6AF in six varieties of oats. *Can. J. Genet. Cytol.* **7,** 268–274.
McKenzie, R. I. H., and Martens, J. W. (1968). Inheritance in the oat strain C.I. 3034 of adult plant resistance to race C 10 of stem rust. *Crop Sci.* **8,** 625–627.
McKenzie, R. I. H., and Martens, J. W. (1974). Breeding for stem rust resistance in oats. *In* "Induced Mutations for Disease Resistance in Crop Plants," pp. 45–48. IAEA, Vienna.
McKenzie, R. I. H., Fleischmann, G., and Green, G. J. (1965). A close association of stem rust and crown rust resistance in 'Ukraine' and 'Rosen's Mutant' oats. *Crop Sci.* **5,** 551–552.
McKenzie, R. I. H., Martens, J. W., Fleischmann, G., and Samborski, D. J. (1968). An association of stem rust and crown rust resistance in Jostrain oats. *Can. J. Genet. Cytol.* **10,** 190–195.
McKenzie, R. I. H., Martens, J. W., and Rajhathy, T. (1970). Inheritance of oat stem rust resistance in a Tunisian strain of *Avena sterilis. Can. J. Genet. Cytol.* **12,** 501–505.
McKenzie, R. I. H., Martens, J. W., Mallough, E. D., and Fleischmann, G. (1976). Registration of Hudson Oats. *Crop Sci.* **16,** 740–741.
McKenzie, R. I. H., Martens, J. W., Brown, P. D., Harder, D. E., Nielsen, J., and Boughton, G. R. (1981). Registration of Fidler Oats. *Crop Sci.* **21,** 632–633.
McKenzie, R. I. H., Brown, P. D., Martens, J. W., Harder, D. E., Neilsen, J., Gill, C. C., and Boughton, G. R. (1984). Registration of Dumont oats. *Crop Sci.* **24,** 207.

Mac Key, J., and Mattsson, B. (1972). Breeding for race specific resistance against Scandinavian oat stem rust. *Sver. Utsaedesfoeren. Tidskr.* **82,** 186–203.

Martens, J. W. (1971). Stem rust of oats in Canada in 1970. *Can. Plant Dis. Surv.* **51,** 11–13.

Martens, J. W. (1973). Competitive ability of oat stem rust races in mixtures. *Can. J. Bot.* **51,** 2233–2236.

Martens, J. W. (1978). Stem rust of oats in Canada in 1977. *Can. Plant Dis. Surv.* **58,** 51–52.

Martens, J. W. (1981). Incidence and virulence of *Puccinia graminis* f. sp. *avenae* in Canada in 1980. *Can. J. Plant Pathol.* **3,** 231–234.

Martens, J. W., and McKenzie, R. I. H. (1979). Virulence dynamics of *Puccinia graminis* f. sp. *avenae* in Canada. *Can. J. Bot.* **57,** 952–957.

Martens, J. W., McKenzie, R. I. H., and Green, G. J. (1967). Thermal stability of stem rust resistance in oat seedlings. *Can. J. Bot.* **45,** 451–458.

Martens, J. W., McKenzie, R. I. H., and Fleischmann, G. (1968). The inheritance of resistance to stem and crown rust in Kyto oats. *Can. J. Genet. Cytol.* **10,** 808–812.

Martens, J. W., McKenzie, R. I. H., and Green, G. J. (1970). Gene-for-gene relationships in the *Avena:Puccinia graminis* host-parasite system in Canada. *Can. J. Bot.* **48,** 969–975.

Martens, J. W., Green, G. J., and McKenzie, R. I. H. (1976). International oat stem rust virulence survey. *Plant Dis. Rep.* **60,** 525–528.

Martens, J. W., Burnett, P. A., and Wright, G. M. (1977). Virulence in *Puccinia coronata* f. sp. *avenae* and *P. graminis* f. sp. *avenae* in New Zealand. *Phytopathology* **67,** 1519–1521.

Martens, J. W., Roelfs, A. P., McKenzie, R. I. H., Rothman, P. G., Stuthman, D. D., and Brown, P. D. (1979). System of nomenclature for races of *Puccinia graminis* f. sp. *avenae. Phytopathology* **69,** 293–294.

Martens, J. W., McKenzie, R. I. H., and Harder, D. E. (1980). Resistance to *Puccinia graminis avenae* and *P. coronata avenae* in the wild and cultivated *Avena* populations of Iran, Iraq, and Turkey. *Can. J. Genet. Cytol.* **22,** 641–649.

Martens, J. W., Rothman, P. G., McKenzie, R. I. H., and Brown, P. D. (1981). Evidence for complementary gene action conferring resistance to *Puccinia graminis avenae* in *Avena sativa. Can. J. Genet. Cytol.* **23,** 591–595.

Mehta, K. C. (1940). "Further Studies on Cereal Rusts in India," Sci. Monogr. No. 14. Govt. of India.

Murphy, H. C., Zillinsky, F. J., Simons, M. D., and Grindeland, R. (1958). Inheritance of seed color and resistance to races of stem and crown rust in *Avena strigosa. Agron. J.* **50,** 539–541.

Newton, M., and Johnson, T. (1944). Physiologic specialization of oat stem rust in Canada. *Can. J. Res., Sect. C* **22,** 201–216.

Orjuela, N. J., Thurston, H. D., and Krull, C. F. (1962). Physiologic specialization of *Puccinia graminis avenae* in Colombia. *Plant Dis. Rep.* **46,** 866–871.

Paradies, M., Sisto, D., and Vallega, J. (1976). Behaviour of some oats to races of "stem rust" found in Italy in 1974 and 1975. *Proc.—Eur. Mediterr. Cereal Rusts Conf., 4th, 1976,* pp. 85–86.

Pavek, J. J., and Myers, W. M. (1965). Inheritance of seedling reaction to *Puccinia graminis* Pers. f. sp. *avenae* race 13A in crosses of oat strains with four different reactions. *Crop Sci.* **5,** 501–504.

Quisenberry, K. S., and Reitz, L. P. (1974). Turkey wheat: The cornerstone of an empire. *Agric. Hist.* **48,** 98–114.

Rajhathy, T., and Thomas, H. (1974). "Cytogenetics of oats (*Avena* L.)," Misc. Publ. No. 2. Genet. Soc. Can. (Ottawa).

Roberts, B. J. (1962). "The Effect of Temperature on Seedling Reaction of Oats to *Puccinia graminis* var. *avenae*." University Microfilms, Inc., Ann Arbor, Michigan.

Roelfs, A. P. (1978). Estimated losses caused by rust in small grain cereals in the United States—1918–1976. *Misc. Publ.—U.S., Dep. Agric.* **1363,** 1–85.

Roelfs, A. P., and Long, D. L. (1980). Analysis of recent oat stem rust epidemics. *Phytopathology* **70,** 436–440.

Roelfs, A. P., Casper, D. H., and Long, D. L. (1978). Races of *Puccinia graminis* f. sp. *avenae* in the United States during 1977. *Plant Dis. Rep.* **62,** 600–604.

Roelfs, A. P., Long, D. L., and Casper, D. H. (1980). Races of *Puccinia graminis* f. sp. *avenae* in the United States during 1979. *Plant Dis.* **64,** 947–949.

Roelfs, A. P., Long, D. L., and Casper, D. H. (1982). Races of *Puccinia graminis* f. sp. *avenae* in the United States and Mexico during 1980. *Plant Dis.* **66,** 208–209.

Sebesta, J. (1973a). The physiological races of *Puccinia graminis* Pers. f. sp. *avenae* Erikss. et Henn. in Czechoslovakia in the years 1967 and 1968. *Ochr. Rostl. Zemed. Velkovyrobe, Ref. Konf.* **9,** 1–6.

Sebesta, J. (1973b). On the relation between virulence and aggressiveness of oat stem rust. *Ochr. Rostl. Zemed. Velkovyrobe, Ref. Konf.* **9,** 155–161; *Rev. Appl. Plant Pathol.* **53,** 2154.

Sebesta, J., and Zwatz, B. (1980). Virulence of central European race populations of oat stem rust (*Puccinia graminis* Pers. f. sp. *avenae* Erikss. et Henn.) with special regard to the effectiveness of resistance genes. *Pflanzenschutzberichte* **46,** 1–42.

Shikina, O. M. (1974). Specialization of *Puccinia graminis* Pers. in the northern Caucasus. *Mikol. Fitopatol.* **8,** 359–361; cited in *Rev. Plant Pathol.* **54,** 2158.

Simons, M. D., Martens, J. W., McKenzie, R. I. H., Nishiyama, I., Sadanaga, K., Sebesta, J., and Thomas, H. (1978). Oats: A standardized system of nomenclature for genes and chromosomes and catalog of genes governing characters. *U.S., Dep. Agric., Agric. Handb.* **509,** 1–40.

Stakman, E. C., Levine, M. N., and Bailey, D. L. (1923). Biologic forms of *Puccinia graminis* on varieties of *Avena* spp. *J. Agric. Res. (Washington, D.C.)* **24,** 1013–1018.

Stakman, E. C., Stewart, D. M., and Loegering, W. Q. (1962). Identification of physiologic races of *Puccinia graminis* var. *tritici*. *U.S., Agric. Res. Serv., ARS* **E617,** 1–53.

Stewart, D. M., and Roberts, B. J. (1970). Identifying races of *Puccinia graminis* f. sp. *avenae:* A modified international system. *U.S., Dep. Agric., Tech. Bull.* **1416,** 1–23.

Suzdalskaya, M. V., Bukanova, V. K., Gorbunova, Y. U., and Koroleva, L. A. (1978). Virulence genes in populations of the oat stem rust pathogen, (*Puccinia graminis* Pers. f. sp. *avenae* Eriks. and E. Henn.) on the European territory of the USSR and in the Transcaucasus. *Mikol. Fitopatol.* **12,** 415–418.

Sztejnberg, A., and Wahl, I. (1967). A new and highly virulent race of oat stem rust in Israel. *Plant Dis. Rep.* **51,** 967–970.

Sztejnberg, A., and Wahl, I. (1976). Mechanisms and stability of slow stem rusting resistance in *Avena sterilis*. *Phytopathology* **66,** 74–80.

Teo, C., and Baker, E. P. (1975). Mutagenic effects of ethyl methanesulphonate on the oat stem rust pathogen (*Puccinia graminis* f. sp. *avenae*). *Proc. Linn. Soc. N.S.W.* **99,** 166–173.

Tozzetti, G. T. (1952). V. Alimurgia: True nature, causes and sad effects of the rust, the bunt, the smut, and other maladies of wheat, and of oats in the field. 139 pp. *In* "Phytopathological Classics" (L. R. Tehon, transl.) No. 9. Am. Phytopathol. Soc., St. Paul, Minnesota (originally published, 1767).

Wahl, I., Dinoor, A., Gerechter-Amitai, Z. K., and Sztejnberg, A. (1966). Physiologic specialization, host range and development of oat stem rust in Israel. *Proc. Cereal Rusts Conf., 1964* pp. 242–253.

Waterhouse, W. L. (1929). Australian rust studies. *Proc. Linn. Soc. N.S.W.* **54,** 615–680.

Waterhouse, W. L. (1952). Australian rust studies. IX. Physiologic race determinations and surveys of cereal rusts. *Proc. Linn. Soc. N.S.W.* **77,** 209–258.

Watson, I. A. (1981). Wheat and its rust parasites in Australia. *In* "Wheat science—Today and Tomorrow" (L. T. Evans and W. J. Peacock, eds.), pp. 129–147. Cambridge Univ. Press, London and New York.

Welsh, J. N., and Johnson, T. (1951). The source of resistance and the inheritance of reaction to 12 physiologic races of stem rust, *Puccinia graminis avenae* (Erikss. and Henn.). *Can. J. Bot.* **29,** 189–205.

Welsh, J. N., Green, G. J., and McKenzie, R. I. H. (1961). New genes for resistance to races of oat stem rust. *Can. J. Bot.* **39,** 513–518.

5

Crown Rust

Marr D. Simons

Agricultural Research Service, U.S. Department of Agriculture, Department of Plant Pathology, Iowa State University, Ames, Iowa

THE CEREAL RUSTS,
Volume II

ISBN 0-12-148402-5

I. Introduction

Crown rust of oats (*Avena sativa* L.), which is caused by *Puccinia coronata* Cda., is the most widespread and damaging disease of this cereal. The fungus has an almost innumerable array of pathogenic variants, and has repeatedly demonstrated its ability to adapt to constraints imposed by man as control measures.

This chapter will briefly review our knowledge of the disease, with emphasis on aspects having relevance to potential control measures.

II. Geographic Distribution

The crown rust fungus occurs nearly worldwide on oats. The distribution even includes islands far from any major land mass.

The aecial stage has been reported from all major oat-producing areas of the Northern Hemisphere where Rhamnaceous hosts occur in proximity to oats, including the Middle East where both hosts and the fungus presumably originated (Wahl, 1970). Susceptible species of *Rhamnus* are rare or nonexistent in South America and Australia, and thus the aecial stage does not occur there. It is also rare in some areas of relatively mild climate, such as the southern United States, even where susceptible hosts exist. This is due to the presumed requirement of low temperatures to break teliospore dormancy, although there are some unanswered questions on this point.

III. Economic Importance

A. History

It is likely that *P. coronata* has caused serious losses in oats ever since the crop was first cultivated. Over two centuries ago Tozzetti (1767) was aware of crown rust, as distinct from stem rust, and wrote that crown rust was very destructive to oats in some years (see Tozzetti, 1952). Generally, however, specific references to crown rust, as distinguished from stem rust and other diseases, are rare in the early literature. This is particularly true with regard to damage from the disease. It was not until late in the nineteenth century that Cornu (1880) wrote of severe damage from crown rust in Europe. At about the same time, Sivers (1887) reported three successive years of complete failure of the oat crop in the Baltic area because of crown rust.

In North America, Peck (1872) specifically listed *P. coronata* as a species new to New York in 1872, and it was collected in Iowa in 1874 (Arthur, 1924). Damage to oats in Connecticut was reported in 1890 (Thaxter, 1890).

B. LOSSES IN YIELD

Since 1900 there have been many reports of losses in yield of oats due to crown rust. Such reports are worldwide, and a representative sample follows.

In Uruguay, winter oats were sometimes killed by crown rust before they reached heading (Gassner, 1916). If flowering occurred, the resulting panicles were mostly barren. An epidemic of crown rust in 1940 in Portugal was so severe that barely enough seed for the next year's sowing was saved (d'Oliveira, 1942). In southeastern Europe, heavy attacks of crown rust were reported in 1959 (Kostic, 1959), and in Israel some cultivars were badly damaged (Wahl and Schreiter, 1953). Severe damage has occurred in coastal areas of Australia, with crops ruined and wild oats killed in years favorable to the disease (Waterhouse, 1952).

Severe losses of both grain yield and quality are not uncommon in both the northern and southern sections of the United States (Murphy *et al.*, 1942). A series of years in the 1940s and 1950s in which severe losses occurred in the United States were associated with the rapid increase in races of *P. coronata* attacking cultivars with the Bond resistance that occurred in conjunction with the increase in popularity of these cultivars. In the state of Iowa, losses were estimated at 12, 18, 20, 8, and 30% in 1949, 1950, 1951, 1952, and 1953, respectively (Sherf, 1954).

In recent years, losses due to crown rust have been light nationwide, but often are substantial over relatively large geographic areas and can be severe in individual fields. In 1979, losses were estimated to have reduced the potential crop by 2% in Minnesota, and by 3% in both North and South Dakota (D. L. Long, unpublished). Such losses in these major oat-producing states represent substantial dollar amounts.

C. METHODS OF ASSESSING LOSSES

There is no doubt that *P. coronata* significantly reduces potential oat yields. As with other economically important plant diseases, however, expressing such losses in terms of quantity or value, especially over large areas, is a vexing problem. An early attempt to make a quantitative estimate of losses caused by *P. coronata* was made by Pammel, who estimated that a severe epidemic in 1907 had resulted in a loss of 50% of the crop (Melhus *et al.*, 1942).

Immer and Stevenson (1928) compared yields of a group of 200 strains of oats that ranged from 5 to 70% infected. Each 1% increase in disease reduced yield by about one-third of a bushel per acre after allowances were made for differences in maturity.

Murphy *et al.* (1940) found that later maturing oats were more severely damaged by *P. coronata.* Studies of the severe epidemic of 1938 showed that the coefficient of infection (an index that combines percentage of leaf area covered by uredia with the size and other characteristics of the uredia) was more highly correlated with damage from *P. coronata* than was percentage of leaf area covered alone.

Quantitative estimates of losses caused by *P. coronata* for the state of Illinois in the moderately severe crown rust year of 1957 were published by Endo and Boewe (1958). Since some of the oat cultivars being grown were resistant to the crown rust races then present, performance of susceptible and resistant cultivars could be compared in nurseries grown in different areas of the state. Data were adjusted to reflect differences in acreage of each major cultivar in each area. Adjustments for inherent differences in cultivar yields were calculated from earlier years in which crown rust had been insignificant. Total losses averaged over cultivars ranged from 10 to 40% among the areas, with a statewide loss of 20%. Grain quality (test weight) of susceptible cultivars was also greatly reduced. A study in the mild rust year of 1956 indicated that crown rust destroyed over 43,000 metric tons of grain, showing that significant losses can result from small amounts of disease.

Fungicides can be used in estimating losses from *P. coronata* (Martens *et al.*, 1972). In Manitoba, yields of control plots, maintained rust-free with a fungicide, were compared with yields of plots exposed to natural infection over a period of years. In 1968 and 1969, the two years in which *P. coronata* developed with some degree of severity and was the only disease of consequence, yields were reduced in the cultivar Kelsey by 26 and 31%, respectively. Seed quality was also significantly reduced.

D. OTHER LOSSES

Puccinia coronata can have other adverse effects on its host in addition to reducing grain yield and quality. One of these is reduction of yield and quality of forage. Severe damage to oats being grown for forage has been reported from South Africa (Sawer, 1909), Germany (Straib, 1937b), and Australia (Miles and Rosser, 1954).

In two New Zealand locations, *P. coronata* reduced the dry-matter yield of forage oats during late winter and early spring from 13.6 to 11.4 and 6.66 to 2.85 tons per hectare (Eagles and Taylor, 1976). The grazing season is shortened by as much as 4–6 weeks by crown rust infection in Texas (Reyes and Futrell, 1962). In the first 2 years of a study of the effect of *P. coronata* on forage oats in Texas, the cultivar Moregrain outyielded Suregrain. Both cultivars were resistant to the prevalent forms of *P. coronata* present in these years. In the third year, forms of the fungus appeared that moderately infected Moregrain. Suregrain was resistant and it outyielded Moregrain significantly.

Oat straw has important economic value in many areas, and *P. coronata* can significantly reduce straw yield of susceptible cultivars (Simons, 1980). The use of controls maintained free of rust by a fungicide showed that straw yields are reduced by as much as half by severe epidemics. Lesser amounts of disease resulted in proportionally smaller losses.

The harvest index is the ratio of grain yield to total yield of grain and straw, and this ratio is sometimes an important consideration in oat breeding. It has been shown (Simons, 1980) that harvest indices of different cultivars can be differentially affected by *P. coronata*. Harvest index, or percentage of grain, of some cultivars was increased, while other cultivars showed a decrease in harvest index as a result of infection.

Winter hardiness is a concern in many areas where winter oats are grown. *Puccinia coronata* reduces the ability of young oat plants to harden at low temperatures, with the loss of cold resistance proportional to the severity and length of exposure to the disease (Murphy, 1939). In the field, greater winter injury was observed in sections where the disease had been heavy, and resistant oats had less winter injury than susceptible strains (Rosen *et al.*, 1942).

The relative protein percentage of oat grain is related to the feeding value of the grain, and attempts have been made to determine the effect of crown rust on protein percentage (Sebesta, 1974; Singleton *et al.*, 1979; Simons *et al.*, 1979b). No single answer has been found, but generally, moderately severe crown rust slightly reduces protein percentage. However infection has caused an increase in protein percentage of certain cultivars. Of greater importance is the significant reduction in total protein produced per acre due to the direct effect of a severe infection on reducing grain yield.

IV. Taxonomy

A. SPECIES LEVEL

The fungus responsible for crown rust of oats and other grasses was first described in the aecial stage by Persoon, as *Aecidium rhamni* (Gmelin, 1791). The telial stage was described by Corda (1837), who listed the rush *Luzula albida* as the host. He named it *Puccinia coronata* due to the projections on the apical end of the teliospore. Castagne (1845) was the first to recognize the fungus as a grass rust. He observed it on *Avena sativa, A. fatua,* and *Festuca arundinacea,* and named it *Solenodonta graminis.* Klebahn (1892) found, as had Nielsen (1875) earlier, that crown rust occurred in two forms, one that parasitized *Rhamnus frangula* and certain grasses, and one that parasitized *R. cathartica,* oats, and certain other grasses. He regarded the form on *R. cathartica* and oats as a separate species and designated it *P. coronifera,* while retaining the name *P. coronata* for the form on *R. frangula.* Later, others divided it into several species by the reactions of various species of *Rhamnus* and genera of grasses, resulting in some controversey regarding the number of species (Simons, 1970).

The controversy has continued, and the status of the specific names *P. coronifera* and *P. lolii* is still not settled. Azbukina (1976) contended that *P. coronifera* should be maintained as a species separate from *P. coronata,* noting they differ somewhat in morphology and markedly in aecial and uredial hosts, and that *P. coronata* and *P. coronifera* do not cross. Nevertheless, most researchers follow Cummins (1956) and Cunningham (1964), who regarded all species differentiated on the basis of pathogenicity as synonyms of *P. coronata.* They recognized, however, that forms of the fungus show considerable morphological diversity, which might indicate a need for subspecific taxa based on morphology.

B. SUBSPECIFIC

Eriksson and Henning (1894) subdivided *P. coronifera,* the species on *R. cathartica,* into two categories, which they designated "f." for form. The form on cultivated oats was designated f. *avenae* and the other, on the grass *Alopecurus pratensis,* f. *alopecuri.* Eriksson (1894) showed that urediospores from one grass genus did not infect species of other grass genera. The term *formae speciales* (f. sp.) was introduced to describe such pathogenic strains, and the term *P. coronata* f. sp. *avenae* became generally accepted for isolates of the fungus that parasitized wild and cultivated oats. Many experiments were done in the next 20 years and were summarized by Muhlethaler (1911). A number of additional special forms were described and, recognizing some exceptions, it was felt that special forms were restricted in pathogenicity to a single grass genus or even to certain species in a genus. However, Melhus *et al.* (1922) felt obliged to stress that some of the forms induced infection in species of several genera in addition to the ones for which they were named.

Klebahn (1895) noted that specialization of the fungus aptly described the situation. Nevertheless, he disliked the term *forma specialis* as it added another unit to the concept of species and varieties and it had no recognized rank. Stakman (1929), aware of these shortcomings of the term *forma specialis,* suggested that the term variety be used to designate subdivisions of species based on pathogenic specialization at the level of host species and genera. Such varieties might differ somewhat in size and shape of spores but were mainly characterized by pathogenicity. Each variety could parasitize several species of one or more genera of grasses but could not parasitize members of other genera, which might, however, be susceptible to other varieties of the fungus.

Fraser and Ledingham (1933), working in Canada, followed Stakman in the use of variety to designate subdivisions of *P. coronata.* Morphological features and rhamnaceous hosts were important parts of their scheme. Isolates of *P. coronata* parasitizing oats were designated *P. coronata* var. *avenae,* and this term was subsequently recognized by many investigators.

Eshed and Dinoor (1980) compared pathogenicity of cultures of *P. coronata* isolated from the grass genera *Agrostis, Alopecurus, Arrhenatherum, Phalaris, Festuca,* and *Lolium,* and nine cultures from cultivated oats. Seedlings of 22 grass species in eight genera and one oat cultivar were inoculated with each culture. The results showed that differentiation of *P. coronata* into *formae speciales* is largely meaningless. The fact that a culture from one host infected a second host did not necessarily mean that another culture originally from the second host would infect the first one. The culture isolated from *Festuca* was infectious to hosts in six other genera, while *Festuca* itself was susceptible to only one additional culture. At the other extreme, the culture isolated from *Arrhenatherum* was innocuous to any of the other host genera while *Arrhenatherum* was susceptible to at least some extent to all the forms. The nine cultures originally isolated from oats should belong to form *avenae.* They were classified into seven races on differential oat cultivars (see Section V,A), but reactions of the grass hosts grouped them into four different forms. There was no greater similarity between these four forms than between forms from the other host genera. Thus, it seemed that there was no essential difference between a race as defined on oat cultivars and a form as defined on grass species and genera. They concluded that forms and races are parallel taxonomic entities, and races are not subordinate to forms in a taxonomic hierarchy. Thus, I find it difficult to justify the use of Latin trinomials such as *P. coronata avenae* and follow the suggestion of Eshed and Dinoor (1980) that the subspecific divisions of the species *P. coronata* be abandoned.

V. Pathogenic Specialization

A. On Oats

Pathogenic specialization of *P. coronata* at the cultivar level is also the basis of cultivar resistance. Surveys of pathogenicity, or virulence, have covered various areas over different periods of time. The data have been interpreted in relation to the presence or absence of the alternate host, the susceptibility or resistance of the cultivars grown, and the reactions of oats having potential value as sources of resistance. Until recently, most surveys have been reported in terms of races (defined below), as this summarized the data in a form that was readily understandable.

Proof of specialization within *P. coronata* at the cultivar level was first published by Hoerner (1919). Seven cultivars of oats were inoculated with 30 isolates of *P. coronata* collected in widely scattered locations in the U.S. and Canada. The reactions of two of the cultivars, Ruakura and Green Russian, clearly illustrated pathogenic specialization at the cultivar level. Isolates of *P. coronata* characterized by such specific patterns of pathogenicity on designated groups of oat cultivars are usually designated as races.

Murphy (1935) summarized the work of Hoerner and those who followed him in the area of pathogenic specialization of *P. coronata* and established a system of classification of infection types into reaction classes that is still in general use:

Immune (I): no macroscopic evidence of infection.
Nearly immune (O): no uredia; necrotic areas or chlorotic flecks present.
Highly resistant (1): uredia few, small; some necrotic areas without uredia.

Moderately resistant (2): plentiful small to medium-sized uredia; necrotic area seldom without uredia.
Mesothetic (M): a combination of two or more types in varying proportions.
Moderately susceptible (3): abundant medium-sized uredia in chlorotic areas; no necrosis.
Completely susceptible (4): abundant large uredia without necrosis or chlorosis.

Races were characterized on the basis of only three reaction classes: resistant (classes I, 0, 1, and 2), mesothetic, and susceptible.

After extensive testing, Murphy (1935) established a standard set of 11 different cultivars (later enlarged to 13) for purposes of race identification. Thirty-three races were identified among 533 isolates collected throughout the United States as well as from Mexico and Canada during the years 1927–1932. They varied in versatility from very aggressive to very restricted forms. Some were common and present annually, and others were rare and appeared occasionally. Certain races overwintered on fall-sown oats in the winter-oat region; others were dependent on the alternate host for initial dissemination in the spring. More races were identified from 100 isolates collected in the north than in the south, indicating that hybridization and segregation of races on *Rhamnus* in the north resulted in a more diversified *P. coronata* population than occurred in the asexual population in the south.

In Canada, Peturson (1935) identified 11 races among 544 isolates of *P. coronata,* most from the prairie provinces of Canada. Straib (1937b), working in Germany, established a set of 15 differential cultivars, including seven of those used in North America. He identified 142 races from 144 isolates, and concluded that specialization of the fungus in Germany was stronger than in North America. Since almost every isolate represented another race, he felt there was little point in additional testing. Straib used finer distinctions of infection types than had the North American workers, and he also noted, as had others, that different environmental conditions, particularly temperature during the incubation period, could make a difference in reaction class. Subsequent race surveys in Northern Europe have generally indicated that *P. coronata* in this area is markedly specialized. For example, Hermansen (1961) isolated five races from six samples.

In South America, Pessil (1960) and Vallega (1951) showed that North and South America had certain prevalent races in common, but some South American races parasitized the cultivars Landhafer and Trispernia, which were resistant to North American races. Races that parasitized the diploid cultivar Saia were rare in North America but were common in Brazil.

The most resistant oat cultivars in the Soviet Union were Victoria,

Landhafer, Santa Fe, and Bondvic (Zhemchuzhina, 1978), while in North America they were generally susceptible.

The 13 standard differentials were used worldwide from 1935 to the early 1950s. Descriptions of the 112 races identified were given by Simons and Murphy (1955). The cultivars of the standard set chosen in the mid-1930s had lost much of their usefulness by 1950 because new or formerly unimportant races of *P. coronata* had become prevalent. In 1950 and 1951, Simons and Murphy tested *P. coronata* isolates made in the United States on additional cultivars. As a result, a second set of 10 cultivars was established as a new standard for identification of races of *P. coronata.* Races identified on it by Simons and Murphy (1955) were combined with races identified in South America and Canada and 59 races were initially described. These were numbered starting with number 201, to prevent confusion with races identified on the old set of cultivars. The new set of cultivars was widely used for the next two decades, and a total of 294 races were identified worldwide by 1976 (Michel and Simons, 1977).

The current trend in the study of pathogenic specialization of *P. coronata* at the host cultivar level is away from the use of standard sets of differential cultivars. Individual investigators in different areas now commonly survey and evaluate the virulence of *P. coronata* in their area in relation to available sources of resistance or resistance found in commercial cultivars grown locally. Examples of such studies have been reported by Martens *et al.* (1977) in New Zealand and Harder (1980) in Canada.

B. ROLE OF *RHAMNUS*

The origin of pathogenic diversity of *P. coronata* at the host cultivar level has important practical implications in the development of methods to control the disease. One way by which new combinations of pathogenicity, or races, arise is by recombination of existing genes for pathogenicity in the sexual cycle. Populations of *P. coronata* originating on *Rhamnus* have been compared with populations that exist in the uredial stage where *Rhamnus* is absent or unimportant. In a study of *P. coronata* isolates from the northern and southern United States, Murphy (1935) showed that certain races seemed to overwinter in the south in the uredial stage, while others appeared to be dependent on spread from *R. cathartica.* A relatively large number of rather rare races were collected from *Rhamnus,* or from oats near *Rhamnus,* suggesting that *Rhamnus* functioned effectively to generate pathogenic diversity.

Surveys over many years show that more races of *P. coronata* occur in eastern Canada than in the west (Fleischmann *et al.*, 1963). There is also less tendency for a few races to predominate in the east. These differences were explained by the relatively frequent occurrence of the alternate host in the oat-growing areas of eastern Canada. It has also been shown in Canada that more races can be identified from *Rhamnus* than in equal numbers of collections made from oats (Fleischmann, 1965). The diversity of races of *P. coronata* in Yugoslavia was attributed to the abundance of aecia on the alternate host (Kostic, 1965).

In 28 isolates of *P. coronata* from a *Rhamnus* hedge in Minnesota, there were 13 known races and four new races (Saari and Moore, 1963). Some could be subdivided by the use of supplementary differential cultivars. It was suggested that planting resistant oat selections next to buckthorn would measure the pathogenic potential of the fungus, while providing a severe test of the value of the resistance being considered. Material from this same *Rhamnus* nursery was compared by Simons *et al.* (1979a) for pathogenicity on 24 differential oat cultivars with *P. coronata* isolates collected in Texas, where *Rhamnus* does not function as a host of the fungus. As expected, more virulence patterns (races) occurred in material from the buckthorn nursery. However, there was very little difference between the two populations in numbers of virulence genes. This emphasizes that *Rhamnus* serves only to segregate virulence genes, and plays no role in their origin. Such segregation conceivably could expedite the development of "super races" having a wide virulence but the data failed to support that possibility.

The general pattern of race succession in the United States in recent years substantiates the subordinate role of *Rhamnus* (Michel and Simons, 1977). For example, race 290 and similar races that dominated in the 1960s were supplanted in the 1970s by race 264B, with a much wider range of virulence. Race 290, however, remained an important component in collections from *Rhamnus*, suggesting that *P. coronata* on *Rhamnus* follows, rather than leads, the evolution of virulence of the fungus.

C. METHODOLOGY

Studies of the occurrence and distribution of virulence in populations of *P. coronata* have advanced our knowledge of the disease in the areas of epidemiology and fundamental genetics. Their primary objective, however, was to help in developing resistant oat cultivars. Information needed for this major objective included (1) year-to-year prevalence of the various races and (2) early detection of new and potentially dangerous races. Unbiassed data on the relative prevalence of common races can best be obtained from material collected on universally susceptible cultivars. This prevents the sampling bias resulting when material is collected from cultivars that have resistance that screens out part of the fungus population. The appearance of new and rare forms can best be detected in collections made from oats with the important resistance genes (Simons, 1955).

Browder (1969) considered the various objectives of surveying virulence of the rusts, and made the important point that all such work must ultimately be grounded in the gene-for-gene theory of parasitism (Flor, 1956). Browder felt that virulence data presented in terms of races tended to obscure important information. He suggested that it would be more meaningful if presented on the basis of virulence to single host cultivars or, better, as virulence corresponding to single genes for resistance. However, there should be some way of recognizing associations of virulence in reporting survey data.

As a compromise, Simons and Michel (1959) utilized temporary differential cultivars that were changed from year to year, or as the need arose, in addition to the standard differential cultivars. Virulence of the fungus isolates on the standard differentials was reported in terms of races, and on the temporary differentials, in terms of virulence on the individual cultivars. Some information is lost using this system because it is not possible to show the range of virulence of individual isolates on the temporary differentials. However, it makes the data understandable in terms of practical breeding. Fleischmann (1967) presented survey data both as races and as virulence for cultivars.

VI. Genetics of *Puccinia coronata*

A. MUTATION

Mutation is the only source of basic new virulence in the fungus. Mutation at a given locus is a rare event, and critical experimentation is limited. However, Zimmer *et al.* (1962) observed measurable rates of mutation of races 202 and 290 for virulence toward cultivars resistant to these races. The mutation rate of race 202 for virulence on Ascencao was estimated at one in 2200 infections; for virulence on Ukraine, one in 6450; and for virulence of race 290 on Ukraine, one in 7900.

The appearance of a new race of *P. coronata* in Canada appeared to exemplify the process of natural mutation in the origin of new virulent races (Fleischmann, 1963b). This race, numbered 332, differed from the then very common race 216 only in its ability to parasitize the cultivar Saia. The appearance of the new race, and its increase to where it was common enough to be detected in the survey, occurred in the absence on any known Saia-type resistance in cultivars being grown commercially, suggesting the operation of selective forces other than host resistance.

B. SEGREGATION

The artificial hybridization of isolates of *P. coronata* on *Rhamnus* could throw light on the relative importance of *Rhamnus* in the appearance of new races. Unfortunately, the problems of producing telia in isolation and inducing telia to germinate have proven to be a deterent. However, some studies are available. Eriksson (1908) inoculated *Rhamnus* with spores from *Alopecurus* and found that the aeciospores parasitize cultivated oats that were not hosts for the original culture. Similarly, a culture from *Festuca* attacked *Lolium* after passage through *Rhamnus*.

Eshed and Dinoor (1976) noted that the overlapping host ranges of different forms of *P. coronata* facilitated genetic studies, because a common host could serve as a propagating host for the F_1 and F_2 generations. A selected group of 32 species of grasses, representing 18 genera, were inoculated with forms of *P. coronata* parasitic on *Avena* and *Phalaris* and with the F_1 and F_2 progeny cultures. Some of the F_2 isolates were unable to parasitize either *Avena* or *Phalaris*. Others had a widened host range to include additional virulence on oats, and included another *Phalaris* species that was resistant to the original cultures. The increased virulence on oat cultivars was assumed to be due to inhibitors. Interactions between parental, F_1, and F_2 cultures and oat cultivars could be explained by 30 loci, while 44 loci were needed to explain interactions on the wild-grass hosts. Segregation within hosts, as well as gradations in reactions to many cultures, pointed to the possibility that despite satisfactory classical genetic explanations of many of the results, pathogenicity might be polygenically controlled in some cases. Biali and Dinoor (1972) showed that the genetics of pathogenicity in *P. coronata* does not always follow the usual pattern with virulence recessive. Selfing an isolate of race 276, virulent on many oat cultivars, showed that virulence was dominant for some loci.

Murphy (1935) identified urediospores of races 1 and 3 from the cultivar Hawkeye and used telial material from the same plants to inoculate *Rhamnus*. Most of the resulting cultures were identified as the previously unknown race 18, but a few were races 1 or 3 or another new race, 19. Hawkeye was resistant to races 18 and 19, and this indicated that races 1 and 3 were highly heterozygous for pathogenicity. Zimmer *et al.* (1965) selfed a relatively homogeneous natural population of *P. coronata* and obtained pathogenically diverse S_1 progenies. Segregation for pathogenicity to 10 of 16 oat cultivars occurred in 52 cultures derived from this selfed population. A high degree of heterozygosity was also found by Dinoor *et al.* (1968), who intercrossed nine pycnia representing five pathogenic races and identified at least 25 races among the progeny. Among the progeny was an array of recombinants ranging from those with limited virulence to those with wide virulence, including virulence that the parental cultures had not exhibited. Nof and Dinoor (1981) crossed two isolates of *P. coronata* and two oat cultivars, and studied segregating populations for virulence and resistance, respectively. They found the oat–*P. coronata* system is under gene-for-gene control, as had been demonstrated for other rust fungi earlier (Flor, 1956).

C. Heterokaryosis via Anastomosis

Hyphal anastomosis of *P. coronata* may be a possible source of pathogenic variation. This was shown by Bartoš *et al.* (1967), who mixed urediospores from two single-spore isolates of *P. coronata* that differed in pathogenicity. The mixture was used to inoculate a susceptible oat cultivar, and single-pustule isolates resulting were identified. The majority of these were of the parental types. However, two new races were found. Reassociation of parental nuclei following germ-tube fusions or hyphal anastomoses apparently was responsible for the occurrence of the new races.

D. Linkage

Whether the virulence of isolates of *P. coronata* is linked to traits like aggressiveness is important both theoretically and practically. Brodny *et al.* (1979) showed that isolates of race 276 usually had greater infection ability and capacity to develop infection than did isolates of race 263. At or above 25°C, there was a greater tendency for early telia formation among isolates of race 263 than race 276.

In New Zealand, Martens *et al.* (1977) found that most isolates of *P. coronata* carried apparently unnecessary genes for virulence that seemed to confer no competitive disadvantage. Such data do not support stabilizing selection as a force in *P. coronata*.

VII. Signs and Symptoms

The uredial stage of *P. coronata* occurs mainly on the leaf blades of the oat plant, but to some extent on the sheaths and floral structures. On susceptible cultivars, the uredia appear as bright orange-yellow, round-to-oblong pustules that are up to 5 mm or more in length when infection is light. Cultivars with different degrees and types of resistance may show reactions ranging from small, light-colored flecks through small to medium-sized pustules surrounded by generally well-defined chlorotic or necrotic areas. Telia usually appear after uredia are well established and sometimes form in rings around the uredia. They are black or dark brown and remain covered for some time by the epidermis. Pycnial and aecial stages occur mainly on the leaves, often with hypertrophy, but also to some extent on the petioles, young stems, and floral structures of susceptible species of *Rhamnus*. The pycnia appear early in the spring as small, round, orange-yellow, slightly raised structures, usually on the upper surface of the leaf. Aecia, which usually form on the underside of the leaf beneath the pycnia, follow the pycnia and appear as round or somewhat irregular, tightly packed clusters of small orange-yellow cups. Aecia may be up to 5 mm or larger in diameter.

VIII. Life History of *Puccinia coronata*

Puccinia coronata is a typical heteroecious, long-cycle rust, with its repeating dikaryotic uredial stage occurring on oats more or less throughout their active growing period (Simons, 1970). As the season advances and as the plants start to mature, telia are formed in and around the uredia, and these serve to overwinter the fungus. Meiotic reduction occurs in the teliospores, and germination of the teliospores results in haploid basidiospores. These infect young leaves of susceptible species of *Rhamnus*. DeBary (1867) first demonstrated this connection between *P. coronata* on grasses and on buckthorn when he used "sporidia" from an unspecified grass to inoculate *R. frangula*. Both "spermogonia" and aecia resulted, but attempts to inoculate oats, rye, and wheat with the aeciospores were unsuccessful. The function of the haploid pycnia and pycniospores that result directly from infection by basidiospores was not discovered until many years later, when Craigie (1928) showed that the diploid condition was restored by the transfer of the self-incompatible pycniospores to flexuous hyphae of other pycnía. Dikaryotic aeciospores then appear in the aecia.

Heteroecism of *P. coronata* was proved by Nielsen (1875) when he obtained uredial infection on *Lolium perenne* from inoculation with aeciospores from *R. cathartica.* A few years later, Cornu (1880) used aeciospores from *R. cathartica* and *R. oleoides* to obtain heavy infection on young oat plants.

In climates where the winters are mild, the fungus may live indefinitely in the uredial stage on cultivated, volunteer or wild oats.

IX. Epidemiology

A. ROLE OF *RHAMNUS*

The relative importance of *Rhamnus* in areas of the world where it functions in the epidemiology of the disease varies greatly. *Puccinia coronata* on oats is invariably associated with *Rhamnus* in Siberia (Wahl *et al.,* 1960). Since the fungus can not overwinter in the uredial stage in the severe climate and there is no source of wind-borne urediospores, the disease is dependent on the presence of *Rhamnus.*

In northern Europe, *P. coronata* does not overwinter in the uredial stage, and *Rhamnus* is the primary source of inoculum (Straib, 1937b). However, wind-borne urediospores may move from the south.

Much of the oat acreage of the world is where *P. coronata* does not overwinter in the uredial stage and where both *Rhamnus* and wind-borne urediospores may be important. The relative importance of aeciospores and urediospores varies annually, and depends on the amount of disease on oats in the overwintering area, velocity and direction of the wind at critical times, progress of *Rhamnus* eradication programs, etc. The north central United States and prairie provinces of Canada exemplify such areas. The progress of a limited epidemic initiated by *Rhamnus* in Iowa was documented by Dietz (1923), who made detailed observations on the progress of several such epidemics. In one of these, the first pycnia appeared on May 13, aeciospores were mature on May 16, and uredia first appeared on nearby oats on May 22. He assumed that uredia found on oats during the next week resulted directly from aeciospore infection. Viable aeciospores were produced up to June 10 on *Rhamnus,* and resulted in no infection beyond 2.4 km from the *Rhamnus* plants. Thus, the direct spread of the disease from aeciospore infection covered 1,300 hectares. Urediospore infection had spread over 8,500 hectares by June 4, 32,600 by June 7, and 165,000 by June 10, at which time infection occurred at a maximum distance of 85 km from the hedge. No other *Rhamnus* bushes were found in the 32,600 hectares in which infection occurred by June 7. Oats in fields adjoining this hedge were a total loss, and yields were significantly lowered in all fields within the area of aeciospore infection.

Simons (1970) summarized work of Dietz (1926) and others that showed that a large number of the known species of *Rhamnus* (Wolf, 1938) are susceptible to the forms of *P. coronata* that parasitize oats. However, only a small number are of economic importance. In North America and northern Europe, *R. cathartica* is the most important species involved in the infection of oats. It was introduced into North America in pioneer times and was well adapted to northern United States and Canada where oats are grown. Because of its desirable qualities as a hedge plant and shrub, it still is widely grown, and has escaped cultivation in many places.

Rhamnus frangula, also native to Europe and common in the northern United States as an introduced shrub, apparently plays no significant role in the epidemiology of crown rust of oats on either continent, in spite of an occasional report of susceptibility to strains of *P. coronata* that parasitize oats. *Rhamnus lanceolata* is a common shrub native to much of the central United States and is susceptible to forms of *P. coronata* capable of infecting oats, but is of little importance as it occurs only infrequently or in wooded areas away from oats.

The situation elsewhere is poorly documented, but it appears likely that a relatively small number of *Rhamnus* species are of economic importance. In Israel, four *Rhamnus* species are present, but *P. coronata* occurs on only two, and only *R. palaestina,* appears to be economically important in the epidemiology of the disease.

B. CROWN RUST WITHOUT *RHAMNUS*

Rhamnus shrubs and hedges are commonly responsible for severe local epidemics of crown rust in northern climates, but general, widespread epidemics in the north central United States and Canada seem to result from infection from airborne urediospores from infected oats planted in the fall or earlier in the spring in milder southern climates.

In most years this spore movement occurs as a stepwise process in which spores are blown relatively short distances from field to field as the season advances. Occasionally, however, disease and climatic conditions are such that spores may be transported in large quantities over long distances and be deposited on oats growing far to the north at a relatively early stage of growth. The severe epidemic that occurred in Iowa and adjoining states in 1953 was good example (Sherf, 1954). In early May, 1953, *P. coronata* was common in maturing oat fields in Texas and Oklahoma. From May 12 to May 14, strong, steady south winds blew across Oklahoma and northward at 10,000 feet and deposited a relatively heavy shower of urediospores over much of Iowa and areas of adjoining states. Initial infection thus occurred on these plants at an early stage in their development, and the weather for the remainder of the season was near ideal for development of subsequent generations of the fungus. The resulting epidemic was one of the most destructive ever recorded, causing an estimated loss of 30% of the potential oat crop in Iowa.

Rhamnus generally does not function in the epidemiology of crown rust in areas where winters are relatively mild and where summers are long and hot. Oats are produced under such conditions in many places in the world, including the southern United States, South America, Australia, etc. The epidemiology of the disease in these situations has not been investigated as intensively as in the north, but it is known that the oats are rusted during at least part of the cool or winter season. It is not always clear how the fungus survives the long, hot summers in the uredial stage to provide initial inoculum in the fall or winter. Generally the fungus survives on volunteer or wild oats that grow through the summer and fall. Observations made by Forbes (1939) showed that this is not true in Louisiana. Urediospores did not survive over the summer on straw piles or on dead leaves, and no volunteer oats or susceptible grasses could be found during the summer. It was suggested that the uredial stage was maintained in the northern states during the fall on volunteer oats and that wind-borne urediospores from these plants gradually spread southward to reach Louisiana in the winter. Such southward movement of *P. coronata* has been reported by Atkins and McFadden (1947) in Texas. The fungus became established on volunteer oats in the fall in the northern part of the state, where winters are too severe for *P. coronata* to survive in the uredial stage, and then spread southward across the state to winter oat areas as the season advanced.

Migrating birds may be a factor in the epidemiology of *P. coronata* and other cereal rusts. Warner and French (1970) noted that large flocks of blackbirds, as well as other species, regularly passed the winter in central and southern Mexico in areas where urediospores are continuously produced by several species of rust fungi on cultivated cereals. Here, they may become heavily contaminated with urediospores and carry these spores with them when they migrate to the northern United States and Canada in the spring. Urediospores could be carried to the south in the fall by the same process. Warner and French showed experimentally that starlings were quite efficient in picking up urediospores from infected oat plants and transferring them in a viable condition to healthy plants in another location.

X. Control

A. Resistance

1. *History*

Of the measures that are available to control *P. coronata,* the use of resistant cultivars is the most important. In one of the earliest known tests of cultivar reaction, Sivers (1887), after three successive years of crop failure due to crown rust, planted 41 cultivars in an effort to find one that would be sufficiently resistant to permit cultivation of oats. Most of them were rated as moderately or strongly infected, but infection was rated as weak on eight and there was almost no infection on one, the Russian Orel oat.

Observations clearly aimed at distinguishing between different reactions to *P. coronata* were made by Norton in 1907 in the United States. At about the same time, Sawer (1909) reported *P. coronata* to be a serious problem on oats in South Africa. He noticed differences in susceptibility among cultivars, observing that three of the red oat group were the most resistant.

Another important milestone in the utilization of resistance was Parker's (1920) demonstration that resistance could be found in progeny from crosses between resistant and susceptible cultivars. Parker (1918) was also one of the first to carry out extensive comparisons of reactions of cultivars in the seedling and later growth stages. This set the stage for the division of resistance to *P. coronata* into two categories: (1) specific (roughly equivalent to seedling, vertical, oligogenic, etc.) and (2) general (adult plant, horizontal, polygenic, etc.).

2. *Specific Resistance*

a. Inheritance. The first clear-cut demonstration of Mendelian inheritance of resistance to *P. coronata* was done by Davies and Jones (1927) studying a cross between the susceptible Scotch Potato oat and the resistant Red Rustproof.

The introduction of the Victoria oat from South America in 1927 and its subsequent recognition as a valuable source of resistance prompted many studies. Murphy *et al.* (1937) showed that Victoria had a single partially dominant gene for resistance, but Smith (1934) found resistance to be dominant only in some crosses. Chang and Sadanaga (1964) found that the resistance of Victoria to race 290 was conditioned by two genes, only one of which conditioned resistance to race 203.

It appears that Victoria has several genes for crown rust resistance. One gene was probably the one reported by Chang and Sadanaga (1964) to condition resistance to race 203. This gene was used in developing the large number of Victoria-derived cultivars that were released in the 1940s that later proved to be susceptible to the destructive Victoria blight, caused by *Helminthosporium victoriae* (Murphy and Meehan, 1946). Victoria also has one or more additional genes conditioning a high degree of resistance to certain races of crown rust that were not associated with susceptibility to *H. victoriae.* Additional genes conditioning moderate or low levels of resistance to many races of crown rust are also present.

The cultivar Landhafer was recognized during the 1930s as having potentially valuable resistance to crown rust and was subsequently used in the development of many important cultivars. Litzenberger (1949) reported its resistance to be controlled by a single dominant gene, which was verified by most other studies.

In the late 1950s the discovery of the great potential value of wild *Avena sterilis* from Israel and other parts of the Mediterranean and Near East (Wahl, 1970) led to investigations of this resistance. An early study (McKenzie and Fleischmann, 1964) of crosses between such wild-oat strains and a susceptible cultivar showed that each strain carried a single gene for resistance. In addition to these major genes for seedling resistance, the strains of *Avena sterilis* each had one or more genes for field resistance. Such results are typical of studies by other investigators.

A synopsis of studies of inheritance of resistance in terms of genes described has been compiled by Simons *et al.* (1978). This catalog lists 61 genes for crown rust resistance. Most of those reported since the early 1960s have been genes carried by *A. sterilis.* Most of the genes listed have, or had, some special characteristic that makes them of value in breeding resistant oat cultivars. It is now apparent that there is little point in studying the genetics of resistance to *P. coronata* unless the resistance has some potential practical usefulness.

The failure of oat cultivars with single genes for resistance to give lasting control of crown rust prompted combining resistance genes in a cultivar. This had the obvious advantage of providing combined resistance to a greater range of rust races than could be achieved with a single gene. Also, such "stacking" of genes in a single cultivar would force the pathogen to mutate simultaneously at two loci to overcome the resistance of the cultivar. There was also a possibility that combined genes might be additive in effect (Finkner, 1954). The significance of such transgressive segregation under field conditions has not been investigated.

A knowledge of dominance is useful in transferring resistance genes to cultivars that are superior agronomically. The majority of studies of inheritance of resistance to *P. coronata* have reported resistance to be dominant or partially dominant. A lack of dominance is not rare, and Simons and Murphy (1954) illustrated the complexity of the situation. The degree of dominance exhibited seems dependent on the genic background involved. Dominance also varies with different races (Sebesta, 1979).

The possibility of linkage between resistance genes and genes governing other traits was studied by Osler and Hayes (1953). They found no linkage between resistance genes and genes for traits such as stem rust resistance, date of heading, number and length of basal hairs, percentage of lower florets awned, strength of awns, or plumpness of seed.

Allelism of genes for resistance is of importance as the resistances conferred by allelic genes cannot be combined in the same plant. The majority of studies have shown that genes for resistance in unrelated oats are generally at different loci. Exceptions, however, have been found (Upadhyaya and Baker, 1965).

b. Breeding. The history of breeding for resistance to *P. coronata* was reviewed from its beginnings until about 1960 by Coffman *et al.* (1961). The first resistant cultivar developed through hybridization was started in 1919, when Dietz crossed the stem-rust-resistant Richland with Green Russian. The cultivar Hawkeye, moderately resistant to *P. coronata,* resulted from this cross. A cross made in 1928 by Coffman between the susceptible Markton and the moderately resistant Rainbow resulted in Marion, which was the first oat cultivar to combine resistance to *P. coronata* with resistance to other major pathogens.

The introduction of Victoria from Uruguay in 1927 provided a much better resistance than had been available to American breeders previously. By 1945, about 90% of the oat acreage of the Corn Belt and 75% of the acreage of the entire United States was planted to Victoria derived cultivars (Murphy, 1952). Then, the sudden appearance and devastating impact of *Helminthosporium victoriae* (susceptibility to which is linked to the *P. coronata* resistance of Victoria) rendered cultivars with the Victoria resistance virtually worthless in large areas of the country. Their decline in popularity was almost as rapid and complete as had been their original acceptance. They had, nevertheless, yielded large benefits during their brief period of popularity (Murphy, 1946) because of their resistance to *P. coronata.*

Races parasitizing Victoria were rare components of the fungus population in the 1930s (Murphy and Levine, 1936), and they did not dramatically increase in prevalence when the Victoria cultivars were being widely grown. Possibly they lacked fitness. They increased to dominate the population in the 1950s. This may have been associated with the growing of cultivars in the South that were derived from Victoria and that still made up an appreciable part of the acreage (Simons and Michel, 1958). The increase in prevalence of the Victoria races in the 1950s also was evident in Canada, where it was correlated with widespread commercial production of the cultivars Garry and Rodney, which were resistant to the older races but susceptible to the Victoria races (Fleischmann, 1963a). Some Victoria-derived cultivars had other Victoria genes. Some of these cultivars were widely grown after the Victoria cultivars susceptible to *H. victoriae* had disappeared. This was particularly true in Canada (Welsh *et al.*, 1953), but important cultivars of this type were also developed in the United States (Poehlman and Kingsolver, 1950). In general, these cultivars had less resistance than the *H. victoriae*-susceptible cultivars, but under field conditions they held up well where susceptible cultivars were severely damaged by *P. coronata.*

The cultivar Bond was introduced from Australia at about the same time as Victoria. The potential of its resistance to *P. coronata* was not recognized until later, but development of cultivars adapted to all the oat-growing areas of the United States carrying the Bond resistance were available to replace the Victoria cultivars. In 1950, the single Bond cultivar Clinton was grown on 75% of the oat acreage of the United States, and other cultivars with the same resistance made up much of the remaining acreage (Murphy, 1965). This very rapid increase of the Bond cultivars provided a striking example of the effect of resistant cultivars on the *P. coronata* population. Races capable of parasitizing Bond were discovered in the 1930s, but remained very rare until the Bond cultivars appeared in commercial fields. At that point the increase in races attacking Bond and the consequent decrease in avirulent races paralleled the increase in acreage of Bond cultivars. The Bond-virulent races made up a trace of the population in 1943 when Bond cultivars were grown only in nurseries, but comprised over 90% of the fungus population in 1949 and 98% in 1950. A similar change was observed in Canada (Peturson, 1951).

Landhafer, an important source of resistance that was used in the United States in the 1950s because of its resistance to the Bond races, was found in Uruguay by Gassner (1916). It was a native strain grown in Uruguay for many decades, during which time it had become adapted to conditions of that region, including crown rust. Other cultivars were also known that were resistant to the races that attacked the Bond derivatives (Coffman *et al.*, 1961). Santa Fe, an introduction from South America, was the source of Clintafe, the first cultivar resistant to *P. coronata* to be developed by systematic backcrossing with the objective of developing a cultivar identical with an existing one, Clinton, except for resistance. Landhafer was used in the same way to develop the popular Clintland and related cultivars.

The appearance and subsequent spread of race 290 and similar races, which effectively parasitized both Landhafer and Santa Fe, largely nullified the value of the Landhafer and Santa Fe derivatives as resistant cultivars (Michel and Simons, 1966; Fleischmann *et al.*, 1963). Since then, no one source of resistance has dominated in plant breeding. This is partly because of the recognition of the futility of using a single resistance gene over a large geographic area, and because of the discovery that the wild *A. sterilis* of the Mediterranean and Near East comprised a rich and diverse source of genes for resistance. With many genes to choose from, there was no need to concentrate on a single source of resistance (Dinoor and Wahl, 1963), and strains of the wild *A. sterilis* have served as the primary sources of resistance in breeding programs since the late 1950s. In the relatively small area of the country of Israel alone, both seedling and adult plant resistance to races with wide virulence were prevalent and widespread (Wahl, 1970). Some form of resistance was found in over 30% of the wild plants collected in 446 locations. Resistance was much more common in the northern areas of the country, where *P. coronata* is common, than in the more arid southern areas where the fungus is rare.

Resistance of wild oat populations from other Mediterranean and

Near Eastern countries was also assayed. Martens *et al.* (1980) screened over 1400 oat accessions from Iran, Iraq and Turkey. Plants of the eight *Avena* species represented were tested as both seedlings and adult plants. Resistance was common in *A. barbata* from Turkey and in *A. sterilis* from all three countries.

Unfortunately, a linkage between resistance genes and yield may occur. Simons (1979) crossed a strain of *A. sterilis* known to carry a gene for resistance to *P. coronata* with a susceptible cultivated oat. In the disease-free plots, the mean yield of resistant selections was 23% less than the susceptible lines, suggesting a linkage of resistance and low yield. Individual lines combining resistance and high yield, however, were found. Frey and Browning (1971), in the process of developing isolines to be used as components of multiline cultivars, found resistance genes from two strains of *A. sterilis* to be associated with significant yield increases in the absence of *P. coronata.*

3. *Diploid and Tetraploid Oats*

Strains of diploid and tetraploid oats with potentially valuable resistance to *P. coronata* are easy to find (Simons, 1959). Such resistance has been used to a very limited extent due to the difficulty of transferring it to cultivated oats.

Simons *et al.* (1959) found each of three diploid strains to have a different single dominant gene. The resistance of a tetraploid oat was conditioned by a single gene without dominance. Another tetraploid strain derived from crossing a resistant diploid with a susceptible tetraploid (Zillinsky *et al.*, 1959) carried a single dominant gene. Other investigators (Marshall and Myers, 1961) found the resistance of diploid and tetraploid strains of oats generally conditioned by single genes.

One diploid line in particular, Saia, has been the subject of intensive study in an effort to transfer genes for resistance from diploid to cultivated oats. Zillinsky *et al.* (1959) crossed Saia with a susceptible tetraploid. Continued backcrossing to cultivated oats led to true-breeding hexaploid oats with the resistance of the diploid (Sadanaga and Simons, 1960). In the Iowa multiline program, lines with this resistance yielded slightly less than the recurrent parents, and they were dropped (Frey and Browning, 1971). Cytogenetic studies showed the presence of 21 pairs of chromosomes plus two fragments with the gene for resistance located on the fragments. Transmission of the fragments was somewhat irregular, and lines without them were susceptible and yielded as well as the recurrent parent (Dherawattana and Sadanaga, 1973; Brinkman *et al.*, 1978).

By using thermal neutron radiation, Sharma and Forsberg (1977) were able to transfer resistance from Zillinsky's tetraploid to cultivated oats. The resistance gene was transferred from an alien chromosome to one in the hexaploid complement, and showed normal transmission.

4. *General Resistance*

a. Inheritance. The conspicuous failure of specific resistance stimulated interest in the use of general resistance. A knowledge of the inheritance of general resistance would be useful, but a very limited amount of information is available.

In segregating progenies from a cross between a resistant and a susceptible oat cultivar, Parker (1920) found that multiple factors were responsible for the resistance. Red Rustproof is typical of a group of cultivars having what we call general resistance. Luke *et al.* (1975) found that the resistance of Red Rustproof was controlled by a small number of genes showing slight partial dominance for susceptibility. Heritability was high, with a broad sense value of 87% making selection for this resistance possible.

Simons (1975) found heritability values of the resistance of four unadapted oat strains to range from 46 to 86% when measured in terms of yield reduction attributable to *P. coronata,* and 65 to 92% in terms of reduction in seed weight. The relationship of yield to resistance in the absence of rust was generally negative, and none of the lines combined maximum yield with maximum resistance. Thus manipulation of general resistance in breeding programs will require large populations in spite of the heritability values.

Kiehn *et al.* (1976) showed that the resistance of two strains of *A. sterilis* was controlled by a number of minor recessive genes having additive effects, a type of inheritance commonly associated with general resistance.

b. Pathogenic Specialization. General resistance to *P. coronata* is thought to be less subject to the vagaries of pathogenic specialization than is specific resistance. Because of the difficulty of testing this hypothesis, few studies have been done. Peturson (1944) found that five oat cultivars, susceptible in the seedling stage, differed in resistance in the adult stage. Some were resistant to all races tested.

Others were resistant to some and susceptible or moderately susceptible to others. Simons (1961) found that cultivars having useful sources of general resistance are clearly susceptible to certain races. It now appears unlikely that a source of general resistance will be discovered that will be effective against all forms of *P. coronata.*

c. Manner of Expression. In contrast to specific resistance, general resistance to *P. coronata* is expressed in many different ways. Durrell and Parker (1920) observed that resistance was expressed as a low percentage of infection from a given quantity of spores, and by a long incubation period. Differences among cultivars for receptivity—that is, the relative numbers of uredia resulting from equal amounts of inoculum—can be striking. In Australia, the number of pustules produced on the cultivars Algerian, Garry, and a strain of *A. sterilis* were 124, 76, and 9, respectively (Kochman and Brown, 1975). Luke *et al.* (1981) regarded low receptivity as a major component of slow rusting, and showed that low receptivity was expressed at low but not at higher levels of infection.

Heagle and Moore (1968) compared the highly susceptible Coachman cultivar with Portage, which has some degree of general resistance. Isolates of *P. coronata* virulent on seedlings of both cultivars did not produce epidemics in pure stands of Portage, but did in Coachman. When the infection processes on adult plants were compared, there were fewer penetrations, hyphal growth rates were slower, onset of sporulation was delayed, and fewer spores were produced per pustule in Portage than in Coachman.

Adult plants of cultivars having general resistance exhibit variation in reaction to *P. coronata;* the younger tissues and the younger plants appear most susceptible, and the older tissues and older plants most resistant (Murphy, 1935; Newton and Brown, 1934).

The epidemiological significance of general resistance was shown by Berger and Luke (1979). They compared the cultivars Fulghum and Burt (susceptible) with Red Rustproof-14 (slow-rusting). The average apparent infection rates on Fulghum, Burt, and Rustproof-14 were 0.4, 0.35, and 0.2 units/day, respectively. When progress of the disease was measured in terms of isopath movement, the rates for Fulghum, Burt, and Red Rustproof-14 were 0.9, 0.4, and 0.35 m/day, respectively.

d. Use in Breeding. General resistance is of potential value in the control of *P. coronata,* but the complexity of its inheritance coupled with difficulties associated with evaluating it delay practical application. To determine the feasibility of manipulating it in breeding programs, Simons (1981b) crossed, and backcrossed, 14 lines of *A. sterilis* known to have general resistance with the susceptible cultivar Clinton. Lines derived from F_2, Bc_1, and Bc_2 plants were selected for cultivated plant type. In terms of reduction in grain weight attributable to disease, 71 of the lines derived from F_1 plants, 52 from Bc_1 plants, and 27 from Bc_2 plants were siginificantly more resistant to *P. coronata* than was the parental Clinton. Lines combining the yield of the cultivated Clinton with a statistically significant improvement in resistance appeared in all but one population.

5. *Multilines*

There is considerable interest in the use of multiline cultivars of various crops to achieve diversity and thereby longer lasting genetic protection from various plant diseases. Much of the work has been done with the crown rust disease. Jensen (1952) noted that a risk of serious loss from *P. coronata* had been created because the fungus could move freely from oat field to oat field without being checked by genetic barriers in the form of resistant cultivars. He suggested a form of intravarietal diversity in which a cultivar would consist of a blend of lines having different resistance genes. The component lines could then be changed from year to year to meet changes in the *P. coronata* population. Borlaug (1965) proposed that such composite cultivars be produced by backcross methods. The component lines would be developed by backcrossing a current commercial cultivar to a number of different types of resistance. The resulting lines would be increased separately, and then mixed to form the cultivar. Such a multiline cultivar would be morophologically uniform, but as the rust races change, individual lines could be changed.

Cournoyer (1970) tested the multiline hypothesis experimentally by using large plots planted to oats consisting of various mixtures of resistant near-isogenic lines. Spores were trapped daily adjacent to the plots, and the final cumulative spore count showed that incorporation of resistant plants decreased the number of spores produced. A greater proportion of resistant plants resulted in fewer spores being produced. Since spore production and disease severity are directly correlated, Cournoyer concluded that her results supported the multiline hypothesis that mixtures of near-isogenic lines effectively buffer the oat population against the pathogen population.

The first commercial multiline cultivars were developed in Iowa and released in 1968 (Browning and Frey, 1981). In the years following, a total of eight early and five midseason multiline cultivars were released. They were well received by farmers and were widely grown for several years before declining in popularity due to the scarcity of *P. coronata* during the period, and to the appearance of new cultivars agronomically superior to the multilines. As this is written, a new multiline is in the final stages of development in Iowa. The cultivar Lang, which is high-yielding and resistant to barley yellow dwarf virus, is the recurrent parent.

Politowski and Browning (1978) found that multiline cultivars developed much less disease than susceptible pureline cultivars. There was only slightly more disease on the multiline cultivars than on cultivars that were resistant to the *P. coronata* isolates that were present. The amount of reduction in yield and seed weight attributable to infection corroborated the observations on incidence of disease.

The early multiline studies were done in Iowa, where the disease season is severe but short. Oats are grown over a much longer disease season in the southern United States. A study under long disease seasons compared the performance of multiline and pureline cultivars at two locations on the southern coastal plain of Texas, Browning and Frey (1981). There, *P. coronata* overwinters in the uredial stage on oats, and severe epidemics developed in the large isolated plots that were used. Relative yields of spores from multilines and from susceptible isoline checks were the same in Texas as in the North, showing that multilines will protect oats from *P. coronata* in long as well as short disease seasons. Disease was less severe at one of the locations, but even there, susceptible pureline check cultivars were killed prematurely by the fungus. Both the multiline and an artificial cultivar mixture in the ratio of one resistant to two susceptible developed no more rust than did the fungicide-sprayed control plots.

Indirect evidence substantiating the value of a relatively small proportion of resistance in a multiline comes from a study of mass selection for resistance (Tiyawalee and Frey, 1970). The frequency of resistance genes was increased from 0.21 in the F_3 to about 0.35 in the F_{10}. Most of the increase occurred in the first three cycles of selection, suggesting that damage from *P. coronata* was negligible when about one-third of the plants carried resistance genes.

The mechanism by which multilines are protected from damage is not completely understood, but it is assumed to result from any plant being resistant to some of the spores that land on it. Thus a portion of the inoculum that would contribute to build-up of the epidemic in a pureline cultivar is removed from contention, resulting in slower development of disease. A given plant in the multiline also serves as a mechanical barrier to the spread of spores to which it is resistant. Tani *et al.* (1980) demonstrated a cross-protection effect from inoculation of oat cultivars with spores of incompatible isolates of *P. coronata*. Since plants in a multiline cultivar are regularly exposed to inoculum of incompatible isolates, they may benefit from this cross-protection effect.

6. *Gene Deployment*

The geographical deployment of genes for resistance has been considered as a means of controlling *P. coronata*. Jensen and Kent (1963) noted that fields of oat cultivars susceptible to *P. graminis* were not severely damaged by this pathogen when they were interspersed among more numerous fields of resistant cultivars. They suggested that a series of cultivars representing a diversity of genes for resistance to *P. coronata* be developed and used. These cultivars would be randomly planted in different fields. Most fields in a given area would thus contain plants resistant to the majority of the spores that might fall on them.

Browning *et al.* (1969) noted that with the eradication of buckthorn in the north central United States, *P. coronata* was initiated mainly by wind-borne urediospores from the South. *Puccinia coronata* in the South, in turn, is initiated, they assumed, by spores produced further north during the summer. They suggested that available genes for resistance be deployed so that those genes used in cultivars in the northern states and Canada would differ from those used in the southern states and Mexico. Thus when spores were blown from one region to the other they would encounter only cultivars on which they would be avirulent, and the cycle would be broken. The geographical deployment of resistance genes also figured importantly in a comprehensive management scheme to control *P. coronata* in the United States and Canada that was described by Frey *et al.* (1980).

7. *Pyramiding of Resistance Genes*

The combination of two or more resistance genes in the same cultivar, sometimes referred to as stacking or pyramiding resistance genes, to control *P. coronata* is theoretically attractive. Mutations in the fungus population for virulence toward a resistance gene are rare events, and the occurrence of two simultaneous mutations necessary to overcome the resistance of such a cultivar would be unlikely. An examination of races that have appeared in the past, however, suggests that pyramiding genes for resistance in the same cultivar may not guarantee long-lasting protection (Simons *et al.*, 1957). During the early 1950s resistance genes carried by Landhafer and Santa Fe were widely used in breeding programs. At that time, no race was known that could parasitize either of them. In the mid-1950s races appeared that were virulent on the combination, but no race was ever found that could parasitize only one. Thus, combination of genes for resistance would not have prevented the breakdown of the resistance.

8. *Telia and Resistance*

In areas where the fungus is dependent on the alternate host, telia are obviously essential to survival. In many of the major oat-growing regions of the world, the alternate host has either little or no role in the epidemiology of the disease. The production of telia under such circumstances merely ends the repeating uredial phase of the disease. Thus plant pathologists tend to regard the formation of telia as an indication of cultivar resistance.

Parker (1918) believed that telia were a sign of resistance as they did not form on highly susceptible cultivars. In Murphy's (1935) more detailed observations, the early development of telia was positively correlated with host resistance. Races showing restricted virulence generally develop telia more readily.

Zimmer and Schafer (1960), however, found that rapidity of telia formation was not correlated with range of virulence, specific virulence of the pathogen, or with maturity of the host. They believed telia formation was a manifestation of a specific relationship between a cultivar and an isolate of the fungus. They noted that some cultures did not produce telia under any of the experimental conditions they worked with, and suggested that some isolates of the rust may have lost the capacity to produce telia under any conditions.

9. *Tolerance*

Tolerance to *P. coronata* was defined by Caldwell *et al.* (1958) as that capacity of a susceptible plant that enabled it to endure severe attack without sustaining severe losses in yield or quality. A clear distinction was made between tolerance and intermediate or lesser degrees of resistance. Theoretically, tolerance should be more stable than resistance. If a pathogenic race of the fungus arises to which the previously tolerant cultivar was not tolerant, the result would be greater injury to the host, but with no consequent increase in the relative rate of increase of the new race. This eliminates the screening mechanism whereby the new race gains an advantage over the established races. Caldwell *et al.* (1958) used two pairs of oat cultivars (Clinton-59 and

Clintland, and Benton and Bentland) identical except that Clintland and Bentland had a resistance gene. In the severe epidemic of 1957, infection reached 100% on the susceptible Benton and Clinton-59, while only a trace of *P. coronata* developed on Bentland and Clintland. Under these conditions Benton yielded 14% less than Bentland, and Clinton-59 yielded 54% less than Clintland. Since both Benton and Clinton-59 were equally infected, the difference between the 14 and 54% losses reflected the greater relative tolerance of Benton.

Simons (1966a) discussed the difficulties involved in the use of tolerance in breeding programs and concluded that the reduction in seed weight attributable to *P. coronata* was the best measure of tolerance, with reduction in yield a useful adjunct. To separate inherent differences in seed weight and yield from differences due to tolerance, field tests were planted in a split-plot design in which half the plot was infected and the other half protected by a fungicide. Data were expressed as ratios of diseased to nondiseased plot pairs. This system, even with hill plots, was effective for measuring small differences in tolerance among unrelated cultivars, and it was equally useful for measuring intermediate or lesser degrees of resistance that are difficult to evaluate visually.

Wahl (1958) observed that although *A. sterilis* was widespread and abundant in Israel and was often heavily infected by *P. coronata,* it did not seem to suffer from the disease. Simons (1972) crossed three strains of *A. sterilis* that were susceptible with the highly susceptible cultivars Richland and Clinton. Cultivated-type, seemingly rust-susceptible segregates selected from the resulting populations were evaluated for tolerance. The amount of tolerance transmitted by the different *A. sterilis* parents varied significantly as estimated by the mean tolerance values of the lines derived from each parent. Most lines derived were more tolerant than the cultivated parent. Heritability of tolerance was estimated to be 76%, and most lines were lower in yield, under rust-free conditions, than the cultivated parents.

Politowski and Browning (1978) further refined the concept of tolerance and showed that dilatory resistance—i.e., resistance that delays the progress of an epidemic—may not be visually apparent. To differentiate between dilatory resistance and tolerance, they compared final cumulative spore counts of *P. coronata* from large plots with host yield and seed weight data from hill plots. Yield and seed weight of Otter and Cherokee were depressed by about the same amount. Otter, however, produced more spores than did Cherokee, and thus was regarded as more tolerant. Both Otter and Cherokee had a degree of dilatory resistance when compared with the susceptible check, which had higher spore counts and whose yield and kernel weight were reduced more than were those of Otter or Cherokee. Singleton *et al.* (1982) found that the yields of some moderately resistant cultivars equalled or exceeded the yields of highly resistant cultivars under severe epidemic conditions, even though the moderately resistant cultivars were more heavily infected.

Genetic studies (Simons, 1969) showed that tolerance was inherited as a complex quantitative trait in two crosses. Heritability of yield and seed weight response to infection was sufficiently high to permit effective selection in breeding programs.

10. *Mutation*

All host genes for resistance to *P. coronata* presumably originated as mutations. Rosen (1955) found a single disease-free plant in a field of oats that was otherwise heavily infected with *P. coronata.* Circumstances and the characteristics of the resistance of this plant were such that it appears mutation is its likely source.

Most of what is known about mutation in oats for resistance to *P. cornoata* was discovered as a result of applying artificial mutagenic agents to susceptible oat cultivars. Frey (1954) X-rayed the susceptible cultivar Huron and obtained a few strains with increased field resistance to *P. coronata.* Luke *et al.* (1958) radiated Floriland with thermal neutrons and isolated lines with field resistance to the virulent race 264. These lines were used in developing the cultivars Florad and Florida 500 (Sechler and Chapman, 1965).

Atkins *et al.* (1964) used both neutrons and X-rays in varying dosages to induce mutations for resistance to *P. coronata,* and one cultivar, Alamo-X, resulted.

Ethyl methane sulfonate (EMS) was used to induce tolerance in susceptible cultivars by Simons (1971). A few lines were obtained that were superior to the control in both yield and seed weight response to infection, but in the absence of *P. coronata* these lines yielded less than the control. Seeds derived from EMS-treated seed of Clintland-60 were retreated with EMS (Simons, 1981a). Lines from this treatment were retested for tolerance. Data suggested that tolerance could be obtained with high yield.

B. ERADICATION OF *RHAMNUS*

The role of the alternate host in the initiation of infection by *P. coronata* on oats early in the spring and its role in the origin of new races through genetic recombination are well known. Thus, eradication of *Rhamnus* is a measure to help control the disease. Historically, an early record of such a specific recommendation was published by von Thumen (1886). He believed that eradication of *Rhamnus* would be an effective but incomplete control measure. Sivers (1887) stated that eradication of *Rhamnus* probably would not be carried out, and indeed in some areas would not even be feasible.

Dietz (1930) surveyed species of *Rhamnus* in the north central United States. He estimated that there were several hundred thousand bushes growing in the upper Mississippi valley, and obviously, the simple physical problem of locating and destroying them would be expensive. A limited goal of eradicating buckthorn hedges adjacent to fields in which oats might be planted is often economically feasible. Several states of the United States and a number of other countries have enacted legislation aimed at promoting eradication of *Rhamnus*. Exact figures are not available, but the combination of legislation and educational programs to make farmers aware of the benefits of eradication has resulted in a great reduction in *Rhamnus* in areas such as Iowa.

C. ERADICATION OF OTHER GRAMINEOUS HOSTS

The susceptibility of wild and volunteer oats and certain grass species to strains of *P. coronata* that infect cultivated oats make them a frequent source of inoculum. Straib (1937a) believed that such infected plants were reservoirs of infection and acted as the primary source of inoculum in certain situations in Germany, and there are other areas in the world where eradication of such plants would be beneficial.

Grasses that are susceptible to *P. coronata* are rarely important in the initiation of *P. coronata* on oats, and therefore eradication of them normally would not be recommended.

D. DATES OF PLANTING AND MATURITY

Early-planted oats often ripen before the fungus has time to affect them seriously. Sivers (1887) was clearly aware of the value of early seeding, and states that early seeding was the best means available of avoiding damage from *P. coronata*. Simons (1966b) showed that the effect of *P. coronata* on yield and grain quality increased significantly as planting data was delayed. This effect was most pronounced with highly susceptible oat cultivars.

Fleischmann and McKenzie (1965) found that under conditions of artificially initiated crown rust, early- and late-seeded plantings of the commercially important cultivar Garry suffered 28 and 50% yield losses, respectively.

Early-maturing oat cultivars tend to be less damaged by *P. coronata* than do later cultivars. This is probably because the early cultivar ripens before the disease has had time to severely damage it. This correlation of relative maturity to damage caused by *P. coronata* was demonstrated by Simons and Michel (1968).

E. FERTILIZATION

Pammel (1892) noted that where oats were rank and thick, rust was severe, but where thin the rust was less severe. Gassner and Hassebrauk (1934) showed that nitrogen tends to favor crown rust whereas potassium reduces it. At present, the feeling is that any kind of a fertilization program resulting in vigorously growing high-yielding oats will permit severe damage to susceptible cultivars when other conditions favor the disease.

F. FUNGICIDES

Puccinia coronata on oats has been used as the experimental system in several studies of the potential usefulness of modern fungicides (Chin *et al.*, 1975; Prusky *et al.*, 1981). Currently, the relatively low acre value of oats, coupled with the infrequent occurrence of severe damage from *P. coronata,* makes the use of fungicides on commercial fields generally impractical. Fungicides may be used effectively in special situations such as experimental work and production of seed.

References

Arthur, J. C. (1924). The uredinales (rusts) of Iowa. *Proc. Iowa Acad. Sci.* **31,** 229–255.

Atkins, I. M., and McFadden, E. S. (1947). Oat production in Texas. *Tex., Agric. Exp. Stn. [Bull.]* **691,** 1–66.

Atkins, I. M., Futrell, M. C., Raab, Q. J., Lyles, W. E., Pawlisch, P. E., Rivers, G. W., and Norris, M. J. (1964). Studies of the progenies of irradiated oats. *Tex., Agric. Exp. Stn. [Misc. Publ.] MP* **MP-742,** *1–12.*

Azbukina, Z. M. (1976). Further investigations about taxonomy of certain species of rust fungi on cereals. Proc.—Euro. Mediterr. Cereal Rusts Conf. 4th, 1976, pp. 69–71.

Bartoš, P., Fleischmann, G., Green, G. J., and Samborski, D. J. (1967). Nuclear reassociation in *Puccinia coronata* f. sp. *avenae. Phytopathology* **57,** 803 (abstr.).

Berger, R. D., and Luke, H. H. (1979). Spatial and temporal spread of oat crown rust. *Phytopathology* **69,** 1199–1201.

Biali, M., and Dinoor, A. (1972). Genetics of virulence in *Puccinia coronata. Proc.—Eur. Mediterr. Cereal Rusts Conf., 3rd, 1972,* pp. 103–108.

Borlaug, N. E. (1965). Wheat, rust, and people. *Phytopathology* **55,** 1088–1098.

Brinkman, M. A., Frey, K. J., and Browning, J. A. (1978). Influence of an extra pair of fragment chromosomes on grain yield in a hexaploid oat. *Crop Sci.* **18,** 147–148.

Brodny, U., Rotem, J., and Wahl, I. (1979). Competition among physiological races of *Puccinia coronata avenae* on oat. *Phytoparasitica* **7,** 59 (abstr.).

Browder, L. E. (1969). Pathogenic specialization in cereal rust fungi, especially *Puccinia recondita* f. sp. *tritici:* Concepts, methods of study, and application of knowledge. *U.S., Dept. Agric., Tech. Bull.* **1432,** 1–51.

Browning, J. A., and Frey, K. J. (1981). The multiline concept in theory and practice. *In* "Strategies for the Control of Cereal Diseases" (J. F. Jenkyn and R. T. Plumb, eds.), pp. 37–46. Blackwell, Oxford.

Browning, J. A., Simons, M. D., Frey, K. J., and Murphy, H. C. (1969). Regional deployment for conservation of oat crown rust resistance genes. *Iowa, Agric. Exp. Stn., Spec. Rep.* **64,** 49–56.

Caldwell, R. M., Schafer, J. F., Compton, L. E., and Patterson, F. L. (1958). Tolerance to cereal leaf rusts. *Science* **128,** 714–715.

Castagne, L. (1845). "Catalogue des plants qui croissent naturellement aux environs de Marseille," pp. 202–204. Imprimerie de Nicot et Pardigon, Pont-Moreau, Aix.

Chang, T. D., and Sadanaga, K. (1964). Crosses of six monosomics in *Avena sativa* L. with varieties, species, and chlorophyll mutants. *Crop Sci.* **4,** 589–593.

Chin, M. Y., Edgington, L. V., Bruin, G. C. A., and Reinbergs, E. (1975). Influence of formulation on efficacy of three systemic fungicides for control of oat leaf rust. *Can. J. Plant Sci.* **55,** 911–917.

Coffman, F. A., Murphy, H. C., and Chapman, W. H. (1961). Oat breeding. *In* "Oats and Oat Improvement" (F. A. Coffman, ed.), pp. 263–329. Am. Soc. Agron., Madison, Wisconsin.

Corda, A. C. J. (1837). Icones Fungorum Hucusque Cognitorum. 1:6. Abbdildungen der Pilze und Schwaemme. Apud J. G. Calve, Prague.

Cornu, M. M. (1880). Notes sur quelques parasites des plantes vivantes: Generations alternantes; Pezizes a sclerotes. *Bull. Soc. Bot. Fr.* **27,** 209–210.

Cournoyer, B. M. (1970). Crown rust epiphytology with emphasis on the quantity and periodicity of spore dispersal from heterogeneous oat cultivar-rust race populations. Ph.D. Thesis, Iowa State University Library, Ames.

Craigie, J. H. (1928). On the occurrence of pycnia and aecia in certain rust fungi. *Phytopathology* **18,** 1005–1015.

Cummins, G. B. (1956). Host index and morphological characterization of the grass rusts of the world. *Plant Dis. Rep., Suppl.* **237,** 1–52.

Cunningham, J. L. (1964). Variation in the crown rusts. *Phytopathology* **54,** 891 (abstr.).

Davies, D. W., and Jones, E. T. (1927). Further studies on the inheritance of resistance to crown rust (*P. coronata,* Corda) in F_3 segregates of a cross between Red Rustproof (*A. sterilis*) and Scotch Potato oats (*A. sativa*). *Welsh J. Agric.* **3,** 232–235.

DeBary, A. (1867). Neue Untersuchungen uber Uredineen. *Monatsber. K. Preuss. Akad. Wiss. Berlin, 1866,* pp. 205–216.

Dherawattana, A., and Sadanaga, K. (1973). Cytogenetics of a crown rust-resistant hexaploid oat with 42 + 2 fragment chromosomes. *Crop Sci.* **13,** 591–594.
Dietz, S. M. (1923). The role of the genus *Rhamnus* in the dissemination of crown rust. *U.S., Dep. Agric., Bull. 1162,* 1–18, pp.
Dietz, S. M. (1926). The alternate hosts of crown rust, *Puccinia coronata* Corda. *J. Agric. Res. (Washington, D.C.)* **33,** 953–970.
Dietz, S. M., and Leach, L. D. (1930). Methods of eradicating buckthorn (*Rhamnus*) susceptible to crown rust (*Puccinia coronata*) of oats. *U.S., Dep. Agric., Circ.* **133,** 1–15.
Dinoor, A., and Wahl, I. (1963). Reaction of non-cultivated oats from Israel to Canadian races of crown rust and stem rust. *Can. J. Plant Sci.* **43,** 263–270.
Dinoor, A., Khair, J., and Fleischmann, G. (1968). Pathogenic variability and the unit representing a single fertilization in *Puccinia coronata* var. *avenae. Can. J. Bot.* **46,** 501–508.
d'Oliveira, B. (1942). A estacao agronomica e os problemas nacionais de fitopatologia. *Rev. Agron.* **30,** 414–438.
Durrell, L. W., and Parker, J. H. (1920). Comparative resistance of varieties of oats to crown and stem rusts. *Iowa, Agric. Exp. Stn., Res. Bull.* **62,** 25–56d.
Eagles, H. A., and Taylor, A. O. (1976). Forage oat varieties for the North Island with emphasis on disease resistance. *Proc. Agron. Soc. N. Z.* **6,** 31–35.
Endo, R. M., and Boewe, G. H. (1958). Losses caused by crown rust of oats in 1956 and 1957. *Plant Dis. Rep.* **42,** 1126–1128.
Eriksson, J. (1894). Ueber die Specialisirung des Parasitismus bei den Getreiderostpilzen. *Ber. Dtsch. Bot. Ges.* **12,** 292–331.
Eriksson, J. (1908). Neue Studien uber die Spezielisierung der grasbewohnenden Kronenrostarten. *Ark. Bot.* **8,** No. 3, 1–26.
Eriksson, J., and Henning, E. (1894). Die Hauptresultate einer neuen Untersuchung uber die Getreideroste. IV. *Z. Pflanzenkr.* **4,** 257–262.
Eshed, N., and Dinoor, A. (1976). Widening the host range and exposure of new virulence genes by intervarietal crosses in crown rust. *Proc.–Eur. Mediterr. Cereal Rusts Conf., 4th, 1976,* pp. 30–31.
Eshed, N., and Dinoor, A. (1980). Genetics of pathogenicity in *Puccinia coronata:* Pathogenic specialization at the host genus level. *Phytopathology* **70,** 1042–1046.
Finkner, V. C. (1954). Genetic factors governing resistance and susceptibility of oats to *Puccinia coronata* Corda var. *avenae,* F. and L. race 57. *Iowa, Agric. Exp. Stn., Bull.* **411,** 1040–1063.
Fleischmann, G. (1963a). Crown rust of oats in Canada in 1963. *Can. Plant Dis. Surv.* **43,** 168–172.
Fleischmann, G. (1963b). The origin of a new physiologic race of crown rust virulent on the oat varieties Victoria and Saia. *Can. J. Bot.* **41,** 1613–1615.
Fleischmann, G. (1965). Variability in the physiologic race populations of oat crown rust isolated from aecia and uredia. *Plant Dis. Rep.* **49,** 132–133.
Fleischmann, G. (1967). Virulence of uredial and aecial isolates of *Puccinia coronata* Corda f. sp. *avenae* identified in Canada from 1952 to 1966. *Can. J. Bot.* **45,** 1693–1701.
Fleischmann, G., and McKenzie, R. I. H. (1965). Yield losses in Garry oats infected with crown rust. *Phytopathology* **55,** 767–770.
Fleischmann, G., Samborski, D. J., and Peturson, B. (1963). The distribution and frequency of occurrence of physiologic races of *Puccinia coronata* Corda f. sp. *avenae* Erikss. in Canada, 1952 to 1961. *Can. J. Bot.* **41,** 481–487.

Flor, H. H. (1956). The complementary genic systems in flax and flax rust. *Adv. Genet.* **8,** 29–54.

Forbes, I. L. (1939). Factors affecting the development of *Puccinia coronata* in Louisiana. *Phytopathology* **29,** 659–684.

Fraser, W. P., and Ledingham, G. A. (1933). Studies of the crown rust, *Puccinia coronata* Corda. *Sci. Agric. (Ottawa)* **13,** 313–323.

Frey, K. J. (1954). Artificially induced mutations in oats. *Agron. J.* **46,** 49.

Frey, K. J., and Browning, J. A. (1971). Association between genetic factors for crown rust resistance and yield in oats. *Crop. Sci.* **11,** 757–760.

Frey, K. J., Browning, J. A., and Simons, M. D. (1980). Management systems for host genes to control disease loss. *Indian J. Genet. Plant Breed.* **39,** 10–21.

Gassner, G. (1916). Die Getreideroste und ihr Auftreten im subtropischen ostlichen Sudamerika. *Zentralbl. Bakteriol., Parasitenkd., Infektionskr. Hyg., Abt. 2* **44,** 305–381.

Gassner, G., and Hassebrauk, K. (1934). Zweijährige Feldversuche über den Einfluss der Düngung auf die Rostanfälligkeit von Getreidepflanzen. *Phytopathol. Z.* **7,** 53–61.

Gmelin, J. F. (1791). "Caroli a Linne, Systema naturae," Tomus 2, p. 1472. Lipsiae.

Harder, D. E. (1980). Virulence and distribution of *Puccinia coronata avenae* in Canada in 1979. *Can. J. Plant Pathol.* **2,** 249–252.

Heagle, A. S., and Moore, M. G. (1968). Effect of moderate adult resistance on penetration, hyphal growth rates, and urediospore production by crown rust of oats. *Phytopathology* **58,** 1053 (abstr.).

Hermansen, J. E. (1961). Studies on cereal rusts in Denmark. I. Physiological races in 1958–1959. *Arsskr.—K. Vet.- Landbohoe.jsk. (Copenhagen),* pp. 99–105.

Hoerner, G. R. (1919). Biologic forms of *Puccinia coronata* on oats. *Phytopathology* **9,** 309–314.

Immer, F. R., and Stevenson, F. J. (1928). A biometrical study of factors affecting yield in oats. *Agron. J.* **20,** 1108–1119.

Jensen, N. F. (1952). Intra-varietal diversification in oat breeding. *Agron. J.* **44,** 30–34.

Jensen, N. F., and Kent, G. C. (1963). New approach to an old problem in oat production. *Cornell Univ. Farm Res.* **29** (2), 4–5.

Kiehn, F. A., McKenzie, R. I. H., and Harder, D. E. (1976). Inheritance of resistance to *Puccinia coronata avenae* and its association with seed characteristics in four accessions of *Avena sterilis. Can. J. Genet. Cytol.* **18,** 717–726.

Klebahn, H. (1892). Kulturversuche mit heterocischen Uredineen. *Z. Pflanzenkr. Gallenkd.* **2,** 332–343.

Klebahn, H. (1895). Kulturversuche mit heterocischen Rostpilzen. III. *Z. Pflanzenkr. Gallenkd.* **5,** 149–156.

Kochman, J. K., and Brown, J. F. (1975). Host and environmental effects on post-penetration development of *Puccinia graminis avenae* and *P. coronata avenae. Ann. Appl. Biol.* **81,** 33–41.

Kostic, B. (1959). The cereal rusts in the south-eastern part of Yugoslavia in 1958 and 1959. *Robigo* **9,** 8–12.

Kostic, B. (1965). Fizioloski rase *Puccinia coronata* Cda. var. *avenae* Fraser et Led. I stepen osetljivosti nekih sortata i hibrida Ovsa prema njima. *Zast. Bilja* **83,** 99–108.

Litzenberger, S. C. (1949). Inheritance of resistance to specific races of crown and stem rust, to Helminthosporium blight, and of certain agronomic characters in oats. *Iowa, Agric. Exp. Stn., Res. Bull.* **370,** 453–496.

Luke, H. H., Chapman, W. H., Wallace, A. T., and Pfahler, P. L. (1958). Oat selections resistant to certain Landhafer-attacking races of crown rust. *Plant Dis. Rep.* **42,** 1250–1253.

Luke, H. H., Barnett, R. D., and Pfahler, P. L. (1975). Inheritance of horizontal resistance to crown rust in oats. *Phytopathology* **65,** 631–632.

Luke, H. H., Pfahler, P. L., and Barnett, R. D. (1981). Influence of disease severity and environmental conditions on low receptivity of oats to crown rust. *Plant Dis.* **65,** 125–127.

McKenzie, R. I. H., and Fleischmann, G. (1964). The inheritance of crown rust resistance in selections from two Israeli collections of *Avena sterilis. Can. J. Genet. Cytol.* **6,** 232–236.

Marshall, H. G., and Myers, W. M. (1961). A cytogenetic study of certain interspecific *Avena* hybrids and the inheritance of resistance in diploid and tetraploid varieties to races of crown rust. *Crop Sci.* **1,** 29–34.

Martens, J. W., Fleischmann, G., and McKenzie, R. I. H. (1972). Effects of natural infections of crown rust and stem rust on yield and quality of oats in Manitoba. *Can. Plant Dis. Surv.* **52,** 122–125.

Martens, J. W., Burnett, P. A., and Wright, G. M. (1977). Virulence in *Puccinia coronata* f. sp. *avenae* and *P. graminis* f. sp. *avenae* in New Zealand. *Phytopathology* **67,** 1519–1521.

Martens, J. W., McKenzie, R. I. H., and Harder, D. E. (1980). Resistance to *Puccinia graminis avenae* and *P. coronata avenae* in the wild and cultivated *Avena* populations of Iran, Iraq, and Turkey. *Can. J. Genet. Cytol.* **22,** 641–649.

Melhus, I. E., Dietz, S. M., and Willey, F. (1922). Alternate hosts and biologic specialization of crown rust in America. *Iowa, Agric. Exp. Stn., Res. Bull.* **72,** 211–236.

Melhus, I. E., Sheperd, D. R., and Corkle, M. A. (1942). Diseases of cereals and flax in Iowa. *Proc. Iowa Acad. Sci.* **49,** 217–247.

Michel, L. J., and Simons, M. D. (1966). Pathogenicity of isolates of oat crown rust collected in the USA, 1961–65. *Plant Dis. Rep.* **50,** 935–938.

Michel, L. J., and Simons, M. D. (1977). Aggressiveness and virulence of *Puccinia coronata avenae* isolates, 1971–1975. *Plant Dis. Rep.* **61,** 621–625.

Miles, L. G., and Rosser, D. (1954). Bovah—A new grazing oat for Queensland. *Queensl. Agric. J.* **78,** 311–315.

Muhlethaler, von F. (1911). Infektionsversuche mit *Rhamnus* befallenden Kroenrosten. *Zentralbl. Bakteriol., Parasitenkd., Infektionskr. Hyg., Abt. 2* **30,** 386–419.

Murphy, H. C. (1935). Physiologic specialization in *Puccinia coronata avenae. U.S., Dep. Agric., Tech. Bull.* **433** 1–48.

Murphy, H. C. (1939). Effect of crown and stem rusts on the relative cold resistance of varieties and selections of oats. *Phytopathology* **29,** 763–782.

Murphy, H. C. (1946). A study of the parasitism of the Rusts, Smuts, and other diseases affecting oats. *Iowa, Agric. Exp. Stn., Annu. Rep.* 1946. pp. 167–170.

Murphy, H. C. (1952). Problems involved in breeding oats for disease resistance. *Phytopathology* **42,** 482–483 (abstr.).

Murphy, H. C. (1965). Protection of oats and other cereal crops during production. *In* "Food Quality: Effects of Production Practices and Processing" (G. W. Irving, ed.), Publ. No. 77, pp. 99–113. Am. Assoc. Adv. Sci., Washington, D.C.

Murphy, H. C., and Levine, M. N. (1936). A race of crown rust to which the Victoria oat variety is susceptible. *Phytopathology* **26,** 1087–1089.

Murphy, H. C., and Meehan, F. (1946). Reaction of oat varieties to a new species of *Helminthosporium. Phytopathology* **36,** 407 (abstr.).

Murphy, H. C., Stanton, T. R., and Stevens, H. (1937). Breeding winter oats resistant to crown rust, smut, and cold. *Agron. J.* **29,** 622–637.

Murphy, H. C., Burnett, L. C., Kingsolver, C. H., Stanton, T. R., and Coffman, F. A.

(1940). Relation of crown-rust infection to yield, test weight, and lodging of oats. *Phytopathology* **30,** 808–819.
Murphy, H. C., Stanton, T. R., and Coffman, F. A. (1942). Breeding for disease resistance in oats. *Agron. J.* **34,** 72–89.
Newton, M., and Brown, A. M. (1934). Studies on the nature of disease resistance in cereals. I. The reactions to rust of mature and immature tissues. *Can. J. Res.* **11,** 564–581.
Nielsen, P. (1875). De for Landbruget farligste Rustarter og Midlerne imod dem. *Ugeskr. Landmaend.* **4,** 549–556.
Nof, E., and Dinoor, A. (1981). The manifestation of gene-for-gene relationships in oats and crown rust. *Phytoparasitica* **9,** 240 (abstr.).
Norton, J. B. (1907). Notes on breeding oats. *Am. Breeders' Assoc. Rep.* **3,** 280–285.
Osler, R. D., and Hayes, H. K. (1953). Inheritance studies in oats with particular reference to the Santa Fe type of crown rust resistance. *Agron. J.* **45,** 49–53.
Pammel, L. H. (1892). Some diseases of plants common to Iowa cereals. *Iowa, Agric. Exp. Stn., Bull.* **18,** 488–505.
Parker, J. H. (1918). Greenhouse experiments on the rust resistance of oat varieties. *U.S., Dep. Agric., Bull.* **629,** 1–16.
Parker, J. H. (1920). A preliminary study of the inheritance of rust resistance in oats. *Agron. J.* **12,** 23–38.
Peck, C. H. (1872). Report of the Botanist. *N.Y. State Mus. Nat. Hist., Annu. Rep.* **24,** 41–108.
Pessil, P. (1960). Racas fisiologicas de *Puccinia coronata avenae,* identificadas no periodo de 1952 a 1958 no Rio Grande do Sul. *Rev. Esc. Agron. Vet., Univ. Rio Grande do Sul* **3,** 53–56.
Peturson, B. (1935). Physiologic specialization in *Puccinia coronata avenae. Sci. Agric. (Ottawa)* **15,** 806–810.
Peturson, B. (1944). Adult plant resistance of some oat varieties to physiologic races of crown rust. *Can. J. Res., Sect. C* **22,** 287–289.
Peturson, B. (1951). Recent changes in the relative prevalence of physiologic races of crown rust in Canada. *Phytopathology* **41,** 29 (abstr.).
Poehlman, J. M., and Kingsolver, C. H. (1950). Disease reaction and agronomic qualities of oat selections from a Columbia × Victoria-Richland cross. *Agron. J.* **42,** 498–502.
Politowski, K., and Browning, J. A. (1978). Tolerance and resistance to plant disease: An epidemiological study. *Phytopathology* **68,** 1177–1185.
Prusky, D., Dinoor, A., and Jacoby, B. (1981). The fungicide or heat induced hypersensitive reaction of oats to crown rust: relations between various treatments and infection type. *Physiol. Plant Pathol.* **18,** 181–186.
Reyes, L., and Futrell, M. C. (1962). Loss in oats grown for forage in Texas due to occurrence of new races of crown rust. *Plant Dis. Rep.* **46,** 835–837.
Rosen, H. R. (1955). New germ plasm for combined resistance to *Helminthosporium* blight and crown rust of oats. *Phytopathology* **45,** 219–221.
Rosen, H. R., Weetman, L. M., and McClelland, C. K. (1942). Winter injury as related to fall and winter growth and crown rust infection in oat varieties and their hybrids. *Bull.—Arkansas, Agric. Exp. Stn.* **418,** 1–17.
Saari, E. E., and Moore, M. B. (1963). Measuring the pathogenic potential of *Puccinia coronata* var. *avenae. Phytopathology* **53,** 887 (abstr.).
Sadanaga, K., and Simons, M. D. (1960). Transfer of crown rust resistance of diploid and tetraploid species to hexaploid oats. *Agron. J.* **52,** 285–288.
Sawer, E. R. (1909). The cereals in South Africa. *Cedara Mem. S. Afr. Agric.* **1,** 264–272.

Sebesta, J. (1974). The effect of stem rust and crown rust on feeding value of oats. *Ved. Pr. Vysk. Ustavu Rastl. Vyroby Praze-Ruzyni* **18,** 87–91.
Sebesta, J. (1979). Complete or incomplete dominance of resistance of the oat cultivar Delphin as a function of crown rust culture. *Euphytica* **28,** 807–809.
Sechler, D., and Chapman, W. H. (1965). Florida 500 oats. *Circ.—Fla., Agric. Exp. Stn.* **S-166,** 1–12.
Sharma, D. C., and Forsberg, R. A. (1977). Spontaneous and induced interspecific gene transfer for crown rust resistance in *Avena. Crop Sci.* **17,** 855–860.
Sherf, A. F. (1954). The 1953 crown and stem rust epidemic of oats in Iowa. *Proc. Iowa Acad. Sci.* **61,** 161–169.
Simons, M. D. (1955). An examination of the present status and proposed modifications of the annual crown rust race survey in the United States. *Plant Dis. Rep.* **39,** 956–959.
Simons, M. D. (1959). Variability among strains of noncultivated species of *Avena* for reaction to races of the crown rust fungus. *Phytopathology* **49,** 598–601.
Simons, M. D. (1961). Testing oats in the field with specific races of crown rust. *Proc. Iowa Acad. Sci.* **68,** 119–123.
Simons, M. D. (1966a). Relative tolerance of oat varieties to the crown rust fungus. *Phytopathology* **56,** 36–40.
Simons, M. D. (1966b). Relationship of date of planting of oats to crown rust damage. *Phytopathology* **56,** 41–45.
Simons, M. D. (1969). Heritability of crown rust tolerance in oats. *Phytopathology* **59,** 1329–1333.
Simons, M. D. (1970). "Crown Rust of Oats and Grasses," Monogr. No. 5. Am. Phytopatol. Soc., Worcester, Massachusetts.
Simons, M. D. (1971). Modification of tolerance of oats to crown rust by mutation induced with ethyl methanesulfonate. *Phytopathology* **61,** 1064–1067.
Simons, M. D. (1972). Crown rust tolerance of *Avena sativa*-type oats derived from wild *Avena sterilis. Phytopathology* **62,** 1444–1446.
Simons, M. D. (1975). Heritability of field resistance to the oat crown rust fungus. *Phytopathology* **65,** 324–328.
Simons, M. D. (1979). Influence of genes for resistance to *Puccinia coronata* from *Avena sterilis* on yield and rust reaction of cultivated oats. *Phytopathology* **69,** 450–452.
Simons, M. D. (1980). Effect of *Puccinia coronata* on straw yield and harvest index of oats. *Phytopathology* **70,** 604–607.
Simons, M. D. (1981a). Effect of recurrent chemical mutagen treatment on tolerance of oats to crown rust *Puccinia coronata* Cda. *Prot. Ecol.* **3,** 131–140.
Simons, M. D. (1981b). Transfer of crown rust field resistance from *Avena sterilis* to cultivated oats by backcrossing. *Phytopathology* **71,** 904 (abstr.).
Simons, M. D., and Michel, L. J. (1958). Physiologic races of crown rust of oats identified in 1957. *Plant Dis. Rep.* **42,** 1246–1249.
Simons, M. D., and Michel, L. J. (1959). A comparison of different methods used in conducting surveys of races of the crown rust fungus. *Plant Dis. Rep.* **43,** 464–469.
Simons, M. D., and Michel, L. J. (1968). Oat maturity and crown rust response. *Crop Sci.* **8,** 254–256.
Simons, M. D., and Murphy, H. C. (1954). Inheritance of resistance to two races of *Puccinia coronata* Cda. var. *avenae* Fraser and Led. *Proc. Iowa Acad. Sci.* **61,** 170–176.
Simons, M. D., and Murphy, H. C. (1955). A comparison of certain combinations of oat varieties as crown rust differentials. *U.S., Dep. Agric., Tech. Bull.* **1112,** 1–22.
Simons, M. D., Luke, H. H., Chapman, W. H., Murphy, H. C., Wallace, A. T., and Frey, K.

J. (1957). Further observations on races of crown rust attacking the oat varieties Landhafer and Santa Fe. *Plant Dis. Rep.* **41,** 964–969.

Simons, M. D., Sadanaga, K., and Murphy, H. C. (1959). Inheritance of resistance of strains of diploid and tetraploid species of oats to races of the crown rust fungus. *Phytopathology* **49,** 257–259.

Simons, M. D., Martens, J. W., McKenzie, R. I. H., Nishiyama, I., Sadanaga, K., Sebesta, J., and Thomas, H. (1978). Oats: A standardized system of nomenclature for genes and chromosomes and catalog of genes governing characters. *U.S., Dep. Agric., Agric. Handb.* **509,** 1–40.

Simons, M. D., Rothman, P. G., and Michel, L. J. (1979a). Pathogenicity of *Puccinia coronata* from buckthorn and from oats adjacent to and distant from buckthorn. *Phytopathology* **69,** 156–158.

Simons, M. D., Youngs, V. L., Booth, G. D., and Forsberg, R. A. (1979b). Effect of crown rust on protein and groat percentages of oat grain. *Crop Sci.* **19,** 703–706.

Singleton, L. L., Stuthman, D. D., and Moore, M. B. (1979). Effect of crown rust on oat groat protein. *Phytopathology* **69,** 776–778.

Singleton, L. L., Moore, M. B., Wilcoxson, R. D., and Kernkamp, M. F. (1982). Evaluation of oat crown rust disease parameters and yield in moderately resistant cultivars. *Phytopathology* **72,** 538–540.

Sivers, M. N. (1887). Ein Probeanbau verschiedener Hafersorten. *Balt. Wochenschr. Landwirt. Sch. Gewerbefl. Handel, Dorpat* No. 39; pp. 390–391; *Bot. Jahresb.* **15** (Abt. 2); 325.

Smith, D. C. (1934). Correlated inheritance in oats of reaction to diseases and other characters. *Minne., Agric. Exp. Stn., Tech. Bull.* **102,** 1–38.

Stakman, E. C. (1929). Physiological specialization in plant pathogenic fungi. *Leopoldina* **4,** 263–289.

Straib, W. (1937a). Flughafer als Zwischentrager des Kronenrostes. *Nachrichtenbl. Dtsch. Pflanzenschutzdienstes* **17,** 89; *(Braunschweig) Rev. Appl. Mycol.* **17,** 234 (abstr.).

Straib, W. (1937b). Die Bestimmung der physiologischen Rassen von *Puccinia coronata* Cda. auf Hafer in Deutschland. *Arb. Biol. Reischanst. Land- Forstwirtsch., Berlin-Dahlem* **22,** 121–157.

Tani, T., Yamashita, Y., and Yamamoto, H. (1980). Initiation of induced nonhost resistance of oat leaves to rust infection. *Phytopathology* **70,** 39–42.

Thaxter, R. (1890). Report of mycologist. Miscellaneous notes on fungus diseases. *Conn., Agric. Exp. Stn., Annu. Rep., 1889,* pp. 161–164.

Tiyawalee, D., and Frey, K. J. (1970). Mass selection for crown rust resistance in an oat population. *Iowa State J. Sci.* **45,** 217–231.

Tozzetti, G. T. (1952). "True Nature, Causes and Sad Effects of the Rust, the Bunt, the Smut, and other Maladies of Wheat, and of Oats in the Field" (L. R. Tehon, transl.), Phytopathol. Classic, No. 9, Am. Phytopathol. Soc., Washington, D. C. (originally published, 1767).

Upadhyaya, Y. M., and Baker, E. P. (1965). Studies on the inheritance of rust resistance in oats. III. Genetic diversity in the varieties Landhafer, Santa Fe, Mutica Ukraine, Trispernia, and Victoria for crown rust resistance. *Proc. Linn. Soc. N.S.W.* **90,** 129–151.

Vallega, J. (1951). Herencia de la resistencia a *"Puccinia coronata avenae"* y *"P. graminis avenae." Rev. Invest. Agric.* **4,** 523–539.

von Thumen, F. (1886). "Die Bekampfung der Pilzkrankheiten unserer Culturgewachse." Versuch einer Pflanzentherapie, Wien.

Wahl, I. (1958). Studies on crown rust and stem rust on oats in Israel. *Bull. Res. Counc. Isr. Sect. D* **6,** 145–166.

Wahl, I. (1970). Prevalence and geographic distribution of resistance to crown rust in *Avena sterilis. Phytopathology* **60,** 746–749.

Wahl, I., and Schreiter, S. (1953). A highly virulent physiological race of crown rust on oats in Israel. *Bull. Res. Conc. Isr.* **3,** 256–257.

Wahl, I., Dinoor, A., Halperin, J., and Schreiter, S. (1960). The effect of *Rhamnus palaestina* on the origin and persistence of oat crown rust races. *Phytopathology* **50,** 562–567.

Warner, G. M., and French, D. W. (1970). Dissemination of fungi by migratory birds: Survival and recovery of fungi from birds. *Can. J. Bot.* **48,** 907–910.

Waterhouse, W. L. (1952). Australian rust studies. IX. Physiologic rust determinations and surveys of cereal rusts. *Proc. Linn. Soc. N.S.W.* **77,** 209–258.

Welsh, J. N., Carson, R. B., Cherewick, W. J., Hagborg, W. A. F., Peturson, B., and Wallace, H. A. H. (1953). Oat varieties—past and present. *Publ.—Can. Dep. Agric.* **891,** 1–51.

Wolf, C. B. (1938). The North American species of *Rhamnus. Rancho Santa Ana Bot. Gard. Monogr.* **1,** 1–136.

Zhemchuzhina, I. (1978). The strain composition of oat crown rust in the European part of the USSR from 1971–1975. *Mikol. Fitopatol.* **12,** 496–498.

Zillinsky, F. J., Sadanaga, K., Simons, M. D., and Murphy, H. C. (1959). Rust-resistant tetraploid derivatives from crosses between *Avena abyssinica* and *A. strigosa. Agron. J.* **51,** 343–345.

Zimmer, D. E., and Schafer, J. F. (1960). Variability of telial formation of *Puccinia coronata. Proc. Indiana Acad. Sci.* **70,** 91–95.

Zimmer, D. E., Schafer, J. F., and Patterson, F. L. (1962). Spontaneous mutations for virulence in *Puccinia coronata. Phytopathology* **52,** 34 (abstr.).

Zimmer, D. E., Schafer, J. F., and Patterson, F. L. (1965). Nature of fertilization and inheritance of virulence in *Puccinia coronata. Phytopathology* **55,** 1320–1321.

6

Barley Leaf Rust

B. C. Clifford
Welsh Plant Breeding Station,
Aberystwyth, Wales, United Kingdom

I. Introduction

Leaf (brown) rust of barley, although widespread where barley is grown, has generally been considered unimportant in economic terms. However, in the last 10–15 years, changes in cropping practices, particularly in northwest Europe, have resulted in a general increase in the disease with some severe local outbreaks. In particular, there has been an increase in the intensive cultivation of barley in areas in which the climate is particularly conducive to development of the causal organism (*Puccinia hordei* Otth.), and this has been exacerbated by the

THE CEREAL RUSTS,
Volume II

ISBN 0-12-148402-5

widespread cultivation of highly susceptible cultivars. As a consequence of this, there has been a considerable increase in research and development work, particularly in relation to disease control through the use of host resistance and the recently developed "systemic" fungicides. Because little effort had been made prior to this period to control barley leaf rust through the use of host resistance, cultivars were essentially susceptible to the disease and the stage was set to apply the newly emerging philosophies of breeding for resistance and in particular the ideas propounded by Vanderplank (1963). Their application, and the concurrent availability of greatly improved systemic fungicides, has resulted in a steady decline in the disease in northwest Europe up to the time of writing. Additionally, and perhaps more significantly, fundamental studies of the host–pathogen system have resulted in theoretical models and new techniques that have practical application and value for the control of other cereal rust diseases and will no doubt stimulate research and development in these areas.

II. The Pathogen

A. DESCRIPTION

Puccinia hordei Otth., syn. *P. anomala* Rostr., *P. simplex* Eriks. et Henn.

Pycnia and aecia on *Ornithogalum* spp. Pycnia amphigenous, numerous, in groups or scattered, spherical to ellipsoid, 86–144 × 99–158 μm, honey-colored, then blackish. Pycniospores pear-shaped to elongate, 2 × 3–5 μm. Aecia amphigenous, scattered, cupulate, yellow, 200–300 μm diameter, peridial cells hyaline, polygonal, external walls smooth, 6–8 μm thick, internal wall coarsely verrucose, 3–4 μm thick. Aeciospores globoid or ellipsoid, 18–26 × 16–33 μm, subhyaline, walls minutely and thickly verrucose. Uredia and telia on *Hordeum* spp. Uredia amphigenous, scattered, minute, cinnamon-brown. Urediospores subgloboid or ellipsoid, yellow, 21–34 × 15–24 μm, wall 1.5–2 μm thick, echinulate, pores 8–10, scattered, indistinct. Telia amphigenous or on culms, blackish-brown, oblong, confluent up to 3 mm, 6.5 mm on culms, remaining covered by epidermis, sori surrounded by or divided into compartments by clusters of brown paraphyses, flattened at tip. Teliospores angularly oblong or clavoid, 15–24 × 34–56 μm, truncate or obtuse, narrowed below, slightly constricted at septum, wall chestnut-brown, smooth, 1.5 μm thick at sides, 4–8 μm at apex, pedicels short, brownish. Mesospores abundant, variable, 25–45 × 15–28 μm, walls slightly thickened at apex, 4–6 μm (Wilson and Henderson, 1965; Critopoulos, 1956) (Fig. 1).

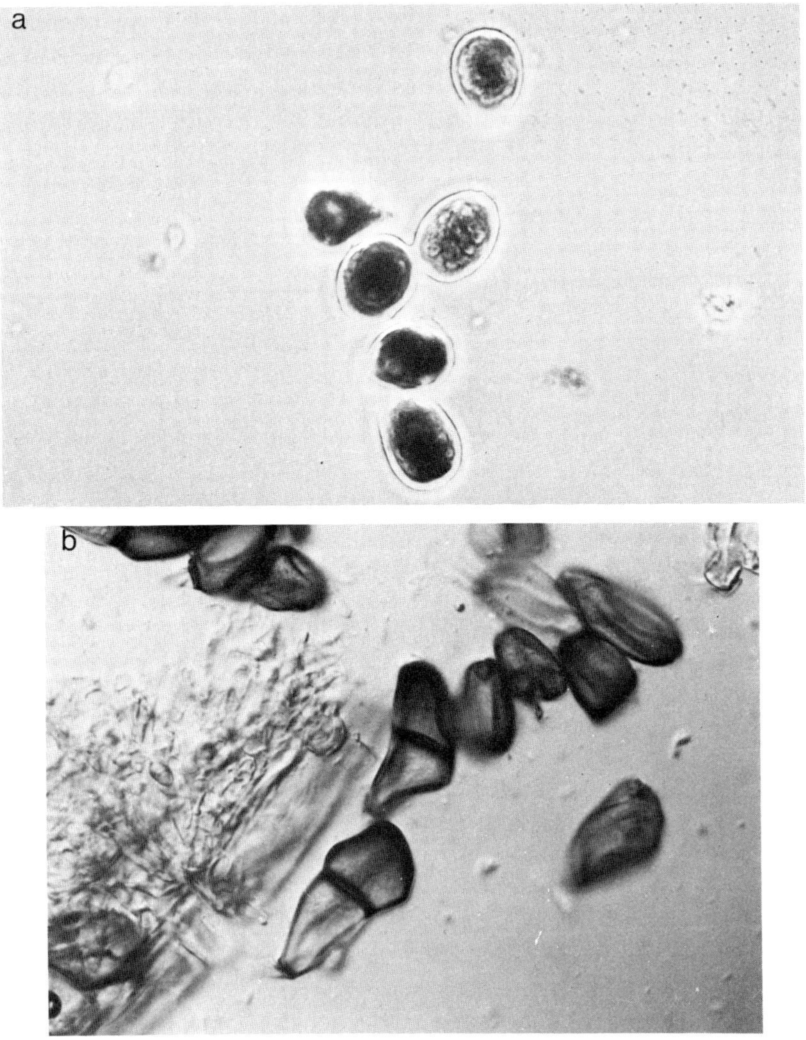

Fig. 1. Spore forms: (a) urediospores (×400), (b) teliospores and mesospores (×400).

B. TAXONOMY, LIFE CYCLE, AND HOST RANGE

Puccinia hordei Otth. is a macrocyclic, heteroecious rust, the uredia and telia of which occur on wild and cultivated *Hordeum* spp. with aecia on *Ornithogalum, Leopoldia,* and *Dipcadi* spp. in the Liliaceae. The uredial and telial stages occur widely on cultivated barley (*Hordeum vulgare* L.) and commonly on the wild species *H. spontaneum* C. Koch and *H. bulbosum* L. in Israel (Anikster and Wahl, 1979). The uredial stage has also been reported in *H. murinum* L. in Norfolk, England, by Ellis (1934). The forms on *H. murinum* and *H. bulbosum* have been considered an autonomous species, *P. hordei-murini* Buch., but Anikster *et al.* (1971) present evidence against this classification and suggest that these forms are cospecific with *P. hordei.*

Tranzschel (1914) first implicated *Ornithogalum* spp. as the alternate host of *P. hordei,* and this has subsequently been confirmed by workers in Austria, England, France, Germany, Hungary, Israel, Portugal, Switzerland, the United States, and the Soviet Union. D'Oliveira (1960) demonstrated that 32 species of *Ornithogalum,* together with *Dipcadi serotinum* (L.) Medic., were hosts and showed that the distribution of the alternate host partly overlaps that of *H. spontaneum,* a fact of significance for the co-evolution of the rust on the two host families. In Israel, the *Ornithogalum* flora coexists with wild *Hordeum* spp., and the alternate host is essential for the survival of the pathogen and for the generation of pathogenic variability in the uredial stage (Anikster *et al.,* 1971, 1975, 1976). Critopoulos (1956) similarly emphasised the importance of *Ornithogalum* spp. in Greece. In other parts of the world—e.g., central Europe—the alternate host is unimportant because teliospore germination is not synchronized with the growth of *Ornithogalum.* In North America, *Ornithogalum* spp. are widely distributed, and, although aecia do occur (Mains and Jackson, 1924), they are unimportant. In England, Dennis and Sandwith (1948) reported aecia occurring naturally on *O. pyrenaicum* L. (= *O. narbonense* Auct.) at one location, but, as in other parts of northwest Europe, the sexual stage is unimportant. Pycnia and aecia have been cultured by Lumbroso *et al.* (1977) on detached leaves or segments of fleshy bulb scales of *O. barbonense* or *O. brachystachys* C. Koch, and in a recent study, Anikster (1982) has successfully infected *O. brachystachys, O. trichophyllum* Boiss. et Heldr., *Dipcadi erythraeum* Webb et Bert., and *Leopoldia eburnea* Eig et Feinbr. with basidiospores derived from either *H. vulgare, H. spontaneum, H. bulbosum,* or *H. murinum.* Pycnial fertilizations were successful between all combinations from the four alternate host species, but aeciospores would only infect the *Hordeum* species of origin, with the exception of *H. spontaneum* and *H. vulgare,* implying that pathogenicity on the alternate hosts is less specialized.

C. *UROMYCES* RUSTS OF BARLEY

Anikster and Wahl (1966) have described a number of morphologically and biologically well-defined *Uromyces* rusts of the section *Angulati* that parasitize wild *Hordeum* spp. in Israel and that complete their life cycles on *Bellevalia, Muscari,* and *Scilla* spp. However, they recognized that speciation in the genus *Uromyces* was controversial. They described *Uromyces christensenii, U. reicherti,* and *U. viennot-bourginii* as new species first found in Israel and occurring naturally on the wild barleys *Hordeum spontaneum* C. Koch. and *H. bulbosum* L., with the latter rust species infecting cultivated barley. Anikster *et al.* (1977a) further described *U. viennot-bourginii* as infecting cultivars of *H. vulgare* that carry a wide range of *Pa* genes (see Section IV, D,1) for resistance to *P. hordei.*

Cummins (1971) recognized *U. turcomanicum* Katajev as the species that parasitizes cultivated barley and listed *U. iranensis* V.-Bourgin, *U. boissierae* V.-Bourgin, *U. prismaticus* V.-Bourgin, *U. viennot-bourginii* Wahl and Anikster, and *U. christensenii* Anikster and Wahl as synonyms. The following description of *U. turcomanicum* is condensed from Cummins (1971).

Aecia on *Bellevalia* and *Muscari* spp. Uredia and telia on *Boissiera pumilo* (Trin.) Hack., *Festuca ovina* L., *Hordeum bulbosum* L., *H. spontaneum* Koch, *H. violaceum* Regel, *H. vulgare* L., *Secale montanum* Guss. Uredia amphigenous, with distinctive colorless, saccate, collapsing paraplyses. Urediospores 24–32 × 19–25 μm, wall pale yellow to nearly colorless, echinulate; germ pores 7–11, scattered. Telia amphigenous, loosely covered by epidermis or exposed, always pulverulent, chocolate-brown. Teliospores one-celled, 18–24 × 14–20 μm, variable, often angular, tending to dimorphism with paler spores more angular, walls often ridged, occasionally punctate, otherwise smooth. The species is distributed from southern Russia to Israel, Iraq, and Iran.

III. The Disease

A. Distribution and Economic Importance

Leaf rust is the most important rust disease of barley and is widely distributed where the crop is grown. It does not cause severe losses on a widespread and regular basis but it is locally important, particularly in the cool temperature regions of barley cultivation. It is reported as potentially damaging in North America (Newton *et al.*, 1945; Levine and Cherewick, 1956; Reinhold and Sharp, 1982; Mathre, 1982), Argentina (Vallega *et al.*, 1955), the Netherlands (Wilten, 1953), Kenya

(R. Little, personal communication), New Zealand (Arnst *et al.*, 1979), and the United Kingdom (Johnson, 1970; Jenkins *et al.*, 1972; Melville *et al.*, 1976; King, 1977), with yield losses of up to 30% occurring in field trials incorporating fungicide treatments. Data on actual losses in field crops is scarce, but losses in the order of 10–20%, at least partly attributable to leaf rust, have been recorded in New Zealand (Arnst *et al.*, 1979) and in England (Jenkins *et al.*, 1972).

Intensification of barley cultivation in the cool, temperate regions has resulted in a considerable increase in the disease in the last 10–15 years. In the United Kingdom it was the most important disease of spring barley in 1970 and was also at high levels in 1971. These epidemics were associated with highly susceptible spring barley varieties that were being grown in southwest England at the time (King, 1972, 1977; Chamberlain and Doodson, 1972; Melville and Lanham, 1972; Priestley, 1978). In New Zealand, the disease was widespread but unimportant up to 1970 (Arnst and Fenwick, 1973) but has subsequently increased, this being attributed to changing rotational practices and the cultivation of susceptible varieties (Arnst, 1976), and resulting in losses in the order of 20% in the 1974–1975 season in Canterbury province (Arnst *et al.*, 1979), where late-sown crops are particularly at risk. In Denmark, as in other parts of northwest Europe, *P. hordei* overwinters in the uredial state on autumn-sown crops and volunteers (Hermansen, 1968). The problems that winter barley created in acting as a disease reservoir led to it being banned in Denmark, but in the 1970s there has been a very large swing to autumn sowing in northwest Europe generally. This, combined with the increased number of barley crops in the rotation and high-nitrogen regimes that favor the disease (Widdowson *et al.*, 1976; Udeogalanya and Clifford, 1982), is leading to a much greater potential threat from leaf rust. A shift to the cultivation of susceptible cultivars and curtailment of fungicide usage for economic reasons would have potentially dangerous consequences.

B. Signs and Symptoms

On the barley host, uredial infections appear as small (up to 0.5 mm) orange-brown pustules that darken with age. Chlorotic halos are usually associated with the pustules, which are scattered, mainly on the upper but also on the lower surface of leaf blades and also on leaf sheaths. With severe infections late in the season, some stem, glume, and awn infection can occur, and there is often general tissue chlorosis and eventual necrosis associated with these late infections. Late in the season, blackish-brown telia are formed. These often occur in stripes, particularly on leaf sheaths, and they are long-covered by the epidermis; they also occur on stems, heads, and leaf blades.

Effects on the host depend on the duration and severity of infection, but it is in the nature of biotrophy that there are general adverse effects on photosynthesis, respiration, transport of nutrients, and water relationships resulting in a general debilitation. Spring barley, especially if late-sown, is particularly affected as it is at risk when the pathogen is actively developing. Early, severe infections can thus result in reduced root and shoot growth, which gives rise to stunting and to reduction in fertile tiller numbers and numbers of grains per ear (Lim and Gaunt, 1981; Udeogalanya and Clifford, 1982). Epidemics tend to occur late, and consequently the most common effect is on grain size and quality (Newton *et al.*, 1945; Teng and Close, 1977b; Johnson and Wilcoxson, 1979a; Lim and Gaunt, 1981; Udeogalanya and Clifford, 1982). Grain characteristics of importance in brewing can also be affected (Newton *et al.*, 1945). Heavily infected plants tend to ripen (senesce) prematurely, and these general effects are exacerbated by other plant stress factors such as low fertility, drought, and excessively high temperatures.

C. ETIOLOGY AND EPIDEMIOLOGY

In the major barley-growing areas, the alternate host (*Ornithogalum* spp.) is unimportant in the survival and development of the pathogen (Mathre, 1982). Where endemic, the fungus perennates on either volunteer plants or autumn-sown crops. This survival has been documented in Germany (Gassner and Pieschel, 1934; Tan, 1976), the Netherlands (Parlevliet and van Ommeren, 1976, 1981), and the United Kingdom (d'Oliveira, 1939; Simkin and Wheeler, 1974b). Survival and development over winter are limited by both temperature and survival of host tissue, there being a general decline in the pathogen throughout the winter months (Simkin and Wheeler, 1974b; Parlevliet and van Ommeren, 1976). Cultivar susceptibility and competition with other leaf pathogens also affects survival (Parlevliet and van Ommeren, 1981), and competition studies (Simkin and Wheeler, 1974c; Round and Wheeler, 1978) have shown that prior inoculation with *Erysiphe graminis hordei* results in reduced development of *P. hordei*.

Free moisture, usually satisfied by nighttime dew, is essential for germination and penetration (Simkin and Wheeler, 1974a). Germination occurs over the temperature range of 5° to 25°C and is high from 10° to 20°C (see Table I and Joshi *et al.*, 1959; Simkin and Wheeler, 1974a). Germination is normally by a single branched germ tube, which grows at right angles to the long axis of the leaf and terminates with a simple cushion-shaped appressorium over a host stomate (Fig. 2). Appressorial formation is most rapid and greatest at 15°C and high from 10° to 20°C but curtailed above 25°C (B. C. Clifford, unpublished). A cigar-shaped substomatal vesicle is then formed (Fig. 3) that produces hyphae from either end (Clifford, 1972) to initiate the mycelial colony, which gains nutrients from the host cells via simple round-to-reniform haustoria. Colonization is limited by temperature and increases to a maximum in the range 5°–25°C (Simkin and Wheeler, 1974b; Teng and Close, 1978). Under optimum conditions just described, which have been summarized by Polley and Clarkson (1978), sporulation begins in 6–8 days after infection, but can take up to 60 days at 5°C (Simkin and Wheeler, 1974a). The sporulation (infectious) period is similar in the range 10°–20°C but is reduced at 25°C, and uredial size, generation time, and sporulation period are all reduced as uredia density increases (Teng and Close, 1978). Spore release is by wind, and the subsequent survival of spores is affected by environment. Teng and Close (1980) found that spores lost viability rapidly when exposed to sunlight during warm summer days in New Zealand, but in simulated cloudy weather, spores survived for 38 days.

The above data show that the uredial stage can survive and develop under winter conditions prevalent in cool temperate regions, and this emphasizes the importance of the autumn-sown crop as a "green bridge." Rapid disease development only occurs in warm summer weather and when free moisture is available overnight. Temperature during the day period is critical in the field, and R. W. Polley (personal

Table I

Urediospore Germination and Penetration of Barley Leaves after a 16-Hour Dark Dew Period[a]

Event	Temperature (°C)						
	0	5	10	15	20	25	30
Ungerminated spores	100	26.0	4.8	3.6	6.0	11.4	100
Germ tubes	0	29.2	21.2	11.0	13.6	55.6	0
Appressoria	0	34.6	14.2	6.0	18.8	4.2	0
Penetrations	0	10.2	59.6	79.4	60.4	4.6	0

[a]Mean of five leaves, 100 events per leaf. From B. C. Clifford (unpublished).

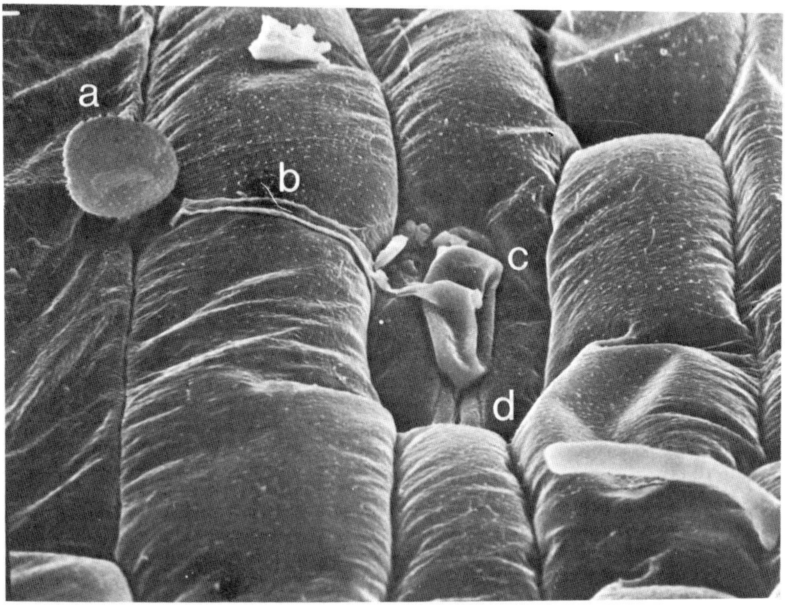

Fig. 2. Urediospore germination and penetration of barley leaf stoma: (a) urediospore, (b) germ tube, (c) appressorium, (d) stoma, (a–d) × 1600.

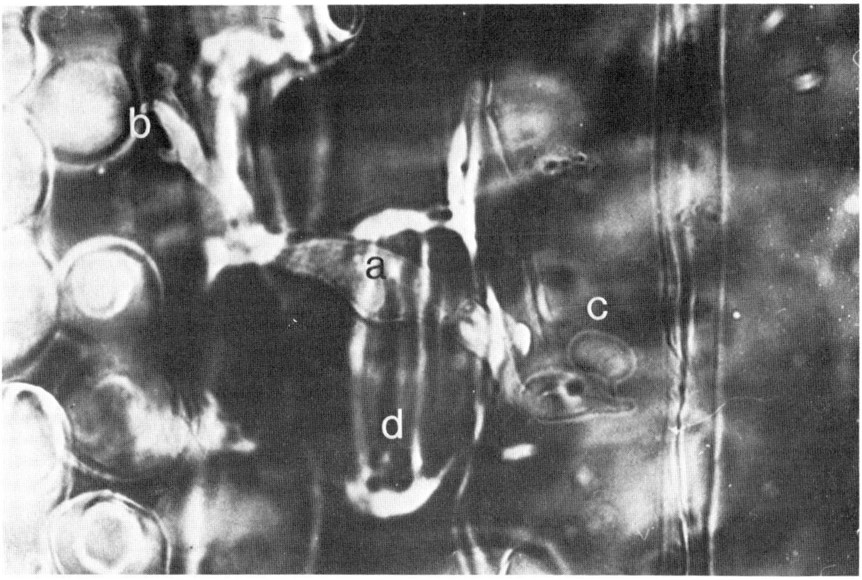

Fig. 3. Early postpenetration growth of uredial stage on barley. (a) substomatal vesicle, (b) primary infection hypha, (c) haustorium, (d) stoma, (a–d) × 700.

communication) considers that at least 9 hr of surface wetness are required at 7°C. These conditions are best satisfied by anticyclonic weather patterns during crop growth.

IV. Disease Control

A prerequisite for the initiation of any disease control program is an evaluation of the importance of the target disease in economic terms, but, as we have just discussed, such information is limited for leaf rust of barley. However, the severe epidemics in England in 1970 and 1971 resulted in a considerably increased effort in breeding and related research in northwest Europe. Although some rust-specific fungicides were also developed during this period, leaf rust has not generally been the main target disease of cereal fungicides, but several currently available systemic chemicals have a broad spectrum of activity that includes *P. hordei.*

A. CROP LOSS APPRAISAL AND FORECASTING

Several attempts have been made to relate infection levels to losses in yield, and in such trials fungicides have been used to give different disease development profiles and different final disease levels. Amount of disease has then been related to yield losses. In one such study, Melville *et al.* (1976) determined that each 1% increment of rust assessed on the flag leaf at GS75 (Zadoks *et al.,* 1974) resulted in a yield loss of 0.77%. In a similar experiment, a yield loss value of 0.6% was obtained by King and Polley (1976), but if disease on the leaf below the flag was assessed, a lower yield loss estimate of 0.4% was obtained. Where disease on whole plants was assessed at GS75, a yield loss estimate of 0.6% for each 1% increment of rust was obtained by Udeogalanya and Clifford (1982), but under a low nitrogen regime a much higher loss (1.5%) was observed. This latter high figure indicates that loss assessments must take into account the physiological state of the plant as, under stress, the effect of rust infection is relatively greater. Although the other loss values are similar to each other, differences have been partly explained by King and Polley (1976) as being due to discrepancies in disease assessment. In these experiments, a single disease assessment was made at a specific or critical point in the growth of the plant (GS75), but this may not be the best procedure for leaf rust. Teng *et al.* (1979) have compared a number of yield loss models based on (1) a single assessment of disease (critical point, CP), (2) several assessments of disease (multiple point, MP), or (3) measurement of the disease profile from the area under the disease progress curve (AUDPC). They concluded that all models were satisfactory with various CP models explaining 45.3–62.2% of the total variation in yield loss and AUDPC models explaining 58.2–74.7%. The MP models were best and explained over 90% of the total variation in yield loss. Even two-point models reached this standard, and the best was one based on two assessments on two leaves (flag and the one below).

The need for accurate disease assessment in estimating yield losses is apparent. Whole-plant or two-leaf assessments are better than single-leaf evaluations, but the criteria of assessment are also important. For example, King and Polley (1976) considered that the lower loss values that they obtained compared with Melville *et al.* (1976) were partly due to Melville *et al.* including in their disease assessment figures of chlorosis associated with uredial pustules. Further work is required to determine the relationships between visible signs (uredial pustules), cryptic infection (intercellular and intracellular growth), and the physiological effects on the host. More accurate assessments of fungal biomass may be possible by assays of fungal mannan and chitin (Whipps *et al.*, 1980) or remote sensing of spectral reflectance, especially in the infrared region (Teng and Close, 1977a). Evaluations of the effects of infection on the plant are also required, especially in relation to varietal responses. In general, a direct linear relationship between infection and yield loss appears to occur, but there are cultivar differences in response. For example, Newton *et al.* (1945) reported that the cultivar Mensury was virtually unaffected by a severe rust infection and was considered to be tolerant to the disease. By extrapolation, it may be assumed that cultivars differ in their sensitivity to infection, and this must be considered when estimating crop losses and benefits from control procedures.

B. DISEASE FORECASTING AND SIMULATION MODELLING

The application of fungicidal sprays is expensive and must therefore be made as cost-effective as possible. Although the widespread practice of program or timetable spraying has many attractions to management, there are obvious economic advantages to applying chemicals only in response to warnings or forecasts of disease build-up. Within this framework, persistent and effective systemic chemicals with prophylactic and therapeutic properties can be applied with the greatest degree of flexibility, whereas nonpersistent, less biologically active chemicals need to be applied precisely for maximum effect.

The first reported attempt to rationalize fungicide control of leaf rust was by King and Polley (1976), who concluded, from spore trapping and disease monitoring experiments, that there is a high risk of infection when 5 hr of dew or rain follows a day with a maximum temperature of over 15°C. From this, the best time for a single spray in England was judged to be mid-June, when weather is generally favorable and before rapid increase in rust (Polley and Clarkson, 1978). These workers acknowledged that many more data inputs are required to improve the value of such predictive models, including pathological data, host factors that affect response to the pathogen, and short-term weather data. Such a system simulation model has been described by Teng *et al.* (1977) for general rust–host interactions with the stated objectives of investigating the relationships between the components of an epidemic, studying the progress of an epidemic, and predicting yield loss. From this, specific models for barley leaf rust have been developed (Teng *et al.*, 1978), resulting in what the authors claim to be the first reported operational disease management scheme based on a computer simulation model. This is called BARSIM-1 (Teng *et al.*, 1980), standing for barley rust simulation model, version 1.

Teng and co-workers stated that the objective of the model was to provide a means for quantitative projection of disease leading to a cost–benefit basis for crop protection decisions. They argued that, compared with mathematical (fixed) models, simulations are "live" and responsive to the continually changing pattern of disease development: they thus offer the only realistic means of analysing and understanding epidemics. Such models are limited by knowledge of the pathogen dynamics in epidemics, by host response and crop loss data, and by the efficacy and economics of fungicide control measures. Much of this information is available or obtainable but, in admitting that a major constraint is the prediction of weather in the short term, they point out that there is a "window of response" governed by the latent period of *P. hordei*—in the order of 7–10 days under early summer field conditions. Of course, this response requires the availability of effective, persistent chemotherapeutants. However, it appears paradoxical to the author that the more effective the chemical is—and the trend is very much in this direction—the less need there is for sophisticated prediction and response models.

C. FUNGICIDE CONTROL

Relatively little information is available on fungicidal control of leaf rust because of its relative unimportance as a target disease on a world wide basis. Although the older contact eradicant fungicides such as zineb and sulfur are active against *P. hordei* (Jenkins *et al.*, 1972; Melville *et al.*, 1976), repeated applications were necessary, which rendered them uneconomic. However, the favorable economics of cereal growing in Europe in the past decade have resulted in the common agricultural usage of fungicides to control disease. During this period a number of "systemic" fungicides have been developed that are translocated from one part of the plant to another to give control of infection at a distance from the site of application. Some of the first produced systemics showed specific activity against *Basidiomycetes* including the cereal rusts. Carboxin, which is transported acropetally in the xylem (von Schmeling and Kulka, 1966), is active against *P. hordei* (Udeogalanya and Clifford, 1982), as is the related compound, benodanil (Frost *et al.*, 1973; Widdowson *et al.*, 1976; Teng and Close, 1977b; Lim and Gaunt, 1981; Udeogalanya and Clifford, 1982). Both of these compounds, which are carboxylic acid anilides, are absorbed by roots and green plant parts and are translocated apoplastically (Udeogalanya, 1979). Another of these earlier systemics, benomyl, is also reported as having some activity against *P. hordei* (Evans *et al.*, 1969), and it has a broad spectrum of activity. The most recently produced systemics are active against a wide range of fungi including *P. hordei.*

Triadimefon applied as a foliar spray gives good control of leaf rust (Teng and Close, 1977b), as does the related compound triadimenol when applied as a seed dressing (Lim and Gaunt, 1981). Seed dressing of triadimenol has been reported as remaining effective in the plant for 51 days, but late epidemics of leaf rust require an additional foliar spray for control (Moore *et al.*, 1979). Both compounds are prophylactic, therapeutic, and transported acropetally but triadimefon owes some of its effect to vapor activity. This is highly advantageous in a crop situation but can lead to problems of interplot interference in glasshouse and field experiments. Other compounds with similar properties and that are active against *P. hordei* include diclobutrazol (Skidmore, 1980) and propiconazole (Urech *et al.*, 1980). Of the four fungicides currently approved for control of leaf rust of barley by the United Kingdom Ministry of Agriculture, Fisheries and Food (Anonymous, 1982), propiconazole, triadimefon, and prochloraz are all grouped together as having the same mode of action through an inhibition of ergosterol biosynthesis. The fourth chemical, fenpropimorph, also inhibits sterol biosynthesis but apparently at a point different from that of the other group of compounds. Such site specificity has led to the evolution of fungicide-insensitive strains of some plant pathogens but there is no evidence for the emergence of such strains of *P. hordei*.

D. HOST RESISTANCE

In common with other cereal–rust interactions, resistance of barley to *P. hordei* can be expressed in a number of ways. Two main types have been recognized and studied sufficiently for their characteristics to be summarized (Table II); these characteristics are not mutually exclusive, and there is considerable overlap between types. However, evidence shows that there are real distinctions and that the separation is not just one of convenience. The writer has chosen not to use any of the particular descriptive terms to be found in the literature, as there is no consent as to which should be adopted; the commonly used synonyms are included in Table II.

Table II

Characterization of Two Types of Resistance to *Puccinia hordei* Otth. in *Hordeum* spp.

Type I Resistance

Expression: "Hypersensitive" host response. Host cell death giving immune response or, more commonly, visible chlorotic or necrotic flecks. Sporulation may occur, but uredia are usually associated with host tissue necrosis or chlorosis.

Genetic control: Major or oligogenes. Usually single completely or partially dominant factors, which may be temperature sensitive in expression.

Stability: Commonly completely overcome by matching virulence in the pathogen resulting in host susceptibility in the field.

Descriptive terms: Hypersensitive; major gene; race-specific; differential, vertical.

Type II Resistance

Expression: Quantitative reduction in compatibility with little or no host cell necrosis. Reduced numbers of uredia that are slower to develop and are reduced in size and sporulation capacity.

Genetic control: Quantitative, recessive major gene and/or limited numbers of modifying minor genes.

Stability: Pathogen variation demonstrated experimentally but effective in widespread agricultural use to date.

Descriptive terms: Nonhypersensitive; partial; slow-rusting; polygenic; minor gene; horizontal.

1. *Type I Resistance*

This resistance is expressed macroscopically as small chlorotic or necrotic flecks, the latter often appearing tan with a brown halo of pinhead size. A varying amount of sporulation may be associated with the lesions depending on environment, homozygosity of the locus, and the dominance of the gene governing the response. These genes (designated *Pa* genes) commonly condition a necrotic reaction when in the dominant state. Histological studies of such interactions governed by gene Pa_2 in the cultivars Batna, Peruvian, and Ricardo indicate that resistance operates after penetration of the host but allows fairly normal colony development up to the time of sporulation. At this point, host-cell necrosis becomes extensive and sporulation is prevented or severely retarded. This development is quantitatively and qualitatively modified by the genotype of the host (Clifford, 1973, 1974a; Clifford and Roderick, 1978a). The slow development of fungal colonies in the cultivar Cebada Capa, which carries gene Pa_7 and which results in limited numbers of necrotic flecks (Clifford and Roderick, 1981), is governed by a number of minor genes which are epistatic to Pa_7 (Parlevliet, 1980).

The expression of *Pa*-gene governed resistance is also affected by the degree of dominance, but no recessively expressed *Pa*-genes appear to have been identified. Some genes are also temperature labile. For example, Vallega *et al.* (1955) indicated that some cultivars only expressed resistance under specific environmental conditions. More recently it has been shown that resistance governed by Pa_7 is not expressed at 5°C and that of Pa_9 (in CI 1243) becomes ineffective with increasing temperature in the range 5°–25°C (Clifford and Udeogalanya, 1976; Udeogalanya and Clifford, 1978). The resistance of Estate, which carries gene Pa_3, is expressed as an immune or chlorotic fleck over the entire range of temperature.

Type I resistance to *P. hordei* was first identified by Waterhouse (1927) who, in later summarizing his results (Waterhouse, 1948), concluded that all genes studied were at the same locus. Henderson (1945) identified two loci governing reaction that he designated *Pa* loci from the old name *Puccinia anamola.* Other workers (Watson and Butler, 1947; Zloten, 1952; Starling, 1956; Roane, 1962; Moseman and Greeley, 1965) identified host resistance factors, but there was some confusion of nomenclature and genetic relationships. This was largely resolved by genetic studies of Roane and Starling (1967, 1969, 1970), which resulted in the classification of six loci, *Pa*, Pa_2, Pa_3, Pa_4, Pa_5, and Pa_6 (C. W. Roane, personal communication). One anomaly was the identity of the gene in Cebada Capa (Dillard and Brown, 1969), which Roane and Starling (1970) considered to be at the same locus as Pa_5 in Quinn. Nover and Lehmann (1974) used pathogen isolates to differentiate the resistance of Quinn and Cebada Capa, and they suggested the designation Pa_7 for the factor in Cebada Capa. This was subsequently confirmed by genetic studies (Parlevliet, 1976c), and it is now clear that the gene is the one designated Pa_y (Johnson, 1970) and that occurs in CI 12201, Dabat (= Rabat), Forrajera Klein, Gondar, H2212, and La Estanzuela 75a (Macer *et al.*, 1966; Johnson, 1970; Dillard and Brown, 1969; Frecha, 1970, 1971; Brückner, 1971; Parlevliet, 1976c; Udeogalanya and Clifford, 1978). Pa_8 has been proposed for the locus in Egypt 4 (Tan, 1977a) and Pa_9 for the locus in CI 1243 (Tan, 1977b). This latter gene governs a temperature-sensitive response to avirulent isolates (Clifford and Udeogalanya, 1976), with compatibility increasing with increasing temperature. The gene is also apparently present in the widely grown East German cultivar Trumpf (Triumph) (Walther and Lehmann, 1980). The only other known factor governing a temperature-sensitive response is Pa_7, where compatibility is expressed at very low temperatures (Clifford and Udeogalanya, 1976). Other unclassified loci have been detected in EP 75 (Parlevliet, 1976c), in Austral, Chilean D, Cruzat, and Berg (Tan, 1977b), and in lines selected from *Hordeum spontaneum* collections (Moseman *et al.*, 1979). Mutations for resistance have also been artificially induced (Grunewaldt and Grunewaldt, 1977). A summary of cultivars known to carry specific *Pa* genes either singly or in combinations has been made by C. W. Roane (personal communication), and some genes have been described (Roane, 1976).

Some linkage relationships have been determined. P_4 in Franger has been shown to be linked to the *Mla* locus conditioning reaction to *Erysiphe graminis hordei* with a recombination value of 16 ± 1.5% (Moseman and Reid, 1961), and a similar value was obtained (17 ± 3.5%) by McDaniel and Hathcock (1969). Both studies identified chromosome 5 as the carrier of these loci. N. A. Tuleen and M. E. McDaniel (personal communication) confirmed this and also located *Pa* on chromosome 2 and Pa_5 and Pa_7 on chromosome 3. Tan (1978) calculated linkage with a recombination value of 7.6 ± 1.4% between Pa_5 and Pa_7 and located the genes on chromosome 3. From trisomic analysis he concluded that Pa_3 was on either chromosome 1 or 2. The observation of Macer *et al.* (1966) that resistances to *P. hordei* and *P. striiformis* were associated in Forrajera Klein and La Estanzuela 75a was shown by Udeogalanya and Clifford (1978) to be due to linkage between Pa_7 and the locus governing reaction to *P. striiformis,* with a recombination value of between 12.1 and 18.3%.

A number of resistant cultivars, and in particular those carrying gene Pa_2, have been shown to carry additional specific resistance factors (C. W. Roane, personal communication; Reinhold and Sharp, 1980) that have not been classified. Additionally, some Pa_2 carriers vary quantitatively in their degree of susceptibility to pathogen isolates with virulence for Pa_2 (Clifford, 1974a). Parlevliet and Kuiper (1977a) identified a number of minor genes governing this "partial resistance" in Cebada Capa. These minor genes were epistatic to Pa_7 (Parlevliet, 1980) and govern the number of incompatible necrotic infection sites, a phenomenon previously observed for gene Pa_2 carried by Batna, Peruvian, and Ricardo (Clifford, 1974a). The characteristics and genetic control of this partial resistance are described next.

2. *Type II Resistance*

It has commonly been observed that cultivars grown in field nurseries differ in their levels of infection. For example, Vallega *et al.* (1955) screened the world collection of barley and identified cultivars with low average levels of leaf rust. In an attempt to identify stable and effective resistance to *P. hordei,* Clifford (1970) selected a range of cultivars with low recorded leaf rust levels from various international disease nurseries and tested them under conditions of uniform inoculum in the field. He thus identified cultivars that, although susceptible, developed less disease, and these he termed slow rusters, using the convention of the day. Among others, Vada expressed this character to a high degree and was subjected to histological analysis. The resistance elicited essentially no necrotic host response and operated only after penetration of the host. It was expressed by the relatively slow development (longer latent period) of fewer uredia (low infection frequency), which were smaller and produced fewer spores (Clifford, 1972). Other cultivars that show this resistance in the field have been identified (Parlevliet and van Ommeren, 1975; Johnson and Wilcoxson, 1979b), and similar resistance characteristics have been attributed to them (Johnson and Wilcoxson, 1978; Parlevliet and Kuiper, 1977b; Neervoort and Parlevliet, 1978). Further studies have confirmed that prepenetration development of the fungus is unimpaired (Niks, 1981) and that the low infection frequency results from the abortion of colonies early in the infection process (Clifford and Roderick, 1978a, 1981; Niks, 1982), although some colonies do develop further but fail to sporulate (Whipps *et al.,* 1980). These processes are only incidentally associated with collapse of host cells (Niks, 1982). Thus the physiological processes underlying the resistance are not clear from histology.

The resistance is expressed in seedling and juvenile stages of growth (Clifford, 1972; Parlevliet, 1975; Niks, 1982), but the separation between cultivars and the absolute expression of resistance is greater in adult plants and particularly on young flag leaves (Parlevliet, 1975). The principal parameters, namely increased latent period and decreased infection frequency, appear to be related (Johnson and Wilcoxson, 1978), but the association is thought not to be complete (Neervoort and Parlevliet, 1978). The resistance, which has been variously described as nonhypersensitive (Clifford, 1972) or partial (Parlevliet, 1975), results in low infectability by the host (Clifford and Clothier, 1974; Niks, 1982) and slow rusting in the field (Clifford, 1970; Johnson and Wilcoxson, 1978). Although environmentally stable, it does not appear to be expressed in the *erectoides* mutant types of spring barley.

The inheritance of Type II resistance in some European cultivars has been studied by Parlevliet (1976a) who followed the inheritance of latent period—defined as the time from penetration to the eruption of 50% of the final number of uredia—in crosses between the highly susceptible (short latent period) genotypes *L92* and *L94* and the highly resistance (long latent period) sister cultivars Vada and Minerva. He concluded that long latent period was governed by the cumulative action of a recessive gene with fairly large effect and some four or five minor genes with small, additive effects. A subsequent diallel analysis (Parlevliet, 1978a) of *L94* and Vada together with the moderately resistant Julia and the susceptible Volla and Sultan identified seven loci, all with additive factors except for the recessive gene in Vada, which had a considerably larger effect than the other loci. Inheritance studies involving Vada and Cebada Capa led Parlevliet and Kuiper (1977a) to conclude that Cebada Capa carries three or four minor genes for long latent period different from those in Vada. In a study involving North American cultivars (Johnson and Wilcoxson, 1979b), the inheritance of Type II resistance was found to be quantitatively expressed. Transgressive segregation occurred in some crosses and heritability values of 53–89% were obtained.

In the European germ plasm studied, minor genes with small and additive effects appear to operate in cultivars classified as susceptible to highly susceptible, whereas the high level of Type II resistance, exemplified by Vada, appears to be under simple genetic control. This conclusion is supported by the observation that the Vada level of resistance is easily recognized and selected for in breeding nurseries.

This, and the association of the resistance with mildew resistances from *Hordeum laevigatum* and Arabische, has resulted in its current widespread distribution in European spring barley cultivars (Clifford, 1981).

3. *Tolerance*

The identification of tolerance of barley to leaf rust is of historical interest in that it appears to be the first recorded observation of the phenomenon. This was by Newton *et al.* (1945), who observed that Mensury, although heavily infected with rust, was little affected in terms of yield and quality compared with other cultivars. A central problem in determining tolerance is the precise assessment of infection and damage, a difficulty acknowledged by Kramer *et al.* (1980), who studied 15 spring barley cultivars in 2 years. Although differences in response to infection, as measured by yield loss, were detected, they found a poor correlation between the 2 years' data and concluded that tolerance was the result of a genotype × environment interaction. More accurate assessments of infection may be possible through assays of fungal biomass (Whipps *et al.*, 1980), although damage to host tissue must also be taken into account, and a technique which may be of value in resolving this is the measurement of infrared reflectance (Teng and Close, 1977a).

E. PATHOGEN VARIATION

Measurement of variation for pathogenicity can only be made in relation to identified resistance in the host. As the main objective of such studies is the identification of biotypes that are pathogenic on host resistance factors of importance to breeders and commerce, the basis of such studies is a set of "differential cultivars" that carry the resistance factors in question. The identification of new resistances and the loss of effectiveness of existing resistances due to increases in corresponding pathogen virulence requires the constant review and updating of differential cultivar sets. It is convenient to consider pathogenic variation in *P. hordei* in relation to the two types of identified resistance discussed above.

1. *Type I Virulence*

Most studies with *P. hordei* have been related to Type I resistance governed by *Pa*-genes. Cultivars carrying particular *Pa*-genes or with unidentified resistance have been used for virulence analysis and these have been changed as knowledge of the genetics of resistance has evolved. A recent revision of the standard set of international differentials has been proposed (Clifford, 1977; Tan, 1977a), and these cultivars, their *Pa*-gene composition, and other carriers of these genes are given in Table III. This information is based on work cited in the following historical and geographical account of virulence analysis.

Early surveys identified variation for pathogenicity in North America (Mains, 1930) and Europe (Straib, 1936; d'Oliveira, 1939), but relationships between resistant cultivars were not clear. Consequently, Levine and Cherewick (1952) reviewed the North American, European, and Australian differentials and proposed a revision to include nine cultivars that they considered critical in differentiating main race groups that they designated as consolidated or unified (UN) races. They concluded that New and Old World races were different, and this still appears to be the case (Reinhold and Sharp, 1982). North American races appear to carry few virulences and in fact race 4, which carries virulence only for Pa_8, was dominant during the 30-year period following Mains (1930) work and virulence for Pa_3 and Pa_7, was not detected (Moseman, 1964; Moseman and Roane, 1959; Moseman and Greeley, 1965; Joshi, 1971). The recent survey in southern Texas where leaf rust is becoming more serious (Reinhold and Sharp, 1982) detected virulence to *Pa*, Pa_2, Pa_4, and Pa_8 but not to any of the other known resistances. Similarly, few virulence factors have been detected in earlier (Vallega *et al.*, 1955) and more recent (Antonelli *et al.*, 1976) surveys in Argentina. In Europe and elsewhere, the situation is rather different.

Formal surveys of virulence were begun in 1968 in the United Kingdom under the auspices of the National Physiological Race Surveys (Cereal Pathogens) (since 1977, the United Kingdom Cereal Pathogen Virulence Survey) by the writer and colleagues. It soon became apparent that the most common biotypes were those that carried a wide range of virulence (Clifford, 1974b), including virulence to *Pa*, Pa_2, Pa_4, Pa_5, Pa_6, and Pa_8. The most effective resistances were conferred by Pa_3 and Pa_7. Similar spectra of virulences were detected in Czechoslovakia (Brückner, 1970a,b), East Germany (Walther and Lehmann, 1980), West Germany (Rintelen, 1976, 1979a; Tan, 1976; Grunewaldt-Stocker, 1981), Italy (Ceoloni, 1979) Israel (Anikster *et al.*, 1975; Anikster, 1982), the Netherlands (Parlevliet, 1976c), and the Soviet Union (Shchelko, 1974), and also in North Africa (Reinhold and Sharp, 1980) and in New Zealand (Arnst *et al.*, 1979). Virulence to Pa_3 has been detected throughout Europe, and more recently, workers in Israel

Table III

Cultivars Carrying Type I Resistance Factors (*Pa* Genes) Used for Pathogen Virulence Analysis

Host gene	Differential cultivar	Other carriers
Pa	Sudan CI 6489	Oderbrucker CI 940
		Speciale CI 7536
		Kwan CI 1016
Pa2	Peruvian CI 935	Juliaca CI 1114
Pa2 +?	—	Ariana CI 2524
		Batna CI 3391
		Club Mariout CI 261
		Reka 1 CI 5051
		Ricardo C I 6306
		Weider CI 1021
Pa3	Ribari	Aim CI 3737
		Estate CI 3410
		Baladi 16
Pa4	Gold CI 1145	Franger CI 8811
		Lechtaler CI 6488
Pa5	—	Magnif CI 13806
Pa5 (+*Pa2*)	Quinn CI 1024	
Pa6 (+*Pa2*)	Bolivia CI 1257	
Pa7	Cebada Capa CI 6193	Dabat
		Gondar
		La Estanzuela 75a
		H2212
Pa8	Egypt 4 CI 6481	
Pa9	CI 1243	Abyssinian Schwarz
		Ab 14
		Triumph (Trumpf)
		Uadera

(Anikster *et al.*, 1977b; Golan *et al.*, 1978; Anikster and Wahl, 1979) have detected virulence to what was hitherto the most effective resistance factor, namely Pa_7; it has also been identified in North Africa (Parlevliet *et al.*, 1981). Essentially, none of the *Pa* genes have been widely used in commerce, the exception being Pa_9, which is present in the widely grown European cultivar Triumph (Trumpf), for which virulence has been detected in East Germany (Walther, 1979) and the United Kingdom (B. C. Clifford, unpublished).

The alternate host, *Ornithogalum* spp., is important in the survival of the fungus, in the evolution of virulence, and in the generation of new virulence combinations in Israel (Anikster *et al.*, 1977b), but overwintering in the uredial state occurs on autumn-sown crops and volunteer barley plants in the major barley-growing areas, and complex combinations of virulence survive this overwintering (Tan, 1976).

Variation in urediospore color has been observed in nature (d'Oliveira, 1939) and has been induced artificially (Falahati-Rastegar, 1981). Attempts at the artificial induction of mutations for virulence were successful at the Pa_3 locus using gamma radiation but was unsuccessful at the Pa_7 locus (Parlevliet and van Eijk-Bos, 1978).

A recent proposal for revising the set of differential cultivars and for standardizing virulence surveying procedures has been made (Clifford, 1977) following a meeting of European leaf rust workers at the European and Mediterranean Cereal Rusts Conference, Interlaken, in 1976. This includes a proposal for a standard set of international differentials with regional supplementary sets and a suggestion for the adoption of an octal system of virulence nomenclature (Gilmour, 1973); this has been instituted for the United Kingdom Cereal Pathogen Virulence Surveys (Jones and Clifford, 1980).

2. *Type II Virulence*

Physiological specialization on genotypes carrying Type II resistance was first reported by Clifford and Clothier (1974). In glasshouse tests of field isolates from the cultivars Vada, Julia, and Sultan, it was found that in general, isolates from a given cultivar were adapted to that cultivar, with the greatest degree of adaptation, as measured by uredia numbers, being with Vada-adapted isolates. When circular field plots of the cultivars were inoculated with selected adapted isolates, no epidemiological advantage over nonadapted isolates could be demonstrated (Clifford, 1975), possibly due to the observation that the latent period aspect of the Vada resistance was unaffected although some adaptation was detectable in nurseries (Clifford *et al.*, 1977). Parlevliet (1976b), however, using latent period as the measure of compatibility, identified three isolates of *P. hordei* adapted to Vada. In a subsequent paper, Parlevliet (1977) identified a specific interaction between the moderately resistant cultivar Julia and isolate 18 of *P. hordei* that was expressed as a shortening of the latent period. From this, and his observation that Julia carries a minor resistance gene not present in other cultivars, he concluded that a specific virulence factor in isolate 18 was interacting with a specific resistance gene in Julia; this formed the basis for his minor gene proposition (Parlevliet, 1978b). However, Niks (1982), in his comparative histological study of Julia infected with isolate 18 together with other genotype–isolate combinations, failed to detect any specific adaptation in terms of abortion of fungal colonies in seedling leaves.

Two Moroccan isolates of *P. hordei* have also been reported as effecting a reduced latent period on the cultivars Peruvian, Bolivia, and Vada (Parlevliet *et al.*, 1981), and dramatic interactions between pathogen isolates and cultivars in current agricultural use in West Germany have been recently reported (Aslam and Schwarzbach, 1980) under glasshouse conditions. Despite the above findings, trap nurseries of Type II resistant cultivars grown in field crops of resistant cultivars have failed to detect adapted field isolates; furthermore, there has been no diminution in the expression of Type II resistance, which has been widely deployed in cultivars grown in Britain up to the present time (Clifford, 1981).

F. BREEDING FOR RESISTANCE

Prior to about 1970, breeding programs were based on Type I resistance. The procedure, in common with the approach to breeding for resistance to other cereal rust diseases, is to screen various "world" and local collections of barley against pathogen isolates common and prevalent in the area. Because the resistance, when effective, is expressed qualitatively in the seedling stage as well as later, selection can be conducted on glasshouse-grown plants as well as in the field. Inoculum should be at a level to avoid disease escape but uniformity is not a requirement for valid testing. Such programs have been carried out over a number of years in North (Levine and Cherewick, 1956) and South America (Vallega *et al.*, 1955; J. H. Frecha and F. L. Mijica, personal communication) and also in Europe (Brückner, 1970a, 1972, 1973; Berbigier *et al.*, 1964; Nover and Mansfeld, 1959; Nover and Lehmann, 1974; Rintelen, 1979b). This approach is still actively pursued by some breeders and, although relatively few cultivars carrying *Pa*-gene resistance have been released, the East German cultivar Triumph (Trumpf) that carries Pa_9 (Walther and Lehmann, 1980) is a notable exception.

Utilization of Type II resistance was begun around 1970 by public- and private-sector breeders in Europe. Its quantitative expression requires the relatively precise techniques of uniformly inoculated spreader-bed nurseries for field selection. Identification of resistance in such nurseries on plants grown singly or in small clumps (hills) is highly effective (Clifford, 1970; Parlevliet and van Ommeren, 1975;

Johnson and Wilcoxson, 1979b; Kostić *et al.*, 1979) when percentage infection is assessed, preferably on two or more occasions in the season. The relative ranking of such nursery entries correlates well with that based on large plot assessments, although differences are reduced and absolute resistance is underestimated (Parlevliet and van Ommeren, 1975; Parlevliet *et al.*, 1980). This means that, in a segregating population, resistant components can be identified, but the absolute level of resistance needs to be determined by other means. Selection for variation in latent period can be effected in monocyclic tests in the glasshouse under conditions of reasonably uniform inoculation of plants in the flag-leaf stage (Neervoort and Parlevliet, 1978; Johnson and Wilcoxson, 1979b). Selection for long latent period or low uredia number can be effected in the seedling stage (Parlevliet, 1976d), although the necessary uniformity of inoculation required for the latter character is difficult to achieve on large plant populations. The expression of resistance is stable to environment, and the relative latent period is reported as being unaffected by temperature, photoperiod, or light intensity (Parlevliet, 1975). The correlation between uredia number and latent period ensures that selection for either one identifies resistant genotypes, although Niks and Parlevliet (1979) found that, in lines selected for either character in a barley composite, the two components were not strongly correlated.

The relative ease with which Type II resistance can be selected for in breeding programs is demonstrated by the wide range of resistant cultivars produced in northwest Europe since 1971. The resistance can be identified in field trial plots with a reasonable degree of certainty, and this has ensured the recommendation to farmers of cultivars with Type II resistance (Chamberlain *et al.*, 1972). The level of resistance of lines selected from breeding programs as potential cultivars can be assessed in relatively small (1 m^2) replicated and artificially inoculated field plots by reference to standard cultivars with known levels of resistance/susceptibility, a procedure developed at the Welsh Plant Breeding Station (Clifford and Roderick, 1978b).

Although there are some reports of pathogen strains adapted to Type II resistance, it has nevertheless remained stable and effective in widespread agricultural use over a 10-year period. On the other hand, Type I resistance has a history of ephemerality. One general problem with the use of "hypersensitive" resistance is that, when effective, it masks the degree of background resistance; consequently, its breakdown is often associated with the "vertifolia effect" (Vanderplank, 1963). For these reasons it is highly desirable to combine different resistances into one genotype, thus giving a broader spectrum of resistance. This can be achieved conventionally, for example by progeny testing to identify carriers of two or more *Pa*-genes or by selecting Type I resistant plants in segregating F_3 families that show Type II resistance on the susceptible components. In addition, some novel methods have been proposed to achieve these ends.

One such is to assemble lines carrying different resistances into a composite population that goes through cycles of interbreeding through the agency of male sterile genes in the population. Such composites have been assembled (Moseman, 1976; Moseman and Baenziger, 1979) in which different known *Pa*-genes and some unclassified resistances are included. Other schemes that exploit various characteristics of Type I and Type II resistances have been proposed (Udeogalanya and Clifford, 1978; Clifford, 1981). Some make use of the property of temperature sensitivity of certain *Pa* genes, which allows them to be combined by testing at appropriate temperatures. Furthermore, background Type II resistance can be assessed at temperatures that do not allow expression of the sensitive *Pa*-gene. Another scheme uses the observation that the number of visible necrotic flecks elicited by certain *Pa* genes is a function of the background Type II resistance (Clifford, 1974a; Clifford and Roderick, 1981), which is under the control of the Type II resistance genes (Parlevliet, 1980).

V. Conclusions and Future Prospects

Barley leaf rust was considered to be of less importance than other barley diseases up to around 1970. However, the recent increase in the disease, particularly in northwest Europe but also in New Zealand and parts of the United States, has resulted in a corresponding increase both in fundamental research and in the development of disease control measures based both on host resistance and fungicides. Although these latter efforts have been very successful, further work is required to identify new sources of resistance and new fungicides to counter any loss of effectiveness of existing measures resulting from changes in the pathogen population. To this end, programs should be maintained to monitor pathogen variation, particularly in relation to Type II host resistance and modern systemic fungicides. There also appears to be a need for more information on the nature of host resistance to allow predictive statements on its durability. Further studies on the interrelationships between the pathogen, the host, and the environment are necessary for the development of predictive, interactive schemes for disease control, especially with regard to fungicide usage. Further elucidation of the nature of pathogenesis and the relationships between symptomatology, host damage, and crop losses are also required to design the most efficient and effective disease control systems.

Acknowledgments

I am grateful to Professor J. P. Cooper, Director of the Welsh Plant Breeding Station, for permission to carry out this review and to my colleague H. W. Roderick for his help in many ways.

References

Anikster, Y. (1982). Alternate hosts of *Puccinia hordei*. *Phytopathology* **72,** 733–735.

Anikster, Y., and Wahl, I. (1966). Uromyces rusts on barley in Israel. *Isr. J. Bot.* **15,** 91–105.

Anikster, Y., and Wahl, I. (1979). Coevolution of the rust fungi on Gramineae and Liliaceae and their hosts. *Annu. Rev. Phytopathol.* **17,** 367–403.

Anikster, Y., Abraham, C., Greenberger, Y., and Wahl, I. (1971). A contribution to the taxonomy of *Puccinia* brown leaf rust of barley in Israel. *Isr. J. Bot.* **20,** 1–12.

Anikster, Y., Moseman, J. G., and Wahl, I. (1975). Parasitic specialization of *Puccinia hordei* Otth. and sources of resistance in populations of *Hordeum spontaneum* C. Koch. *Proc. Congr. Mediterr. Phytopathol. Union, 4th, 1975,* Summaries, pp. 69–70.

Anikster, Y., Moseman, J. G., and Wahl, I. (1976). Parasite specialization of *Puccinia hordei* Otth. and sources of resistance in *Hordeum spontaneum* C. Koch. Barley Genetics *Proc. Int. Barley Genet. Symp., 3rd, (1975)* pp. 468–469.

Anikster, Y., Moseman, J. G., and Wahl, I. (1977a). *Uromyces viennot-bourginii,* its life cycle, pathogenicity and cytology. *Trav. dédiés à G. Viennot-Bourgin,* pp. 9–17.

Anikster, Y., Golan, T., Moseman, J. G., and Wahl, I. (1977b). Role of *Ornithogalum* plants in evolution of virulence in *Puccinia hordei. Proc. Am. Phytopathol. Soc.* **4,** 213 (abstr.).

Anonymous (1982). "Use of Fungicides and Insecticides on Cereals 1982," Booklet 2257 (82). U. K. Ministry of Agriculture, Fisheries and Food, England.

Antonelli, E. F., Frecha, J. H., and Mujica, F. L. (1976). Presencia en la Argentina de una raza de *Puccinia hordei* virulente sobre el gene Pa_5. *Bol. Genet. Inst. Fitotec. Castelar* **9,** 15–20.

Arnst, B. J. (1976). Brown rust of barley. *Proc. N. Z. Weed Pest Control Conf.* **29,** 225–227.

Arnst, B. J., and Fenwick, J. E. (1973). A survey of barley diseases in New Zealand. *Proc. N. Z. Weed Pest Control Conf.* **26,** 157–160.

Arnst, B. J., Martens, J. W., Wright, G. M., Burnett, P. A., and Sanderson, F. R. (1979). Incidence, importance and virulence of *Puccinia hordei* on barley in New Zealand. *Ann. Appl. Biol.* **92,** 185–190.

Aslam, M., and Schwarzbach, E. (1980). Specificity of quantitative interactions between brown rust isolates and barley varieties. *Proc.—Eur. and Mediterr. Cereal Rusts Conf., 5th, 1980,* pp. 123–127.
Berbigier, A., Chery, J., and Larambergue, de R. (1964). Etude de la resistance à différents parasites d'une collection de varíetés d'orge. *Ann. Amelior. Plant.* **14,** 419–426.
Brückner, F. (1970a). Varietal resistance of summer barley to leaf rust (*Puccinia hordei* Otth.) in Czechoslovakia. *Genet. Slechteni. (Prague)* **6,** 143–151.
Brückner, F. (1970b). The physiological specialization of *Puccinia hordei* Otth. on the Czechoslovak territory in the years 1966 and 1967. *Ochr. Rostl., Zemed. Velkovyrobe, Ref. Konf. 1967* **6,** 13–17.
Brückner, F. (1971). Inheritance of the resistance to barley leaf rust and powdery mildew of cereals in some varieties of barley. *Genet. Slechteni (Prague)* **7,** 95–102.
Brückner, F. (1972). The breeding of barley for resistance to *Puccinia hordei* Otth. *Proc.—Eur. Mediterr. Cereal Rusts Conf., 3rd, 1972,* Vol 1, p. 55 (abstr.).
Brückner, F. (1973). The use of resistance sources from genetically remote forms in the breeding of malting barley. *Genet. Slechteni (Praque)* **9,** 135–140.
Ceoloni, C. (1979). *Puccinia hordei* in Italy: A preliminary survey on the virulence characteristics of the fungus in our country. *Cereal Rusts Bull.* **6,** 11–16.
Chamberlain, N. H., and Doodson, J. K. (1972). Brown rust of barley. *Agriculture (London)* **79,** 302–305.
Chamberlain, N. H., Doodson, J. K., and Meadway, M. H. (1972). A technique for the evaluation of the resistance of barley varieties to infection with brown rust (*Puccinia hordei* Otth.). *J. Natl. Inst. Agric. Bot. (G. B.)* **12,** 440–446.
Clifford, B. C. (1970). Brown rust of barley. *Welsh Plant Breed. Stn. Jubilee Rep. 1919–1969,* pp. 124–126.
Clifford, B. C. (1972). The histology of race non-specific resistance to *Puccinia hordei* Otth. in barley. *Proc.—Eur. Mediterr. Cereal Rusts Conf., 3rd, 1972,* Vol. 1, pp. 75–79.
Clifford, B. C. (1973). 6. Brown rust of barley The relationship between compatible and incompatible infections. *Rep.—Welsh Plant Breed. Stn. (Aberystwyth, Wales)* pp. 97–99.
Clifford, B. C. (1974a). Relation between compatible and incompatible infection sites of *Puccinia hordei* on barley. *Trans. Br. Mycol. Soc.* **63,** 215–220.
Clifford, B. C. (1974b). The choice of barley genotypes to differentiate races of *Puccinia hordei* Otth. *Cereal Rusts Bull.* **2,** 5–6.
Clifford, B. C. (1975). Vulnerability of barley cultivars having the Vada type of resistance to brown rust (*Puccinia hordei* Otth.). *Rep.—Welsh Plant Breed. Stn. (Aberystwyth, Wales),* p. 39.
Clifford, B. C. (1977). Monitoring virulence in *Puccinia hordei:* A proposal for the choice of host genotypes and survey procedures. *Cereal Rusts Bull.* **5,** 34–38.
Clifford, B. C. (1981). Combining different resistances to barley brown rust. *Barley Genet., Proc. Int. Barley Genet. Symp., 4th, 1981,* pp. 479–483.
Clifford, B. C., and Clothier, R. B. (1974). Physiologic specialization of *Puccinia hordei* on barley hosts with non-hypersensitive resistance. *Trans. Br. Mycol. Soc.* **63,** 421–430.
Clifford, B. C., and Roderick, H. W. (1978a). A comparative histology of some barley brown rust interactions. *Ann. Appl. Biol.* **89,** 295–298.
Clifford, B. C., and Roderick, H. W. (1978b). Disease standards for barley foliar diseases. *Rep.—Welsh Plant Breed. Stn. (Aberystwyth, Wales),* pp. 71–73.
Clifford, B. C., and Roderick, H. W. (1981). Detection of cryptic resistance of barley to *Puccinia hordei. Trans. Br. Mycol. Soc.* **76,** 17–24.

Clifford, B. C., and Udeogalanya, A. C. C. (1976). Hypersensitive resistance of barley to brown rust (*Puccinia hordei* Otth.) *Proc.—Eur. Mediterr. Cereal Rusts Conf., 4th, 1976,* pp. 27–29.

Clifford, B. C., Clothier, R., and Roderick, H. W. (1977). The resistance of spring barley cultivars to *Puccinia hordei* Otth. *Rep.—Welsh Plant Breed. Stn. (Aberystwyth, Wales),* pp. 67–69.

Critopoulos, P. (1956). Perpetuation of the brown rust of barley in Attica. *Mycologia* **48,** 596–600.

Cummins, G. B. (1971). "The Rust Fungi of Cereals, Grasses and Bamboos." Springer-Verlag, Berlin and New York.

Dennis, R. W. G., and Sandwith, N. Y. (1948). Aecidia of barley rust in Britain. *Nature (London)* **162,** 461.

Dillard, M. W., and Brown, A. R. (1969). Inheritance of reaction to race 8 of *Puccinia hordei* Otth. in two barley crosses. *Crop Sci.* **9,** 677–678.

d'Oliveira, B. (1939). Studies on *Puccinia anomala* Rost. I. Physiologic races on cultivated barleys. *Ann. Appl. Biol.* **26,** 56–82.

d'Oliveira, B. (1960). Host range of the aecidal stage of *Puccinia hordei* Otth. *Melhoramento* **13,** 161–188.

Ellis, E. A. (1934). Flora of Norfolk, rust fungi. *Trans. Norfolk Nat. Hist. Soc.* **13,** 489–505.

Evans, E., Richard, M., and Whitehead, R. (1969). Effect of benomyl on some diseases of spring barley. *Proc. Br. Insectic. Fungic. Conf., 6th, 1969,* Vol. 3, p. 758.

Falahati-Rastegar, M., Manners, J. G., and Smartt, J. (1981). Effects of temperature and inoculum density on competition between races of *Puccinia hordei. Trans. Br. Mycol. Soc.* **77,** 359–368.

Frecha, J. H. (1970). Inheritance of the resistance to *Puccinia hordei* Otth. in barley. *Bol. Genet. (Engl. Ed.),* **7,** 1–8; summary in *Plant Breed. Abstr.* **42,** 605 (1972).

Frecha, J. H. (1971). Inheritance of resistance to *Puccinia hordei* in barley. *Inf. Tec. Estac. Exp. Reg. Agropecu. Pergamino* **105,** 38–42; summary in *Plant Breed. Abstr.* **44,** 202 (1974).

Frost, A. J. P., Jung, K. U., and Bedford, J. L. (1973). The timing of application of benodanil (BAS 317OF) for the control of cereal rust diseases. *Proc. Br. Insectic. Fungic. Conf., 7th, 1973,* pp. 111–118.

Gassner, G., and Pieschel, E. (1934). Untersuchungen zur Frage der Uredo - überwintering der Getreideroste in Deutchland. *Phytopathol. Z.* **7,** 355–392.

Gilmour, J. (1973). Octal notation for designating physiologic races of plant pathogens. *Nature (London)* **242,** 620.

Golan, T., Anikster, Y., Moseman, J. G., and Wahl, I. (1978). A new virulent strain of *Puccinia hordei. Euphytica* **27,** 185–189.

Grunewaldt, J., and Grunewaldt, G. (1977). Selection of *Puccinia hordei* resistant mutants. *Cereal Rusts Bull.* **5,** 24–26.

Grunewaldt-Stocker, G. (1981). Untersuchungen zur Rassenanalyse von *Puccinia hordei* Otth. und zur Braunrostresistenz der Gerste. Ph.D. Thesis, Universität Hannover.

Henderson, M. T. (1945). Studies of sources of resistance and inheritance of reaction to leaf rust *Puccinia anomola* Rostr. in barley. Ph.D. Thesis, University of Minnesota, Minneapolis.

Hermansen, J. E. (1968). Studies on the spread and survival of cereal rust and mildew diseases in Denmark. *Friesia* **8,** 161–359.

Jenkins, J. E. E., Melville, S. C., and Jemmett, J. L. (1972). The effect of fungicides on leaf diseases and on yield in spring barley in south-west England. *Plant Pathol.* **21,** 49–58.

Johnson, D. A., and Wilcoxson, R. D. (1978). Components of slow-rusting in barley infected with *Puccinia hordei. Phytopathology* **68,** 1470–1474.
Johnson, D. A., and Wilcoxson, R. D. (1979a). Yield losses of fast and slow rusting barleys infected with *Puccinia hordei. Plant Dis. Rep.* **63,**764–768.
Johnson, D. A., and Wilcoxson, R. D. (1979b). Inheritance of slow rusting of barley infected with *Puccinia hordei* and selection of latent period and number of uredia. *Phytopathology* **69,** 145–151.
Johnson, R. (1970). The genetics of resistance of some barley varieties to *Puccinia hordei. Proc.—Eur. Mediterr. Cereal Rusts Conf., 2nd, 1968,* pp. 160–162.
Jones, E. R. L., and Clifford, B. C. (1980). Brown rust of barley. *U. K. Cereal Pathogen Virulence Surv., 1979,* pp. 55–59.
Joshi, L. M. (1971). Barley leaf rust: Sensitivity to external environment and physiologic specialization. *Plant Dis. Rep.* **55,** 1026–1030.
Joshi, L. M., Misra, D. P., Sreekanti, K. R., Lele, V. C., and Kak, D. (1959). Studies of *Puccinia hordei* Otth. the leaf rust of barley in India. *Indian Phytopathol.* **12,** 69–75.
King, J. E. (1972). Surveys of foliar diseases of spring barleys in England and Wales, 1967–70. *Plant Pathol.* **21,** 23–35.
King, J. E. (1977). Surveys of foliar diseases of spring barley in England and Wales, 1972–75. *Plant Pathol.* **26,** 21–29.
King, J. E., and Polley, R. W. (1976). Observations on the epidemiology and effect on grain yield of brown rust in spring barley. *Plant Pathol.* **25,** 63–73.
Kostić, B., Pribaković, M., and Mikić, K. (1979). Reaction of barley varieties to *Erysiphe graminis* f. sp. *hordei* Marchal and to *Puccinia hordei* Otth. in adult stage. *Zast. Bilja* **30,** 119–124 (Engl. summ.).
Kramer, T., Gildemacher, B. H., van der Ster, M., and Parlevliet, J. E. (1980). Tolerance of spring barley cultivars to leaf rust, *Puccinia hordei. Euphytica* **29,** 209–216.
Levine, M. N., and Cherewick, W. J. (1952). Studies on dwarf leaf rust of barley. *U.S., Dep. Agric., Tech. Bull.* **1056,** 1–17.
Levine, M. N., and Cherewick, W. J. (1956). Greenhouse and fieldplot studies on varietal reactions to barley leaf rust. *Can. J. Agric. Sci.* **36,** 127–143.
Lim, L. G., and Gaunt, R. E. (1981). The timing of spray applications against powdery mildew and leaf rust in barley. *Proc. N. Z. Weed Pest Control Conf.* **34,** 195–198.
Lumbroso, E., Anikster, Y., Moseman, J. G., and Wahl, I. (1977). Completion of life cycles of *Puccinia hordei* and *Uromyces scillarum* on detached leaves of their hosts. *Phytopathology* **67,** 941–944.
McDaniel, M. E., and Hathcock, B. R. (1969). Linkage of the Pa_4 and Mla loci in barley. *Crop. Sci.* **9,** 822.
Macer, R. C. F., Johnson, R., and Wolfe, M. S. (1966). *Plant Breed. Inst., Cambridge Annu. Rep.* pp. 119–124.
Mains, E. B. (1930). Host specialization of barley leaf rust, *Puccinia anomala. Phytopathology* **20,** 873–882.
Mains, E. R., and Jackson, H. S. (1924). Aecial stages of the leaf rusts of rye, *Puccinia dispersa* Erikss. and Henn., and of barley, *P. anomala* Rostr. in the United States. *J. Agric. Res. (Washington, D.C.)* **28,** 1119–1126.
Mathre, D. E. (1982). Compendium of barley diseases. *Proc. Am. Phytopathol. Soc.,* pp. 32–34.
Melville, S. C. and Lanham, C. A. (1972). A survey of leaf diseases of spring barley in south-west England. *Plant Pathol.* **21,** 59–66.
Melville, S. C., Griffin, G. W., and Jemmett, J. L. (1976). Effects of fungicide spraying on brown rust and yield in spring barley. *Plant Pathol.* **25,** 99–107.
Moore, M. S., Close, R. C., and Lai, R. (1979). Control of seed-borne and foliar diseases of

barley with new systemic fungicides triadimenol and triadimefon. *Proc. N. Z. Weed Pest Control Conf.* **32,** 267–271.

Moseman, J. (1964). Present status of plant pathological research on barley in the United States. *Barley Genet. 1, Proc. Int. Symp., 1st, 1963,* pp. 250–258.

Moseman, J. G. (1976). Controlled facilitated recurrent selection of barley. *Agron. Abstr.,* p. 57.

Moseman, J. G., and Baenziger, P. S. (1979). Registration of barley composite crosses XXXV, XXXV-A, -B and -C. *Crop Sci.* **19,** 750–751.

Moseman, J. G., and Greeley, L. W. (1965). New pathogenic strains of *Puccinia hordei* among physiologic races identified in United States from 1959 through 1964. *Plant Dis. Rep.* **49,** 575–578.

Moseman, J. G., and Reid, D. A. (1961). Linkage relationship of genes conditioning resistance to leaf rust and powdery mildew in Franger barley. *Crop Sci.* **1,** 425–427.

Moseman, J. G., and Roane, C. W. (1959). Physiologic races of barley leaf rust (*Puccinia hordei*) isolated in the United States from 1956 through 1958. *Plant Dis. Rep.* **43,** 1000–1003.

Moseman, J. G., Tan, B. H., and Baenziger, P. S. (1979). Genes conditioning resistance of *Hordeum spontaneum* to *Puccinia hordei. Agron. Abstr.,* p. 70.

Neervoort, W. J., and Parlevliet, J. E. (1978). Partial resistance of barley to leaf rust, *Puccinia hordei* V. Analysis of the components of partial resistance in eight barley cultivars. *Euphytica* **27,** 33–39.

Newton, M., Peturson, B., and Meredith, W. O. S. (1945). The effect of leaf rust of barley on the yield and quality of barley varieties. *Can. J. Res., Sect. C* **23,** 212–218.

Niks, R. E. (1981). Appressorium formation of *Puccinia hordei* on partially resistant barley and two non-host species. *Neth. J. Plant Pathol.* **87,** 201–207.

Niks, R. E. (1982). Early abortion of colonies of leaf rust, *Puccinia hordei* in partially resistant barley seedlings. *Can. J. Bot.* **60,** 714–723.

Niks, R. E., and Parlevliet, J. E. (1979). Variation for partial resistance to *Puccinia hordei* in the barley composite. XXI. *Cereal Rusts Bull.* **6,** 3–10.

Nover, I., and Lehmann, C. O. (1974). Resistenzeigenschaften im Gerstenund Weizensortiment Gatersleben 18. Prufung von Sommergersten auf ihr Verhalten gegen Zwergrost (*Puccinia hordei* Otth.). *Kulturpflanze* **22,** 25–43.

Nover, I., and Mansfeld, R. (1959). II. Originalarbeiten Resistenzeigenschaften im Gerstenund Weizensortiment Gatersleben, 3. Prufung der Gersten auf ihr Verhalten gegen Zwergrost (*Puccinia hordei* Otth.). *Kulturpflanze* **7,** 29–36.

Parlevliet, J. E. (1975). Partial resistance of barley to leaf rust, *Puccinia hordei.* I. Effect of cultivar and development stage on latent period. *Euphytica* **24,** 21–27.

Parlevliet, J. E. (1976a). Partial resistance of barley to leaf rust, *Puccinia hordei.* III. The inheritance of the host plant effect on latent period in four cultivars. *Euphytica* **25,** 241–248.

Parlevliet, J. E. (1976b). Evaluation of the concept of horizontal resistance in the barley, *Puccinia hordei* host pathogen relationship. *Phytopathology* **66,** 494–497.

Parlevliet, J. E. (1976c). The genetics of seedling resistance to leaf rust *Puccinia hordei* Otth. in some spring barley cultivars. *Euphytica* **25,** 249–254.

Parlevliet, J. E. (1976d). Screening for partial resistance in barley to *Puccinia hordei* Otth. *Proc.—Eur. Mediterr. Cereal Rusts Conf., 4th, 1976,* pp. 153–156.

Parlevliet, J. E. (1977). Evidence of differential interaction in the polygenic *Hordeum vulgare - Puccinia hordei* relation during epidemic development. *Phytopathology* **67,** 776–778.

Parlevliet, J. E. (1978a). Further evidence of polygenic inheritance of partial resistance in barley to leaf rust, *Puccinia hordei. Euphytica* **27,** 369–379.

Parlevliet, J. E. (1978b). Race-specific aspects of polygenic resistance of barley to leaf rust, *Puccinia hordei. Neth. J. Plant Pathol.* **84,** 121–126.

Parlevliet, J. E. (1980). Minor genes for partial resistance epistatic to the Pa_7 gene for hypersensitivity in the barley—*Puccinia hordei* relationship. *Proc.—Eur. Mediterr. Cereal Rusts Conf., 5th, 1980.* pp. 53–57.

Parlevliet, J. E., and Kuiper, H. J. (1977a). Resistance of some barley cultivars to leaf rust, *Puccinia hordei;* polygenic, partial resistance hidden by monogenic, hypersensitivity. *Neth. J. Plant Pathol.* **83,** 85–89.

Parlevliet, J. E., and Kuiper, H. J. (1977b). Partial resistance of barley to leaf rust *Puccinia hordei.* IV. Effect of cultivar and development stage on infection frequency. *Euphytica* **26,** 249–255.

Parlevliet, J. E., and van Eijk-Bos, G. (1978). Induced mutations for virulence to barley in brown rust. *Puccinia hordei* Otth. *Cereal Rusts Bull.* **6,** 3–10.

Parlevliet, J. E., and van Ommeren, A. (1975). Partial resistance of barley to leaf rust, *Puccinia hordei.* II. Relationship between field trials, micro plot tests and latent period. *Euphytica* **24,** 293–303.

Parlevliet, J. E., and van Ommeren, A. (1976). Overwintering of *Puccinia hordei* in the Netherlands. *Cereal Rusts Bull.* **4,** 1–4.

Parlevliet, J. E., and van Ommeren, A. (1981). The development of the leaf rust, *Puccinia hordei* population during winter, spring and early summer in 11 winter barley cultivars. *Neth. J. Plant Pathol.* **87,** 131–137.

Parlevliet, J. E., Lindout, W. H., van Ommeren, A., and Kuiper, H. J. (1980). Level of partial resistance to leaf rust, *Puccinia hordei,* in West European barley and how to select for it. *Euphytica* **29,** 1–8.

Parlevliet, J. E., van der Beek, J. G., and Pieters, R. (1981). Presence in Morocco of brown rust, *Puccinia hordei,* with a wide range of virulence to barley. *Cereal Rusts Bull.* **9,** 3–8.

Polley, R. W., and Clarkson, J. D. S. (1978). Forecasting cereal disease epidemics. *In* "Plant Disease Epidemiology" (P. R. Scott and A. Bainbridge, eds.), pp. 141–150. Blackwell, Oxford.

Priestley, R. H. (1978). The incidence of rust diseases in cereal cultivar trials in England and Wales, 1957–1976. *J. Natl. Inst. Agric. Bot., (G.B.)* **14,** 414–427.

Reinhold, M., and Sharp, E. L. (1980). Some virulence types of *Puccinia hordei* from semi-arid environment. *Proc.—Eur. Mediterr. Cereal Rusts Conf., 5th, 1980,* pp. 167–168.

Reinhold, M., and Sharp, E. L. (1982). Resistance to leaf rust of barley in Southern Texas. *Cereal Rusts Bull.* **10,** 4–10.

Rintelen, J. (1976). Physiological specialization of *Puccinia hordei* Otth. and reactions of barley varieties and cultivars to physiological races. *Barley Genet. 3, Proc. Int. Barley Genet. Symp., 3rd, 1975.* pp. 464–467.

Rintelen, J. (1979a). Survey on the present races of brown rust of barley (*Puccinia hordei*) in the Federal Republic of Germany. *Z. Pflanzenkr. Pflanzenschutz* **86,** 173–179.

Rintelen, J. (1979b). Verfügen unsere Gerstensorten über spezifische Resistenzen gegen physiologische Rassen des Zwergrostes? *Bayer. Landwirtsch. Jahrb.* **56** Jahrg 4, 393–397 (English summary).

Roane, C. W. (1962). Inheritance of reaction to *Puccinia hordei* in barley. I. Genes for resistance among North American race differentiating varieties. *Phytopathology* **52,** 1288–1295.

Roane, C. W. (1976). Description of genes resistant to *Puccinia hordei* Otth. (barley leaf rust) BGS 70, 121, 122, 218. *Barley Genet. Newsl.* **6,** 120–122, 129.

Roane, C. W., and Starling, T. M. (1967). Inheritance of reaction to *Puccinia hordei* in barley. II. Gene symbols for loci in differential cultivars. *Phytopathology* **57,** 66–68.

Roane, C. W., and Starling, T. M. (1969). Genes conditioning reaction to *Puccinia hordei* in barley cultivars Cebada Capa and Franger. *Phytopathology* **59,** 1046 (abstr.).

Roane, C. W., and Starling, T. M. (1970). Inheritance of reaction to *Puccinia hordei* in barley. III. Genes in the cultivars Cebada Capa and Franger. *Phytopathology* **60,** 788–790.

Round, P. A., and Wheeler, B. E. J. (1978). Interactions of *Puccinia hordei* and *Erysiphe graminis* on seedling barley. *Ann. Appl. Biol.* **89,** 21–35.

Shchelko, L. G. (1974). Rasovaia differentsiatsia vozbuditelia i istochrika immuniteta iachmenia Karlikovoi rhovchine. *Tr. Prikl. Bot. Genet. Sel.* **53,** 105–112 (Engl. summ.).

Simkin, M. B., and Wheeler, B. E. J. (1974a). The development of *Puccinia hordei* on barley cv. Zephyr. *Ann. Appl. Biol.* **78,** 225–235.

Simkin, M. B., and Wheeler, B. E. J. (1974b). Overwintering of *Puccinia hordei* in England. *Cereal Rusts Bull.* **2,** 2–4.

Simkin, M. B., and Wheeler, B. E. J. (1974c). Effects of dual infections of *Puccinia hordei* and *Erysiphe graminis* on barley, cv. Zephyr. *Ann. Appl. Biol.* **78,** 237–250.

Skidmore, A. M. (1980). Diclobutrazol, a new systemic fungicide for control of cereal rusts and other diseases. *Proc.—Eur. Mediterr. Cereal Rusts Conf., 5th, 1980,* pp. 243–247.

Starling, T. M. (1956). Sources, inheritance, and linkage relationships of resistance to race 4 of leaf rust (*Puccinia hordei* Otth.) race 9 of powdery mildew (*Erysiphe graminia hordei* El. Marchal.), and certain agronomic characters in barley. Ph.D. Thesis, Iowa State College, Ames, 1955; also *Iowa State Coll. J. Sci.* **30,** 438–439.

Straib, W. (1936). Die bestimmung der physiologischen rassen des gerstenzwergrostes, *Puccinia simplex* (Kcke.) Erikss. et Henn. *Arb. Biol. Reichsanst. Land- Forstwirtsch. Berlin-Dahlem* **22,** 43–63.

Tan, B. H. (1976). Recovery and identification of physiologic races of *Puccinia hordei* from winter barley. *Cereal Rusts Bull.* **4,** 36–39.

Tan, B. H. (1977a). Evaluating host differentials of *Puccinia hordei. Cereal Rusts Bull.* **5,** 17–23.

Tan, B. (1977b). A new gene for resistance to *Puccinia hordei* in certain Ethiopian barleys. *Cereal Rusts Bull.* **5,** 39–43.

Tan, B. H. (1978). Verifying the genetic relationships between three leaf rust resistance genes in barley. *Euphytica* **27,** 317–323.

Teng, P. S., and Close, R. C. (1977a). Spectral reflectance of healthy and rust-infected barley leaves. *Aust. Plant Pathol. Soc. Newsl.* **6,** 7–9.

Teng, P. S., and Close, R. C. (1977b). A preliminary comparison of benodanil and MEB 6447 for control of leaf rust of barley. *Aust. Plant Pathol. Soc. Newsl.* **6,** 55–57.

Teng, P. S., and Close, R. C. (1978). Effect of temperature and uredinium density on urediniospore production, latent period, and infectious period of *Puccinia hordei* Otth. *N. Z. J. Agric. Res.* **21,** 287–296.

Teng, P. S., and Close, R. C. (1980). Effect of solar radiation on survival of *Puccinia hordei* uredospores in New Zealand. *Cereal Rust. Bull.* **8,** 23–29.

Teng, P. S., Blackie, M. J., and Close, R. C. (1977). A simulation analysis of crop yield loss due to rust disease. *Agric. Syst.* **2,** 189–198.

Teng, P. S., Blackie, M. J., and Close, R. C. (1978). Simulation modelling of plant diseases to rationalize fungicide use. *Outlook Agric.* **9,** 273–277.

Teng, P. S., Close, R. C., and Blackie, M. J. (1979). Comparison of models for estimating yield loss caused by leaf rust (*Puccinia hordei*) on Zephyr barley in New Zealand. *Phytopathology* **69**, 1239–1244.

Teng, P. S., Blackie, M. J., and Close, R. C. (1980). Simulation of the barley leaf rust epidemic: structure and validation of BARSIM-1. *Agric. Syst.* **5**, 85–103.

Tranzschel, W. (1914). Kulturversuche mit uredineen in den Jahren 1911–1913. *Vorlauf. Mitt. Mycol. C1.* **4**, 70–71.

Udeogalanya, A. C. C. (1979). Some relationships between *Puccinia hordei*, its barley host and the fungicides oxycarboxin and benodanil. Ph.D. Thesis, University of Wales.

Udeogalanya, A. C. C., and Clifford, B. C. (1978). Genetical, physiological and pathological relationships of resistance to *Puccinia hordei* and *P. striiformis* in *Hordeum vulgare*. *Trans. Br. Mycol. Soc.* **71**, 279–287.

Udeogalanya, A. C. C., and Clifford, B. C. (1982). Control of barley brown rust, *Puccinia hordei* Otth., by benodanil and oxycarboxin in the field and the effects on yield. *Crop Prot.* **1**, 299–308.

Urech, P. A., Pietà, S. D., and Speich, J. (1980). The use of CGA 64250 against cereal rust diseases. *Proc.—Eur. Mediterr. Cereal Rust Conf., 5th, 1980*, pp. 249–254.

Vallega, J., Cenoz, H. P., Favret, E. A., Sarasola, J. A., de Sarasola, M. D. R. C., Tessi, J. L., and Frecha, J. H. (1955). Compartamento de algunas Cebadas con respecto a la raza 14 de *Puccinia hordei* Otth. *Rev. Invest. Agri.* **9**, 187–200.

Vanderplank, J. E. (1963). "Plant Diseases: Epidemics and Control." Academic Press, New York.

von Schmeling, B., and Kulka, M. (1966). Systemic fungicidal activity of 1,4-oxathiin derivatives. *Science* **152**, 659–660.

Walther, U. (1979). Die Virulenz-und Resistenzgensituation bei *Puccinia hordei* West. *Arch. Zuechtungsforsch.* **9**, 49–54.

Walther, U., and Lehmann, C. O. (1980). Resistenzeigenshaften im Gersten- und Weizensortiment Gatersleben 24. Prüfung von Sommerund Wintergersten auf ihr Verhalten gegenüber Zwergrost (*Puccinia hordei* Otth.). *Kulturpflanze* **28**, 227–238.

Waterhouse, W. L. (1927). Studies in the inheritance of resistance to leaf rust *Puccinia anomala* Rostr. in crosses of barley. I. *Proc. R. Soc. N. S. W.* **61**, 218–247.

Waterhouse, W. L. (1948). Studies in the inheritance of resistance to rust of barley. II. *Proc. R. Soc. N. S. W.* **81**, 198–205.

Watson, I. A., and Butler, F. C. (1947). Resistance to barley leaf rust (*Puccinia anomala* Rost.). *Proc. Linn. Soc. N. S. W.* **72**, 379–386.

Whipps, J. M., Clifford, B. C., Roderick, H. W., and Lewis, D. H. (1980). A comparison of development of *Puccinia hordei* Otth. on normal and slow rusting varieties of barley (*Hordeum vulgare* L.) using analyses of fungal chitin and mannan. *New Phytol.* **85**, 191–199.

Widdowson, F. V., Jenkyn, J. F., and Penny, A. (1976). Results from two barley experiments at Saxmundham, Suffolk, measuring effects of the fungicide benodanil on three varieties, given three amounts of nitrogen at two times 1973–4. *J. Agric. Sci.* **86**, 271–280.

Wilson, M., and Henderson, D. M. (1965). "British Rust Fungi." Cambridge Univ. Press, London and New York.

Wilten, W. (1953). Het bestrijden van dwergroest (*Puccinia simplex*) in zomergerst. *Zeventiende Jaarb. Natl. Com. Brouwgerst.* **17**, 72–79.

Zadoks, J. C., Chang, T. T., and Konzak, C. F. (1974). A decimal code for the growth stages of cereals. *Weed Res.* **14**, 415–421.

Zloten, R. R. (1952). Inheritance of reaction of leaf rust in barley. M.Sc. Thesis, University of Manitoba.

7

Corn and Sorghum Rusts

*Arthur L. Hooker**

DeKalb-Pfizer Genetics, St. Louis, Missouri

*Present address: DeKalb-Pfizer Genetics, 3100 Sycamore Road, DeKalb, Illinois 60115.

THE CEREAL RUSTS,
Volume II

ISBN 0-12-148402-5

I. Introduction

Corn (*Zea mays* L.) and sorghum [*Sorghum bicolor* (L.) Moench], along with most grasses, are hosts to rust fungi. *Puccinia sorghi* Schw., *P. polysora* Underw., and *Physopella zeae* (Mains) Cumm. and Ramochar occur on corn and *Puccinia purpurea* Cke. occurs on the sorghums.

A. HISTORICAL PERSPECTIVES

Corn (maize) is an American crop by origin. It has been grown as a food crop for thousands of years in Central America and the adjacent areas in South and North America. All three corn rusts occur here. As corn culture expanded into the United States Corn Belt, Europe, Africa, and elsewhere in the world, *P. sorghi* spread with the crop. However, *P. polysora* remained in the warmer areas of Central, South, and North America until about 1949, when it was accidentally introduced into northwest Africa. From here it spread eastward through all of tropical Africa, the humid areas in and surrounding the Indian Ocean, southeast Asia, the Philippines, and eventually Taiwan (Nattrass, 1952, 1953; Hou *et al.*, 1978). This is a remarkably long distance spread for an obligate parasite. *Physopella zeae,* with the exception of a limited outbreak in Florida followed by its eradication there, has remained confined to Central America.

Sorghum, by contrast, is an African crop by origin, and the greatest variability of sorghum types exists in that area. *Puccinia purpurea* occurs in Africa and apparently has spread with the crop as sorghum was introduced into other areas of the world.

B. CURRENT AREAS OF INTEREST AND CENTERS OF RESEARCH

Research on various aspects of corn and sorghum rusts has been done in different centers throughout the world. Early work on morphology, life cycles, and taxonomy was done at Purdue University in Indiana. Dickson, at Wisconsin, and his students initiated an extensive program on common rust caused by *P. sorghi.* This research was extended and continued in Central America, Iowa, Illinois, and India. Much of this research involved various facets of host–parasite interactions, with considerable attention given to the genetics of resistance. Recently there has been a series of papers from Cambridge, England relating to host and environmental effects on infection, and studies in Hawaii on host resistance and disease losses.

When *P. polysora* spread into Africa, maize research institutes were established in West Africa and in East Africa to study the disease and to bring it under control by resistance. Recent interests in *P. polysora* have centered on fundamental studies in Maryland; epidemiology, disease losses, and host resistance in Mississippi; and physiologic specialization and breeding for resistance in Nigeria.

Research in India has involved both *P. sorghi* on corn and *P. purpurea* on sorghum. The comprehensive research program in Texas on sorghum diseases has included sorghum rust. Other studies have been made elsewhere in the world.

Currently, there are no laboratories giving exclusive and broad attention to any of the corn or sorghum rusts.

II. Common Corn Rust

A. THE PATHOGEN

Puccinia sorghi was first described by Schweinitz from plant tissue he thought to be sorghum. While the species name would suggest sorghum is a host, *P. sorghi* does not infect *Sorghum* spp. The uredial and telial hosts are corn, annual teosinte [*Euchlaena* (*Zea*) *mexicana* Schrad.], and perennial teosinte [*E.* (*Zea*) *perennis* Hitch.]. The aecial hosts are several species of *Oxalis,* mainly *Oxalis corniculata* L., *O. stricta* L., and *O. europaea* Jord. (Mains, 1934; Robert, 1962; Ullstrup, 1977).

Uredia are without paraphyses, circular to elongate, cinnamon-brown, and produced on both surfaces of the leaf. As the dusty spores are released, sections of host epidermis are visible at the margins of uredia. Finely to moderately echinulate cinnamon-brown urediospores are globoid to ellipsoid, 23–29 × 26–32 μm and have three to four equatorial pores.

Open telia are circular to elongate and become brownish-black toward plant maturity as urediospores are replaced by teliospores. Teliospores are two-celled, chestnut-brown, slightly constricted at the septum with a thickened apex and 16–23 × 29–54 μm. The side wall, apical wall, and pedicel measure 1–2, 5–7, and 50–80 μm, respectively (Cummins, 1941, 1956, 1971; Ullstrup, 1977).

The aecial infection is not known to occur naturally in the United States. However, infection is common in Switzerland (Zogg, 1949),

South Africa (Pole-Evans, 1923), India (Misra, 1963; Misra and Sharma, 1964), and Mexico (Borlaug, 1946).

Several investigators have reported on the cytology, life cycle, and heterothallism of *P. sorghi* (Allen, 1933, 1934; Arthur, 1904; Cummins, 1931; Mains, 1934; Pavgi, 1975; Pavgi *et al.*, 1960; Pole-Evans, 1923; Zogg, 1949). The fungus has been grown through its full life cycle on detached leaves of its hosts (Hooker and Yarwood, 1966).

The disease cycle of *P. sorghi* differs in various parts of the world. In Hawaii and other areas where corn is grown continuously, the pathogen simply reproduces on living corn in the uredial stage and spreads from old to newer planted corn throughout the year. In the United States mainland, the fungus spreads northward during mid-season into the Corn Belt and east from the south, where it had overwintered presumably in Mexico. A similar cycle prevails in southern Europe. In Switzerland, South Africa, highland Mexico, India, and perhaps elsewhere, where the aecial stage is needed for survival from one season to the next, the fungus passes through the winter as teliospores. In highland Mexico and in South Africa, it is not uncommon to see pycnial, aecial, uredial, and telial stages of the fungus together in the same field.

B. ENVIRONMENT

Infection and spread of *P. sorghi* are influenced by the environment (Mahindapala, 1978a,b,c,d; McKeen, 1951; Mederick and Sackston, 1972; Melching, 1975; Pavgi and Dickson, 1961; Smith, 1926; Syamananda and Dickson, 1959; Weber, 1922). Mature teliospores have a dormancy that must be broken by low temperatures and weathering before germination occurs. Dry urediospores remain viable for several days at moderate temperatures and store for long periods at 10° to −26°C (von Meyer, 1963a). Urediospores have a self inhibitor, and spore dispersal is required for germination. The optimum temperature for urediospore germination and germ tube growth is in the range of 15°–25°C. The minimum is near 4°C and the maximum is under 31°C. Germination is maximum at 100% relative humidity, and most spores do not germinate at 98% or lower. Light favors urediospore germination. A minimum of about 4 hr is required for infection, and the rate of infection increases after 6–12 hr. Generation time is about 16, 10, 7, and 5 days at 10, 15, 20, and 25°C, respectively. Spore production is higher at 20° than at 15° and least at 10°C. Both temperature and light can influence seedling reactions. In general, field spread by urediospores occurs best when air is humid and temperatures drop overnight, resulting in dew formation. Rains tend to wash spores from the leaves and air into the soil.

Common corn rust can be found from mid-season on in most temperate areas of the world where corn is grown. In tropical areas it occurs both on seedlings and older plants but mainly in highland areas at 1000–1400 m altitude and above (Renfro and Ullstrup, 1976; Schieber, 1965, 1971; Schieber *et al.*, 1964).

Seasonal disease development depends on the area of the world. The primary factors seem to be presence of the pathogen, humidity, and temperature. In the United States mainland, rust development in the southern states is greatest during the early cool part of the growing season and subsides when weather becomes warmer. In the Corn Belt and the states to the east and areas north into Canada, rust generally first appears after anthesis and develops most abundantly during the latter part of the growing season. Early-season epidemics are unusual (McKeen, 1951; Wallin, 1951; Wood and Lipscomb, 1956). Often spores are carried through the air and deposited late in the season as showers over fields. When this happens, the upper leaves are infected while the lower ones have few pustules on them. When viewed from a distance the top leaves appear to have been killed by frost. In Hawaii, when corn is grown continuously all year in the same area, rust develops on seedlings and throughout the growing season. Infection often is in bands across the leaf resulting from spores infecting leaves in the whorl position. Zogg (1949) described rust spread during the epidemic years 1945–1947 in the Rheintal valley of Switzerland, where the primary inoculum is aeciospores produced on *Oxalis.*

C. Effect of Disease on the Host

Several approaches have been taken to assess the severity of rust infection on corn plants. Percentage leaf area infected has been estimated for individual leaves at different leaf positions on the plant or at different time intervals during the growing season. Response-surface models have been tried for sweet corn (Teng and Montgomery, 1981). Several investigators have made visual estimates of entire plant leaf area infected late in the growing season but when leaves are still green (Hooker, 1962b; Kim and Brewbaker, 1976a; Russell, 1965). Conceivably, these assessments would be more meaningful if estimates of uninfected tissue were taken at several intervals during the grain-filling period and an index were calculated similar to that developed for corn leaf blights (Hooker, 1979b). Some changes in methodology may be needed because rust infection in late season may be most severe on the upper leaves, rather than as a gradient up the plant from the lower to the upper leaves as is often the situation with fungal leaf blights.

The main effect of common rust is a reduction of grain yield (Hooker, 1962b; Kim and Brewbaker, 1976a; Kushalappa and Hegde, 1970; Martinez, 1977; Russell, 1965). While only small losses are generally experienced in the major corn-growing areas of the world, the disease has the potential to damage susceptible corn. In temperate areas such as Argentina and the United States Corn Belt, yield losses up to 25% have been measured. Several United States experiments employed nearly isogenic resistant hybrids and their susceptible counterparts, so the loss measurements were quite accurate. In these trials the magnitude of yield loss for hybrids were about 4, 6, 15, 21, and 24%, with plants having 10, 30, 50, 60, and 70% leaf area covered with rust pustules 50 days after anthesis, respectively (Hooker, 1962b; Russell, 1965). In Hawaii, where rust infection occurs at early plant growth stages, 80% rust resulted in a 35% average grain yield loss for several hybrids (Kim and Brewbaker, 1976a). Some individual hybrids were reduced in yield by 75%. Much of the yield loss is a reduction in seed size. This seed size reduction is quite important to the hybrid seed corn industry in seed production fields where small-seeded susceptible inbreds are grown.

Rust infection also affects the plant in other ways (Kim and Brewbaker, 1976a; Martinez, 1977). These include a reduction in plant height, fresh plant weight, ear length, ear diameter, oil content, and protein content, and an increase in stalk rot. Grain from infected plants often has less moisture at harvest. Sweet corn fresh weight and quality can be reduced (Rich and Waggoner, 1979). Corn in late planted fields often sustains the greatest loss.

D. PATHOGEN VARIATION

Like most cereal rusts, *P. sorghi* isolates differ in virulence to plants having major alleles for resistance (Dickson *et al.*, 1959; Hooker, 1963, 1979a; Kim and Brewbaker, 1976b; Lee *et al.*, 1963; Payak *et al.*, 1974; Robert, 1962; Roy and Prasada, 1966; Russell and Hooker, 1959; Stakman *et al.*, 1928; and others). Because it did not seem important to do so, no attempt has been made to establish a standard or widely used set of differential host genotypes or to determine the several hundred possible physiologic races that can be identified and described (Hooker, 1969). However, experience has shown that genes for virulence and avirulence in *P. sorghi* are not randomly distributed throughout the world (Hooker, 1979a).

Other forms of variation in the pathogen are known. Biotypes of *P. sorghi* differ in aggressiveness as measured by spore yield during propagation on susceptible seedlings. Biotypes of *P. sorghi* produce different antigens, which can be distinguished by serology (Flangas and Dickson, 1961a,b). Mains (1934) studied host specialization in *P. sorghi* to *Oxalis* species. Pycnia and aecia were produced on *O. corniculata* by all isolates tested, but only certain isolates were able to infect *O. europaea*.

E. EXPRESSION OF RESISTANCE

Resistance to common rust is clearly of two types. Plants of different genotypes show a quantitative variation in number of pustules that form or a qualitative variation in the kind of infection types that result. These two forms, and perhaps others, interact and function together in the adult plant (Hooker, 1967b, 1969, 1973).

Ratings on plant reaction expressed as pustule number are usually taken in the field one or more times after anthesis. This resistance is often called mature-plant resistance. The conventional way to measure it is to estimate the percentage of leaf area infected, using as a guide that 37% of the leaf surface area as pustules represents 100% of the leaf infected (Peterson *et al.*, 1948). When data are taken only once, ratings are most meaningful when made about 4 or 5 weeks after anthesis. The resistance may not persist later in the growing season, and it becomes more difficult to distinguish rust from other leaf diseases. When rust infection is initiated at about anthesis and secondary spread within fields occurs throughout the grain-filling period, infection is relatively uniform among leaves on the same plant. When disease occurs as a result of a heavy late season infection from spores produced in other fields, only the upper leaves may be infected. In addition, later-maturing genotypes, although not always the most susceptible, may show the most infection. Genotypes with only this mature-plant type of resistance are usually fully susceptible as seedlings, although as older plants in the field they may differ widely in reaction.

Ratings taken as infection type are usually used in seedling tests. The resistance continues to function as the plant gets older, but the infection types become less distinct. Those most commonly seen on corn seedlings are tabulated below.

Type	Description
0;	Small chlorotic flecks
1−	Small necrotic spots
1	Small pustules surrounded by necrotic tissue
2	Small pustules surrounded by chlorotic areas
3	Medium sized sporulating pustules without chlorosis
4	Large sporulating pustules

Sometimes a mixture of small and large pustules are interspersed over the leaf. When the dispersion is uniform it is usually designated as an X infection type. An unusual reaction, designated the Z infection type, consists of resistant flecking in the more mature leaf tip but sporulating pustules on the less mature portions of the leaf that were infected in the whorl (Van Dyke and Hooker, 1969a).

Other forms of reaction are also seen but not extensively studied. In the field some genotypes of corn have smaller uredia, have fewer spores, or have teliospores develop within the uredium earlier in the season than other genotypes. The phenomenon of slow rusting has been mentioned (Sharma, 1977; Wilcoxson, 1981), but it and mature-plant resistance have not been clearly distinguished from each other.

The fundamental relationships between the aecial or uredial host and *P. sorghi* are quite well known. Allen (1934) has studied the aecial stage and Hilu (1965) and others (von Meyer, 1963b; Weber, 1922; Wellensiek, 1927) the uredial stage histologically. Hilu (1965) studied in detail compatible reactions in nearly isogenic host seedlings where either the plant had no known genes for resistance or the allele for resistance was ineffective because the plants were infected with a virulent rust biotype. The light-microscope investigations were extended to study the ultrastructure of corn and *P. sorghi* (Van Dyke and Hooker, 1969b) and *Oxalis* and *P. sorghi* (Rijkenberg and Truter, 1974). Little attention has been given to the physiology of host–parasite interactions (Dickson *et al.*, 1959; Kim *et al.*, 1978).

The nature of mature-plant resistance is largely unknown. Uredia on resistant plants are reduced in number or their appearance delayed. Peletz (1971) showed plant genotype, plant age, and leaf position all to have an effect on urediospore germination. Significantly more spores germinated on the leaf surfaces of susceptible than of resistant inbreds or hybrids.

The histology of incompatible reactions has been studied with the light microscope (Hilu, 1965) and with the electron microscope (Van Dyke and Hooker, 1969b). These studies have revealed much detailed information. Host cells die in and near the infection sites, but when this occurs depends upon the host gene for resistance. Ultrastructural studies suggest that injured or dead host cells are unable to provide materials required by the fungus to grow.

F. GENETICS OF HOST–PATHOGEN INTERACTIONS

Two forms of host–pathogen interactions must be considered from a genetic point of view: (1) interactions that vary quantitatively and (2) interactions that vary qualitatively.

Resistance, expressed in the number of uredia per unit of leaf area on adult plants in the field, ranges continuously from plants having only a few uredia to those having most of the leaf area covered with uredia. Inheritance of this form of plant reaction has been studied in Illinois (Hooker, 1962b, 1967a,b, 1969, 1973), Hawaii (Kim and Brewbaker, 1977), and India (Sharma, 1977; Wilcoxson, 1981). While different plant genotypes were involved and different statistical procedures used to analyze the data collected, these studies all revealed the same general results, and similar conclusions were drawn from each of them. All investigators concluded that the mature-plant quantitatively expressed form of resistance to rust was inherited as a polygenic trait. The gene effects were largely additive. There was a high general combining ability for resistance, and heritability estimates were high.

Resistance expressed as infection type on seedlings and older plants shows qualitative variation and discrete segregation. Mains (1926, 1931) was the first to show in corn that this form of resistance was inherited as a single dominant gene. The gene (*Rp*) was subsequently shown to be located in the short arm of chromosome 10 (Rhoades, 1935; Rhoades and Rhoades, 1939; Russell and Hooker, 1962). Now at least six loci on three chromosomes have been identified for resistance to *P. sorghi.*

The distal end of the short arm of chromosome 10 is an interesting region when resistance to any rust is considered. Located here, in addition to gene *Rp*, are *Rp5* and *Rp6* for resistance to *P. sorghi* (Hagan and Hooker, 1965; Wilkinson and Hooker, 1968) and *Rpp9* for resistance to *P. polysora* (Ullstrup, 1965). The total map distance is about 3.0 units. Using conventional genetic procedures and an array of *P. sorghi* biotypes to distinguish genetic differences in plants, 14 alleles for resistance at the *Rp* locus, designated *Rpa* to *Rpn* inclusive and all dominant to *rp,* were identified in corns from several countries (Hagan and Hooker, 1965; Hooker, 1962a,b, 1969, 1973; Hooker and Le Roux, 1957; Hooker and Russell, 1962a,b; Lee *et al.*, 1963; Russell and Hooker, 1959; Wilkinson and Hooker, 1968). Corns from Uruguay and Turkey have *Rp5* and *Rp6* (Hagan and Hooker, 1965; Wilkinson and Hooker, 1968).

What was first thought to be a series of alleles at *Rp* are now known in some instances to be very closely linked genes (Hooker and Saxena, 1971; Saxena and Hooker, 1968). Testcross progenies of up to 19,000 plants were used in detailed studies. Recombination values between *Rpg* and *Rpl, Rpa* and *Rpk, Rpa* and *Rpc, Rpc* and *Rpk,* and *Rpb* and *Rpc* of 0.37, 0.27, 0.22, 0.16, and 0.10%, respectively, were observed. When *Rpd* was tested with other alleles, no recombinations were detected. Reciprocal recombinants of *Rpc* and *Rpk* were studied for reaction to a series of rust cultures (Hooker, 1979a). The resistance of the *Rpc–Rpk* genotype was equivalent to the additive effect of *Rpc* and *Rpk*. There were no lesser or greater effects. These studies suggested that *Rp* is a chromosome region with functional and nonfunctional genes closely linked together. Considering *Rp5, Rp6,* and *Rpp9* in addition to the *Rp* complex, it appears that the terminal end of chromosome 10 is physiologically active in conditioning resistance to rusts and may have evolved through repeated duplication with subsequent structural modifications of individual genes.

Six alleles for resistance have been identified at the *Rp3* locus on chromosome 3 (Hooker, 1963; Russell and Hooker, 1962; Wilkinson and Hooker, 1968). Alleles *Rp3b* and *Rp3c* exhibit a reversal of dominance to two biotypes of *P. sorghi* (Hooker and Saxena, 1967). Conceivably, these genes exhibit a dosage effect, but it is also possible that closely linked dominant and recessive genes occur at *Rp3*. However, such genes were not revealed by techniques similar to that used with *Rp* (Saxena and Hooker, 1974).

Locus *Rp4* on chromosome 4 is represented by two alleles (Hagan and Hooker, 1965; Russell and Hooker, 1962; Wilkinson and Hooker, 1968). Allele *Rp4b* results in the expression of the Z reaction type with an avirulent rust culture (Van Dyke and Hooker, 1969a).

Recessive genes and modifying genes for resistance to *P. sorghi* are also known (Hooker, 1962b, 1969; Malm and Hooker, 1962). Resistance may be conditioned by one, two, or three recessive genes.

Resistance seen as infection type is also expressed by teosinte and *Tripsacum* (Bergquist, 1981; Malm and Beckett, 1962). A dominant gene from *Tripsacum dactyloides* that occurs within 0.3 map units of

Rpd raises the possibility that genes in corn for rust resistance on chromosome 10 may have come from *Tripsacum* (Bergquist, 1981).

The genetics of virulence and avirulence in *P. sorghi* has been considered in preliminary studies (Flangas and Dickson, 1961a,b; Pavgi, 1966). They indicated but did not prove a variance from the gene-to-gene model.

G. CONTROL

The most feasible means of disease control is through host resistance. It has been recommended that mainly the adult-plant (uredia number) form of resistance be used in agriculture (Hooker, 1967a,b, 1969). This form of resistance is thought to be effective against all biotypes of *P. sorghi,* and it is relatively easy through breeding and selection to accumulate a high frequency of alleles for resistance. The use of mature-plant resistance is believed to be the main reason why common corn rust has not been a major disease in the United States Corn Belt. It is recommended that resistance expressed as infection type be used primarily as a supplement to mature-plant resistance and a way of increasing the value of elite but rust-susceptible genotypes for the time they are useful in agriculture.

Avoiding new plantings adjacent to older infected corn and early planting in areas where *P. sorghi* does not overseason help in disease control. Adequate fertility and good weed control do not prevent rust but do reduce plant stress and yield losses when plants are infected.

Protectant chemicals—e.g. dithane M-45, maneb, zineb, and related compounds—are effective (Kushalappa and Hegde, 1970; Martinez, 1977; Mederick and Sackston, 1972; Singh and Musymi, 1979; Teng and Montgomery, 1981; Townsend, 1951) but, mainly for economic reasons, are not widely used as a means of disease control. Chemical foliar sprays have been used to a limited extent in high-value seed and sweet corn production fields.

III. Southern Corn Rust

A. THE PATHOGEN

Puccinia polysora infects *Erianthus alopecuroides* (L.) Ell., teosinte, *Tripsacum dactyloides* L., *T. lanceolatum* Rupr., *T. laxum* Nash., and *T. pilosum* Scrib. and Merr., as well as corn (Cummins, 1971; Robert,

1962; Schieber, 1975; Schieber and Dickson, 1963; Ullstrup, 1977). Both uredia and telia are produced. The aecial stage is unknown.

Uredia are without paraphyses, circular, orange-red, and produced on both sides of the leaf. Echinulate, yellowish-golden-brown urediospores are ellipsoid or ovoid, 23–29 × 29–36 μm, and have four to five equatorial pores. Some urediospores are atypical and smaller in size.

Subepidermal telia are brownish-black, circular to elongate, and are most often found along the midrib on the underside of the leaf but may appear in a circle around individual uredia. Teliospores are two-celled, chestnut-brown, angular to ellipsoid or oblong, and 20–27 × 29–31 μm. The side wall, apical wall, and pedicel measure 1.5, 1.5–2.5, and 30 μm, respectively (Cummins, 1941, 1956, 1971; Ullstrup, 1977). The covered telia and more angular teliospores with only slightly thickened apical wall help distinguish *P. polysora* from *P. sorghi.*

Puccinia polysora propogates itself through the repeating uredial stage. Teliospores are not known to be functional. The pathogen overseasons on its host and depends on wind dispersal of spores to spread into areas where it cannot persist between seasons.

B. ENVIRONMENT

Hot humid conditions are necessary for the survival and spread of *P. polysora.* The urediospores show a narrow optimum temperature range from about 23°–28°C for germination (Melching, 1975), and they do not remain viable long at temperatures below 20°C (von Meyer, 1963a). Melching (1975, 1981) has studied the factors favoring infection with artificial inoculations.

Puccinia polysora is primarily tropical or subtropical in distribution, mainly because of its temperature sensitivity, but under favorable conditions and high inoculum load can spread from tropical into temperate regions. In the United States, southern rust is confined mainly to the lower Mississippi River valley, but the pathogen in some years has spread as far north as Illinois and Wisconsin (Hooker, 1961; Pavgi and Flangas, 1959). Epidemics occurred in the United States in 1972, 1973, 1974, 1979, and 1981 (Futrell, 1975; A. L. Hooker, personal observation; Leonard, 1974; Sim, 1980).

In areas where corn is grown continuously the pathogen can spread from infections on old corn plants to young plants and infect them at any growth stage, but in most areas the disease attains its full severity at or about flowering time for the corn plant (Futrell, 1975; Storey *et al.*, 1958). Infections occur mainly on the leaves but can occur on leaf sheaths, husks, and sometimes on the stalk. Primary inoculum and secondary spread is by airborne urediospores (Cammack, 1958a). Futrell (1975) associated the 1972, 1973, and 1974 epidemics in the lower Mississippi River valley with a cropping practice change. Corn growers tried to achieve two crops of corn per year to supply the increased demand for grain by livestock and poultry producers in the region. This was achieved by planting in February and harvesting in June and replanting in late June and harvesting in October. Because the growing season was short for each crop, early maturing hybrids were introduced from the more northern Corn Belt states. These hybrids were highly susceptible to *P. polysora,* and they allowed an early buildup of rust inoculum, which subsequently spread northward. Southerly winds, warm humid conditions, and susceptible plants together resulted in large areas of rust-infected plants. The source of primary inoculum to the Gulf Coast is unknown, but it most probably is blown in from the Caribbean area.

C. Effect of Disease on the Host

Several methods of disease assessment are possible. Disease severity has been rated by Rodriguez-Ardon and associates (1980) on a scale ranging from 0 to 10 where 0 is equivalent to no uredia and 10 equivalent to leaves killed by rust. They also took similar ratings again 10 days after anthesis on the lower, middle, and upper third positions of the leaf canopy. Futrell (1975) followed the usual cereal rating system of percentage leaf area infected and the type of uredia present. Ratings similar to that developed by Hooker (1979b), as discussed in Section II,C, could be developed for southern rust.

Southern rust is a serious disease and economically important in warm humid areas unless controlled. *P. polysora* can kill corn plants 7–10 days after infection appears. Leaves fire rapidly, probably because of increased respiration rates in infected leaves. This is followed by rapid utilization of available food materials in the leaves and stalks and a rapid loss of moisture.

When *P. polysora* was introduced into Africa and spread through the continent, yield losses the first few years the disease was in a country were substantial and often in excess of 50% over large areas (Cammack, 1959a; Ellis, 1954; Hemingway, 1955; Rhind *et al.*, 1952; Vanderplank, 1968; Wood and Lipscomb, 1956). In the Philippines, 80–

84% losses were reported on some susceptible corn cultivars (Reyes, 1953). Cultivars with high yield potential sometimes had a greater yield loss when diseased than cultivars with a low yield potential. In all areas, losses were observed to be greatest when infection occurred early in the growing season or on late-planted corn.

Yield losses have also been high in United States trials and observations made under conditions less favorable for rust than prevail in the tropics. With inoculation, losses of about 50% have been reported in a greenhouse test and 24–37% in a field test (Melching, 1975). Futrell (1975) calculated about a 45% yield loss with natural infection in the field. Yield losses were studied in Mississippi trials using inoculated, nearly isogenic, resistant and susceptible hybrids at three different planting dates (Rodriguez-Ardon *et al.*, 1980). The resistant hybrids had almost no rust, and the susceptible hybrids had about 75% of the leaf surface covered with uredia. Yield losses ranged from 4 to 45%, depending on planting date. Part of the loss was due to a reduction in kernel number and part to a reduction in kernel size. The authors stated they believed corn plants, under the conditions of their tests, would have been killed prior to producing grain had the plants been inoculated as early as the seven-leaf stage.

Other effects of rust infection have been noted (Futrell, 1975; Melching, 1975; Sim, 1980). Forage yields and quality are reduced, and infected plants may die prematurely from stalk rot and may lodge more.

D. PATHOGEN VARIATION

Puccinia polysora varies in virulence to corn lines having major genes for resistance. This has been observed in East Africa (Ryland and Storey, 1955), West Africa (Lallmahomed and Craig, 1968), Central America (Schieber and Dickson, 1963), and in the United States (Robert, 1962; Ullstrup, 1965). Several physiologic races have been described.

Urediospores also differ in morphology. Typical spores are symmetrical with four to five equatorial pores. Occasionally some smaller spores occur that are flattened, with usually two pores on each face (D. B. O. Savile, personal communication). Apparently, only the symmetrical spores have been observed in Africa but both types occur in Cuba, Jamaica, North Borneo, and the Philippines (Cammack, 1958b; Savile, Volume I, Chapter 3, Section V,A). The significance of the smaller flattened spores and their distribution is unknown.

E. EXPRESSION OF RESISTANCE

Resistance can be expressed in two main ways, which interact with each other.

One important way resistance is expressed is in terms of amount of tissue not infected (uredia number), but there is little published literature on this aspect. This resistance is quantitative and ranges from nearly rust-free tissue to that heavily rusted. Since rust infection may be mainly on the upper portion of the leaf canopy, it is often desirable to make ratings of percentage uninfected tissue at different positions within the canopy. Multiple ratings and the calculation of an index, as discussed in Section II,C, may be desirable.

Generally four different infection types are seen when seedlings of various corn genotypes are inoculated with different isolates of *P. polysora* (Lallmahomed and Craig, 1968; Robert, 1962; Ryland and Storey, 1955; Storey and Howland, 1957). These represent the other main type of resistance. Authors have given different symbols to the classes, but the plants can be described as follows:

Highly resistant—Chlorotic to necrotic flecks but no sporulating uredia.
Resistant—Necrotic lesions with a small uredium or a small uredium containing a few spores.
Mesothetic or intermediate—Mixed reaction containing highly resistant or resistant infection types interspersed among large sporulating uredia.
Susceptible—Open uredia with or without chlorosis and containing moderate to many spores.

Infection and establishment of the fungus in susceptible and resistant plants is similar to that seen with other cereal rusts (Cammack, 1959b; Melching, 1975; von Meyer, 1963b; Santiago-Oro and Exconde, 1974).

F. GENETICS OF HOST–PATHOGEN INTERACTIONS

While the author is unaware of published data, amount of tissue not infected, as in *P. sorghi*, is probably a quantitative character with polygenic inheritance.

Based on seedling data, 11 major genes that condition resistance to *P. polysora* measured by infection type have been identified or implied to exist in different areas of the world. Proper genetic tests have not been made to fully determine the number of genes or loci involved. Rather, the genes have been identified mainly in relation to the races of *P.*

polysora to which they condition resistance. *Rpp1* and *Rpp2* were identified in Kenya from Central American corn (Storey and Howland, 1957, 1959). *Rpp1* is fully dominant and conditions resistance to race EA1 of *P. polysora* but not to race EA2. *Rpp2* conditions an intermediate reaction to both races and can be modified to near-complete susceptibility by other host genes. The two genes are linked with a recombination value of about 12.23%. Genes *Rpp3* and *Rpp8* inclusive were implied to exist in corn lines that expressed resistance to *P. polysora* races 3–8 (Robert, 1962). Gene *Rpp9* conditions resistance to race 9 in Indiana, and this gene is linked to *Rpd* for resistance to *P. sorghi* by about 1.6 crossover units (Ullstrup, 1965). Inbred B1138T also has a single dominant gene for resistance to the race 9 isolate of the pathogen that occurred in Mississippi (Futrell *et al.*, 1975). Other work in Kenya resulted in the identification of genes *Rpp10* and *Rpp11* from Colombian and Mexican corn (Storey and Howland, 1967). *Rpp10* is fully dominant and conditions a highly resistant reaction to races EA1 and EA2 and is, therefore, different from *Rpp1*. *Rpp11* conditions resistance to both of these races but the reaction is a large necrotic area and the resistance is not fully dominant. There was no evidence for linkage between *Rpp10* and *Rpp11*.

G. CONTROL

Where southern rust continually occurs, the main means of disease control is by resistant cultivars or hybrids. Breeding for resistance is part of all corn breeding programs in the tropics (Fajemisin, 1976). Numerous sources of resistance expressed by seedlings have been identified (Futrell *et al.*, 1975; Robert, 1962; Schieber and Dickson, 1963; Stanton and Cammack, 1953; Storey and Howland, 1957), and major genes for hypersensitive resistance occur with low frequency in Caribbean and Mexican corns. However, the field resistance or tolerance of these corns does not entirely depend upon these genes; they are reinforced by a polygenic system.

At first, major genes for hypersensitive resistance were used in breeding in East and West Africa following the introduction and widespread distribution of *P. polysora* (Storey *et al.*, 1958). However, before these resistant cultivars became widely distributed in agriculture, the local cultivars had become resistant through mass selection by native farmers (Robinson, 1976; Vanderplank, 1968). The gene pools in local cultivars in tropical Africa changed substantially from high susceptibility to resistant in 10–15 crop generations. Currently, there are three main types of corn germ plasm used in tropical Africa: (1) native cultivars with increased rust resistance, (2) cultivars improved for yield and rust resistance, and (3) a mixture of native and improved cultivars.

Vanderplank (1968) makes the interesting statement that resistance to *P. polysora* must be disadvantageous to the plant in areas where resistance is not needed. He cites as evidence for this the African experience where prior to 1949 all native cultivars were susceptible to rust and that in the initial trials to locate sources of resistance (Stanton and Cammack, 1953) none was found in corns from areas of the world where *P. polysora* did not occur. In addition, some of the most productive corns in the world, those of the United States Corn Belt, were susceptible to southern rust (Futrell *et al.,* 1975). They are usually killed by *P. polysora* when planted in tropical Africa during the wet season.

Several cultural practices help reduce disease. These include adjusting planting date to avoid humid conditions during plant growth, maintaining good soil fertility, and avoiding late planting that puts the crop at its most vulnerable plant growth stages during the time when abundant inoculum is available from earlier planted fields.

Several kinds of protectant chemicals—e.g., HOE 2873, HOE 6052, HOE 6053, sulfur, and zineb—have been used on experimental bases to control *P. polysora* infection (Ellis, 1954; Onofeghara and Kapooria, 1975; Storey *et al.,* 1958). All delayed rust development and substantially increased grain yields under conditions favorable for rust development.

IV. Tropical Corn Rust

Physopella zeae infects corn, teosinte, *Tripsacum dactyloides, T. laxum, T. lanceolatum,* and *T. pilosum* (Robert, 1962; Ullstrup, 1977).

Uredia are round to oval, often small and beneath the epidermis, and tend to occur in groups. Frequently, the margin of the uredium is black in color while the center appears white to pale yellow. The uredia are without paraphyses. Echinulate hyaline-yellowish urediospores are elliptical or ovoid, 12–20 × 18–30 μm, and have about five equatorial pores. Telia remain covered and are black or brown. Teliospores are unicellular, golden to light chestnut-brown, cuboid or oblong in chains of two or four spores, and 10–18 × 12–20 μm. The side and apical walls measure 1.5–2 and 3–4 μm, respectively. There is no pedicel (Cummins, 1941, 1956; Cummins and Ramachar, 1958; Mains, 1938).

The disease is restricted to the tropical areas of Central America from Mexico to Colombia (Schieber and Dickson, 1963; Ullstrup, 1977). Outbreaks of this rust are sporadic, and most widely found in cool open country or mountainside fields. Spores are released when the host tissue splits or disintegrates. Secondary spread is by urediospores. No alternate host of the fungus is known. Tropical rust appears to be of no economic importance (Renfro and Ullstrup, 1976; Ullstrup, 1977). Little attention has been given variation in *P. zeae.* Robert (1962) noted that a resistance corn selection from Turkey showed small pinpoint uredia when inoculated with a Peruvian isolate but large uredia when inoculated with Nicaraguan isolates of the pathogen.

The forms of rust resistance expressed by corn to *P. zeae* include chlorotic flecks, small uredia, and mixed or mesothetic (type X) infection types (Robert, 1962; Schieber and Dickson, 1963). Genetic studies of resistance have not been made. In their studies of host reaction, Schieber and Dickson (1963) observed segregation for reaction type within some of the Guatemala lines. No specific control measures are needed. Sources of resistance have been identified (Robert, 1962; Schieber and Dickson, 1963), and breeding for resistance would be possible.

V. Sorghum Rust

A. THE PATHOGEN

Puccinia purpurea infects only species of *Sorghum* (Tarr, 1962), although an isolate from Sudan grass in Wisconsin was reported to produce infection on *Oxalis corniculata* and the aeciospores that then developed produced infections on corn (Le Roux and Dickson, 1957). However, subsequent studies showed this isolate (race 13) was morphologically similar to *P. sorghi* (Pavgi, 1972). It was concluded that *P. purpurea* was a valid species and morphologically distinct from the rust species that infect corn. Johnston and Mains (1931) inoculated several cultivars of corn with *P. purpurea* but did not achieve infection. The host range of *P. purpurea* includes *S. biocolor* (L.) Moench (grain sorghums, sweet sorghums, and broomcorns), *S. sudanense* (Piper) Staph., (sudangrass and other grass sorghums), *S. halpense* (L.) Pers. (johnsongrass), *S. almum* (columbusgrass), *S. virgatum* (tunisgrass), *S. nitidum, S. verticillifloru, S. nitens,* and *S. arundinaceum* (Johnston and Mains, 1931; Tarr, 1962).

Uredia are produced in purplish or tan spots, depending on the ability of the host plant to produce purple pigments, and on both sides of the leaf. Echinulate, cinnamon-brown urediospores are spherical to ovate-ellipsoid, 23–29 × 30–40 μm, and have five to eight scattered pores. The uredia, especially near the margin, contain numerous clavate-capetate paraphyses, which usually are curved and have a brownish-purple wall.

The dark-chocolate-brown open telia are often larger than the uredia. Teliospores are two-celled, chestnut-brown, ellipsoid-oblong, and 24–30 × 40–50 μm. The side wall, apical wall, and pedicel measure 3–3.5., 4–5, and 95 μm, respectively (Cummins, 1956; Johnston and Mains, 1931; Tarr, 1962).

Urediospore and teliospore germination is typical of other long-cycle heteroecious rusts. The aecial stage is unknown, although it has been inconclusively shown that *Oxalis corniculata* might be a host (Le Roux and Dickson, 1957; Pavgi, 1972).

Tarr (1962) lists other rusts infecting *Sorghums*, but none are of economic importance.

B. ENVIRONMENT

Sorghum rust is favored by warm humid weather, and under these conditions the pathogen spreads rapidly (Miller and Cruzado, 1969; Tarr, 1962). Late-summer rains resulting in wet soils followed by light rains or heavy dews and high humidity are ideal conditions.

The disease is widely distributed in the warm temperate and tropical portions of the world and occurs in most areas where sorghum is grown (Frederiksen, 1980; Johnston and Mains, 1931; Tarr, 1962). It is more severe in the tropics than in the drier temperate areas of the world. In the United States, sorghum rust is more important in Hawaii and the gulf coast region than in the main sorghum-producing areas of Texas, Oklahoma, Kansas, and Nebraska.

The pathogen overseasons in the uredial stage on living host tissue. This is most probably on grassy sorghums such as johnsongrass in the United States (Leukel *et al.*, 1951). Successive plantings of sorghum are made in some areas of Asia and Africa, and here green tissue of its host is available to the fungus all year. Secondary spread by means of air-borne urediospores occurs as soon as pustules are present. In humid areas the disease develops throughout the season. However, in the drier areas most of the infection occurs late in the growing season and near crop maturity. Teliospores are not known to play a part in disease development.

C. EFFECT OF DISEASE ON THE HOST

Sorghums vary in reaction to rust, but apparently no special methods have been developed to measure disease intensity and damage to the host (Mains, 1929).

In most major grain-sorghum-producing regions, the disease does not appear until seed is well developed. Hence, grain yield losses are relatively slight when compared to the usual damage caused by cereal rusts (Leukel *et al.*, 1951; Tarr, 1962). However, in warm humid areas of India (Patil-Kulkarni *et al.*, 1972) and in similar areas elsewhere in the world, heavy losses in grain yield occur (Frederiksen, 1980; Miller and Cruzado, 1969).

On many, but not all, sorghums, a characteristic purpling of the infected area occurs. The specific epithet "purpurea" refers to this symptom (Tarr, 1962). Cultivars that lack the *Pp* locus for purple pigmentation do not show purpling.

The rust also has other effects. Forage value of the crop is reduced when rust is abundant. Under these conditions leaves dry and break off (Leukel *et al.*, 1951). In sorgo (sweet sorghum), severe rust infection reduces the sugar content of the juice (Coleman and Dean, 1961).

D. PATHOGEN VARIATION

Rust biotypes differing in virulence to resistant cultivars or hybrids has not complicated breeding for resistance (Frederiksen, 1980; Frederiksen and Rosenow, 1972, 1974, 1979; Johnston and Mains, 1931). Two races were described in the Hawaiian Islands on cultivars that have single dominant genes for resistance (Bergquist, 1974). The possibility of physiologic races in India has also been mentioned (Rana *et al.*, 1976).

E. EXPRESSION OF RESISTANCE

Like other rusts, sorghum reaction can be measured in several different ways. All forms of resistance interact under field conditions. According to Frederiksen (1980; Frederiksen and Rosenow, 1972, 1979), there are three main types of reaction expressed by sorghums to rust. Some cultivars are essentially free from rust under all conditions except for an occasional large uredium near a midvein. Other groups of cultivars characteristically have either smaller or fewer uredia, which develop more slowly than on susceptible hosts. The third group of cultivars have large uredia but have fewer uredia per unit of leaf area.

Differences in amount of uninfected tissue seem to be expressed most clearly in the less favorable environments for rust (Johnston and Mains, 1931; Sharma and Jain, 1978). When the inoculum load is high and the environment favorable to rapid fungal growth, these differences tend to disappear (Miller and Cruzado, 1969). Bergquist (1971) used four reaction classifications (highly resistant, resistant, moderately resistant, and susceptible) for field studies in Hawaii. Balasubramanian *et al.* (1975) observed "green island" formation and concluded that it could be used to assess reaction to *P. purpurea.* Coleman and Dean (1961) classified plants with uredia as susceptible and those with necrotic flecks as resistant. Slow rusting is believed to be a desirable form of resistance (Frederiksen, 1980; Frederiksen and Rosenow, 1979).

Plant growth stage has a bearing on rust reaction. When 4-week-old seedlings were artificially inoculated, those that expressed resistance were also resistant when exposed to inoculum in tropical field conditions, and those that were susceptible remained susceptible at later growth stages (Bergquist, 1971). Johnston and Mains (1931) evaluated cultivars in temperate field conditions and in the greenhouse. When cultivars were highly resistant in the seedling stage, they showed little or no infection at later growth stages in the field, but cultivars with susceptible reactions as seedlings varied widely in reaction in the field from a trace to 100% tissue infected. In this respect sorghum is like corn and common rust.

The host–pathogen relationships are believed to be similar in sorghum to those encountered in other rust diseases, but few studies have been made with sorghum. Only a few hours are required for spore germination when conditions are favorable, and uredia appear in 10–14 days (Dalmacio, 1969).

F. GENETICS OF HOST–PATHOGEN INTERACTIONS

In sorgo (sweet sorghum), crosses between the susceptible cultivar Planter and the resistant MN 960 segregated in ratios expected for a single dominant gene (Coleman and Dean, 1961). Susceptible plants had sporulating uredia in the field in Louisiana and were given the genotype *pupu.* Resistant plants had no sporulating pustules and were given the genotype *PuPu.* Rust resistance was not linked to the *Pp* locus that conditions purple pigmentation.

There are different allelic interactions at the *Pu* locus (Miller and Cruzado, 1969). Allele *Pu* (from Rio or PI267474) functions as a dominant in Puerto Rico throughout the season when the opposite allele is pu^r (from Combine Shallu) but fails to condition resistance late in the season when the opposite allele is *pu* (from B406). Plants with the genotype *Pupu* were resistant only during the early part of the season, and plants with the genotype *PuPu* were resistant throughout the season.

In two studies in India with grain sorghum, susceptibility to rust was dominant. Segregation from a cross between the highly resistant cultivar Combine Shallu and the highly susceptible GM 2-3-1 ranged from immune (no uredia) to highly susceptible (uredia large and coalescing), with intermediate classes of highly resistant (uredia very small and few surrounded by chlorotic margin), semiresistant (uredia small but many), and susceptible (uredia large, scattered, many) (Patil-Kulkarni *et al.*, 1972). It was concluded that more than a single pair of genes were segregating. In another study, resistance was believed to be due to three major genes (Rana *et al.*, 1976).

G. CONTROL

The most feasible means of control is through rust-resistant cultivars and hybrids. Numerous sources of resistance are available, many of which originated in various areas of south and east Africa (Bergquist, 1971; Broadhead and Coleman, 1974; Frederiksen, 1980; Frederiksen and Rosenow, 1979; Johnston and Mains, 1931; Leukel *et al.*, 1951; Lopes *et al.*, 1975). Resistance varies within the major types of sorghum, and the slow-rusting trait is very useful in breeding and disease control.

Early planting helps plants to escape the disease in some areas. Other factors, such as eliminating sources of inoculum, i.e., grassy sorghums, volunteer plants, or limited plantings of the crop itself, could have some effect on slowing disease development in areas where sorghum culture is interrupted by a crop-free season.

Spraying with protectant chemicals or seed treatments has been tried with success in India (Agrawal and Kotasthane, 1973). Fungicides are not now labeled for use as a control of sorghum rust in North America (Frederiksen, 1980). While protectant fungicides are potentially valuable in controlling rust, they are not widely used in the world (Frederiksen, 1980; Tarr, 1962).

VI. Future Outlook

The corn and sorghum rusts seem to be stabilized in the world, and while there are no urgent problems or needs, several interesting areas of research can be mentioned. Corn and sorghum genotypes grown where rust resistance is not needed usually are susceptible to their rusts. Does this mean that susceptible genotypes are inherently more productive than their resistant counterparts in the absence of disease? A greater understanding of the nature of slow-rusting and mature-plant types of resistance would be desirable. Better methods for measuring the effects of rust on plant development would be helpful in assessing the importance for disease control and the degree of resistance or other control needed. Chemicals specific to individual rusts would offer an alternative means of control and be useful in other respects. In corn the structure of the short arm of chromosome 10 poses some interesting questions. What is its composition? How did it evolve? What is the relationship of closely linked genes for rust resistance on a chromosome to combined virulence in rust biotypes? Can complex clusters of loci be developed by breeding or genetic engineering and do they have value in disease control? A greater understanding of pathogen variation in aggressiveness and other features would be helpful in predicting the long-term value of the slow-rusting mature-plant types of resistance. If there are no penalties to unneeded resistance, then continual efforts should be extended in all aspects of host resistance and breeding.

References

Agrawal, S. C., and Kotasthane, S. R. (1973). Efficacy of systemic fungicides and antibiotics in checking the rust of *Sorghum vulgare* L. *Sci. Cult.* **39,** 235–236.

Allen, R. F. (1933). The spermatia of corn rust, *Puccinia sorghi. Phytopathology* **23,** 923–925.

Allen, R. F. (1934). A cytological study of heterothallism in *Puccinia sorghi. J. Agric. (Washington, D.C.) Res.* **49,** 1047–1068.

Arthur, J. C. (1904). The aecidium of maize rust. *Bot. Gaz. (Chicago)* **38,** 64–67.

Balasubramanian, K. A., Dixit, L. A., and Ajagannavar, L. S. (1975). Is green island in sorghum a light mediated reaction? *Curr. Sci.* **44,** 514–515.

Bergquist, R. R. (1971). Sources of resistance in sorghum to *Puccinia purpurea* in Hawaii. *Plant Dis. Rep.* **55,** 941–944.

Bergquist, R. R. (1974). The determination of physiologic races of sorghum rust in Hawaii. *Proc. Am. Phytopathol. Soc.* **1,** 67 (abstr.).

Bergquist, R. R. (1981). Transfer from *Tripsacum dactyloides* to corn of a major gene locus conditioning resistance to *Puccinia sorghi. Phytopathology* **71,** 518–520.

Borlaug, N. E. (1946). *Puccinia sorghi* on corn in Mexico. *Phytopathology* **36,** 395 (abstr.).

Broadhead, D. M., and Coleman, O. H. (1974). Registration of Brandes sweet sorghum (Reg. No. 116). *Crop Sci.* **14,** 494.

Cammack, R. H. (1958a). Factors affecting infection gradients from a point source of *Puccinia polysora* in a plot of *Zea mays. Ann. Appl. Biol.* **46,** 186–197.

Cammack, R. H. (1958b). Studies on *Puccinia polysora* Underw. I. The world distribution of forms of *P. polysora. Trans. Br. Mycol. Soc.* **41,** 89–94.

Cammack, R. H. (1959a). Studies on *Puccinia polysora* Underw. II. A consideration of the method of introduction of *P. polysora* into Africa. *Trans. Br. Mycol. Soc.* **42,** 27–32.

Cammack, R. H. (1959b). Studies on *Puccinia polysora* Underw. III. Description and life cycle of *P. polysora* in West Africa. *Trans. Br. Mycol. Soc.* **42,** 55–58.

Coleman, O. H., and Dean, J. L. (1961). The inheritance of resistance to rust in sorgo. *Crop Sci.* **1,** 152–154.

Cummins, G. B. (1931). Heterothallism in corn rust and effect of filtering the pycnial exudate. *Phytopathology* **21,** 751–753.

Cummins, G. B. (1941). Identity and distribution of three rusts of corn. *Phytopathology* **31,** 856–857.

Cummins, G. B. (1956). Host index and morphological characterization of the grass rusts of the world. *Plant Dis. Rep., Suppl.* **237,** 1–52.

Cummins, G. B. (1971). "The Rust Fungi of Cereals, Grasses, and Bamboos." Springer-Verlag, Berlin and New York.

Cummins, G. B., and Ramachar, P. (1958). The genus *Physopella* (Uredinales) replaces *Angiopsora. Mycologia* **50,** 741–744.

Dalmacio, S. C. (1969). Notes on the penetration and infection of *Puccinia purpurea* Cke. *Philipp. Agric.* **53,** 53–59.

Dickson, J. G., Syamananda, R., and Flangas, A. L. (1959). The genetic approach to the physiology of parasitism of the corn rust pathogens. *Am. J. Bot.* **46,** 614–620.

Ellis, R. T. (1954). Tolerance to the maize rust *Puccinia polysora* Underw. *Nature (London)* **174,** 1021.

Fajemisin, J. M. (1976). Potentials for stable resistance to *Puccinia polysora* in local (Nigerian) and exotic maize varieties. *Cereal Rusts Bull.* **4,** 5–8.

Flangas, A. L., and Dickson, J. G. (1961a). Complementary genetic control of differential compatibility in rusts. Theoretical application to analysis of host-obligate parasite interaction. *Q. Rev. Biol.* **36,** 254–272.

Flangas, A. L., and Dickson, J. G. (1961b). The genetic control of pathogenicity, serotypes, and variability in *Puccinia sorghi. Am. J. Bot.* **48,** 275–285.

Frederiksen, R. A. (1980). Sorghum rust. *In* "Sorghum Diseases: A World Review" (R. J. Williams, R. A. Frederiksen, L. K. Mughogho, and G. D. Bengston, eds.), pp. 240–242. ICRISAT, Andhra Pradesh, India.

Frederiksen, R. A., and Rosenow, D. T. (1972). Sorghum rust, a naturally stabilized disease in North America. *Phytopathology* **62,** 757.

Frederiksen, R. A., and Rosenow, D. T. (1974). A model for evaluation of genetic vulnerability of sorghum to disease. *Proc. Am. Phytopathol. Soc.* **1,** 57 (abstr.).

Frederiksen, R. A., and Rosenow, D. T. (1979). Breeding for disease resistance in sorghum. *In* "Biology and Breeding for Resistance to Arthropods and Pathogens in Agricultural Plants" (M. K. Harris, ed.), MP-1451, pp. 137–167. Texas A&M University, College Station.

Futrell, M. C. (1975). *Puccinia polysora* epidemics on maize associated with cropping practice and genetic homogeneity. *Phytopathology* **65,** 1040–1042.

Futrell, M. C., Hooker, A. L., and Scott, G. E. (1975). Resistance in maize to corn rust, controlled by a single dominant gene. *Crop Sci.* **15,** 597–599.

Hagan, W. L., and Hooker, A. L. (1965). Genetics of reaction to *Puccinia sorghi* in eleven corn inbred lines from Central and South America. *Phytopathology* **55,** 193–197.

Hemingway, J. S. (1955). Effects of *Puccinia polysora* rust on yield of maize. *East Afr. Agric. J.* **20,** 191–194.

Hilu, H. M. (1965). Host–pathogen relationships of *Puccinia sorghi* in nearly isogenic resistant and susceptible seedling corn. *Phytopathology* **55,** 563–569.

Hooker, A. L. (1961). Occurrence of *Puccinia polysora* in Illinois. *Plant Dis. Rep.* **45,** 236.
Hooker, A. L. (1962a). Additional sources of resistance to *Puccinia sorghi* in the United States. *Plant Dis. Rep.* **46,** 14–16.
Hooker, A. L. (1962b). Corn leaf diseases. *Proc. 17th Annu. Hybrid Corn Ind. Res. Conf.,* pp. 24–36.
Hooker, A. L. (1963). A second major gene locus in corn conditioning resistance to *Puccinia sorghi. Phytopathology* **53,** 221–223.
Hooker, A. L. (1967a). Inheritance of mature plant resistance to rust in corn. *Phytopathology* **57,** 815.
Hooker, A. L. (1967b). The genetics and expression of resistance in plants to rusts of the genus *Puccinia. Annu. Rev. Phytopathol.* **5,** 163–182.
Hooker, A. L. (1969). Widely based resistance to rust in corn. *In* "Disease Consequences of Intensive and Extensive Culture of Field Crops" (J. A. Browning, ed.), Spec. Rep. No. 64, pp. 28–34. Iowa Agric. Home Econ. Exp. Stn., Ames.
Hooker, A. L. (1973). Maize. *In* "Breeding Plants for Disease Resistance" (R. R. Nelson, ed.), pp. 132–154. Pennsylvania State Univ. Press, University Park.
Hooker, A. L. (1979a). Breeding for resistance to some complex diseases of corn. *In* "Rice Blast Workshop," pp. 153–181. Int. Rice Res. Inst., Los Baños, Laguna, Philippines.
Hooker, A. L. (1979b). Estimating disease losses based on the amount of healthy leaf tissue during the plant reproductive period. *Genetika* **11,** 181–192.
Hooker, A. L., and Le Roux, P. M. (1957). Sources of protoplasmic resistance to *Puccinia sorghi* in corn. *Phytopathology* **47,** 187–191.
Hooker, A. L., and Russell, W. A. (1962a). Development of nearly isogenic rust-resistant lines of corn. *Phytopathology* **52,** 14.
Hooker, A. L., and Russell, W. A. (1962b). Inheritance of resistance to *Puccinia sorghi* in six corn inbred lines. *Phytopathology* **52,** 122–128.
Hooker, A. L., and Saxena, K. M. S. (1967). Apparent reversal of dominance of a gene in corn for resistance to *Puccinia sorghi. Phytopathology* **57,** 1372–1374.
Hooker, A. L., and Saxena, K. M. S. (1971). Genetics of disease resistance in plants. *Annu. Rev. Genet.* **5,** 407–424.
Hooker, A. L., and Yarwood, C. E. (1966). Culture of *Puccinia sorghi* on detached leaves of corn and *Oxalis corniculata. Phytopathology* **56,** 536–539.
Hooker, A. L., Sprague, G. F., and Russell, W. A. (1955). Resistance to rust (*Puccinia sorghi*) in corn. *Agron. J.* **47,** 388.
Hou, H., Tseng, J., and Sun, M. (1978). Occurrence of corn rusts in Taiwan. *Plant Dis. Rep.* **62,** 183–186.
Johnston, C. O., and Mains, E. B. (1931). Relative susceptibility of varieties of sorghum to rust, *Puccinia purpurea. Phytopathology* **21,** 525–543.
Kim, S. K., and Brewbaker, J. L. (1976a). Effects of *Puccinia sorghi* rust on yield and several agronomic traits of maize in Hawaii. *Crop Sci.* **16,** 874–877.
Kim, S. K., and Brewbaker, J. L. (1976b). Sources of general resistance to *Puccinia sorghi* on maize in Hawaii. *Plant Dis. Rep.* **60,** 551–555.
Kim, S. K., and Brewbaker, J. L. (1977). Inheritance of general resistance in maize to *Puccinia sorghi* Schw. *Crop Sci.* **17,** 456–461.
Kim, S. K., Brewbaker, J. L., and Hasegawa, Y. (1978). Peroxidase activity associated with *Puccinia sorghi* infection in maize. *Korean J. Plant Prot.* **17,** 193–199.
Kushalappa, A. C., and Hegde, R. K. (1970). Studies on maize rust (*Puccinia sorghi*) in Mysore state. III. Prevalence and severity on maize varieties and impact on yield. *Plant Dis. Rep.* **54,** 788–792.
Lallmahomed, G. M., and Craig, J. (1968). Races of *Puccinia polysora* in Nigeria. *Plant Dis. Rep.* **52,** 136–138.

Lee, B. H., Hooker, A. L., Russell, W. A., Dickson, J. G., and Flangas, A. L. (1963). Genetic relationships of alleles on chromosome 10 for resistance to *Puccinia sorghi* in 11 corn lines. *Crop Sci.* **3,** 24–26.

Leonard, K. J. (1974). Foliar pathogens of corn in North Carolina. *Plant Dis. Rep.* **58,** 532–534.

Le Roux, P. M., and Dickson, J. G. (1957). Physiology, specialization, and genetics of *Puccinia sorghi* on corn and of *Puccinia purpurea* on sorghum. *Phytopathology* **47,** 101–107.

Leukel, R. W., Martin, J. H., and Lefebvre, C. L. (1951). Sorghum diseases and their control. *Farmers' Bull.* **1959.**

Lopes, T. A., Lam-Sanches, A. A., Casagrande, A. A., Mendonca, J. R., dos Santos, F. F., and Pane, F. A. (1975). Competicao de variedades de sorgo (*Sorghum bicolor* L. Moench) no Municipio de Jaboticabal, Sao Paulo, durante o ano de 1972. I. Reacôes varietais á antracnose e ferrugem. *Cienc. Cult. (Sao Paulo)* **27,** 1244.

McKeen, W. E. (1951). An exceptionally early outbreak of corn rust in Canada. *Plant Dis. Rep.* **35,** 367.

Mahindapala, R. (1978a). Occurrence of maize rust, *Puccinia sorghi,* in England. *Trans. Br. Mycol. Soc.* **70,** 393–399.

Mahindapala, R. (1978b). Host and environmental effects on the infection of maize by *Puccinia sorghi.* I. Prepenetration development and penetration. *Ann. Appl. Biol.* **89,** 411–416.

Mahindapala, R. (1978c). Host and environmental effects on the infection of maize by *Puccinia sorghi.* II. Post-penetration development. *Ann. Appl. Biol.* **89,** 417–421.

Mahindapala, R. (1978d). Epidemiology of maize rust, *Puccinia sorghi. Ann. Appl. Biol.* **90,** 155–161.

Mains, E. B. (1926). Studies in rust resistance. *J. Hered.* **17,** 313–325.

Mains, E. B. (1929). Relative susceptibility of various varieties of sorghum to rust, *Puccinia purpurea. Phytopathology* **19,** 104.

Mains, E. B. (1931). Inheritance of resistance to rust, *Puccinia sorghi,* in maize. *J. Agric. Res. (Washington, D.C.)* **43,** 419–430.

Mains, E. B. (1934). Host specialization of *Puccinia sorghi. Phytopathology* **24,** 405–411.

Mains, E. B. (1938). Two unusual rusts of grasses. *Mycologia* **30,** 42–45.

Malm, N. R., and Beckett, J. B. (1962). Reactions of plants in the tribe Maydeae to *Puccinia sorghi* Schw. *Crop Sci.* **2,** 360–361.

Malm, N. R., and Hooker, A. L. (1962). Resistance to rust, *Puccinia sorghi* Schw., conditioned by recessive genes in two corn inbred lines. *Crop Sci.* **2,** 145–147.

Martinez, C. A. (1977). Effects of *Puccinia sorghi* on yield of flint corn in Argentina. *Plant Dis. Rep.* **61,** 256–258.

Mederick, F. M., and Sackston, W. E. (1972). Effects of temperature and duration of dew period on germination of rust urediospores on corn leaves. *Can. J. Plant Sci.* **52,** 551–557.

Melching, J. S. (1975). Corn rusts: Types, races, and destructive potential. *Proc. 30th Annu. Corn Sorghum Res. Conf.,* pp. 90–115.

Melching, J. S. (1981). The effect of inoculum density on urediospore germination and infection of corn by *Puccinia polysora,* the cause of southern corn rust. *Phytopathology* **71,** 769.

Miller, F. R., and Cruzado, H. J. (1969). Allelic interactions at the *Pu* locus in *Sorghum bicolor* (L.) Moench. *Crop Sci.* **9,** 336–338.

Misra, D. P. (1963). Natural occurrence of the aecial stage of *Puccinia sorghi* Schw., on *Oxalis corniculata* Linn., in Nepal. *Indian Phytopathol.* **16,** 8–9.

Misra, D. P., and Sharma, S. K. (1964). Natural infection of *Oxalis corniculata* L., the alternate host of *Puccinia sorghi* Schw. in India. *Indian Phytopathol.* **17,** 138–141.

Nattrass, R. M. (1952). Preliminary notice of the occurrence in Kenya of a rust (*Puccinia polysora*) on maize. *East Afr. Agric. J.* **18,** 39–40.

Nattrass, R. M. (1953). Occurrence of *Puccinia polysora* Underw. in East Africa. *Nature (London)* **171,** 527.

Onofeghara, F. A., and Kapooria, R. G. (1975). Effects of systemic fungicides on corn rust. *Ghana J. Sci.* **15,** 89–92.

Patil-Kulkarni, B. G., Puttarudrappa, A., Kajjari, N. B., and Goud, J. V. (1972). Breeding for rust resistance in sorghum. *Indian Phytopathol.* **25,** 166–168.

Pavgi, M. S. (1966). Observations on the morphology and pathogenic reactions of two races of *Puccinia sorghi* and their inbred recombinants on a susceptible host. *Plant Dis. Rep.* **50,** 119–122.

Pavgi, M. S. (1972). Morphology and taxonomy of the Puccinia species on corn and sorghum. *Mycopathol. Mycol. Appl.* **47,** 207–220.

Pavgi, M. S. (1975). Teliospore germination and cytological aberrations in *Puccinia sorghi* Schw. *Cytologia* **40,** 227–235.

Pavgi, M. S., and Dickson, J. G. (1961). Influence of environmental factors on development of infection structures of *Puccinia sorghi. Phytopathology* **51,** 224–226.

Pavgi, M. S., and Flangas, A. L. (1959). Occurrence of southern corn rust in Wisconsin. *Plant Dis. Rep.* **43,** 1239–1240.

Pavgi, M. S., Cooper, D. C., and Dickson, J. G. (1960). Cytology of *Puccinia sorghi. Mycologia* **52,** 608–620.

Payak, M. M., Sharma, R. C., Singh, B. M., and Lilaramani, J. (1974). Performance of maize lines carrying single gene resistance to *Helminthosporium turcicum* and *Puccinia sorghi* in India. *Indian J. Genet. Plant Breed.* **34,** 31–35.

Peletz, B. F. (1971). Variability of uredospore germination of *Puccinia sorghi* in relation to mature-plant resistance in *Zea mays.* Master's Thesis, University of Illinois, Urbana-Champaign.

Peterson, R. F., Campbell, A. B., and Hannah, A. E. (1948). A diagrammatic scale for estimating rust intensity on leaves and stems of cereals. *Can. J. Res.* **26,** 496–500.

Pole-Evans, M. (1923). Rusts in South Africa. II. A sketch of the life cycle of the rust on mealie and *Oxalis. Union S. Afr., Div. Bot. Sci., Bull.,* pp. 1–2.

Rana, B. S., Tripathi, D. P., and Rao, N. G. P. (1976). Genetic analysis of some exotic × Indian crosses in sorghum. XV. Inheritance of resistance to sorghum rust. *Indian J. Genet. Plant Breed.* **36,** 244–249.

Renfro, B. L., and Ullstrup, A. J. (1976). A comparison of maize diseases in temperate and in tropical environments. *PANS* **22,** 491–498.

Reyes, G. M. (1953). An epidemic outbreak of the maize rust in Eastern and Central Visayas, Philippines. *Philipp. J. Agric.* **18,** 115–128.

Rhind, D., Waterston, J. M., and Deighton, F. C. (1952). Occurrence of *Puccinia polysora* Underw. in West Africa. *Nature (London)* **169,** 631.

Rhoades, M. M., and Rhoades, V. H. (1939). Genetic studies with factors in the tenth chromosome in maize. *Genetics* **24,** 302–314.

Rhoades, V. H. (1935). The location of a gene for disease resistance in maize. *Proc. Natl. Acad. Sci. U.S.A.* **21,** 243–246.

Rich, S., and Waggoner, P. E. (1979). Growth and rust (caused by *Puccinia sorghi*) of sweet corn cultivars in Connecticut. *Plant Dis. Rep.* **63,** 1012–1015.

Rijkenberg, F. H. J., and Truter, S. J. (1974). The ultrastructure of sporogenesis in the pycnial stage of *Puccinia sorghi. Mycologia* **66,** 319–326.

Robert, A. L. (1962). Host ranges and races of the corn rusts. *Phytopathology* **52,** 1010–1012.

Robinson, R. A. (1976). "Plant Pathosystems." Springer-Verlag, Berlin and New York.

Rodriguez-Ardon, R., Scott, G. E., and King, S. B. (1980). Maize yield losses caused by southern corn rust. *Crop Sci.* **20,** 812–814.

Roy, M. K., and Prasada, R. (1966). Physiologic specialisation of maize rust *Puccinia sorghi* Schw., in India. *Indian Phytopathol.* **19,** 305–307.

Russell, W. A. (1965). Effect of corn leaf rust on grain yield and moisture in corn. *Crop Sci.* **5,** 95–96.

Russell, W. A., and Hooker, A. L. (1959). Inheritance of resistance in corn to rust, *Puccinia sorghi* Schw., and genetic relationships among different sources of resistance. *Agron. J.* **51,** 21–24.

Russell, W. A., and Hooker, A. L. (1962). Location of genes determining resistance to *Puccinia sorghi* Schw. in corn inbred lines. *Crop Sci.* **2,** 477–480.

Ryland, A. K., and Storey, H. H. (1955). Physiological races of *Puccinia polysora* Underw. *Nature (London)* **176,** 655–656.

Santiago-Oro, R., and Exconde, O. R. (1974). Penetration and infection of corn by *Puccinia polysora* Underwent. *Philipp. Agric.* **58,** 50–60.

Saxena, K. M. S., and Hooker, A. L. (1968). On the structure of a gene for disease resistance in maize. *Proc. Natl. Acad. Sci. U.S.A.* **61,** 1300–1305.

Saxena, K. M. S., and Hooker, A. L. (1974). A study on the structure of gene *Rp3* for rust resistance in *Zea mays. Can. J. Genet. Cytol.* **16,** 857–860.

Schieber, E. (1965). Distribution of *Puccinia polysora* and *P. sorghi* in Africa and pathogenicity of these species on corn lines with Latin American germplasm. *Phytopathology* **55,** 1074–1075.

Schieber, E. (1971). Distribution of *Puccinia polysora* and *P. sorghi* in Africa and their pathogenicity on Latin American maize germ plasm. *FAO Plant Prot. Bull.* **19,** 25–31.

Schieber, E. (1975). *Puccinia polysora* rust found on *Tripsacum laxum* in the jungle of Chiapas, Mexico. *Plant Dis. Rep.* **59,** 625–626.

Schieber, E., and Dickson, J. G. (1963). Comparative pathology of three tropical corn rusts. *Phytopathology* **53,** 517–521.

Schieber, H. E., Rodriguez V., A. E., and Fuentes F., S. (1964). Distribution of *Puccinia sorghi* and *P. polysora* in central Mexico. *Plant Dis. Rep.* **48,** 425–427.

Sharma, H. C., and Jain, N. K. (1978). General and horizontal resistance against leaf spot diseases in some sorghum hybrids and varieties. *Indian J. Genet. Plant Breed.* **38,** 220–227.

Sharma, R. C. (1977). Investigations on host resistance, physiologic specialization and loss assessment in *Puccinia sorghi* Schw. Ph.D. Thesis, Aligarh Muslim University, Aligarh, India.

Sharma, R. C., Payak, M. M., and Khan, A. M. (1978). Reactions of *Coix,* teosinte and some maize lines to *Puccinia sorghi. Indian Phytopathol.* **31,** 393–394.

Sim, T., IV (1980). Southern rust of corn recognized in Kansas. *Plant Dis.* **64,** 500.

Singh, J. P., and Musymi, A. B. K. (1979). Control of rusts and powdery mildews by a new systemic fungicide, Bayleton. *Pesticides* **13** (No. 4), 51–53.

Sinha, D. C., Mishra, B., and Misra, A. P. (1974). Reaction of maize cultivars to *Puccinia sorghi* in Bihar. *Indian Phytopathol.* **27,** 253–254.

Smith, M. A. (1926). Infection and spore germination studies with *Puccinia sorghi. Phytopathology* **16,** 69.

Stakman, E. C., Christensen, J. J., and Brewbaker, H. E. (1928). Physiologic specialization in *Puccinia sorghi. Phytopathology* **18,** 345–354.

Stanton, W. R., and Cammack, R. H. (1953). Resistance to the maize rust, *Puccinia polysora* Underw. *Nature* (*London*) **172,** 505–506.

Storey, H. H., and Howland (Ryland), A. K. (1957). Resistance in maize to the tropical American Rust fungus, *Puccinia polysora* Underw. I. Genes *Rpp1* and *Rpp2. Heredity* **11,** 289–301.

Storey, H. H., and Howland, A. K. (1959). Resistance in maize to the tropical American rust fungus, *Puccinia polysora* II. Linkage of genes *Rpp1* and *Rpp2. Heredity* **13,** 61–65.

Storey, H. H., and Howland, A. K. (1967). Resistance in maize to a third East African race of *Puccinia polysora* Underw. *Ann. Appl. Biol.* **60,** 297–303.

Storey, H. H., Howland, A. K., Hemingway, J. S., Jameson, J. D., Baldwin, B. J. T., Thorpe, H. C., and Dixon, G. E. (1958). East African work on breeding maize resistant to the tropical American rust, *Puccinia polysora. Emp. J. Exp. Agric.* **26,** 1–17.

Syamananda, R., and Dickson, J. G. (1959). The influence of temperature and light on rust reaction of inbred lines of corn inoculated with specific lines of *Puccinia sorghi. Phytopathology* **49,** 102–106.

Tarr, S. A. (1962). "Diseases of Sorghum, Sudan Grass, and Broomcorn." Commonwealth Mycological Institute, Kew, Surrey.

Teng, P. S., and Montgomery, P. R. (1981). Response surface models for common rust of corn. *Phytopathology* **71,** 895 (abstr.).

Townsend, G. R. (1951). Control of the leaf blight and rust diseases of sweet corn. *Plant Dis. Rep.* **35,** 368–369.

Tuley, P. (1961). *Puccinia polysora* on *Tripsacum laxum* and *Zea mays. Nature* (*London*) **190,** 284.

Turner, G. J. (1974). Possible transmission of *Puccinia polysora* by bees. *Trans. Br. Mycol. Soc.* **62,** 205–206.

Ullstrup, A. J. (1965). Inheritance and linkage of a gene determining resistance in maize to an American race of *Puccinia polysora. Phytopathology* **55,** 425–428.

Ullstrup, A. J. (1977). Diseases of corn. *In* "Corn and Corn Improvement" (G. F. Sprague, ed.), pp. 391–500. Am. Soc. Agron., Madison, Wisconsin.

Vanderplank, J. E. (1968). "Disease Resistance in Plants." Academic Press, New York.

Van Dyke, C. G., and Hooker, A. L. (1969a). The Z reaction in corn to *Puccinia sorghi. Phytopathology* **59,** 33–36.

Van Dyke, C. G., and Hooker, A. L. (1969b). Ultrastructure of host and parasite in interactions of *Zea mays* with *Puccinia sorghi. Phytopathology* **59,** 1934–1946.

von Meyer, W. C. (1963a). The influence of storage conditions on the longevity of urediospores of *Puccinia polysora* Underw. and *Puccinia sorghi* Schw. *Plant Dis. Rep.* **47,** 614–616.

von Meyer, W. C. (1963b). A histological study of host parasite relations of *Puccinia polysora* and *P. sorghi* on different genotypes of maize. *Proc. Indiana Acad. Sci.* **73,** 89–96.

Wallin, J. R. (1951). An epiphytotic of corn rust in the North Central region of the United States. *Plant Dis. Rep.* **35,** 207–211.

Weber, G. F. (1922). Studies on corn rust. *Phytopathology* **12,** 89–97.

Wellensiek, S. J. (1927). The nature of resistance in *Zea mays* L. to *Puccinia sorghi* Schw. *Phytopathology* **17,** 815–825.

Wilcoxson, R. D. (1981). Genetics of slow rusting in cereals. *Phytopathology* **71,** 989–993.

Wilkinson, D. R., and Hooker, A. L. (1968). Genetics of reaction to *Puccinia sorghi* in ten corn inbred lines from Africa and Europe. *Phytopathology* **58,** 605–608.

Wood, J. I., and Lipscomb, B. R. (1956). Spread of *Puccinia polysora* with a bibliography on the three rusts of *Zea mays. U.S. Dep. Agric., Agric. Res. Serv., Spec. Publ.* **9,** 1–59.

Zogg, H. (1949). Untersuchungen über die epidemiologie des maisrostes *Puccinia sorghi* Schw. *Phytopathol. Z.* **15,** 143–192.

8

Sugarcane Rusts

L. H. Purdy

Department of Plant Pathology, University of Florida, Gainesville, Florida

I. Introduction

Through the years since 1890 when Krüger (1890), cited by Egan (1964), named the pathogen of one of the two presently known sugarcane rusts, the diseases have been considered of minor importance.

THE CEREAL RUSTS,
Volume II

ISBN 0-12-148402-5

Although sugarcane rust caused significant loss in Asia in the late 1940s and into the 1950s, concern in other sugarcane-growing areas was minimal or nonexistent. All this relative complacency gave way to serious concern beginning in July 1978, when an epidemic of very serious proportions began in the Dominican Republic (Presley *et al.*, 1978) in the cultivar B4362. Although the pathogen was misidentified as *Puccinia kuehnii*, a corrected identity was published later (Koike *et al.*, 1979) in which the accepted name was applied to the fungus, *Puccinia melanocephala.*

Puccinia melanocephala spread almost explosively to all sugarcane-growing locations in the pan-Caribbean area where the cultivar B4362 was growing. Within 1 year of its detection in the Dominican Republic, sugarcane rust had been observed throughout the Caribbean islands, northern South America, Central America, Mexico, and in Florida, Louisiana, Mississippi, and Texas in the United States.

The pan-Caribbean epidemic of sugarcane rust is, in itself, a most interesting occurrence. However, sugarcane rust (*P. melanocephala*) was also reported in Australia for the first time in October 1978 in the cultivar Q90 (Egan and Ryan, 1979), and there were some indications that it probably was present in late September. Thus, there were two major, but widely separated, sugarcane rust epidemics underway initiated in mid-1978. Both increased in intensity in an explosive manner and caused significant loss in production.

In contrast to *P. melanocephala, P. kuehnii* remains obscure and relatively inocuous. One might think of this species as mild-mannered and not a vigorous pathogen. In fact, Egan (1964) stated that it occurs sporadically and never reached epidemic proportions, a situation that seems to prevail.

II. History and Distribution

More than 90 years have past since Krüger (1890) described *Uromyces kuehnii,* the pathogen that causes one of the two rust diseases of sugarcane, interspecific hybrids of *Saccharum,* and *Saccharum* spp. According to Egan (1964), Wakker and Went renamed the pathogen *Uredo kuehnii* on the basis of the urediospores. Egan also credits Butler (1918) for first observing teliospores that placed the pathogen in the genus *Puccinia* as *P. kuehnii* Butler.

Padwick and Khan (1944) described *P. erianthi* from *Erianthus* sp., probably *E. ravennae* (Cummins, 1971). Egan (1979a) stated that the same fungus had been described in 1907 by H. and P. Sydow as *P.*

melanocephala on bamboo, but about 1970 the original host was shown to be a species of *Erianthus*. Therefore, as Sathe (1971) and Egan (1979a) pointed out, the accepted name for the pathogen of the second and more damaging sugarcane rust is *Puccinia melanocephala* H. Syd. and P. Syd. Cummins (1971) shows *P. erianthi* as a synonym of *P. melanocephala*.

Distribution of the two sugarcane rusts relates directly to the identity of the pathogens. For example, according to information presented in map 215 of the Distribution Maps of Plant Diseases (Anonymous, 1981a), *P. kuehnii* occurs over a widespread area in Asia, Australia, and Oceania. On the other hand, *P. melanocephala*, as shown in map 462 (Anonymous, 1981b), is present in almost all sugarcane-growing regions of the world with notable exceptions of almost all of South America (except Venezuela and Colombia, where rust appeared in the Cauca Valley in early 1982), Oceania, and much of Asia.

III. Economic Importance

Rust caused by *P. kuehnii* apparently has never been of significant economic importance, probably because of its inability to initiate and sustain epidemics (Egan, 1964). In contrast, *P. melanocephala* has initiated epidemics in India in the past (Egan, 1964), commencing in 1949 and subsequently with regularity in various provences. Egan stated that *P. melanocephala* would probably be of economic importance in other areas if it ever reached them. This prophecy proved to be extremely accurate beginning in 1978, with almost world-wide rejuvenation of *P. melanocephala*.

Methodology for loss assessment in sugarcane resulting from rust has not been established per se. There have been no experiments designed specifically to establish the magnitude of loss. There is loss to be sure, but loss assessment to date has been established by comparing production from years without rust to 1 or 2 years when rust was present in the same cultivar. Osada and Reyes (1980) compared production of B4362 in 1978–1979 before rust with production in 1979–1980 with rust. They reported losses from four areas of Mexico to range from 1.5 to 21.7 tonnes of cane/hectare, with an average of 12.6 tonnes/hectare. In 1978–1979 there were 10,588 hectares of B4362, and in 1979–1980, 10,668 hectares in the four areas used in the comparison. These data show that 1,344,168 tonnes of cane of B4362 were lost because of rust in 1979–1980.

The impact of *P. melanocephala* was substantial in growing areas predominated by B4362 in the Caribbean. Cuba reportedly grew B4362 on 40% of its production area in 1979, and 28% in 1980. Production losses because of sugarcane rust probably were similar to or perhaps greater than losses reported from Mexico for B4362 with rust. Losses of millions of pesos in 1980 were attributed to sugarcane rust in Cuba.

It was good fortune that resulted in a reduction from 40% of the sugarcane acreage in Jamaica being in B4362 in 1966 to 12% in 1979 (Burgess, 1979). Loss was significant but well short of the devastation that might have occurred had the 1966 acreage of B4362 been retained through 1979.

IV. Taxonomy and Nomenclature

As mentioned previously, there are two sugarcane rust pathogens, *P. kuehnii* Butler and *P. melanocephala* H. Syd. and P. Syd. An aecial stage is not known for either of these species. The following description are adapted from Cummins (1971).

Puccinia kuehnii

Uredia cinnamon or yellowish-brown, with pale brownish or hyaline paraphyses that are cylindric or capitate; urediospores (25–)30–43(–48) × 17–26 μm, mostly obovoid or pyriform, wall 1.5–2.5 μm at sides often thickened to 5 μm at apex, golden- or cinnamon-brown, echinulate, four to five equatorial pores. Telia small, blackish, spores 25–40 × 10–18 μm, mostly, oblong-clavate with rounded apex, not thickened apically, smooth, rounded apex, yellowish (? immature), pedicel hyaline, short.

Puccinia melanocephala

Uredia on abaxial leaf surface, cinnamon-brown, with capitate colorless to golden paraphypes; urediospores (25–)28–33(–36) × 18–23(–25) μm, mostly obovoid, wall 1.5 μm thick, cinnamon-brown, echinulate, germ pores four or five equatorial. Telia on abaxial leaf surface, exposed blackish-brown; teliospores (29–)30–43(–54) × (15–)17–21(–23) μm, mostly clavate, wall 1.5–2 μm at sides, 3–4 μm apically, chestnut-brown, smooth; pedicles thin-walled, brown, to 12 μm long.

It appears from descriptions of the species that, because of the smaller urediospores, the absence of thickened apical wall of urediospores, abundant capitate paraphyses, and darker spore wall, *P. melanocephala* can be distinguished from *P. kuehnii.* In actual practice it appears that there is overlap in spore sizes, and spore color may not be reliable. I have observed lemon-yellow urediospores of *P. melanocephala* on CP 65-357 following inoculation with typical cinnamon-brown urediospores. These plants were maintained in a plastic-covered greenhouse, and the color change occurred during the winter months, December–February, when light intensity was low. Temperatures in the greenhouse were also relatively low, about 5°–15°C. Attempts to purify the lemon-yellow type were relatively successful during the winter because they predominated. As temperature and light increased in later months, the lemon-yellow urediospore type produced cinnamon-brown urediospores. If collections from field-grown plants are made when temperatures and light are both limited, there seems to be a probability of obtaining urediospores much lighter than cinnamon-brown, primarily because of an environmental effect. Assessment of the variation of other specific characters seems to be in order, and such an evaluation may reveal a closer relationship between the two rust pathogens than is presently recognized. For example, Butler (1918) stated that both uniformly and apically thickened walls in urediospores occur on some hosts in India, but in Japan they occur in the same uredium.

Other species of *Puccinia* have been named that were found on *Saccharum* sp. or other closely related genera, such as *P. miscanthi* Mira on *Miscanthus sacchariflorus.* Cummins (1971) indicated that *P. miscanthi* has been treated as *P. eulaliea.* Patel *et al.* (1950) established the binomial *P. sacchari* that is considered to be *nomen nudum* by Laundon and Waterson (1964a). *Puccinia pugiensis* Tai may be a doubtful species or at least rare and might be a variant of *P. kuehnii* except for the very long (140 μm) pedicels of teliospores. One other species, *P. rufipes,* has been reported on sugarcane (personal communication from D. B. O. Savile).

V. Symptoms

According to Egan (1964), symptoms induced by *P. kuehnii* are almost identical to those caused by *P. melanocephala.* The earliest symptoms in susceptible cultivars are small, elongated, yellowish spots that are visible on both leaf surfaces. The spots increase in length, turn brown to orange-brown or red-brown in color, and develop a slight but definite, chlorotic halo. The lesions may be 2–10 mm in length but occasionally may reach 30 mm; they are seldom more than 1–3 mm in width. Uredia develop mainly on the lower leaf surface, but certain cultivars may also have some uredia on the upper surface. Uredia are subepidermal, and they rupture the epidermis because of the formation of masses of orange or cinnamon-brown urediospores. Urediospores may be formed over a considerable period of time, depending on weather conditions, but eventually the lesions darken and the surrounding leaf tissues become necrotic. Telia often develop before leaves become necrotic.

When the rust disease is severe, considerable numbers of uredia occur on a leaf, coalescing to form large, irregular, necrotic areas, and resulting in premature death of even young leaves. There is reduction in cane diameter and apparently in the number of canes per stool.

Rust is regarded as a disease of the leaf only, at least in *Saccharum officinarum* or its hybrids, but Srinivasan and Chenulu (1956) reported the presence of uredia on leaf sheaths and occasionally on stalks of some varieties of *S. spontaneum* in the collection at Coimbatore in India.

VI. Inoculum and Epidemics

Onslaught of plant disease epidemics and the results of such are of special concern when the epidemic develops in an area in which the pathogen was not present previously (Egan and Ryan, 1979; Presley *et al.*, 1978), especially so when the resulting damage is significant (Section III). "How did the epidemic get started?" is a question that stimulates thought. Some facts are known about sugarcane rust in the Americas, one of which is, according to Egan (1979a), that the International Society of Sugar Cane Technologists Standing Committee on Sugar Cane Diseases has not shown rust as occurring in any American country (one report of rust from Cuba in 1956 was withdrawn in 1959). The cultivar B4362 had been grown in the Caribbean region for more than 20 years (Whittle and Holder, 1980). The nearest known source of *P. melanocephala* to the Americas in early 1978 was the Camaroons in west Africa, where previous identification of the rust pathogen as *P. kuehnii* were incorrect according to Egan (1979a).

For rust to have developed in B4362 in the Dominican Republic (Presley *et al.*, 1978), two important events took place. Viable urediospores of *P. melanocephala* landed on leaves of B4362 and initiated the disease. There are at least three possible sources of the inoculum that started the epidemic.

1. *Puccinia melanocephala* was always present in the Dominican Republic but in low incidence on other hosts or cultivars of sugarcane. Rust exploded into epidemic proportions as a result of crop management changes, such as happened for *P. striiformis* on wheat in the Pacific Northwest of the United States in 1959 and 1960 (L. H. Purdy, personal observation). All factors considered, this potential source of

inoculum seems quite unlikely because there are no reports of crop management changes in the Dominican Republic from 1977 to 1978. The primary reason for rejecting this option, however, is the demonstrated susceptibility of B4362. If rust had been in the Dominican Republic before 1978, its presence would have been obvious on that very susceptible cultivar.

2. Rust was introduced into the Dominican Republic by man or by implements employed in sugarcane production. The probability exists, of course, that man did, indeed, have a role as suggested. However, the probability is very low because (a) any cane pieces for seed introduced into the Dominican Republic must have had leaves attached for uredia to be present, which is most unlikely; (b) buds would have had to be infected with rust, but bud infection has not been demonstrated; (c) seed cane pieces might have had urediospores on their surfaces, but infection of emerging shoots has also not been demonstrated; (d) urediospores-contaminated clothing and cutting equipment are highly unlikely because there is no transfer of a labor force to the Dominican Republic from areas where rust occurred in 1978 and before, on the African continent and the Mascarene Islands (Egan, 1979b).

3. Viable urediospores arrived in the Dominican Republic as a result of movement by wind that carried the spores from their origin to the infection site, as suggested by Bernard (1980). Prevailing winds in the southern hemisphere are east to west, and Bowden *et al.* (1971) present data that show that coffee rust (*Hemileia vastatrix*) arrived in Brazil from Africa by wind transport. Stover (1962) discusses intercontinental spread of the sigatoka disease of banana (*Mysosphaerella musicola*) and points clearly to the probability that wind carried spores across the Atlantic Ocean. Inoculum production sites for sigatoka, east Africa and the Camaroons, are almost identical to locations in Africa where sugarcane rust occurred in early 1978. Sigatoka moved to the Americas on winds from Africa, and sugarcane rust appears to have followed that same pattern of transport.

Prevailing wind direction does not account for wind movement in other than the prevailing direction. Wind moves in all directions at some time, although certain circumstance of directional change may be frequent in some locations. Cyclonic winds are not uncommon, particularly in areas where sugarcane grows. Weather disturbances associated with cyclonic winds can account for vertical movement of urediospores to high altitudes and net horizontal movement over long distance in directions not normally touched by prevailing winds. Such a happening was, I believe, the means by which urediospores were introduced into the Dominican Republic, and the most likely site of origin of those spores was the African continent, as suggested by Egan and Ryan (1979) and Whittle and Holder (1980).

Localized epidemics develop in the direction of prevailing surface or near-surface winds as well as high-altitude winds, as shown by Schieber (1972) for coffee rust in Brazil. From the Dominican Republic, sugarcane rust spread to adjacent islands, southward as far as Venezuela, westward to Jamaica, Cuba, Central America, Mexico, and to Florida, Louisiana, and Texas—all by movement of urediospores by wind. Within 1 year, sugarcane rust had attained its present distribution in the Americas.

In addition, D. B. O. Savile (letter to L. H. Purdy, 1981) observed that the range of variation in dimensions of *P. melanocephala* in a mass collection from Mexico is essentially that of the whole species. He suggested that it is possible that the outbreak of sugarcane rust in the Americas stemmed from one introduction. Furthermore, Whittle and Holder (1980) compared measurements of urediospores from 36 collections from locations around the world and found that there were clear population differences based on these measurements. They found that the dimensions of urediospores from Australia fell within the limits of the range of variation of the West Indian samples. Thus, they concluded that introduction of rust into Australia was by airborne spores from the Americas (Whittle and Holder, 1980).

In contrast to Whittle and Holder (1980) Egan and Ryan (1979) state that a common source of inoculum for the outbreaks of sugarcane rust in the Americas and Australia seems most unlikely. They suggest that the Australian epidemic resulted from urediospores arriving there in mid-1978 on the winds of the 1978 northwest monsoonal inflow from the Indian Ocean areas. Such an origin is as probable as is the speculation of airborne spores from the Americas arriving in Australia prior to September 1978.

VII. Factors That Influence Urediospore Germination

A. TEMPERATURE

1. Urediospore Germination

Urediospores of *P. melanocephala* germinate over a wide range of temperatures when spores are deposited on the surface of leaves. Hsieh *et al.* (1977) report germination over a range of 14°–34°C with an optimum of 25°C range from 16°–29°C and an optimum of 20°–25°C for both *P. kuehnii* and *P. melanocephala.* Vasudeva (1958) indicated an optimal range of 18°–25°C for germination of urediospores of *P. kuehnii.* Sahni and Chona (1965) reported urediospore germination at 10°C for what might have been both *P. kuehnii* and *P. melanocephala.*

I. A. Sotomayor and L. H. Purdy (1982 unpublished data) observed that urediospores of *P. melanocephala* produce a germ tube at least the length of the longest dimension of the urediospore over a temperature range of 5°–30°C. From 15° to 30°C, germination percentages reached 96–99 after 12 hr, with percentages of 85 and 78 at 25° and 30°C, respectively, after 1 hr. Urediospores germinated slowly at 5° and 10°C with minimal criteria for germination being reached after 4 hr; however, only 52% of the urediospores germinated at 5° and 10°C after 12 hr. Thus, our observations are in general agreement with respect to optimal temperatures for urediospore germination, but germination at 5°C is lower than reported previously.

2. *Teliospore Germination*

Although teliospores are produced by both rust fungi, reports of their germination are few. Antoine and Perombelon (1964) reported teliospore germination in *P. kuehnii,* but Egan (1980) rejects *P. kuehnii* in favor of *P. melanocephala* for the rust of sugarcane in Mauritius. Vasudeva (1958) indicated that teliospores of *P. kuehnii* germinated optimally at 18° to 22°C. It is probable that identification of the rust as *P. kuehnii* was correct, although *P. melanocephala* also occurs in India (Egan, 1980). Sahni and Chona (1965) reported germination at 15°–20°C for teliospores from sugarcane and *Erianthus* spp.

B. MOISTURE

Egan (1964), Lambhate *et al.* (1976), Vasudeva (1958), Sreeramulu (1971), Srinivasan and Chenulu (1956), and others all refer to rust being most severe during warm weather, but severity and incidence decreased with the onset of hot weather. Also, high humidity was indicated by almost all as being prerequisite for germination.

High humidity, per se, will not support germination of urediospores of rust fungi. However, that free liquid water is prerequisite for germination of urediospores of *Puccinia* spp. is an accepted axiom. High humidity does, indeed, have a role in urediospore germination, but in an indirect manner. The higher the humidity at a so-called warm temperature of 30°C, the fewer are the degrees that the temperature must drop to reach the dew point. Condensation of water vapor (humidity) occurs at the dew point, and liquid water covers plant surfaces. The film of liquid may not be visible, but urediospore germination may take place.

High humidity also reduces transpirational loss and resulting loss in leaf cell turgor. Sporulation almost certainly is affected by the humidity of the air, and may even increase during periods of dew deposition. Sreeramulu (1971) reported that urediospore content of air was highest during daytime hours. This suggests that spores were produced during periods of leaf wetness, and as they dried and winds increased more urediospores became airborne.

C. LONGEVITY OF UREDIOSPORES

According to Egan (1964), urediospores lose their viability after 5 weeks when conditions are cool but lose viability rapidly during hot weather. Urediospores have remained viable for more than 5 months in my laboratory after being placed in a closed glass container and stored at 4°C, and other urediospores have retained their viability for more than 11 months when stored in liquid nitrogen (L. H. Purdy, unpublished data).

VIII. Factors That Influence Infection

A. TEMPERATURE

Temperature effects on urediospore germination, discussed in the previous section, show that germination occurs from 5° to 30°C, with optimal temperatures of 15° to 30°C. A greater influence of temperature on infection is on the formation of appressoria. I. A. Sotomayor and L. H. Purdy (unpublished data, 1982) found that 15°–30°C was optimal for the formation of appressoria, but appressoria developed in sufficient percentages to initiate infection at 5° and 10°C. Thus, one might conclude that if urediospores germinate on sugarcane leaves and favorable conditions of moisture prevail for 8 or more hr, infection will probably occur when temperatures range from 5° to 34°C.

The influence of soil temperatures directly on infection has not been established. One might assume, however, that the indirect effect is on plant growth, especially in areas such as Louisiana, where soil temperature falls below the temperature that allows plant growth.

B. MOISTURE

Atmospheric moisture is, in a sense, the most influential environmental parameter that not only affects but governs infection. It is a plus or minus phenomenon, in that liquid water (dew, fog, light rain) is either on leaf surfaces or it is not. The additional parameter of the length of time that liquid water is present also exerts great influence on infection. Information in the literature requires extrapolation from data on urediospore germination and appressoria formation for the assumed role of liquid water in the infection process. There are no directly determined data on the influence of moisture on infection of sugarcane by the rust fungi. There are, of course, many comments that rust develops on sugarcane when temperatures are cool to warm and humidity is high.

Data accumulated by I. A. Sotomayor and L. H. Purdy (unpublished data, 1982) show that when urediospores are on leaf surfaces, germination and appressorium formation occur over the range of 5°–30°C if liquid water is on the leaf surface for at least 6 hr.

From these laboratory-generated data, extensive field evaluation of the role of leaf wetness on infection seems essential for adequate understanding of sugarcane rust epidemics.

Soil moisture exerts a pronounced influence on plant growth and thereby imposes an indirect influence on infection. Soil moisture might be a limiting factor for plant growth, but infection might still occur if urediospores are present and temperature and leaf wetness are adequate for infection to take place, such as reported by Gargantiel and Barredo (1980) in the Philippines and Ricaud and Autrey (1979) in Mauritius.

The diversified agriculture on the muck soils of Florida, where sugarcane might be grown adjacent to fields where various vegetable crops are growing or will be grown, stands alone in the world of sugarcane as unique. An unusual situation developed in the spring, February to May, 1981, in that a field of CP63-588 was growing in a block of fields that included several fields where vegetables (celery, etc.) had grown recently and would be grown again. Preparation of the fields included flooding, or at least a very high water table, which is manipulated by pumping water out of or into drainage ditches around blocks of fields.

The high water table, very near or at times above the soil surface, resulted in high atmospheric humidity near the soil surface within the canopy of CP63-588. Abundant inoculum was present, temperature was satisfactory, leaf wetness was adequate—all factors supported a localized epidemic within that particular field. In another field 1 km away, the same grower planted CP63-588 from the same seed source during the same week that the rusted field was planted. The second field had very little rust, a much lower water table, and conditions not favorable for infection. The only difference between the two fields was the location of the water table, which resulted in different leaf wetness conditions.

C. STAGE OF PLANT DEVELOPMENT

A so-called mature-plant resistance has been suggested as one means by which seasonal sugarcane rust epidemics subside. There are, however, no data to support the existence of mature-plant resistance that can be distinguished from the suppressive influence that increasing seasonal temperature has on rust development. Observations of field situations have been reported in which mature-plant resistance has been suggested. Burgess (1979) observed that plants of B4362 older than 6 months show less severe symptoms and that, at the outbreak of rust in Jamaica, older fields of B4362 seemed to escape infection. Liu (1980b) reported that several cultivars gained mature-plant resistance slowly, 5 months after planting, whereas others developed resistance rapidly in only 2 months. Susceptible cultivars planted each month in two different years at Canal Point, Florida, showed that some cultivars seemed to develop mature plant resistance 6 months after planting (B4362) but that others failed to show resistance regardless of age, such as CP78-1735 (personal observations with J. L. Dean). Increasing summer temperatures at Canal Point, Florida, suppressed rust development both years in all cultivars included in the plantings.

To date, clear demonstration of the nature of observed mature-plant resistance has not been presented. Further, the role of environment on the expression of mature plant resistance has not been established.

IX. Host Resistance Development

Egan (1964) pointed out that the only control measures taken against rust have been the withdrawal or withholding of susceptible cultivars of sugarcane from cultivation. Implied, of course, is replacement with resistant cultivars. The amount of resistance in cultivar development programs seems to favor rust resistant cultivars becoming available where they are needed a few years after the onset of rust epidemics. The primary reason seems to be that continual screening throughout the entire development program takes place in the field. There may be certain seasons when enhancement of naturally produced inoculum may be needed, and this might be done as reported by Todd and Summers (1980). As time passes, and with the concomitant introduction of rust-resistant cultivars, rust nurseries may be needed specifically for the production of inoculum to maintain programs to screen for resistance to rust. Perhaps greater use may be made of the detached leaf techniques (Vasudeva, 1958; Egan, 1964) for evaluation of rust resistance under controlled environmental conditions. With near worldwide distribution of *P. melanocephala,* a vigorous pathogen, continued screening for resistance will be essential.

Exchange of germ plasm most certainly can advance significantly the progress made in almost all cultivar development programs. Concerns relating to movement of seed pieces from one area to another should be alleviated to achieve maximal exchange of germ plasm. Whittle (1980a) discussed some of the difficulties in reaching agreement and establishing workable quarantine procedures for exchange of cane. Whittle (1980a) also discussed exchange of buds grown in cultures that are apparently free from certain diseases, at least leaf scald (*Xanthomonas albalineans*), for the situation mentioned. Leu (1978) discussed apical meristem culture to eliminate certain systemic diseases from sugarcane. The future of protoplast isolation and culture holds great promise as a vehicle by which germ plasm might be exchanged. In addition, protoplast culture also has tremendous potential to uncover variability and useful disease resistance for cultivar improvement.

To maximize benefits from germ plasm exchange, methods must be developed to exchange observed results in a form that is understood by the participants of the exchange as well as other interested individuals. A data-recording system for the sugarcane rust–host interaction was proposed by Purdy and Dean (1981) to facilitate recording and exchanging information. They pointed out the importance of the stage of plant development when rust response is determined, as plant growth stage relates to mature-plant resistance. Whatever system is used for data recording, it must be one that conveys information accurately and in a form that can be understood.

X. Pathogenic Specialization

The question of pathogenic races of the sugarcane rusts is poorly defined. Albuquerque and Arakeri (1958) reported that the response of 23 cultivars of sugarcane to *P. kuehnii* suggested the presence of pathogenic races. A report from Lucknow reported two races of *P. melanocephala,* one on Co475, the other on Co410 (Anonymous, 1962). Liu (1980a) suggests strongly that pathogenic races occur, citing PR1000, which is resistant in Puerto Rico but susceptible in Nicaragua. In contrast to these reports of pathogenic races, Whittle (1980b) concludes that sugarcane cultivar development programs are utilizing "horizontal resistance mechanisms." He implies also that pathogenic races of sugarcane rust can not exist, that claims of their presence ignore the host, and that it is highly unlikely that vertical genes for resistance occur in sugarcane. Whittle (1980b) credits Robinson (1976) with the comment that vertical resistance has been postulated for the two rusts of sugarcane solely on the grounds that they are rusts. Future work will determine the controversy over vertical versus horizontal resistance in sugarcane to the rust fungi.

XI. Host Range

Sugarcane cultivars grown today are interspecific hybrids of *Saccharum* species, including *S. officinarum, S. sponteneum, S. robustum, S. barberi* and *S. sinensis.* The susceptibility to rust of the interspecific hybrids, or present-day cultivars, has been mentioned frequently in the literature. For example, Hsieh *et al.* (1977) reported rust on F176 in Taiwan. Similar announcements of cultivar susceptibility can be found, but the response of many cultivars is usually tabulated as part of the information published, e.g., Todd and Summers (1980). Egan (1979b) presented a list of cultivars grown commonly in various countries with the observation that these "indicator" cultivars should be monitored for changes in susceptibility.

Other hosts of sugarcane rusts include species in *Saccharum* and related genera. Hosts of sugarcane rusts as reported by Cummins (1971) and Laundon and Waterston (1964a,b) are for *P. kuehnii: Saccharum arundinaceum* Retz., *S. narenga* Wall., *S. offinicinarum* L., *S. sponteneum* L., and *Sclerostachy fusca* (Roxb.) A. Camus; and for *P. melanocephala: Saccharum officinarum* L., *S. sponteneum* L., *Erianthus ravennae* (L.) Beauv., and *E. rufipilis* (Steud.) Griseb.

XII. Pathological Histology

The absence of information that describes the intimate relationship between the sugarcane rust fungi and their hosts may be testimony to the belief that host resistance will, for the most part, take care of the rust problem. Indeed, host resistance is the only realistic means to combat sugarcane rust at this time. However, are immunity from infection or immunity from disease synonymous? What happens within the leaf that results in either response? Host response has been evaluated macroscopically to select for resistance, but if resistance types are similar macroscopically, are there microscopic differences that can be utilized to enhance selection for resistance? With this as an ultimate objective, initial microscopic evaluation of the response of a susceptible cultivar, CL41-223, established the procedures to be used (I. A. Sotomayor and L. H. Purdy unpublished data, 1982). Their results showed the sequential development of *P. melanocephala* from urediospore germination through the establishment of the fungus within the host, and finally, production of urediospores in uredia (pustules).

Urediospores germinate to produce a germ tube of variable length that advances across the leaf generally perpendicularly to the long axis of epidermal cells. When contact with a stoma (guard cells) occurs, an appressorium forms over the stomatal aperture. A penetration peg develops and passes between the guard cells to enter the substomatal cavity, where a lobed substomal vesicle develops. Infection hyphae, two or more but usually four, develop and begin to ramify intercellularly between mesophyll cells. Upon contact with walls of mesophyll cells, the tip of an infection hypha is cut off by a septum to form a haustorium mother cell. Penetration of the mesophyll cell wall occurs, after which the haustorium develops. The haustorium differs from haustoria of other *Puccinia* spp. in that it is lobed, with several lobes arranged like fingers on a hand. The haustorium occupies almost all of the cell lumen. Haustoria can be observed by 36 hr after inoculation at 25°C in susceptible hosts. The rust fungus continues growth parallel to venation of the leaf and advances in both directions from the point of penetration. After 7 days, sporogous hyphae develop as part of the fungus thallus, and these cells appear similar to parenchyma of plants. Urediospores and paraphyses develop, and subsequently the epidermis is ruptured by the developing urediospore mass.

In the resistant cultivar CP70-1133, a golden fluorescence develops around the point of penetration and in host cells adjacent to fungal structures, suggesting host cell necrosis indicative, perhaps, of an incompatible interaction between host and pathogen, as reported by Rohringer *et al.* (1977). Rust development in CP70-1133 is slower and less encompassing than in CL41-223. However, uredia have been observed in CP70-1133 in controlled environments, but chlorotic specks with necrotic centers are the field symptoms of rust in the cultivar. Other field-resistant cultivars are being evaluated with respect to their microscopic response to *P. melanocephala*.

XIII. Coexistence with Other Diseases

There are very few reports of the coexistence of sugarcane rusts and other diseases. Fors (1979) has observed what appears to be a predisposition to red stripe (*Pseudomonas rubrilineans*) that resulted from infection of B4362 by *Puccinia melanocephala* in Guatemala. Personal observations in Canal Point, Florida, with J. L. Dean have shown that rust increases symptom severity of *Ustilago scitaminea*, and rust apparently provides sufficient necrotic tissue for *Colletotrichum falcatum* to become established on leaves with rust uredia.

XIV. Control

A. Resistant Cultivars

Egan (1964) pointed out that the only control measures taken against rust were withdrawing or withholding susceptible cultivars from cultivation. Rust-resistant cultivars would be used to replant fields formerly occupied by susceptible ones. This practice prevails, and considerable effort has been devoted to replacement of B4362 in the Caribbean basin region of the Americas, CL41-223 in Florida, and, I assume, Q90 in Australia. Resistance seems to be available, and evaluation of resistant selections for agronomic acceptability will yield cultivars that will become replacements for susceptible ones. There is, however, a need to establish criteria for selection that before 1978 did not exist. For example, if uredia develop, will a cultivar be useful in the field in large populations, must resistance equal immunity, and so on? It seems that cultivars that are known to support the development of a small number of uredia, such as CP70-1133, may contribute to the development of new pathogenic races, unless Whittle (1980b) is correct in his assumption that races cannot exist because rust resistance is horizontal in sugarcane.

B. CHEMICAL CONTROL

Results of evaluations of fungicidal chemicals to control or reduce the incidence of sugarcane rust has been reported. Tiwari and Singh (1962) observed that ferban and ziram were most effective of seven chemicals tested. Singh and Muthaujan (1968) indicated that nickel sulfate plus ferban controlled sugarcane rust, and that the combination of both was more effective than either one alone. Bachchhav *et al.* (1978) presented data to show that four applications of ferbam plus nickel sulfate at 21-day intervals reduced the incidence of rust by 37.5% and increased yield by 7.66 tonnes/hectare over the nontreated control.

L. H. Purdy and J. L. Dean (unpublished data) evaluated some potential systemic chemicals and two protectant fungicides in Canal Point, Florida. Even with five applications at 30-day intervals there was no beneficial effect from the chemicals applied. These observations agree with those of E. Reyes in Cordoba, Mexico (personal conversation).

Availability of new-generation systemic fungicides with specificity against sugarcane rust may make the use of chemicals for rust control in sugarcane economically advantageous. Chemicals available now do not seem to offer economic return of sufficient magnitude, so they are not used.

C. CULTURAL AND BIOLOGICAL CONTROL

There is little that might be done culturally to reduce the incidence of rust. Maintaining soil water in the range of ready availability will reduce soil moisture stress that is reported to increase damage from rust (Gargantiel and Barredo, 1980; Ricaud and Autrey, 1979). However, an excessively high water table may contribute to a high incidence of rust, as was observed in Florida in 1981 (L. H. Purdy, personal observation).

Biological control has been suggested by the information presented by Chona *et al.* (1965). They reported that a heat-stable toxin was produced by *Trichothecium roseum* in culture that reduced germination of urediospores. Koike *et al.* (1979) observed *Darluca filum* frequently in pusutles of sugar cane rust from the Caribbean. Ryan and

Wilson (1981) suggest that a hyperparasitic relationship might exist between *Cladosporium uredinicola* and *Puccinia melanocephala*. However, currently no location is known where hyperparasites are adequately controlling the disease naturally.

References

Albuquerque, M. J., and Arakeri, H. R. (1958). Sugarcane rust in Bombay state. *Indian Inst. Sugarcane Res. Dev.* **2,** 199–203; *Hort Abstr.* **30,** 178(1960).

Anonymous (1962). *Indian Inst. Sugarcane Res., Lucknow Agric. Res., New Delhi* **2**(3), 183–187.

Anonymous (1981a). *Puccinia kuchnii* Butler. "Distribution Maps of Plant Diseases," 4th ed., Map No. 215. Commonw. Mycol. Inst., Ferry Lane, Kew, Surrey, England.

Anonymous (1981b). *Puccinia melanocephala* H. & P. Sydow. "Distribution Maps of Plant Diseases," 2nd ed., Map No. 462. Commonw. Mycol. Inst., Ferry Lane, Kew, Surrey, England.

Antoine, R., and Perombelon, M. (1964). Rust. *Annu. Rep. –Mauritius Sugar Ind. Res. Inst.* pp. 63–65.

Bachchhav, M. B., Hapse, D. G., Patil, A. O., and Ghure, T. K. (1978). Chemical control of sugarcane rust. *Sugarcane Pathol. Newsl.* **20,** 32–33.

Bernard, F. A. (1980). Considerations of appearance of sugarcane rust disease in the Dominican Republic. *Proc. Int. Soc. Sugar-Cane Technol.* **17,** 1382–1386.

Bowden, J., Gregory, P. H., and Johnson, C. G. (1971). Possible wind transport of coffee leaf rust across the Atlantic Ocean. *Nature (London)* **229,** 500–501.

Burgess, R. A. (1979). An outbreak of sugarcane rust in Jamaica. *Sugarcane Pathol. Newsl.* **22,** 4–5.

Butler, E. J. (1918). "Fungi and Diseases in Plants." Thacker, Spink & Co., Calcutta.

Chona, B. L., Durgapal, J. C., Gugnani, H. C., and Sohi, H. S. (1965). *Trichothecium roseum* Link. A hyperparasite of sugarcane rust. *Indian Phytopathol.* **18,** 386–387.

Cummins, G. B. (1971). "The Rust Fungi of Cerals, Grasses and Bamboos," pp. 94 and 102. Springer-Verlag, Berlin and New York.

Egan, B. T. (1964). Rust. *In* "Sugar-cane Diseases of the World" (C. G. Hughes, E. V. Abbott, and C. A. Wismer, ed.), Vol. II, pp. 61–68. Am. Elsevier, New York.

Egan, B. T. (1979a). A name change for the rust pathogen. *Sugarcane Pathol. Newsl.* **22,** 1.

Egan, B. T. (1979b). Susceptible indicator varieties for rust disease (*Puccinia melanocephala*). *Sugarcane Pathol. Newsl.* **22,** 10–11.

Egan, B. T. (1980). A review of the world distribution of *Puccinia* spp. attacking sugar cane. *Proc. Int. Soc. Sugar-Cane Technol.* **17,** 1373–1381.

Egan, B. T., and Ryan, C. C. (1979). Sugarcane rust, caused by *Puccinia melanocephala,* found in Australia. *Plant Dis. Rep.* **63,** 822–823.

Fors, A. L. (1979). Rust in Guatemala. *Sugarcane Pathol. Newsl.* **22,** 8.

Gargantiel, F. T., and Barredo, F. C. (1980). Observation on sugarcane rust in Occidental Negros, Philippines. *Sugarcane Pathol. Newsl.* **25,** 4.

Hsieh, W., Lee, C., and Chan, S. (1977). Rust disease of sugarcane in Taiwan: The causal organism *Puccinia melanocephala* Sydow. *Taiwan Sugar* **24,** 416–419.

Koike, H., Pollack, F. G., Lacy, S., and Dean, J. L. (1979). Rust of Sugarcane in the Caribbean. *Plant Dis. Rep.* **63,** 253–255.

Krüger, W. (1890). Mededeel. Proefstatin West-Java, Kagok-Tegal, Deel I.
Lambhate, S. S., Bachchhav, M. B., Shingte, V. V., and Ghure, T. K. (1976). Influence of climatic conditions on the incidence of sugarcane rust in the Deccan tract of Maharashtra State. *J. Maharashtra Agric. Univ.* **1,** 257–259; *Rev. Plant Pathol.* **57,** 2641 (1978).
Laundon, G. F., and Waterston, J. M. (1964a). *Puccinia erianthi. In* "C. M. I. Descriptions of Pathogenic Fungi and Bacteria," No. 9 Commonw. Mycol. Inst., Ferry Lane, Kew, Surrey, England.
Laundon, G. F., and Waterston, J. M. (1964b). *Puccinia kuehnii. In* "C. M. I. Descriptions of Pathogenic Fungi and Bacteria," No. 10. Commonw. Mycol. Inst., Ferry Lane, Kew, Surrey, England.
Leu, L. S. (1978). Apical meristem culture and redifferentiation of callus masses to free some sugarcane systemic diseases. *Plant Prot. Bull., Taiwan* **20,** 77–82.
Liu, L.-J. (1980a). Observations and considerations on sugarcane rust incidence, varietal reaction and possible occurrence of physiological races. *Sugarcane Pathol. Newsl.* **25,** 5–10.
Liu, L.-J. (1980b). Maturity resistance, a useful phenomenon for integrated control of sugarcane rust. *Sugarcane Pathol. Newsl.* **25,** 11–13.
Osada, S., and Reyes, E. (1980). Estimación de perdidas causadas por la roya de la caña de azucar. *Conv. Nac. ATAM, 1980.*
Padwick, G. W., and Khan, A. (1944). Notes on Indian fungi. II. *Mycol. Pap.* **10,** 1–17.
Patel, M. K., Kamat, M. N., and Padhye, Y. A. (1950). A new record of *Puccinia* on sugarcane in Bombay. *Curr. Sci.* **19,** 121–122.
Presley, J. T., Perdomo, R., and Ayats, J. D. (1978). Sugarcane rust found in the Dominican Republic. *Plant Dis. Rep.* **62,** 843.
Purdy, L. H., and Dean, J. L. (1981). A system for recording data about the sugarcane rust/host interactions. *Sugarcane Pathol. Newsl.* **27,** 35–40.
Ricaud, C., and Autrey, J. C. (1979). Identity and importance of sugarcane rust in Mauritius. *Sugarcane Pathol. Newsl.* **22,** 15–16.
Robinson, R. A. (1976). "Plant Pathosystems." Springer-Verlag, Berlin and New York.
Rohringer, R., Kim, W. K., Samborski, D. J., and Howes, N. K. (1977). Calcofluor: An optical brightener for fluorescence microscopy of fungal plant parasites in leaves. *Phytopathology* **67,** 808–810.
Ryan, C. C., and Wilson, J. A. (1981). A possible hyperparasite of sugarcane rust: *Cladosporium uredinicola,* speg. *Sugarcane Pathol. Newsl.* **27,** 31–32.
Sahni, M. L., and Chona, B. L. (1965). Studies on sugarcane rust in India. *Indian Phytopathol.* **18,** 191–203.
Sathe, A. V. (1971). Nomenclatural revision of the common rust fungus affecting sugarcane. *Curr. Sci.* **40,** 42–43.
Schieber, E. (1972). Economic impact of coffee rust in latin America. *Annu. Rev. Phytopathol.* **10,** 491–510.
Singh, K., and Muthaujan, M. C. (1968). Efficacy of fungicides against *Puccinia erianthi* Padw, and Khan causing rust of sugarcane. *Proc. Int. Soc. Sugar-Cane Technol.* **13,** 1203–1207.
Sreeramulu, T. (1971). Aeromycological observations and their implications in the epidemiology of some diseases of sugarcane. *Proc. Indian Natl. Sci. Acad., Part B* **37,** 506–510.
Srinivasan, K., and Chenulu, V. V. (1956). A preliminary study of the reaction of *Sacharum sponteneum* variants to red rot, smut, rust, and mosaic. *Proc. Int. Soc. Sugar-Cane Technol.* **9,** 1097–1107.

Stover, R. H. (1962). Intercontinental spread of banana leaf spot (*Mycosphaerella musicola* Leach). *Trop. Agric. (Trinidad)* **39,** 327–338.

Tiwari, M. M., and Singh, K. (1962). Chemical control of sugarcane rust. *Indian J. Sugarcane Res. Dev.* **6,** 179–180.

Todd, E. H., and Summers, T. E. (1980). Sugarcane rust inoculation technique and varietal resistance ratings in Florida. *Sugar J.* **42,** 17.

Vasudeva, R. S. (1958). Report of the Division of Mycology and Plant Pathology. *Sci. Rep. Agric. Res. Inst. New Delhi, 1956–1957,* pp. 86–100.

Whittle, A. M. (1980a). Cane movement and quarantine in some countries of the Caribbean. *Sugarcane Pathol. Newsl.* **25,** 24–25.

Whittle, A. M. (1980b). Do races exist in pathogens of sugarcane? *Proc. Inter-Am. Sugar Cane Semin., Cane Dis., 1980,* pp. 16–19 (sponsored by Inter-American Transport and Equipment Company, Vanguard, Miami, Florida).

Whittle, A. M., and Holder, D. (1980). The origin of the current rust epidemic in the Caribbean. *Sugarcane Pathol. Newsl.* **24,** 4–7.

PART II

Disease Distribution

9

World Distribution in Relation to Economic Losses

Eugene E. Saari
J. M. Prescott
Centro Internacional de Mejoramiento de Maíz y Trigo (CIMMYT),
Mexico City, Mexico

THE CEREAL RUSTS,
Volume II

ISBN 0-12-148402-5

I. Cereal Crops and Their Allies

Two-thirds of the world's total food supply is comprised of eight major cereal crops: wheat, rice, barley, rye, oats, corn, sorghum, and pearl millet. These crops may include more than one plant species, and the Food and Agriculture Organization (FAO) production statistics are often for total production. The millets are the most diverse group, being made up of several plant genera. In addition to the cereal crops, sugarcane is recognized as a major contributor to the world food supply (Hanson *et al.*, 1982; FAO, 1981) and triticale (× *Triticosecale* Wittmack) is also a promising species [International Maize and Wheat Improvement Center (CIMMYT), 1980; Muntzing, 1979].

A. RUST DISEASES

The food crops just mentioned are all subject to a number of diseases, and many are attacked by one or more species of the rust fungi. Table I comprises a list of the major cereal crops (and other gramineous crops), and the rust fungus (fungi) reported to attack them (Cummins, 1971; ICRISAT, 1980; Kranz *et al.*, 1977; Madumarov and Gorshkov, 1978; Malm and Beckett, 1962; Mathre, 1982; Ramakrishnan, 1963; Shurtleff, 1980; Wiese, 1977; Zillinsky, 1983). It is not the purpose of this chapter to review either the rust fungi or the diseases they cause. Those interested in the fungi are referred to Volume I of this treatise. The minor cereal crops are sparsely grown and information about the host and rust pathogens fungus (fungi) and their importance and distribution is very limited (Anonymous, 1967; Rachie, 1974).

B. WHEAT AND BARLEY CULTURE AND DISTRIBUTION

The authors are familiar with the cultural methods used for wheat and barley, particularly in the developing countries of Africa and Asia. The majority of the observations and comments regarding wheat and barley culture will be based on this experience, supplemented by reports and communications from colleagues.

Wheat and barley are grown primarily as temperate crops with a small area currently sown in the subtropics. The majority of the world's wheat is produced north of the Tropic of Cancer and south of the Tropic of Capricorn. Barley is grown in similar climates, but usually in the more marginal production environments where limited

Table 1

Cereal Crops, Related Species, and Rust Pathogens

Crop	Species	Rust fungus
Wheat	*Triticum aestivum* L.	*Puccinia graminis* Pers. f. sp. *tritici* Erikss and Henn.
	T. durum Desf.	P. stiiformis Westend. f. sp. *tritici*
	T. dicoccum Schrank.	*P. recondita* Rob. ex Desm. f. sp. *tritici*
	Triticum spp.	
Rice	*Oryza sativa* L.	None
Barley	*Hordeum vulgare* L.	*P. graminis* Pers. f. sp. *tritici* Erikss and Henn.
	H. distichum L.	*P. striiformis* Westend. f. sp. *hordei*
		P. hordei Otth.
Rye	*Secale cereale* L.	*P. recondita* Rob. ex Desm. f. sp. *secalis*
		P. striiformis Westend.
		P. graminis Pers. f. sp. *secalis* Erikss and Henn.
Oats	*Avena sativa* L.	*P. graminis* Pers. f. sp. *avenae* Erikss and Henn.
		P. coronata Corda. var. *avenae* Fraser and Ledingham
Corn	*Zea mays* L.	*P. sorghi* Schw.
		P. polysora Underw.
		Physopella zeae (Mains) Cumm. and Ramochar
Sorghum	*Sorghum bicolor* (L.) Moench	*Puccinia purpurea* Cke.
Pearl millet	*Pennisetum americanum* (L.) Leeke	*P. substriata* Ell. and Barth. var. *penicillarie* (Speg.) Ramachar and Cumm.
Sugarcane	*Saccharum officinarum* L.	*P. melanocephala* H. Syd. and P. Syd.
		P. kuehnii Butl.
Triticale	*X Triticosecale* Wittmack	*P. graminis* Pers.
		P. striiformis Westend.
		P. recondita Rob. ex Desm.
Kodo millet	*Paspalum scrobiculatum* L.	*P. substriata* Ell. and Barth.
Job's tears millet	*Coix lacryma-jobi* L.	*P. operta* Mund. and Thirum.
Foxtail millet	*Setaria italica* (L.) Beauv.	*Uromyces setariae-italicae* Yosh.
Small millet	*Panicum miliare* Lamk.	*U. liniaris* Berk. and Br.
Teff	*Eragostis tef* Trotter	*U. eragrostidis* Tracy
Proso millet	*Panicum miliaceum* L.	None
Finger millet	*Eleusine coracana* (L.) Gaertn.	None
Barnyard millet	*Echinochloa frumentacea* (Roxb.) Link	None

moisture, poorer soils or cooler temperatures prevail (Anonymous, 1981; Barghouti *et al.*, 1978a,b; Saari and Srivastava, 1977). Experimental wheat and barley trials are being conducted in tropical environments (Anonymous, 1982; Danakusuma, 1982; CIMMYT, 1976, 1977, 1978, 1979, 1980; Paganiban, 1980), but these constitute a noncommercial area at the present time.

The FAO lists 82 countries that grow wheat and 75 that grow barley (FAO, 1981). Twenty-seven countries in the developing world sow 100,000 hectares or more of wheat annually (CIMMYT, 1981). Crop distribution maps are available, but they fail to distinguish between areas that are spring- or autumn-sown. There is no distinction between autumn-sown spring wheats, facultative wheats, or the true winter wheats that require a vernalization period. Distribution maps do not reflect differences in moisture availability, or areas where partial or full irrigation is used. Cultivar distribution maps are not available. A similar situation exists regarding the definition of areas and environments in which barley is cultivated. This lack of specific information regarding host distribution tends to limit our ability to explain the distribution of the cereal rusts.

A worldwide map of wheat distribution (Reitz, 1967) shows some obvious areas of concentration, i.e., the Great Plains of North America; the Pampas of South America; western and southern Europe; the Russian New Lands; the Balkan Penninsula; Middle East; North Affrica; the Indo-Gangetic Plain; the North China Plain of Asia; and southeastern Australia. Although each area is itself geographically contiguous, a gradual change from one climatic zone to another may occur, and sharp geographical separations usually occur between major wheat-growing areas. If a crop distribution map could be superimposed onto a relief map, the isolation of the major wheat areas would be more obvious. Different types of wheat and barley cultivars are generally grown in each major geographical region. Similarly, the diseases present on wheat and barley and their importance also varies between regions (Abdel-Hak *et al.*, 1974; Anonymous, 1974, 1975a, 1979a,b; Saari, 1981; Saari and Prescott, 1980); the geographical separation between areas allows for the evolution of distinct pathogen populations. The separation may determine the predominant rust species and the virulence frequencies present (CIMMYT, 1980; Prescott and Saari, 1975; Rajaram and Campos, 1974; Saari and Wilcoxson, 1974).

C. Disease Distribution

The rusts of wheat are still the most important diseases on a global basis. This is shown by analyses of data from disease and breeding nurseries, as well as yield trials (CIMMYT, 1978, 1979; Kampmeijer, 1981, 1982; Kinaci, 1983; Rajaram and Campos, 1974; Saari and Prescott, 1980; Saari *et al.*, 1979; Young *et al.*, 1978), which support the authors' experiences and the general opinion of colleagues (Anonymous, 1975a, 1979b; Barghouti *et al.*, 1977b). The importance of these diseases is due to their wide distribution, their capacity to mutate and become virulent on previously resistant cultivars, their rapid rate of disease increase (epidemic potential), and their ability to remain viable after dispersal over great distances (see Chapters 10–13, this volume).

The rust disease situation for barley is generally less distinct. Most of the barley in the developing world is sown in marginal production environments that are too dry for wheat, thereby reducing the disease pressure (Anonymous, 1981; Barghouti *et al.*, 1977a,b; Saari and Srivastava, 1977). The rusts of barley, particularly leaf rust and stripe rust are considered significant in areas with more favorable moisture (Abdel-Hak and Ghobrial, 1978; Anonymous, 1981; Saari and Prescott, 1978b; Srivastava, 1978; Chapter 6, this volume).

Normally, one or two of the rusts of wheat or barley are found in any given area. The prevailing temperature pattern is the primary factor determining which rust species predominates because the rusts have similar moisture requirements. The effect of resistance in the cultivar(s) grown is often overlooked when considering the distribution of rust diseases and evaluating epidemic potentials.

Data from screening nurseries and from such disease trap nurseries as the Regional Disease Trap Nursery (RDTN) (Prescott, 1976) can be used to establish the dominant rust species and basic virulence spectrum for an area (CIMMYT, 1980; Rajaram and Campos, 1974). The differences between rust pathogens in adjacent areas can then be compared, and if there are sharp and consistent differences, the possibility arises that the pathogen populations are distinct. Caution is advised, however, in making this kind of comparison. Differences in pathogen populations may be as much a function of selection pressure (due to variations in the cultivars grown) as the result of environmental dissimilarities between areas.

II. Epidemiological Zones for the Cereal Rusts

An epidemiological zone is a region that has an endemic focus (foci) for primary rust infection with free movement and exchange of the rust inoculum within the zone. The evolution of new virulences or virulence combinations can occur within the zone, and these new virulences can increase in frequency and become well distributed in the zone. Generally, epidemiological zones are separated by geographic boundaries, such as mountain ranges, deserts, or oceans. The use of nursery data, along with field surveys (Chantarasnit and Luangsodsai, 1981; CIMMYT, 1972, 1973, 1974; Hassan, 1974; Joshi *et al.*, 1974; Khazra and Bamdadian, 1974; Sharif and Bamdadian, 1974), physiologic race identifications (Abdel-Hak, 1970; Abdel-Hak and Kamel, 1972; Abdel-Hak *et al.*, 1974; Annual Report, 1983; Aslam, 1978; Bahadur *et al.*, 1982; Bamdadian, 1973, 1980; Harder *et al.*, 1972; Hussain *et al.*, 1980; Rizvi, 1982; Stubbs, 1974), virulence analysis (Karki, 1980; Khalifa, 1979; Khokhar, 1977; Kirmani, 1980), and experiencc in the geographic area can help define epidemiological zones or pathological regions for the rust diseases.

Figure 1 represents our current concept of rust zones in Asia and Africa. Within any zone there may be subzones, and substantial areas that require exodemic primary inoculum. A pathway or cycling of rust inoculum is often a part of the epidemiological system. The zones outlined in Fig. 1 are based on the physical spread of urediospores by wind. The dispersal pattern within the zone would represent the normal or expected movement of a cereal rust on an annual cycle. A zone may be too large for all three of the rusts because of some other limiting factor. For example, in south India all three rusts of wheat are found in the Niligris Hills at about 1800 m elevation. The wheat crop sown in the plains of peninsular India becomes infected with leaf rust (*Puccinia recondita* Rob. ex Desm. f. sp. *tritici*) and stem rust (*P. graminis* Pers. f. sp. *tritici*) from this primary source each year (Joshi and Palmer, 1973; Joshi *et al.*, 1974, 1976b; Nagarajan and Joshi, 1980). The yellow rust fungus (*P. striiformis* Westend. f. sp. *tritici*) does not spread to the adjacent plains, because of the unfavorably high temperatures.

In all epidemiological zones, urediospores represent the major source of primary inoculum. The aecial host seems to be either nonfunctional or of little importance, except in a few limited locations. Barberry is reported to be a significant source of primary inoculum for stem rust in the Soviet Union Far East, (Azbukina, 1980) and is considered important to the initiation of stem rust epidemics in eastern Europe (Spehar, 1975; Spehar *et al.*, 1976). The alternate host of leaf rust of wheat is considered important in parts of the Mediterrean Basin (Anikster and Wahl, 1979; d'Oliveira and Samborski, 1966; Spehar, 1975; Tommasi *et al.*, 1980). *Isopyrum* spp. have been reported as a source of primary inoculum of leaf rust in the central Siberian region of Russia (Anikster and Wahl, 1979; Chester, 1946). *Ornithogalum* spp. are common in the Mediterrean region, providing primary inoculum for leaf rust of barley (

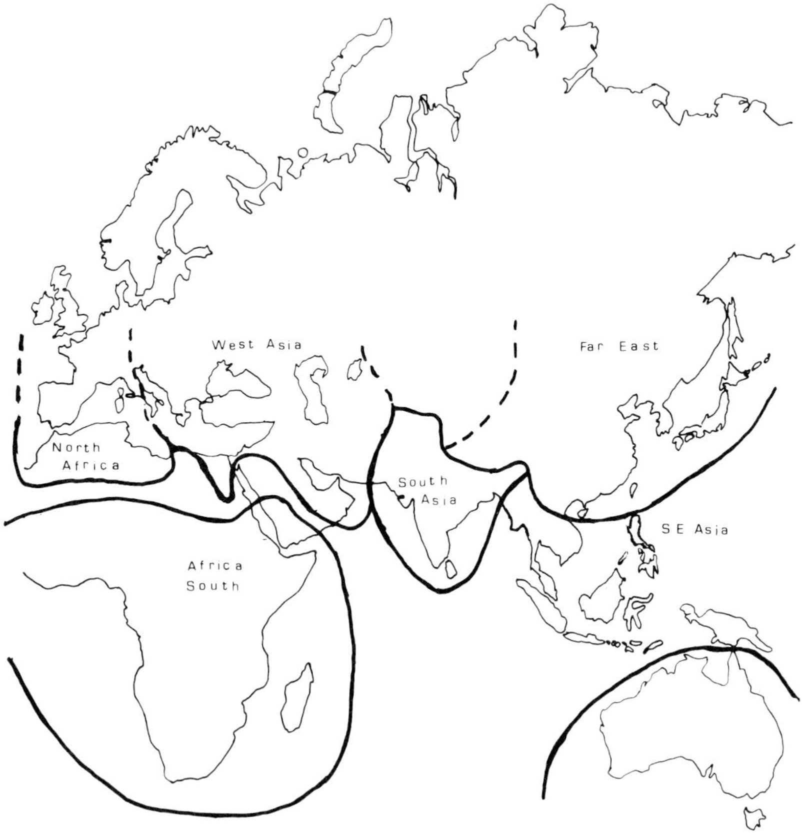

Fig. 1. Major epidemiological zones for the wheat rusts in Asia and Africa.

P. hordei Otth.) (Anikster and Wahl, 1979; Browning, 1974; Browning *et al.*, 1979; Dinoor, 1974).

A. SOUTH ASIA

This area includes India, Pakistan, Nepal, Bangladesh, and southeast Afghanistan. The annual cycle of the rusts has been well documented for many years (Annual Report, 1977, 1978, 1979; Hassan, 1970, 1974, 1978; Hassan *et al.*, 1977; Joshi *et al.*, 1974, 1976a,b, 1977; Nagarajan and Joshi, 1978; Nagarajan *et al.*, 1978, 1980, 1982). Wheat and barley are grown throughout the year in south India and the mountain areas of Nepal, India, and Pakistan. The wheat and barley crops grown at the higher elevations allow the rusts of these two crops to oversummer and provide inoculum for reinfection of the main crop sown from September to December in the plains of Bangladesh, India, Nepal, and Pakistan (CIMMYT, 1978; Saari, 1977, 1981; also see Chapter 12, this volume).

Temperature is the critical factor in the establishment of rust in the plains. The reinfection of stem rust in the northern plains is not common. Sowing is normally done in November when temperatures are ideal for leaf and stripe rust infection. In the southern hills of India, stem and leaf rust inoculum is available during September and October when wheat is sown on the plains of peninsular India. The two foci for leaf rust merge and mix freely each year. Stem rust inoculum primarily comes from the southern hills, and stripe rust spreads from the northern mountains. Alternate hosts are not important in the epidemiology of the South Asia zone (Hassan, 1970; Joshi *et al.*, 1974).

Leaf rust of wheat is widespread, and the authors consider it the most important disease in this zone. A diverse virulence spectrum exists for both bread wheat and durum wheat. In the northern portion of the region, epidemics have occurred on susceptible cultivars in 3 years out of 4. Losses of 30–40% have been recorded in severely affected fields. National losses of 10–20% are possible.

Stripe rust of wheat is important in the northern areas of India, Pakistan, and Nepal, in southeast Afghanistan, and at the higher elevations in the south. Epidemics have occurred about once in every 10 years. Losses of 100% can occur in fields of susceptible cultivars. The pathogen has a narrow spectrum of virulence in this zone on both bread wheat and durum wheat.

Stem rust of wheat is a significant disease in south Pakistan and south and central India, but epidemics are infrequent and have been confined to local land cultivars. When such epidemics occur, losses can reach 100% in individual fields, yet losses are minor on a national basis. The pathogen has a diverse virulence spectrum.

Barley stripe rust is found in the northern plains of Pakistan, India, and Nepal, and in the higher elevations in the northern areas of this zone. The northern mountains serve as a source of inoculum for the annual reinfection of crops in the northern plains. Epidemics can occur when environmental conditions are favorable; losses in individual fields can reach 100%. Most of the barley grown in this zone is found in areas that are marginal for disease development, thereby limiting the impact of stripe rust on a national basis.

Epidemics of stem rust of barley parallel epidemics of stem rust of wheat. Little barley is lost due to its early maturity. Leaf rust of barley is found in the northern areas of the zone, where it is endemic and oversummers at the higher elevations. This disease spreads sporadically into the plains, and losses are minimal.

B. WEST ASIA AND EGYPT

The topography and climate of this large region is very diverse. It is difficult to accurately classify this epidemiological zone and describe the host populations, because it is the center of origin for both wheat and barley (Frankel and Bennett, 1970; Morris and Sears, 1967). There are also numerous related species growing in nature, which further complicates the analysis of any rust studies.

Differences in elevation and latitude within the zone make possible the cultivation of wheat throughout the year. In some of the mountainous areas of Turkey, Iran, and Iraq, a wheat crop may take as long as 13 months to mature due to the long snow cover and cool summer temperatures (Saari and Srivastava, 1977).

The geographic diversity of the zone, along with the numerous wild species of wheat, barley, and related grasses, results in a diverse rust flora. The alternate host of *P. hordei* is present in the region and functions in nature (Anikster and Wahl, 1979). Barberry species are found, but no association with *P. graminis* has yet been confirmed. The numerous foci and diversity of hosts comprises the type of system that reaches an equilibrium between pathogens and hosts (Browning, 1974; Browning *et al.*, 1979; Dinoor, 1974).

A major feature of this system is the diversity at the sources of primary inoculum. This competition or natural equilibrium has probably been beneficial in reducing epidemics and increasing the duration that resistant cultivars retain a functional level of resistance. However, the authors are unaware of any data measuring this type of influence on primary inoculum. The introduction of modern agricultural practices and the extensive cultivation of homogenous cultivars will undoubtedly upset this balance and result in an increase in epidemic potential (Dinoor, 1974; Pinto, 1971; Wodageneh, 1974).

Egypt is a special case. Wheat and barley have been cultivated there for thousands of years, and the rust diseases have also been present. None of the wheat rusts survive in the Nile Valley between crop seasons; and primary inoculum must be reintroduced each year (Abdel-Hak, 1970; El-Tobgy and Taalat, 1971; Saari, 1976). The rust diseases are usually confined to the Delta area. There is a tendency for both rust severity and winter rainfall to decrease southward from the coast. The annual rainfall is 124 mm at the coast, 36 mm in the northern Delta, 15 mm in the southern Delta, and 0.0 mm in middle and upper Egypt (El-Tobgy, 1976; CIMMYT, 1978). The observed disease gradient probably reflects the deposition of primary inoculum rather than moisture (or other factors) for disease establishment. Dew formation is common throughout the irrigated Nile Valley, so artificial inoculations for disease screening develop to severe levels. This would indicate that the arrival of inoculum is the important factor in the development of the rusts on wheat. Stripe rust and stem rust of barley are introduced each year, but seldom become important (Abdel-Hak and Ghobrial, 1978; Saari and Prescott, 1978; Srivastava, 1978). In the case of *P. hordei*, primary inoculum is probably of local origin from the alternate host.

The inoculum source for the initial infection of wheat each year has not been determined, but there is an association between the atmospheric low-pressure systems that bring the rains and the subsequent appearance of rust infections. This would suggest that the Middle East countries are the source. Race analyses support this hypothesis (Abdel-Hak, 1970; Abdel-Hak and Kamel, 1972; Abdel-Hak *et al.*, 1972, 1974). One study has shown that the virulence factors present in the stripe rust population of Egypt are more similar to those of populations collected from different areas of the Middle East than to other populations (Khalifa, 1979). Unfortunately, the sampling was too limited to be conclusive.

The screening effect of the cultivars grown plays an important role in the comparison of virulences. The resistance of Giza 155 was effective against the three wheat rusts for 25 years in Egypt. As the cultivation of new semidwarf cultivars increase in the Middle East, new virulences have been propagated, which has apparently changed the established balance that has prevailed for so long (Abdel-Hak *et al.*, 1974).

Leaf rust of wheat is an important disease where the climate is moderated by bodies of water, such as the Caspian, Black, and Mediterranean Seas and the Persian Gulf. Leaf rust oversummers on wheat and related grasses at higher elevations. Local land-race cultivars can experience a 30% reduction in yield, but early maturity and resistance keep overall losses low. A wide range of virulence exists in the region.

Stripe rust of wheat is the most important rust disease in the region, resulting in serious losses in areas where cooler temperatures prevail and in the northern Nile delta. Epidemics are common on the widely grown land-race cultivars. Stripe rust oversummers on wheat and related grasses at the higher elevations, and a very diverse population exists for virulence.

Stem rust of wheat is present throughout the region and is important on late-maturing, local land-race cultivars. Stem rust oversummers on wheat and related grasses in the higher-elevation valleys. Overall losses are low, but entire fields can be destroyed, particularly in the coastal areas. A range of virulence combinations exists for both bread wheat and durum wheat. Barberry species are found in this zone, except in Egypt, but no association with epidemics has been confirmed.

Stem and stripe rusts of barley occur throughout the region but are considered by the authors to be unimportant. The alternate host of barley leaf rust is present and functional throughout the area. Leaf rust of barley is a serious disease and can result in losses of 100% on susceptible cultivars. The pathogen population is very diverse.

C. SOUTHERN AFRICA AND SOUTHWESTERN ARABIAN PENINSULA

Much of the wheat in this zone is grown above 1000 m (Anonymous, 1975a, 1979a; CIMMYT, 1975, 1976, 1977, 1978, 1979, 1980); but there is a small area sown at lower elevations. The moderate climate associated with the higher elevations in this zone favors the development of stem and stripe rust (Baghadadi, 1983; Cormack, 1974; Dmitriev and Gorshkov, 1980; El Ghouri, 1979). Leaf rust of wheat is considered a minor disease in much of the area, but becomes more important in the south where seasonal differences are more distinct (Edwards, 1975; Olugbemi, 1974).

This epidemiological zone is large and has several subzones. The total wheat production area is relatively small and clustered in widely separated areas. The principal wheat-growing countries are Saudi Arabia, People's Republic of Yemen, Yemen Arab Republic, Sudan, Ethiopia, Kenya, Tanzania, Zimbabwe, Lesotha, Nigeria, Angola, and South Africa. Wheat and barley are traditional crops north of the equator, and they were introduced south of the equator about 100 years ago (Pinto and Hurd, 1970). Wheat is cultivated in most of the countries of south and east Africa, but the area is relatively small, except in South Africa. Barley is grown on a small area and commercially only in Kenya and South Africa. North of the equator, many local land cultivars have developed during the thousands of years of wheat and barley cultivation; many of these cultivars are still in cultivation today.

The moderate climate and the ability to grow wheat continuously throughout the year, due to differences in elevation, creates a local endemic disease cycle (Edwards, 1975; Hogg *et al.*, 1969; Oggema, 1974; Pinto and Hurd, 1970). In Saudi Arabia, People's Republic of Yemen, and Yemen Arab Republic, overlapping crop cycles also result in a local endemic rust population (Anonymous, 1975b; Ba-Angood, 1975; El Ghouri, 1979; El-Saadi, 1974). The rust diseases of both wheat and barley can be found at any time, depending on the elevation.

The Ethiopian highlands are considered a center of diversity for durum wheat and barley (Frankel and Bennett, 1970). Similarly, notable diversity exists in the virulence patterns of the rust populations (Dmitriev and Gorshkov, 1980; Gebeyhou, 1975; Khalifa, 1979; Pinto, 1971; Wodageneh, 1974).

The expansion of wheat cultivation at the lower elevations north of the equator has resulted in the occasional appearance of leaf and stem rust. The source of primary inoculum for Sudan is probably either the Arabian Peninsula or Ethiopia (Ali, 1979; Baghadadi, 1983; Khalifa, 1979). The source of inoculum for Nigeria is uncertain (Andrews, 1968; Olugbemi, 1974).

The three rusts of wheat and barley can be found in Ethiopia, Kenya, and Tanzania. South of Tanzania, only leaf rust and stem rust are reported (Edwards, 1975; CIMMYT, 1978, 1979, 1980; Hogg *et al.*, 1969; Saari and Wilcoxson, 1974). The epidemiology of the area is associated with the Rift Valley, which runs through much of the wheat growing area. The similarities of rust races between Ethiopia and Kenya suggests some interchange of inoculum (Abdel-Hak and Kamel, 1972; Harder *et al.*, 1972). Similarities also exist between the Kenya and Tanzania rust populations (Harder *et al.*, 1972; Hogg *et al.*, 1969; Northwood, 1970). The degree and frequency of inoculum exchange in the area south of Tanzania is not well known, but the probability of its occurrence is illustrated by the documented spread of *P. polysora* Underw. from West Africa to East Africa and subsequently to Southern Africa (Hogg *et al.*, 1969). Wheat-growing areas are isolated in the south, and endemic disease cycles also exist.

Leaf rust of wheat is present in this zone, but it is a minor disease and causes minimal losses nationally. Stripe rust can be serious at higher elevations in this zone, occasionally destroying individual fields. Barley leaf rust occurs sporadically throughout the zone and is a major disease only in Ethiopia. Stem rust of barley is present, sometimes causing serious losses locally. Overall, losses are generally low since most barleys mature before the disease develops. Stripe rust can cause serious losses on barley grown at higher elevations.

D. North Africa

This epidemiological zone includes the countries of Morocco, Algeria, Tunisia, and Libya. The area has a typically Mediterranean climate. Primarily, spring-habit wheat and barley are sown in the autumn, but some autumn-sown winter-habit wheat is grown at higher elevations. The alternate hosts of both *P. recondita* and *P. hordei* are found in the area (Anonymous, 1981; Anikster and Wahl, 1979; d'Oliveira and Samborski, 1966). The importance of the alternate host in the disease epidemiology of this region has not been well documented (Dinoor, 1974). The rusts of wheat and barley occur annually, and the primary inoculum must come either from the alternate hosts, from urediospores from higher elevations (such as in the Atlas mountains) or from exogenous sources (Arthand *et al.*, 1966; Hogg *et al.*, 1969; Spehar, 1975).

Wheat leaf rust is present throughout the zone, causing serious losses on late-maturing local land cultivars. An especially wide virulence spectrum exists for durum wheats. Stripe rust of wheat is serious in coastal areas, with losses as high as 50% occuring in individual fields. Losses are diminished by the dry warming period that occurs in the late stage of plant growth.

Barley leaf rust is present in the coastal areas of Tunisia, Algeria, and Morocco, causing major losses on susceptible cultivars. Stem and stripe rust are not common on barley in this region and are considered minor diseases.

E. FAR EAST

China, Japan, Korea, Mongolia, and the eastern Soviet Union comprise this zone, and the general epidemiological patterns of the region appear to center on the concentration of wheat in China (Azbukina, 1980; Chen *et al.*, 1975; Chester, 1946; Gotoh, 1972). Stem rust of wheat was considered the most important disease in the 1950s, and there was a regular occurrence of the disease in eastern China, starting in the south and moving northward. Today, the use of resistant cultivars and a reduction of wheat cultivation in Fujian Province to reduce inoculum build-up in the south has reduced the losses caused by stem rust (S. M. Chen, personal communication). Stem rust is still a major disease of wheat in the north, northeast, south, and southeast regions of China.

Stripe rust is currently the most important rust disease in the Far East, and three subzones have been identified (Chen *et al.*, 1975; Johnson and Beemer, 1977; Chiu and Chang, 1982). The winter wheat area of northern China between the Yangtze River and the Great Wall is subject to severe stripe rust. This area alone contains about half the total wheat grown in China. The use of resistant cultivars has reduced the frequency of epidemics, but the evolution of new races requires a periodic change in cultivars. The source of inoculum for this area is the high elevations in the northwest spring wheat region: Inner Mongolia, Gansu, and Qinghai Provinces (north and Chiu and Chang, 1982;Johnson and Beemer, 1977).

A second subzone consists of the Upper Yangtze River Valley winter wheat region and the Provinces of Sichun, Guizhou, and Yunnan. Stripe rust could be severe in this subzone every year if resistant cultivars were not used. Host resistance commonly remains effective for only 3 to 5 years. *Puccinia striiformis* oversummers in the region at elevations above 2100 m (Chiu and Chang, 1982).

A third subzone for stripe rust is in the far west in Xijang Province. Epidemics occur frequently, but resistant varieties are being introduced. Wheat has been cultivated here for at least 2000 years, and many *Triticum* species are currently under cultivation.

Leaf rust is found throughout the wheat-growing areas of China. It is considered a minor disease, except for occasional local epidemics. There appears to be a general movement of *P. graminis* inoculum from south to north; inoculum does not appear to return (S. M. Chen, personal communication). The oversummering sites for stem rust have not been identified. Stripe rust inoculum moves in a easterly or northeasterly direction, and leaf rust apparently follows a similar pathway, at least in the northern areas. It has been suggested that China is an inoculum source for the Soviet Union Far East, Japan, and possibly Korea. Exchange between the rust populations of India and western China has been suggested and would appear to be possible through some mountain passes (Nagarajan and Joshi, 1975b), but there is no other data to support this hypothesis.

The alternate host of *P. recondita* has been reported in central Asia (Chester, 1946). Barberry is the alternate host for *P. graminis* and is found in the Soviet Union Far East (Azbukina, 1980).

F. SOUTHEAST ASIA

This zone consists of the Philippines, Sri Lanka, Indonesia, Malaysia, Thailand, and Burma. To date, only leaf rust has been observed in the area, except for Burma and Thailand. The races appear to be primitive with a narrow virulence spectrum, which can attack only the old local land-race cultivars. It suggests that the rust is also present on some noncultivated grass hosts.

The leaf rust population observed in Burma and possibly Thailand provides the exception to the prevailing situation in this zone. The virulence spectrum appears to be distinctly different from those observed in other countries of southeast Asia and Bangladesh. Burma also has a stem rust population that seems to have developed in isolation (U Kyi Maung and A. S. Gomaa, personal communication). The predominant wheat in Burma is Momywa White, which is the old Indian cultivar known variously as IP-4, P-4, or NP-4. This cultivar is susceptible to *P. graminis,* and it has suffered from epidemics of stem rust in the past. Recently, semidwarf cultivars with greater resistance have been introduced, and their use is spreading.

G. NORTH AMERICA

Stem rust has historically been the major disease on wheat in North America. In the nineteenth and early twentieth centuries, local epidemics were common and the primary inoculum came from urediospores overwintering in the south. Since the mid-1950s, however, resistant cultivars have been used to control this disease.

Leaf rust of wheat consistently causes some production losses, but never as severe as the losses that were associated with stem rust. Currently, the commercially grown spring bread wheat and durum wheat cultivars have adequate resistance to prevent severe losses. However, shifts in the virulence patterns of the pathogen continues to be of concern. Most winter wheat cultivars are currently susceptible to at least part of the pathogen population.

Stripe rust is only important in the Pacific Northwest, and even then not every year. The disease is controlled by a combination of adult plant resistance and, in recent years, by chemical fungicides.

Generally, leaf rust of barley is of minor importance in this epidemiological zone, and stem rust on barley has been important only when inoculum is generated on susceptible wheats. The barley cultivars grown in the northern rust area possess the *T* gene for resistance, which has provided adequate resistance for a number of years.

The epidemiology of the cereal rusts in North America are discussed in detail in Chapter 13, this volume. Annual losses due to the rust diseases of small grain cereals in the United States have been estimated (Roelfs, 1978).

H. SOUTH AMERICA

There are two subzones in South America: the Andean countries of Colombia, Ecuador, Peru, and western Bolivia, and a southeastern subzone that includes the Pampas of Argentina, Uruguay, Paraguay, southern Brazil, and lowland Bolivia. Leakage from the Andean subzone has occurred through the southern Andes (see Section III). Another possible point of leakage is where the two areas are separated by only a short distance in Bolivia.

Rust may exist as an endemic population within both subzones. *Puccinia* species often survive locally, except at the southern end of the Andean subzone and probably at both the north and south ends of the eastern subzone. In parts of the Andean subzone, agriculture is of a primitive nature and there are tremendous differences in elevation over short distances. Wild species of *Hordeum* are common and often severely rusted (CIMMYT, 1978; Dubin, 1984). In the eastern subzone, modern agriculture prevails and double cropping is common. Self-sown wheat is common in the summer crop and along the roadsides. The climate is generally mild except in the south end of the region. Planting and harvesting occurs over a period of several months. Thus, the rusts can often cycle between early-planted, late-planted, and self-sown cultivars.

Stripe rust is common in the Andean subzone and can cause serious losses on both wheat and barley in the higher elevations near the southern boundary of the zone. Stripe rust is relatively unimportant in the eastern subzone, except in Argentina near the southern end of the zone. Leaf rust is common in South America and has caused significant losses. Leaf rust resistance has been incorporated into many of the South American wheat varieties.

Stem rust is also common throughout the entire zone. Losses are generally light, except occasionally in localized epidemics. Barberry species are found in the Andean subzone, but they are not a part of the disease cycle in the south and their importance in the north is unknown.

I. AUSTRALIA AND NEW ZEALAND

Wheat-production areas are not contiguous in this zone. Western Australia is over 1000 km distant from the major wheat-producing areas in southern and eastern Australia. New Zealand is about 2000 km southeast of the eastern Australian wheat area. The major wheat-growing region in south and east Australia consists of a narrow geographic band with scattered outlaying areas, especially toward the end of the band (Luig, Chapter 10, this volume, Fig. 1). Most of the wheat is fall-sown and matures before the hot dry summer.

Little is known about rust epidemiology during the summer when wheat is not grown. However, all rusts of wheat survive this period, and self-sown plants must play an important role in disease survival. The geographic isolation of Australia has provided an excellent opportunity to study long-distance movements of rust populations. It has been suggested that at least two introductions of *P. graminis* into Australia have occured from southern Africa. The rust fungi can move to New Zealand from the eastern Australian wheat belt almost annually by wind dispersal (reviewed by Luig, Chapter 10, this volume).

Stem rust of wheat is an important disease causing severe losses on susceptible cultivars. However, few detailed studies or estimates of rust-caused losses are available. Currently, many of the important varieties grown are resistant to all prevalent races (see Luig, Chapter 10). Yet, virulence is evolving in the pathogen population for the resistant cultivars grown. Stripe rust was introduced into Australia in 1979 (see Section III; also Luig, Chapter 10, this volume) and has since become a serious disease of wheat. Intensive efforts are currently underway to select resistant cultivars. Leaf rust of wheat is an important disease in Australia, but causes less severe losses than the other two wheat rusts. Resistant cultivars are being used to minimize losses.

J. EUROPE AND CENTRAL ASIA

This epidemiological zone has great diversity in the host cultivars grown, cultural practices used, and environmental conditions. It stretches eastward across Europe through the Russian Steppes into the New Lands. Along the southern edge of this zone and across the Mediterranean Sea is the North African zone (Section II,D); to the east are the Mediterranean Sea, Black Sea, and Caspian Sea; to the west is the West Asia and Egypt zone (Section II,B). Spore movement in this zone is northward from Spain following a westerly track, and eastward from Yugoslavia and Bulgaria following an eastern track. The western tract is connected with the North Africa zone and the eastern tract with the West Asia zone. Winter-planted cereals are grown through much of the year, and overwintering rust is common (Zadoks and Bouwman, Chapter 11, this volume).

A subzone exists in eastern Europe from the Ukraine and Caucasus across the Russian Steppes south of the mountains into the Russian New Lands. Winter wheats are grown in the Caucasus, and rust spreads eastward, where spring wheat is grown in the New Lands. Considerable effort has been made to breed resistant cultivars and identify pathogen races in the Soviet Union; most of this effort, however, has occurred on a local level. No recent national description of the epidemiology of the cereal rusts was available to the authors.

Stem rust of wheat was an important disease in Europe in the nineteenth and early twentieth centuries, often spreading from barberry. Even though many western European cultivars are susceptible to stem rust, the disease is currently of minor importance in this zone, except along the eastern tract where barberry still may be involved in the disease cycle. Stem rust occurs in the Russian subzones, but generally does not cause serious losses.

Thalictrum and *Anchusa* species are alternate hosts of *P. recondita* and are commonly infected in southwestern Europe. There appears to be specialization for both the alternate host as well as the wheat host (d'Oliveira and Samborski, 1966; Young and d'Oliveira, 1982). Leaf rust of wheat is widespread, but losses are generally light in western Europe. The disease is more severe in eastern Europe and in the Eastern subzone (Chester, 1946; Hogg *et al.*, 1969). Recent epidemics reported in the eastern subzone were in the Ukraine in 1970, 1972, and 1977 (Novokhatka, 1979), in the southern foothills of the Caucasin in 1973 (Minkevich and Zakharova, 1982), and in the Transcaucus in 1975 (Konovalova *et al.*, 1977).

Stripe rust of wheat and barley is the most important cereal rust in western Europe and can result in serious losses. The disease is generally endogenous in western Europe, and it has occurred throughout the zone in many years (Zadoks and Bouwman, Chapter 11, this volume). Susceptible cultivars are often sprayed with fungicides.

Barley leaf rust is an important disease in northern Europe and is controlled there by the use of resistant cultivars and fungicides (see Parlevliet, Chapter 16, and Clifford, Chapter 6, this volume).

The epidemiology of the cereal rusts in Europe is reviewed in detail by Zadoks and Bouwman in Chapter 11, this volume.

III. Long-Distance Dissemination

A major problem with establishing geographical zones for the rust pathogens is the implication that migration from one zone to another is impossible. This is not the intent of the authors in proposing the zones above. There is strong evidence that migration occurs between adjacent zones and even zones that are rather distant (Abdel-Hak *et al.*, 1974; Dinoor, 1974; Dinoor *et al.*, 1976; Hogg *et al.*, 1969; Rajaram and

Campos, 1974; Saari, 1977; Sharma *et al.*, 1972). It should, however, be recognized that exchange between zones is either rare or only occurs sporadically. Exchanges are associated with either special weather circumstances or "leakage" points between zones (CIMMYT, 1980; Nagarajan and Joshi, 1975b). Both types of exchanges are illustrated by the introduction into South America of barley stripe rust race 24.

The first reports of race 24 were recorded in 1975 in Bogota, Colombia (CIMMYT, 1978; Dubin, 1984; Stubbs, Chapter 3, this volume). Race 24 has a distinctive virulence pattern that corresponds to barley stripe rust in Europe. The authors speculate that race 24 was inadvertently introduced by jet-age travel. Introductions of this type are becoming increasingly more common, as more people quickly travel between distant locations. Race 24 subsequently migrated southward throughout the Andean region. It was found in Ecuador in 1977, southern Peru in 1977, northern Chile in 1980, and southern Chile in 1981. (CIMMYT, 1978; Dubin, 1984). In 1981 a severe outbreak of race 24 occurred in south central and coastal Chile (CIMMYT, 1984).

The pattern of spread was a stepwise migration, proceeding from the point of introduction in Colombia along the Andean mountains to southern Chile. Most commercial cultivars or land varieties of the area possessed no resistance to hamper its spread. The intermountain movement appears to have been assisted by the susceptibility of the wild *Hordeum* species that prevail in the area (CIMMYT, 1979; Dubin, 1984). Severe attacks and losses in yield were recorded in all of the barley-growing areas of Colombia, Ecuador, Peru, Bolivia, and Chile (CIMMYT, 1978, 1984; Dubin, 1984). In 1982, race 24 was reported to have "leaked" through from Chile to Argentina via the southern mountain passes (Dubin, 1984; M. Kohli, personal communication). This case is a classic example deserving a more detailed review than possible here.

A similar case involving the introduction of stripe rust of wheat into Australia occurred in 1979 (Wellings and McIntosh, 1981). The introduced race, 104E137, is identical to one that previously was found only in Europe. This single race multiplied, spread rapidly, and survived the off-season. The first stripe rust of wheat observed in New Zealand was reported the next year (Beresford, 1982). The introduction into Australia appears to be human-aided, while the introduction into New Zealand appears to be by way of wind dispersal of urediospores.

A new biotype of race 104E137 that could attack the cultivar Avocet was identified in 1981 (Annual Report, 1982), indicating how quickly changes can occur in the rust fungi. The new biotype was also virulent on local species of *Agropyron* and *Bromus* (Annual Report, 1982), a

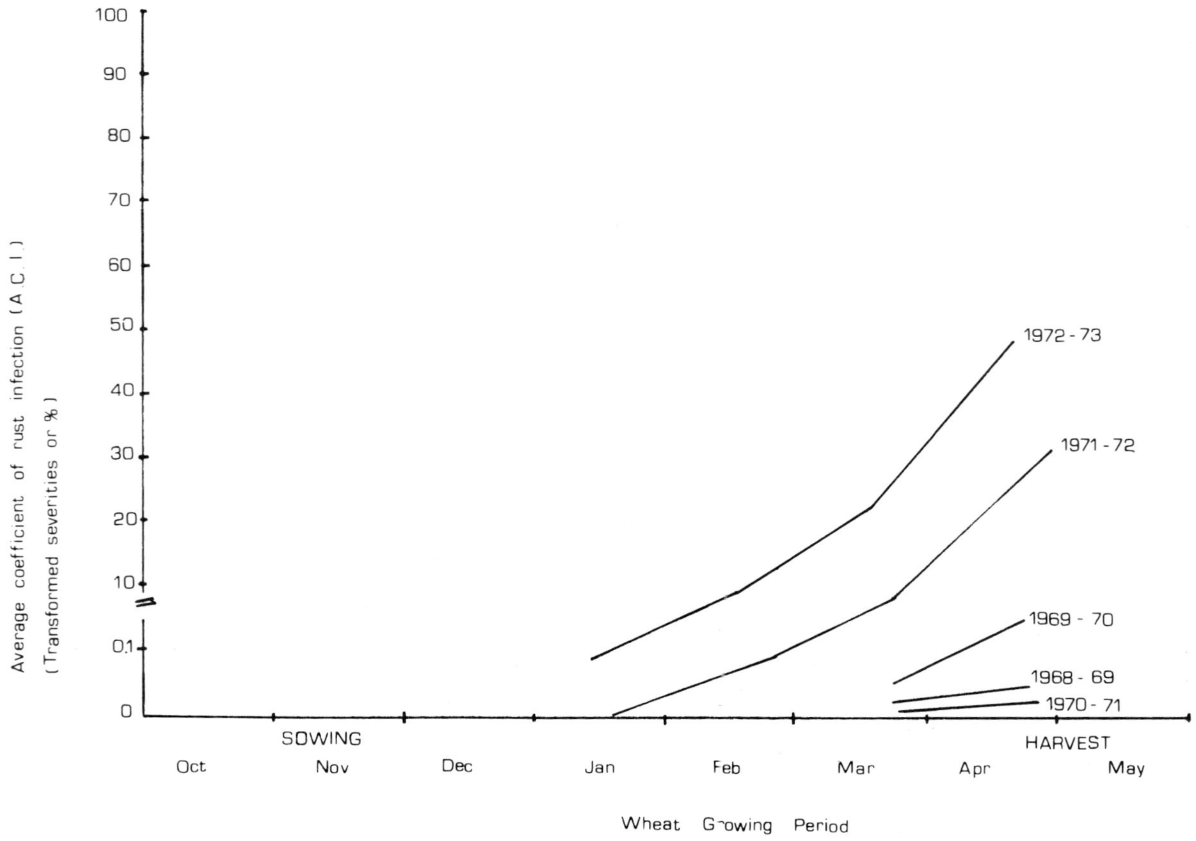

Fig. 2. The change in the virulence frequency of leaf rust to the 8156 cultivars in commercial fields and in trap nurseries in northwest India, 1968–1973.

factor that undoubtedly provides an additional means of survival over the summer season.

Another example of migration versus evolution involves a comparison of leaf rust and stripe rust race developments recorded in India and Pakistan. The 8156 cultivars (Kalyansona in India and Mexipak in Pakistan) were extensively cultivated in both countries in the late 1960s and early 1970s. Eventually, a new race of leaf rust capable of attacking the 8156 cultivars was detected. It increased in frequency until serious losses were incurred in the 1971–1972 and 1972–1973 seasons (Joshi, 1980). This appears to be evolution within the zone. The change is shown in Fig. 2, where the coefficient of infection in frequency is plotted against time (Saari, 1977). The coefficient of infection is obtained by multiplying disease severity by the coded host responses. Values of 0.2, 0.4, 0.8, and 1.0 are assigned to resistant, moderately resistant, moderately susceptible, and susceptible responses, respectively (CIMMYT, 1976).

Stripe rust virulence for 8156 cultivars appears to have migrated to Pakistan and India from the Middle East (see Fig. 3). In the case of the 8156 virulence, the migration from Turkey (Ozkan and Prescott, 1972; Saari, 1977; Sharma *et al.*, 1972) eastward was associated with a meteorological feature called the "western disturbance" (Nagarajan and

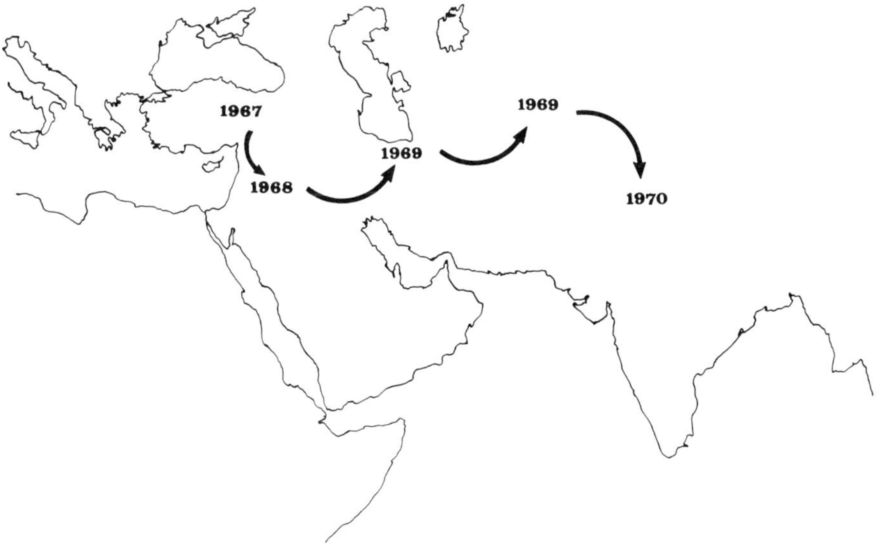

Fig. 3. Years of the first recorded appearance of the *Puccinia striiformis* virulence for 8156 cultivars and its possible migration path.

Joshi, 1978; Pisharoty and Desai, 1956). This weather feature probably provided the means for the sequential movement of the rust spores to Pakistan and India. The western disturbance is a low-pressure system that forms in the Middle East and moves eastward, eventually dissipating in northern Pakistan and India. It seems that the 8156 virulence was introduced into India and Pakistan late in the crop cycle of 1970, and established itself and survived in the high hills of the northern areas (Hassan, 1970; Joshi *et al.*, 1976a,b, 1977). It began increasing in frequency, but at a much slower rate than the 8156 leaf rust virulence.

IV. Epidemics and Yield Losses

The history of wheat and, to some extent, barley cultivar improvement is filled with examples of new virulences arising in the rust fungi and eventually increasing to a level where previously resistant varieties are vulnerable to attack. If early infection occurs and favorable weather prevails, then the classic pattern of virulent pathogen, susceptible host, favorable weather, and time result in an epidemic.

Cereal rust epidemics occurring in Europe, North America, and Oceania have been well documented (Zadoks and Bouwman, Chapter 11; Roelfs, Chapter 13; and Luig, Chapter 10, respectively, this volume). Rust epidemics occurring in South America have been reported in various local publications with limited availability.

Rust epidemics have occurred in Asia and Africa on a regular basis, but documentation is fragmentary. Many references to epidemics are in mimeographed reports or obscure journals. Several languages are involved, so a comprehensive review of epidemics is difficult. When records are available, the recording of disease occurrence may be stated, but these are seldom accompanied by yield loss data. We have tried to summarize the occurrence of rust epidemics for Asia and Africa based on our experience, notes from colleagues, and published papers where available (Joshi and Pathak, 1975; Varughese, 1975).

One serious problem regarding the evaluation of rust losses is the lack of statistical precision. Losses of less than 10% are not measurable under most circumstances. Consequently, disease development must be severe to more accurately measure losses. Furthermore, small-plot evaluations have also been shown to overestimate disease losses (Stucker, 1980). It is also difficult to accurately measure the effects of rust that develops in the late stages of plant growth. In years that are favorable for crop production, yields may be above average and the effects of rust diseases are discounted and/or underestimated.

Stem rust was considered the most important rust disease on a global basis until the late 1950s, when resistant cultivars began appearing in several countries (S. M. Chen, personal communication; El-Tobgy and Taalat, 1971; Hassan, 1978; Hassan *et al.*, 1977; Nagarajan and Joshi, 1975a; Pinto and Hurd, 1970; Saari and Wilcoxson, 1974). The introduction of semidwarf bread wheats into Asia and Africa in the late 1960s changed the relative importance of the three rust diseases of wheat (Dalrymple, 1978; Hanson *et al.*, 1982; Saari and Wilcoxson, 1974). The incorporation of resistance and earliness further reduced the frequency of severe stem rust infections in all areas except east Africa, most notably in Kenya and Ethiopia (Cormack, 1974; CIMMYT, 1979, 1980; Oggema, 1974; Pinto, 1971; Saari and Wilcoxson, 1974). The incorporation of earliness in host cultivars has eliminated the last generation(s) of disease development when inoculum densities are the highest, thereby enhancing the resistance of the semidwarf cultivars.

Stripe rust is also considered an important disease in areas where temperatures are favorable for disease development (Rapilly, 1979). True winter and/or facultative wheats are especially vulnerable to stripe rust because they are grown in areas characterized by extended cold periods in winter and slow warming in the spring. The incorporation of resistance, however, has greatly reduced the frequency of stripe rust epidemics (Bamdadian, 1980; Chen *et al.*, 1975; Nagarajan *et al.*, 1978; Saari and Wilcoxson, 1974). In China, stripe rust is considered one of the most serious diseases, affecting more than 50% of the wheat-growing areas. Serious epidemics were common until resistant cultivars were developed (Chen *et al.*, 1957; Johnson and Beemer, 1977).

In areas where spring wheats are fall-sown, stripe rust has been a serious problem in the past. The introduction of semidwarf wheats into these areas, however, brought in good levels of resistance, which reduced the frequency of severe epidemics (Nagarajan *et al.*, 1978). An additional factor, especially in Pakistan and India, is the later sowing date used with these cultivars (Hanson *et al.*, 1982). The reduced time for stripe rust development, due to the typical drying and warming trend in the late stages of crop development, undoubtedly acts as a constraint under normal crop-production conditions. In areas where the climate remains favorable for stripe rust development, the planting of susceptible cultivars often results in epidemics (Hassan, 1978;

CIMMYT, 1978; Saari, 1978). Losses can be serious if the weather is favorable for infection at the flowering or grain-filling stages.

In Figs. 4 and 5, the recorded levels of stripe rust and stem rust can be seen for four of the zones in Asia and Africa from which data from the Regional Disease Trap Nursery (RDTN) is available. The average level remains relatively low even for the local land-race cultivars or susceptible checks. In contrast, leaf rust has remained a serious problem in south Asia and parts of west Asia and Egypt (zone 2), despite major efforts to incorporate resistance. The variability and flexibility of the leaf rust fungus seems particularly well suited to conditions in the

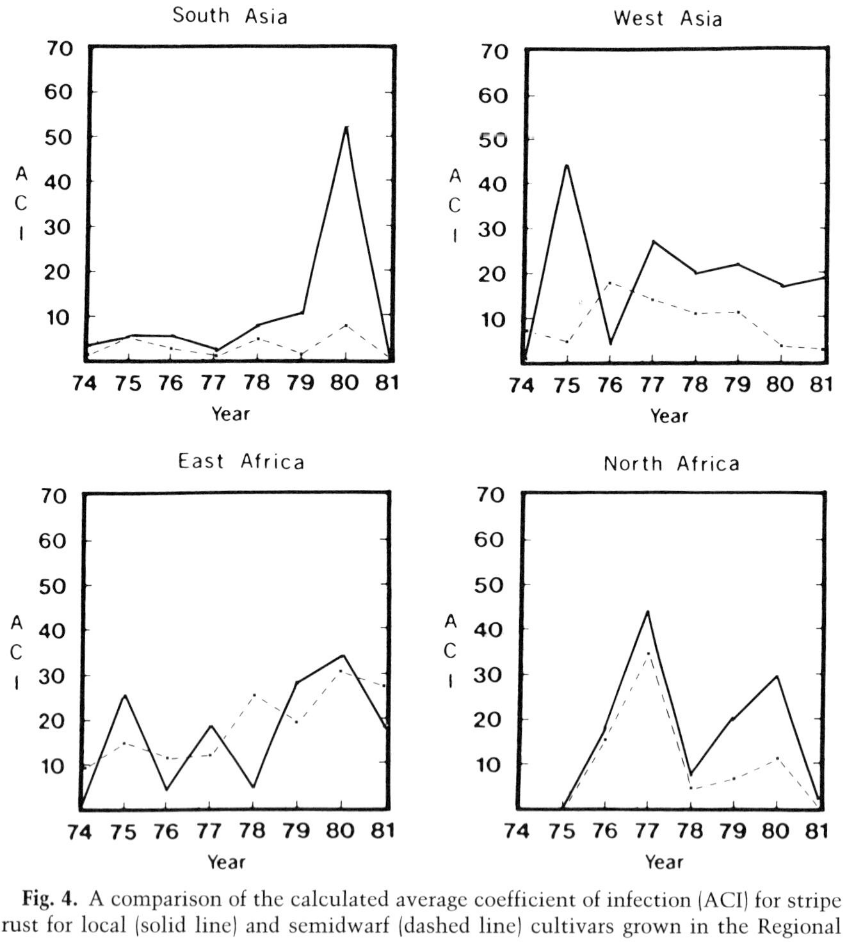

Fig. 4. A comparison of the calculated average coefficient of infection (ACI) for stripe rust for local (solid line) and semidwarf (dashed line) cultivars grown in the Regional Disease Trap Nurseries.

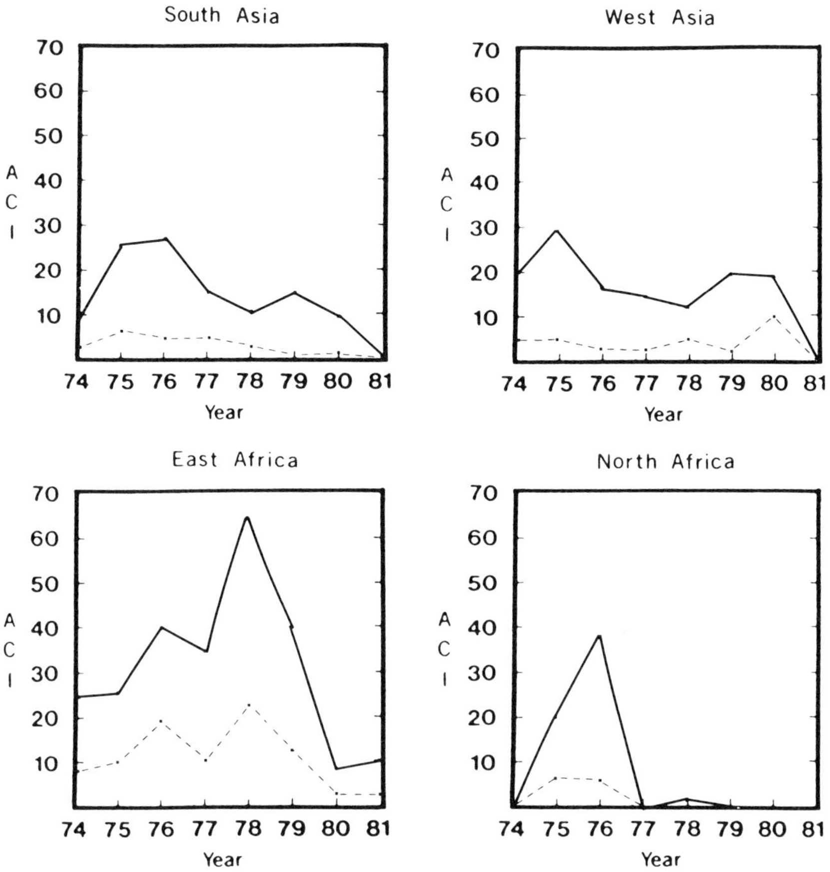

Fig. 5. A comparison of the calculated average coefficient of infection (ACI) for stem rust for local (solid line) and semidwarf (dashed line) cultivars grown in the Regional Disease Trap Nurseries.

wheat areas of the south Asia zone, and where the weather is influenced by large bodies of water (Fig. 6).

The development of leaf rust in India and Pakistan during the 1972 and 1973 seasons caused considerable losses (Hassan *et al.*, 1977; CIMMYT, 1972, 1973, 1978; Joshi, 1980). Survey results from India suggest that reductions of 0.73 and 1.37 million tonnes occurred in 1972 and 1973, respectively (Joshi, 1980). Most of this loss occurred in the Indian states of Punjab, Haryana, and Uttar Pradesh. Following the 1972 and 1973 leaf rust epidemics, the cultivars grown in India were changed rapidly (Nagarajan *et al.*, 1978; Swaminathan, 1978), while in

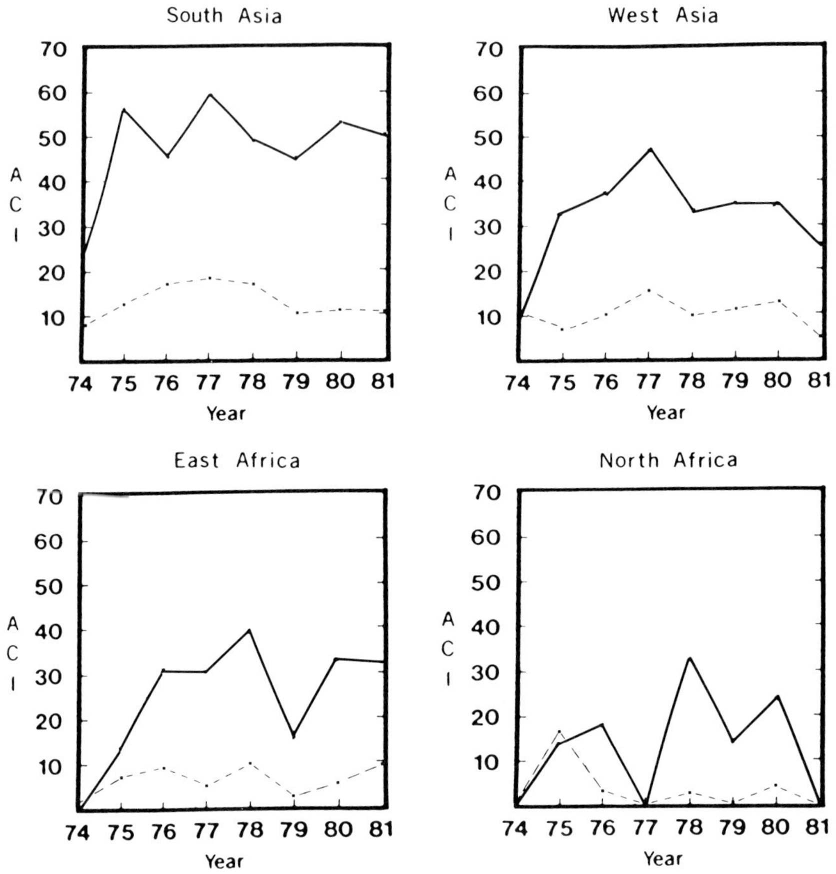

Fig. 6. A comparison of the calculated average coefficient of infection (ACI) for leaf rust for local (solid line) and semidwarf (dashed line) cultivars grown in the Regional Disease Trap Nurseries.

Pakistan they were not changed (Hassan, 1978; CIMMYT, 1978; Saari, 1978). This diversification became important (and was effective) when an accumulation of factors led to epidemic conditions for leaf and stripe rust in the northern regions of Pakistan and India in 1978 (Hassan, 1978; CIMMYT, 1978; Nagarajan and Joshi, 1980; Nagarajan *et al.*, 1980, 1982). The losses in Pakistan appear to have been somewhere between 10 and 20% of the total crop (CIMMYT, 1978; Kidwai, 1979), while India had a record production of wheat. The primary difference between the Punjab of India and the Punjab of Pakistan was in the level of resistance of the cultivars grown (Saari, 1978; Swaminathan, 1978).

Leaf rust has been a problem in Egypt in recent years. Studies indicate that losses of up to 25% are possible with some of the semidwarf wheats. Losses with improved cultivars ranged from 11 to 19%. However, the yield level of the semidwarfs was still sufficient to out-yield the local, tall cultivars (Abdel-Hak *et al.*, 1980). There have been several other recorded epidemics in the area, but they have involved either local land-race cultivars (Bamdadian, 1980; Torabi, 1980) or older semidwarf cultivars that continued to be cultivated after pathogen virulence was known to exist in the rust population (Ghodbane, 1977; Hassan, 1978; Hassan *et al.*, 1977; Joshi, 1980; Joshi *et al.*, 1980; Saari and Wilcoxson, 1974). A lack of adequate disease-screening pressure has been a major factor leading to epidemics on some of the newly released semidwarf cultivars (Saari, 1978).

The evolution of a new race of rust usually allows sufficient time for replacement of a cultivar from the recommended list, but this can only be accomplished if a resistant line and seed production capabilities are present. In much of the developing world, seed production and the infrastructure for distribution of new cultivars is still a major constraint to good disease management in wheat.

The number and significance of the recorded rust epidemics vary greatly. The examples mentioned above represent some of the better documented epidemics and illustrate some of the problems involved in food production and disease control in developing countries. Table II lists major known rust epidemics and the estimated losses they caused.

V. Future Prospects

In the past two decades, much has been accomplished in controlling the cereal rusts, especially those attacking wheat. A principal reason for the progress has been the number of trained individuals working on the problem. This collective effort of plant pathologists and plant breeders in developing resistant cultivars and understanding disease epidemiology has gradually reduced the magnitude and frequency of epidemics. Gains can be measured over a 10-year period only, since the cycle of education, training, and application in a breeding program for disease resistance requires at least a decade to come to fruition.

Progress can be best measured by the developments since World War II. There has been a substantial increase in facilities during this 40-year period; the combination of trained personnel and improved facilities has resulted in a growth in knowledge that is reflected in the number of publications now available. Furthermore, the number of profes-

Table II

Wheat Rust Epidemics on Commercial Areas in Asia and Africa since 1970

Disease	Year	Area	Cultivar(s) losses on	Losses	Estimated effect
Stripe rust	1981	Ethiopia	—	Serious	Losses up to 20–30%
	1979	Central coast of Algeria	Siete Cerros	Moderate	Losses of 20% in yield trials
	1978	East and central coast of Algeria	Siete Cerros	Serious	Loss of 20% in yield trials
		Punjab areas of India and Pakistan	Mexipak, Chenab 70, Punjab 76, Khushal 69, Tarnab 70	Major	National loss of 5–10%
		Far northwest China	—	Serious	Losses of 5–30%
	1977	Coastal Algeria	Siete Cerros	Serious	Losses up to 25% in yield trials
		Central plateau and coastal Turkey	Penjamo 62, local land cultivars of winter wheat	Serious	Losses up to 12% nationally
		Northwest and west Iran	Local land cultivars	Serious	—
		Northern Iraq	—	Serious	—
		Tunisia and Algeria	Soltane, Siete Cerros	Serious	More than a 30% loss
	1976	East Central Coastal Algeria	Siete Cerros	Serious	Losses of up to 18% on Siete Cerros
		Coastal Tunisia	Soltane		On Soltane, losses 20–30%
		Turkey	Local land cultivars	Moderate	About a 10% loss
	1975	Turkey	Local land cultivars	Moderate	About a 10% loss

	1973	North China Plain	—	Minor	Less than 5%
		Kenya highlands	Kenya Nyali, Kenya Mamba	Severe	Losses of 5–10%
	1972	Northwest China	Local land cultivars	Severe	Losses of 5–10%
	1970	Iran	Penjamo 62	Severe	—
Leaf rust	1978	Central and southern Pakistan	Mexipak, Chenab 70, Pak 70, Punjab 76	Serious	National losses of 5–10%
		Egypt	Mexipak, Chenab 70	Moderate	Loss of 20% in yield trials
		Zimbabwe	—	Serious	25% Of area affected
	1977	Egypt	Chenab 70, Giza 155, Mexipak	Serious	Losses of 10–20% in yield trials
	1976	Egypt	Chenab 70, Mexipak	Moderate	Losses up to 20% in yield trials, national loss minor
		Central Pakistan	Mexipak	Minor	—
	1973	Northern India, Pakistan, and Nepal	Kalyansona, Mexipak	Moderate	6% Loss in India, no estimate elsewhere
	1972	Northern India, Pakistan, and Nepal	Kalyansona, Mexipak	Moderate	3% Loss in India, no estimate elsewhere
Stem rust	1979	Ethiopia	Durums	Serious	—
	1978	Kenya	Kenya Nyako	Moderate	—
		Zimbabwe		Serious	Loss of 25%
	1977	Yeman Arab Republic	Local land cultivars	Serious	—
	1976	Zambia	—	Moderate	—
	1975	Southern Tunisia	Local land durum	Serious	—
	1972	Kenya	Trophy, Kenya Kanga	Serious	—

sional conferences and journals has increased dramatically since World War II.

Another major contributor to more effective control of the wheat rusts has been the exchange of improved germ plasm. The International Wheat Rust Nurseries distributed by the United States Department of Agriculture was a major contributor to the exchange of germ plasm on a worldwide basis (Kilpatrick, 1974). The creation of the International Centers for Agricultural Research, such as the International Maize and Wheat Improvement Center (CIMMYT), further stimulated the international exchange of seed (Hanson *et al.*, 1982). The volume and types of seed exchange have also changed dramatically (Dalrymple, 1978; CIMMYT, 1982). A diversity of germ plasm is now available to the wheat improvement programs of the world, and this diversity can be exploited in a number of ways. In established breeding programs, this germ plasm provides a vital source of variation. In new programs or those with limited manpower, the evaluation and selection of advanced lines or cultivars provides an alternative. The variation available today is much broader than would be available without such an exchange of germ plasm and provides opportunities for diversification that would not be possible to a scientist working in isolation. Consequently, the possibilities for reducing the hazards inherent in monocropping have been greatly improved.

There are a number of critics who point to the dangers of such germ plasm exchanges, but their criticisms have not been properly weighed against the benefits already accrued to mankind; nor have these critics adequately addressed the vital question of the viability of alternatives to the free exchange of germ plasm. Increasingly restrictive plant quarantine regulations in many developing countries are bringing this issue to the forefront. The restrictions are often applied to experimental seed intended for scientific uses, but not to commercial trade. This differential application is inconsistent with the increasing international world demand for basic food commodities.

There is a growing interest in the cultivation of wheat in areas where wheat is not a traditional crop. These areas constitute a possible expansion of wheat production into climatic zones normally considered unsuitable for wheat. The countries concerned import wheat for domestic consumption and are interested in the potential for import substitution. Most of this interest in wheat cultivation has resulted because of reductions in the food deficit of other basic food staples, such as rice. Many of these potential wheat-growing countries have subtropical or tropical environments. Preliminary evaluations suggest that there is sufficient variation in the *Triticum* species to select cultivars that can be grown in these environments (CIMMYT, 1980).

There are number of difficulties inherent in cultivating wheat in new areas. There are agronomic or production problems specific to local environments, and the plant protection problems are often different, with important implications for wheat diseases (especially the rusts). In trials grown in the Philippines, Indonesia, Thailand, and Sri Lanka, leaf rust of wheat has been observed. The only cultivars attacked were old tall or local land-race cultivars. Semidwarf cultivars considered highly susceptible to leaf rust in traditional wheat-production areas are highly resistant. It thus appears that the leaf rust races present are primitive and probably come from other grass hosts. The implication is that evolution of the pathogens will result in virulence on the new wheat introductions, and that commercial-scale production may lead to major disease problems.

Stem rust of wheat seems likely to become a serious disease problem in new production areas because of climatic considerations. In south and southeast Asia, stem rust of wheat has not been observed. However, in southern India stem rust is present throughout the year at higher elevations. If wheat cultivation expands in Sri Lanka, the appearance of stem rust there seems highly probable. Certainly, any significant amount of wheat cultivation in Sri Lanka will also have major implications for the leaf rust and stem rust pathways affecting India and Pakistan.

The expansion of wheat production in the Sudan and west Africa has resulted in the appearance of stem rust (Andrews, 1968; Baghadadi, 1983; Khalifa, 1979), and leaf rust has also been observed in western Africa (Andrews, 1968). The expansion of wheat into the more tropical environments of South America has also resulted in the appearance of stem rust (J. Dubin, personal communication). The case of stem rust of wheat in South America, west Africa, and Sudan appears to be an example of extending the "Puccinia Pathway". There are known endemic foci that could serve as a source of urediospores for primary infection. The spread from these areas seems to be the most likely explanation for the appearance of stem rust.

The appearance of stem rust in South America is significant because the expanded acreage is an extension of the existing wheat production area. In West Africa, the appearance of stem rust is sporadic, and timely sowing often provides a means of escape (Andrews, 1968). The wheat in Sudan will probably always be subject to stem rust disease because of nearby sources of inoculum. The evidence suggests that *P. graminis* inoculum is endemic in Ethiopia and the southern and western areas of the Arabian Peninsula, which could serve as sources of primary inoculum for Sudan depending on weather patterns.

In recent years, chemical control of the wheat rusts has become a routine practice in Europe (Obst, 1981; Zadoks, 1980). The use of chemical controls appears to be both economically and strategically feasible in high-yield production environments where the inputs are available and the agricultural infrastructure is well established. For the developing world, however, where the yield potential is lower, where inputs tend to be scarce, and where the agricultural infrastructure is often less well established, the potential for effective chemical control of the cereal rusts is minimal, at least for the forseeable future. Breeding for resistance remains the most promising and viable means of controlling the cereal rusts in developing countries, and this will require renewed and, in some cases, expanded commitments by various groups and institutions within the international scientific community. The need for the free exchange of information and germ plasm on a global basis can only increase in the future. The extent to which this cooperative effort to increase food production around the world can be facilitated and made more efficient is a major challenge facing agricultural scientists today.

References

Abdel-Hak, T. M. (1970). Importance of leaf rust in the Near East. *Proc. FAO/Rockefeller Found. Wheat Semin., 3rd, 1970,* pp. 239–246.

Abdel-Hak, T. M., and Ghobrial E. (1978). The barley disease situation in the Near East with special reference to sources of resistance. *Proc. Reg. Winter Cereals Workshop—Barley, 4th, 1977,* Vol. 2, pp. 311–319.

Abdel-Hak, T. M., and Kamel, A. H. (1972). Present status of wheat stem rust in the Near East Region. *In* "Disease," Proc. Reg. Wheat Workshop, Vol. 1 (mimeo.). Ford Found., Beirut, Lebanon.

Abdel-Hak, T. M., Stewart, D. M., and Kamel, A. H. (1972). The current stripe rust situation in the Near East Region. *In* "Disease," Proc. Reg. Wheat Workshop, Vol. 1 (mimeo.). Ford Found., Beirut, Lebanon.

Abdel-Hak, T. M., Stewart, D. M., and Kamel, A. H. (1974). Wheat diseases and their relevance to the improvement and production programs in the Near East. *Proc. FAO/Rockefeller Found. Wheat Semin., 4th, 1973,* pp. 287–291.

Abdel-Hak, T. M., El-Sherif, N. A., Bassiouny, A. A., Shafik, I. I., and El-Dauadi, Y. (1980). Control wheat leaf rust by systemic fungicides. *Proc.—Eur. Mediterr. Cereal Rusts Conf., 5th, 1980,* pp. 255–266.

Ali, F. M. (1979). Constraints to production and possible solutions in Sudan. *In* "The Gap Between Present Farm Yields and the Potential," Proc. 5th Cereal Workshop, Vol. 1, pp. 82–87. Minist. Agric., ICARDA and CIMMYT, El Batan, Mexico.

Andrews, D. J. (1968). Wheat cultivation and research in Nigeria. *Niger. Agric. J.* **5**, 67–72.

Anikster, Y., and Wahl, I. (1979). Coevolution of the rust fungi on Gramineae and Liliaceae and their hosts. *Annu. Rev. Phytopathol.* **17**, 367–403.

Annual Report (1977). "National Wheat Development Program." Dept. of Agric., Kathmandu, Nepal.
Annual Report (1978). "National Wheat Development Program." Dept. of Agric., Kathmandu, Nepal.
Annual Report (1979). "National Wheat Development Program." Dept. of Agric., Kathmandu, Nepal.
Annual Report (1982). "Plant Breeding Institute." University of Sydney, Sydney, Australia.
Annual Report (1983). "Indian Agricultural Research Institute." Regional Station, Flowerdale, Simla.
Anonymous (1967). "The Millets: A Bibliography of the World Literature Covering the Years 1930–1963." Scarecrow Press, Metuchen, New Jersey.
Anonymous (1974). "Proceedings of the Fourth FAO/Rockefeller Foundation Wheat Seminar." FAO, Rome.
Anonymous (1975a). "Proceedings of the Third Regional Wheat Workshop." International Maize and Wheat Improvement Center, El Batan, Mexico.
Anonymous (1975b). Wheat production in the Kingdom of Saudi Arabia. *Proc. Reg. Wheat Workshop, 3rd, 1975,* pp. 355–360.
Anonymous (1979a). "The Gap Between Present Farm Yield and the Potential," 5th Cereals Workshop, Vol. I. Minist. Agric., ICARDA and CIMMYT, El Batan, Mexico.
Anonymous (1979b). "The Gap Between Present Farm Yield and the Potential," 5th Cereals Workshop, Vol. II. Minist. Agric., ICARDA and CIMMYT, El Batan, Mexico.
Anonymous (1981). "Proceeding of the Barley Diseases and Associated Breeding Methodology Workshop." Rabat, Morocco USAID-Montana State Univ., ICARDA and CIMMYT, El Batan, Mexico.
Anonymous (1982). "Wheat Excels in the Dry Zone," Res. Highlights, Vol. 10, pp. 2–4. Dept. Agric., Peradeniya, Sri Lanka.
Arthand, J., Guyot, L., and Malencon, G. (1966). Comparative biometric studies of the formae specialis of black rust (*Puccinia graminis* Pers.) living on wild grasses of the Moroccan Atlas. *Proc. Cereal Rusts Conf., 1964,* pp. 204–206.
Aslam, M. (1978). Genetic analysis of rust races for breeding resistant cultivars of wheat. *In* "Technology for Increasing Food Production" (J. C. Holmes, ed.), pp. 424–425. FAO, Rome.
Azbukina, Z. (1980). Economical Importance of aecial hosts of rust fungi of cereals. *Proc.—Eur. Mediterr. Cereal Rusts Conf., 5th, 1980,* pp. 199–201.
Ba-Angood, S. A. A. (1975). Wheat improvement in the Peoples Democratic Republic of Yemen. *Proc. Reg. Wheat Workshop, 3rd, 1975,* pp. 352–354.
Baghadadi, A. M. (1983). Black stem rust disease of wheat in the Sudan—A preliminary report. *Cereal Rusts Bull.* **11,** 1–5.
Bahadur, P., Nayar, S. K., Srivastava, M., Goel, L. B., Sharma, S. K., and Meena, K. L. (1982). Distribution and frequency of physiologic races of wheat rusts in India during 1979–80. *Indian Phytopathol.* **35,** 478–482.
Bamdadian, A. (1973). Physiologic races of *Puccinia recondita* in Iran (1968–1972). *Cereal Rusts Bull.* **1,** 45–47.
Bamdadian, A. (1976). The studies on chemical control of *Puccinia striiformis* West. with three systemic fungicides in Iran. *Proc.—Eur. Mediterr. Cereal Rusts Conf., 4th, 1976,* pp. 93–95.
Bamdadian, A. (1980). Epidemiology of yellow rust of wheat (*Puccinia striiformis*) in Iran. *Proc.—Eur. Mediterr. Cereal Rusts Conf., 5th, 1980,* pp. 219–222.
Barghouti, S., Saari, E. E., Srivastava, J. P., and Chancellor, G., eds. (1978a). "Proceedings

of the Fourth Regional Winter Cereals Workshop—Barley," Vol. 1. ICARDA--CIMMYT, Aleppo, Syria.

Barghouti, S., Saari, E. E., Srivastava, J. P., and Chancellor, G., eds. (1978b). "Proceedings of the Fourth Regional Winter Cereals Workshop—Barley," Vol. 2. ICARDA--CIMMYT, Aleppo, Syria.

Beresford, R. M. (1982). Stripe rust (*Puccinia striiformis*), a new disease of wheat in New Zealand. *Cereal Rusts Bull.* **10,** 35–41.

Browning, J. A. (1974). Relevance of knowledge about natural ecosystems to development of pest management programs for agro-ecosystems. *Proc. Am. Phytopathol. Soc.* **1,** 191–199.

Browning, J. A., Frey, K. J., McDaniel, M. E., Simons, M. D., and Wahl, I. (1979). The biologic of using multilines to buffer pathogen populations and prevent disease loss. *Indian J. Genet. Plant Breed.* **39,** 3–9.

Chantarasnit, A., and Luangsodsai, H. (1981). A review of five years research on the diseases of wheat and barley. *In* "Proceedings of the Highland Cereal Crops Workshop," pp. 233–236. North. Agric. Dev. Cent., Chiang Mai, Thailand.

Chen, S. M., Chow, C. P., Lee, S. P., Wang, K. N. Ou-Yang, Y., Hung, S. W., Lu, S. I., Yang, T. M., and Wu, W. C. (1957). Studies on the epidemiology of stripe rust of wheat in north China. *Acta Phytopathol. Sin.* **3,** 63–85 (Engl. Summ).

Chester, K. S. (1946). "The Nature and Prevention of Cereal Rusts as exemplified in Leaf Rust of Wheat." Chronica Botanica, Waltham, Massachusetts.

Chiu, W. F., and Chang, Y. H. (1982). Advances of science of plant protection in the People's Republic of China. *Annu. Rev. Phytopathol.* **20,** 71–92.

Cormack, M. W. (1974). Wheat diseases in Kenya. *Proc. FAO/Rockefeller Found. Wheat Semin., 4th, 1973,* pp. 306–308.

Cummins, G. B. (1971). "The Rust Fungi of Cereals, Grasses and Bamboos." Springer-Verlag, Berlin and New York.

Dalrymple, D. G. (1978). Development and spread of high yielding varieties of wheat and rice in the less developed nations. *Foreign Agric. Econ. Rep.* **95,** 1–134.

Danakusuma, T. (1982). "Performance of Wheat Varieties Ex India and Pakistan" (mimeo. rep.). Food Crops Res. Inst., Sukamandi, Indonesia.

Dinoor, A. (1974). Role of wild and cultivated plants in the epidemiology of plant diseases in Israel. *Annu. Rev. Phytopathol.* **12,** 413–436.

Dinoor, A., Prusky, D., Biali, M., and Neubaner, J. (1976). Chemical control of wheat leaf rust; a regional approach. *Proc.—Eur. Mediterr. Cereal Rusts Conf., 4th, 1976,* pp. 96–97.

Dmitriev, A. P., and Gorshkov, A. K. (1980). The results of some rusts investigation in Ethiopia. *Proc.—Eur. Mediterr. Cereal Rusts Conf., 5th, 1980,* pp. 157–159.

d'Oliveira, B., and Samborski, D. J. (1966). Aecial stage of *Puccinia recondita* on Rammunculaceae and Boraginaceae in Portugal. *Proc. Cereal Rusts Conf., 1964,* pp. 133–150.

Dubin, H. J. (1984). *Puccinia striiformis* sp. *hordei:* Cause of barley yellow rust epidemic in South America. *Phytopathology.* **74,** 820 (abstr.).

Edwards, I. B. (1975). The wheat situation in Rhodesia. *Rhod. Agric. J.* **72,** 49–54.

El Ghouri, M. W. (1979). Constraints to cereal production and possible solution in the Yemen Arab Republic. *In* "The Gap Between Present Farm Yields and the Potential," Proc. 5th Cereals Workshop, pp. 90–96. CIMMYT, El Batan, Mexico.

El-Saadi, M. (1974). Saudi Arabia. *Proc. FAO/Rockefeller Found. Wheat Semin., 4th, 1973,* pp. 101–103.

El-Tobgy, H. A. (1976). "Contemporary Egyptian Agriculture," 2nd ed. Ford Found., New York (Egyptian Book House, Cairo).

El-Tobgy, H. A., and Taalat, E. (1971). Wheat improvement and production in Egypt. *In* "Proceedings of the First International Wheat Workshop" (R. A. Fischer and D. Bork, eds.), pp. 25–27. International Maize and Wheat Improvement Center, El Batan, Mexico.
Food and Agriculture Organization (FAO) (1981). "Production Yearbook," Vol. 35. Food Agric. Organ. U.N., Rome.
Frankel, O. H., and Bennett, E. (1970). "Genetic Resources in Plants. Their Exploration and Conservation." Davis, Philadelphia, Pennsylvania.
Gebeyhou, G. (1975). Durum wheat production and research in Ethiopia. *Proc. Reg. Wheat Workshop, 3rd, 1975,* pp. 341–344.
Ghodbane, A. (1977). Tunisia: Stripe rust on wheat. *FAO Plant Prot. Bull.* **25,** 212.
Gotoh, T. (1972). "Winter Wheat Improvement Problems in Far East." Printed Rep. Tohoky Natl. Agric. Exp. Stn., Morioka, Japan.
Hanson, H., Borlaug, N. E., and Anderson, R. G. (1982). "Wheat in the Third World." Westview Press, Boulder, Colorado.
Harder, D. E., Mathenge, G. R., and Mwaura, K. (1972). Physiologic specialization and epidemiology of wheat stem rust in East Africa. *Phytopathology* **62,** 166–171.
Hassan, S. F. (1970). Cereal rusts situation in Pakistan. *Proc. Cereal Rusts Conf., 1968,* pp. 124–125.
Hassan, S. F. (1974). Wheat diseases and their relevance to improvement and production in Pakistan. *Proc. FAO/Rockefeller Found. Wheat Semin., 4th, 1973,* pp. 284–286.
Hassan, S. F. (1978). Rusts problem in Pakistan. *In* "Wheat Research and Production in Pakistan" (M. Tahir, ed.), Vol. I. pp. 90–93. Pak. Agric. Res. Counc. Islamabad, Pakistan.
Hassan, S. F., Hussain, M., and Rizvi, S. A. (1977). Investigations on rust of wheat in Pakistan. *Cereal Rusts Bull.* **5,** 4–10.
Hogg, W. H., Hounam, C. E., Mallik, A. K., and Zadoks, J. C. (1969). Meteorological factors affecting the epidemiology of wheat rusts. WMO, *Tech. Note* **99,** 1–143.
Hussain, M., Hassan, S. F., and Kirmani, M. A. S. (1980). Virulences in *Puccinia recondita* Rob. Ex. Desm. f. sp. *tritici* in Pakistan during 1978 and 1979. *Proc.—Eur. Mediterr. Cereal Rusts Conf., 5th, 1980,* pp. 179–184.
ICRISAT (1980). "Proceedings of the International Workshop on Sorghum Diseases, Hyderabad, India." ICRISAT-Patancherw, Andhra Pradesh, India.
International Maize and Wheat Improvement Center (CIMMYT) (1970). "Report on Wheat Improvement." El Batan, Mexico.
International Maize and Wheat Improvement Center (CIMMYT) (1971). "Report on Wheat Improvement." El Batan, Mexico.
International Maize and Wheat Improvement Center (CIMMYT) (1972). "Report on Wheat Improvement." El Batan, Mexico.
International Maize and Wheat Improvement Center (CIMMYT) (1973). "Report on Wheat Improvement." El Batan, Mexico.
International Maize and Wheat Improvement Center (CIMMYT) (1974). "Report on Wheat Improvement." El Batan, Mexico.
International Maize and Wheat Improvement Center (CIMMYT) (1975). "Report on Wheat Improvement." El Batan, Mexico.
International Maize and Wheat Improvement Center (CIMMYT) (1976). "Report on Wheat Improvement." El Batan, Mexico.
International Maize and Wheat Improvement Center (CIMMYT) (1977). "Report on Wheat Improvement." El Batan, Mexico.
International Maize and Wheat Improvement Center (CIMMYT) (1978). "Report on Wheat Improvement." El Batan, Mexico.

International Maize and Wheat Improvement Center (CIMMYT) (1979). "Report on Wheat Improvement." El Batan, Mexico.

International Maize and Wheat Improvement Center (CIMMYT) (1980). "Report on Wheat Improvement." El Batan, Mexico.

International Maize and Wheat Improvement Center (CIMMYT) (1981). "World Wheat Facts and Trends; Report One: An Analysis of Changes in Production, Consumption, Trade and Prices over the Last Two Decades." CIMMYT, El Batan, Mexico.

International Maize and Wheat Improvement Center (CIMMYT) (1982). "CIMMYT Review." El Batan, Mexico.

International Maize and Wheat Improvement Center (CIMMYT) (1984). "Report on Wheat Improvement, 1981." El Batan, Mexico.

Johnson, V. A., and Beemer, H. L., Jr., ed. (1977). "Wheat in the People's Republic of China," CSCPRC Rep. No. 6. Nat. Acad. Sci., Washington, D.C.

Joshi, L. M. (1980). The disease surveillance system in India. *In* "Assessment of Crop Losses due to Pests and Diseases" (H. C. Govindu, G. K. Veeresh, P. T. Walker, and J. F. Jenkyn, eds.), UAS Tech. Ser. No. 3, pp. 211–217. Univ. Agric. Sci., Hebbal, Bangalore.

Joshi, L. M., and Palmer, L. T. (1973). Epidemiology of stem, leaf and stripe rusts of wheat in India. *Plant Dis. Rep.* **57**, 8–12.

Joshi, L. M., and Pathak, K. D. (1975). Loss in yield due to brown rust (*Puccinia recondita* Rob ex Desm) on late sown wheat crop. *Indian J. Agron.* **20**, 79–81.

Joshi, L. M., Saari, E. E., Gera, S. D., and Nagarajan, S. (1974). Survey and epidemiology of wheat rusts in India. *In* "Current Trends in Plant Pathology" (S. P. Raychandhuri and J. P. Verma, eds.), pp. 150–159. Dept. of Botany, University of Lucknow, Lucknow.

Joshi, L. M., Goel, L. B., and Sinha, V. C. (1976a). Role of the Western Himmalayas in the annual occurrence of yellow rust in Northern India. *Cereal Rusts Bull.* **4**, 27–30.

Joshi, L. M., Nagarajan, S., and Srivastava, K. D. (1976b). Epidemiology of leaf rust of wheat (*Puccinia recondita tritici*) in India. *Proc.—Eur. Mediterr. Cereal Rusts Conf., 4th, 1976,* pp. 52–55.

Joshi, L. M., Nagarajan, S., and Srivastava, K. D. (1977). Epidemiology of brown and yellow rusts of wheat in North India. I. Place and time of appearance and spread. *Phytopathol. Z.* **90**, 116–122.

Joshi, L. M., Srivastava, K. D., Singh, D. V., and Ramanujam, K. (1980). Wheat rust epidemics in India since 1970. *Cereal Rusts Bull.* **8**, 17–21.

Kampmeijer, P. (1981). Differ: A procedure to find new differential varieties in large cultivar-isolate reaction matrices. *Cereal Rusts Bull.* **9**, 913.

Kampmeijer, P. (1982). "EPIDAT, Data Analysis for Disease Nurseries." Rep. Res. Inst. Plant Prot. (IPO), Wageningen, The Netherlands.

Karki, C. B. (1980). Report on "Evaluation of Nepalese Wheat and Barley Varieties in the Seedling Stage on their Resistance to Yellow Rust." Res. Inst. Plant Prot. (IPO), Wageningen, The Netherlands.

Khalifa, M. M. (1979). "Evaluation of the Resistance of Egyptian Durum and Bread Wheat Varieties against Yellow Rust in Seedling Stage." Rep. Res. Inst. Plant Prot. (IPO), Wageningen, The Netherlands.

Khazra, H., and Bamdadian, A. (1974). The wheat diseases situation in Iran. *Proc. FAO/Rockefeller Found. Wheat Semin., 4th, 1973,* pp. 292–299.

Khokhar, L. K. (1977). "Yellow (Stripe) Rust in Wheat and Barley." Rep. Res. Inst. Plant Prot. (IPO). Wageningen, The Netherlands.

Kidwai, A. (1979). Pakistan reorganizes agriculture research after harvest disaster. *Nature (London)* **277**, 169.

Kilpatrick, R. A. (1974). International rust nurseries. *Proc. FAO/Rockefeller Found. Wheat Semin., 4th, 1973,* pp. 19–21.

Kinaci, E. (1983). The importance of the Turkish Trap Nursery (TTN) for monitoring wheat and barley diseases. *Cereal Rusts Bull.* **11,** 36–38.

Kirmani, M. A. S. (1980). "Comparative Evaluation of Wheat Varieties from Pakistan against Stripe Rust in Seedling Stage." Res. Inst. Plant Prot. (IPO), Wageningen, The Netherlands.

Konovalova, N. E., Suzdalskaya, M. V., Semionova, L. P., Sorokina, G. K., Gorbunova, Yu. V., Zhemchuzhina, A. I., Arutiunian, E. A., Akhmerov, R. A., Bazhenova, V. M., Volkova, V. T., Gogava, T. I., Zhdanov, V. P., Zakarian, M. A., Kryzhanovskaya, M. S., Kulikova, G. N., Kurbatova, V. Sh., Lekomtseva, S. N., Paichadze, L. V., Sarkisian, D. D., Simonian, L. Kh., Solotchina, G. F., Khachaturian, G. A., and Yaremenko, Z. I. (1977). Occurrence of cereal rust causal agent races in the U.S.S.R. in 1975. *Mycol. Phytopatol.* **11,** 499–503 (D. H. Casper, transl.).

Kranz, J., Schmutterer, H. S., and Koch, W. (1977). "Diseases, Pests and Weeds in Tropical Crops." Parey, Berlin.

Madumarov, T. M., and Gorshkov, A. K. (1978). Alternate hosts of rust fungi, Puccinia species and *Uromyces eragrostitis,* which infect cereals in Ethiopia. *Ethiop. J. Sci.* **1,** 123–126.

Madumarov, T. M., and Gorshkov, A. K. (1979). Survey of wheat and teff rust distribution and evaluation of yield loss in 1977. *Proc. Natl. Crop Improve. Comm. Meet., Inst. Agric. Res., 11th, 1979,* pp. 67–71. Addis Ababa, Ethiopia.

Malm, N. R., and Beckett J. B. (1962). Reactions of plants in the tribe Maydeae to *Puccinia sorghi* Schw. *Crop Sci.* **2,** 360–361.

Mathre, D. E. (1982). "Compendium of Barley Diseases." Am. Phytopathol. Soc., St. Paul, Minnesota, and Montana State Univ., Bozeman.

Minkevich, I. I., and Zakharova, T. I. (1982). Effects of weather conditions on the development of rust epiphytotics. *Mycol. Phytopatol.* **16,** 351–357 (D. H. Casper, transl.).

Morris, R., and Sears, E. R. (1967). The cytogenetics of wheat and its relatives. *In* "Wheat and Wheat Improvement" (K. S. Quisenberry and L. P. Reitz, eds.), pp. 19–87. Am. Soc. Agron. Madison, Wisconsin.

Muntzing, A. (1979). "Triticale: Results and Problems." Parey, Berlin.

Nagarajan, S., and Joshi, L. M. (1975a). A historical account of wheat rust epidemics in India and their significance. *Cereal Rusts Bull.* **3,** 29–33.

Nagarajan, S., and Joshi, L. M. (1975b). Presence of wheat rust uredospores over the Rohtang pass (3,954 m) in the interior of the Himalayas. *Cereal Rusts Bull.* **3,** 34–35.

Nagarajan, S., and Joshi, L. M. (1978). Epidemiology of brown and yellow rusts of wheat over north India. II. Associated meteorological conditions. *Plant Dis. Rep.* **62,** 186–188.

Nagarajan, S., and Joshi, L. M. (1980). Further investigations on predicting wheat rusts in Central and Peninsula India. *Phytopathol. Z.* **98,** 84–90.

Nagarajan, S., Joshi, L. M., Srivastava, K. D., and Singh, D. V. (1978). Epidemiology of brown and yellow rust in North India. III. Importance of varietal change. *Plant Dis. Rep.* **62,** 694–698.

Nagarajan, S., Saari, E. E., and Kranz, J. (1980). Conditions that led to the 1978 epidemic of *Puccinia striiformis* over Northwestern parts of the Indian Subcontinent. A preliminary report. *Proc.—Eur. Mediterr. Cereal Rusts Conf., 5th, 1980,* pp. 213–216.

Nagarajan, S., Kranz, J., Saari, E. E., and Joshi, L. M. (1982). Analysis of wheat rusts epidemic in the Indo-Gangetic Plains 1976–78. *Indian Phytopathol.* **35,** 473–477.

Northwood, P. J. (1970). Wheat production in Tanzania. *World Crops* **22,** 226–230.

Novokhatka, V. G. (1979). Epiphytotics of *Puccinia recondita* Rob. ex. Desm. f. sp. *tritici* on winter wheat in the forest-steppe zone (Ukraine) of the USSR. *Mycol. Phytopatol.* **13,** 488–493 (D. H. Casper, transl.).

Obst, A. (1981). Chemical control in intensive wheat cultivation. *In* "Integrated Plant Protection for Agricultural Crops and Forest Trees" (T. Kommedahl, ed.), Proc. Symp. Int. Congr. Plant Prot., 9th, Vol. II, pp. 448–451. Burgess, Minneapolis, Minnesota.

Oggema, M. W. (1974). Kenya. *Proc. FAO/Rockefeller Found. Wheat Semin. 4th, 1973,* pp. 86–88.

Olugbemi, L. B. (1974). Nigeria. *Proc. FAO/Rockefeller Found. Wheat Semin., 4th, 1973,* pp. 92–96.

Ozkan, M., and Prescott, J. M. (1972). Cereal rusts in Turkey. *Proc.—Eur. Mediterr. Cereal Rusts Conf., 3rd, 1972,* Vol. 2, pp. 183–185.

Paganiban, D. F. (1980). The prospects of wheat production in the Philippines. *Grains J.* **1,** 38–42.

Pinto, F. F. (1971). Current wheat (disease) situation in Ethiopia. *In* "Proceedings of the First Wheat Workshop" (R. A. Fischer and D. Bork, eds.), pp. 21–24. International Maize and Wheat Improvement Center, El Batan, Mexico.

Pinto, F. F., and Hurd, E. A. (1970). Seventy years with wheat in Kenya. *East Afr. Agric. For. J.* **26,** Spec. Issue, 1–24.

Pisharoty, R. R., and Desai, B. N. (1956). "Western Disturbances" and Indian weather. *Indian Meteorol. J. Geophys.* **7,** 333–338.

Prescott, J. M. (1976). Regional disease monitoring. *Turk. Phytopathol.* **5,** 1–5.

Prescott, J. M., and Saari, E. E. (1975). Major disease problems of durum wheat and their distribution within the region. *Proc. Reg. Wheat Workshop, 3rd, 1975,* pp. 104–114.

Rachie, K. O. (1974). "The Millets and Minor Cereals; A Bibliography of the World Literature on Millets, pre-1930 and 1964–69, and of all Literature on other Minor Cereals." Scarecrow Press, Metuchen, New Jersey.

Rajaram, S., and Campos, A. (1974). Epidemiology of wheat rusts in the western hemisphere. *CIMMYT Res. Bull.* **27,** 1–27.

Ramakrishnan, T. S. (1963). "Diseases of Millets." ICAR, New Delhi.

Rapilly, F. (1979). Yellow rust epidemiology. *Annu. Rev. Phytopathol.* **17,** 59–73.

Reitz, L. P. (1967). World distribution and importance of wheat. *In* "Wheat and Wheat Improvement" (K. S. Quisenberry and L. P. Reitz, eds.), pp. 1–18. Am. Soc. Agron., Madison, Wisconsin.

Rizvi, S. S. A. (1982). "Current Status of Wheat Leaf Rust Investigations in Pakistan," Nat. Wheat Prod. Semin. (mimeo. pap.). Pak. Agric. Res. Council, Islamabad, Pakistan.

Roelfs, A. P. (1978). Estimated losses caused by rust in small grain cereals in the United States—1918–76. *Misc. Publ.—U.S., Dep. Agric.* **1363,** 1–85.

Saari, E. E. (1976). Long distance transportation and expansion of wheat rusts. *Turk. Phytopathol.* **5,** 7–12.

Saari, E. E. (1977). Wheat disease surveillance and its role on genetic vulnerability. *In* "Proceedings of the Workshop on Crop Surveillance" (B. H. Zandstra, ed.), pp. 145–155. East-West Center and University of Hawaii, Honolulu.

Saari, E. E. (1978). Rust problems and how to fight them. *In* "Wheat Research and Production in Pakistan" (M. Tahir, ed.), pp. 82–89. Pak. Agric. Res. Conc., Islamabad, Pakistan.

Saari, E. E. (1981). Wheat in the developing countries from the Atlantic to the Pacific. *In* "Integrated Plant Protection for Agricultural Crops and Forest Trees" (T. Kommedahl, ed.), Proc. Symp. Int. Congr. Plant Prot., 9th, Vol. II, pp. 437–441. Burgess, Minneapolis, Minnesota.

Saari, E. E., and Prescott, J. M. (1978). Barley diseases and their surveillance in the region. *Proc. Reg. Winter Cereal Workshop—Barley, 4th, 1977,* Vol. 2, pp. 320–330.

Saari, E. E., and Prescott, J. M. (1980). Disease problems of rainfed wheat. *In* "Proceedings of the Third International Wheat Conference," pp. 453–465. University of Nebraska, Lincoln.

Saari, E. E., and Srivastava, J. P. (1977). Improved and stabilization of production of winter cereals: Potentials in a single-crop system. *In* "Middle East and Africa Agricultural Seminar" (mimeo.). Ford Found., New York.

Saari, E. E., and Wilcoxson, R. D. (1974). Plant disease situation of high-yielding dwarf wheats in Asia and Africa. *Annu. Rev. Phytopathol.* **12,** 49–68.

Saari, E. E., Prescott, J. M., and Kamel, A. H. (1979). The significance of diseases and insects in cereal production. *In* "The Gap Between Present Farm Yields and the Potential," Proc. 5th Cereal Workshop, Vol. II, pp. 149–256. Minist. Agric., ICARDA and CIMMYT, El Batan, Mexico.

Sharif, G., and Bamdadian, A. (1974). Importance and situation of wheat and barley diseases in Iran. *Proc. FAO/Rockefeller Found. Wheat Semin., 4th, 1973,* pp. 300–305.

Sharma, S. K., Joshi, L. M., Singh, S. D., and Nagarajan, S. (1972). New virulence of yellow rust on Kalyansona variety of wheat. *Proc.—Eur. Mediterr. Cereal Rusts Conf., 3rd, 1972,* Vol. 1, pp. 263–265.

Shurtleff, M. C. (1980). "Compendium of Corn Diseases," 2 ed. Am. Phytopathol. Soc., St. Paul, Minnesota.

Spehar, V. (1975). Epidemiology of wheat rust in Europe. *Proc. Int. Winter Wheat Conf., 2nd, 1975,* pp. 435–440.

Spehar, V., Vlahovic, V., and Koric, B. (1976). The role of *Berberis* sp. on appearance of virulent physiologic races of *Puccinia graminis* f. sp. *tritici. Proc.—Eur. Mediterr. Cereal Rusts Conf., 4th, 1976,* pp. 63–67.

Srivastava, J. P. (1978). Barley production, utilization and research in the Afro-Asian region. *Proc. Reg. Winter Cereal Workshop—Barley, 4th, 1977,* Vol. 2, pp. 242–259.

Stubbs, R. W. (1974). Significance of stripe rust in wheat production and the role of the European-Mediterranean cooperative program. *Proc. FAO/Rockefeller Found. Wheat Semin., 4th, 1973,* pp. 258–266.

Stucker, R. E. (1980). Statistical problems in measurement of yield and loss. *Misc. Publ.—Minn., Agric. Exp. Stn.* **7,** 23–27.

Swaminathan, M. S. (1978). Wheat revolution, the next phase. *Indian Farm.* **27,** 7–11.

Tommasi, F., Siniscalo, A., and Parodies, M. (1980). Aecia of an unidentified rust on *Thalictrum flavum* L. in Southern Italy. *Proc.—Eur. Mediterr. Cereal Rusts Conf., 5th, 1980,* pp. 191–198.

Torabi, M. (1980). Factors affecting and epidemic of yellow rust on wheat in the North-Western regions of Iran. *Proc.—Eur. Mediterr. Cereal Rusts Conf., 5th, 1980,* pp. 217–222.

Varughese, G. (1975). Needs of durum wheat in the Third North African and Near the Middle East Regions. *Proc. Reg. Wheat Workshop, 3rd, 1975,* pp. 53–67.

Wellings, C. R., and McIntosh, R. A. (1981). Stripe rust—a new challenge to the wheat industry. *Agric. Gaz. N.S.W.* **92,** 2–4.

Wiese, M. V. (1977). "Compendium of Wheat Diseases." Am. Phytopathol. Soc., St. Paul, Minnesota.

Wodageneh, A. (1974). Ethiopia. *Proc. FAO/Rockefeller Found. Wheat Semin., 4th, 1973,* pp. 64–68.

Young, H. C., Jr., and d'Oliveira, B. (1982). A further study of race populations of *Puccinia recondita* f. sp. *tritici. García de Orta, Sér. Est. Agron., Lisboa* **9,** 37–52.

Young, H. C., Jr., Prescott, J. M., and Saari, E. E. (1978). Role of disease monitoring in preventing epidemics. *Annu. Rev. Phytopathol.* **16,** 263–285.

Zadoks, J. C. (1980). Yield losses and costs of crop protection. Three views with special reference to wheat growing in the Netherlands. *Misc. Publ.—Minn., Agric. Exp. Stn.* **7,** 17–22.

Zillinsky, F. J. (1983). "Common Diseases of Small Grain Cereals: A guide to Identification." International Maize and Wheat Improvement Center, El Batan, Mexico.

PART III

Epidemiology

10

Epidemiology in Australia and New Zealand

N. H. Luig

Department of Agricultural Genetics and Biometry, The University of Sydney, Sydney, New South Wales, Australia

THE CEREAL RUSTS,
Volume II

ISBN 0-12-148402-5

I. Introduction

Australasia, comprising mainland Australia, Tasmania, and New Zealand, constitutes an epidemiologic area, isolated from other cereal-growing areas. Urediospore movement occurs readily in the eastern and southern Australian wheat belts, Tasmania, and both islands of New Zealand (Waterhouse, 1938; Watson and Cass Smith, 1962; Watson and Luig, 1966; Luig and Watson, 1970; Luig, 1977). While the development of epiphytotics in such a vast expanse is an interesting study, Australasia also offers other features, namely, (1) the occasional introduction of new species or new strains of rust pathogens, and (2) the virtual absence of sexual cycles. Thus, the spread and survival of new types can be monitored and their development can be compared with that of the established strains. Furthermore, stepwise mutational changes and somatic hybridization events in the rust flora are readily observable.

II. Wheat-Growing Regions

Wheat is the most important crop in Australia, and about 10 million hectares are sown annually. Oats occupy $2\frac{1}{2}$ million hectares, barley 2 million hectares, and rye about 50,000 hectares. Rye is grown as a pasture, for erosion control and grain.

The wheat area in New Zealand (North and South Islands) is small (Fig. 1), but since conditions for pathogen growth and survival are much more favorable than in Australia, it is important epidemiologically.

Australasia has been divided into regions based on differences in selection pressure on the wheat stem rust pathogen, *Puccinia graminis* Pers. f.sp. *tritici* Eriks. and Henn. (*Pgt*) (Luig and Watson, 1970) (Figs. 1 and 2).

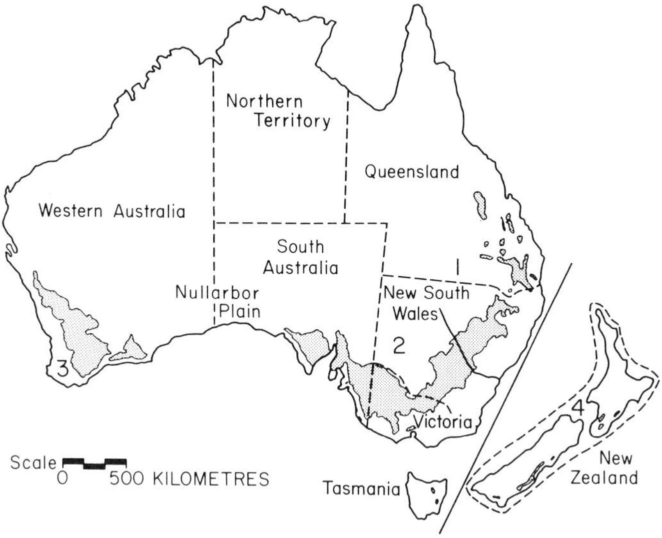

Fig. 1. The wheat-producing area (shaded) of Australia (10,249,000 hectares) and New Zealand (91,000 hectares) in 1978. Thousands of hectares per region were as follows: 1 = 2047 (747 hectares in Queensland and 1300 in northern New South Wales); 2 = 4519 (1866 in southern New South Wales, 1345 in Victoria, and 1308 in South Australia; 3 = 3728, all in Western Australia; and 4 = 91 (5 on the North Island and 85 on the South Island).

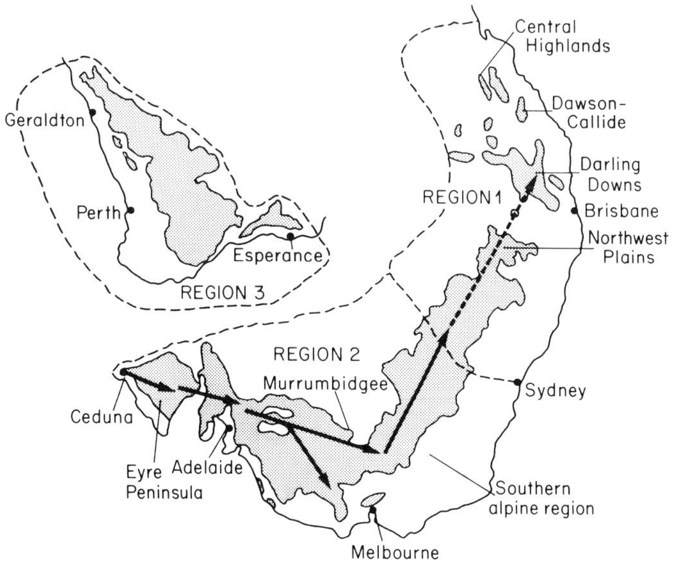

Fig. 2. The sequential movements of the 1973–1974 rust epidemic: →, major and -->, minor movement.

A. REGION 1

This is the predominantly summer rainfall area of eastern Australia and traditionally was the rust-liable part of the wheat belt. Heavy postharvest rains are common, allowing rust survival on volunteer plants and native grasses, especially *Agropyron scabrum* Beauv. (rough wheat grass) (Rees, 1972a,b). Areas in Queensland include the Central Highlands and the Dawson-Callide Valley, and further south the fertile Darling Downs. The southern part of region 1 comprises the northern areas of New South Wales and the northwestern plains (Fig. 2).

B. REGION 2

Until the 1970s, few stem rust-resistant cultivars were grown in region 2 (southern New South Wales, Victoria, South Australia, and Tasmania). Notable exceptions were Glenwari (*Sr17*) and Insignia 49 (*Sr11*). Rainfall is predominantly in the fall and winter, and temperatures are lower during the first part of the growing season (late autumn to early spring) than in region 1. In the Murrumbidgee region of southern New South Wales, many fields are irrigated.

C. REGION 3

The wheat areas of Western Australia make up region 3. The eastern edge of this area, Esperance, is separated from region 2 by approximately 1300 km of desert. Inoculum is occasionally exchanged between the two regions, mainly in a west to east direction.

D. REGION 4

In New Zealand over 90% of the wheat area is on the South Island, where low summer temperatures favor leaf rust. Stem rust survives readily, but the comparatively cool spring weather delays increase. Thus initial inoculum is low, and genes conferring a low degree of resistance could have a significant effect on rust incidence. Since the growing season is much longer than in Australia, tolerance to stem rust could also result in much higher yields.

On the North Island, stem rust usually starts at a very low level but can develop rapidly, and epidemics may be severe. On both islands, pathogen survival is facilitated by sowing times ranging from May to August.

Wheat areas of New Zealand are isolated from those of eastern Australia by about 2000 km. Survey data suggest that in most years both islands receive urediospores from regions 1 and 2. The earlier-maturing crops in Queensland can provide urediospores for those in the North Island, while the later-maturing crops in region 2 are probably a source of inoculum for the South Island. Thus, the build-up of a new strain in region 1 is frequently followed by the detection of the same strain in the North Island (McEwan, 1966), while strains from region 2, with less genes for virulence, will be found on the South Island (Close, 1967).

III. Terminology

The most common terms used in this chapter are:

1. Virulence and avirulence. Both are characteristics of the pathogen. An isolate of the pathogen is referred to as virulent only in relation to a specific host or host gene.
2. Aggressiveness. Refers to the competitive ability of pathogen biotypes when in mixtures.
3. Race. Refers to biotypes of the rust pathogens that have been identified on the international differentials.
4. Strain. Refers to biotypes that have been separated on supplemental differentials.

IV. Race Survey

While physiological race surveys in many countries are mainly aimed at establishing the presence or absence of certain virulence genes or combinations, the Australian surveys for *Pgt* also attempt to assess the total variability of the rust flora. Thus, the Australian data is better suited for epidemiological studies. In this context, all the host genes used in the Australian differential set for *Pgt* are not of equal value. Those present in common cultivars frequently cause major shifts in *Pgt* populations. Monitoring the virulent strains establishes a relationship with the cultivars with which these strains are associated. Where a single gene (*A*) for resistance is present in part of a host population and where the other cultivars are susceptible to all prevalent field strains, a simple correlation between the area sown to the resistant cultivar(s) and the percentage frequency of the strain(s) that can attack it can be demonstrated (Watson, 1958). When other genes for resistance become important, the data for gene *A* cannot be interpreted on a gene-for-gene basis as the simple equilibria no longer exist. Under Australian conditions, it is not uncommon for the strain(s) that mutated for gene *A* to disappear despite the continued cultivation of the cultivar(s) with the gene. Normally in such a situation a more important mutation for virulence has occurred in a strain(s) avirulent for gene *A*. The main genes that have been exploited in breeding for stem rust resistance in Australia are *Sr5, Sr6, Sr8, Sr9b, Sr9g, Sr11, Sr12, Sr17, Sr30,* and *Sr36* (= *SrTt1*), each of which has had a significant influence on the epidemiology.

The way samples were obtained for annual race surveys often biased the results. In some years, too many samples were collected from cultivars that had been previously resistant. The strain(s) identified from such samples usually had more genes for virulence than those collected from other susceptible wheats. Samples from susceptible cultivars in the vicinity of those reacting differentially also biased the results in favor of those strains with additional virulences. Further, the similarities between the frequency patterns of particular strains in different regions are, in some instances, due to the cultivation of the same hosts (Luig and Watson, 1977).

Since Person (1959) discussed host : parasite relationships, there has been a tendency to emphasize the significance of stocks carrying single genes for resistance. However, for field work, combined resistances are more important. It is essential to know whether cultivars with more than a single gene for resistance are endangered. When studying rust development in a breeding nursery, it must be possible to determine which are the major and minor strains in the pathogen populations. In most instances, single gene lines do not reveal the frequencies of strains with two or more genes for virulence. Generally, combined resistances are not utilized adequately in breeding nurseries to convey this information.

Likewise, in rust surveys, more use could be made of combined resistances. For instance, a host line combining two genes, one conferring a high degree of resistance [Infection type (I.t.) 0 or;1], but ineffective against most field isolates, and a second conditioning an intermediate resistance (I.t. 2), could replace the two respective single-gene testers.

Where the field inoculum mostly comprises strains virulent on host genotypes with combined resistances, they can be included with the susceptible genotype for increasing inocula. Subsequent grouping of the rust accessions would eliminate the need to test a large number of differentials in each case.

V. Wheat Stem Rust Epiphytotics

A. ENVIRONMENT

The arid Australian climate often leads to moisture stress, and such conditions are unfavorable for host and rust pathogen (Nix, 1975). In contrast to stripe rust and leaf rust, stem rust develops at higher temperatures. At low temperatures stem rust largely hibernates in myceliar form around the infection sites. Brown and Shipton (1964) concluded that maximum infection occurs when cool nights, conducive to dew formation, are followed by warm days. Disease development had an optimal temperature range of 25°–30°C. These temperature regimes are frequently achieved in the wheat belt during the middle of spring. Higher-than-normal rainfall can also delay the maturing of the crop. This enables the parasite to undergo an extra generation. Much greater yields can be expected in Australia when grain filling takes place under cool conditions, but heavy losses from stem rust can negate these gains.

B. CROP STAGE AT DISEASE ONSET

In Australia a main factor determining the severity of wheat stem rust is the growth stage of the crop at the onset of high temperatures. In the late 1880s, W. Farrer observed that earlier-maturing cultivars suffered less from this disease, and his early-maturing cultivars were a major contribution to the Australian wheat industry. This allowed an expansion of the wheat belt into hotter and drier districts, which were less favorable for rust. Since the planting season extends from late April to early August, the importance of the planting date of a susceptible cultivar cannot be overstated. For instance, during the early 1980s in Queensland, Oxley (*Sr5, Sr6, Sr8, Sr12*) was susceptible to the predominant pathotypes, but only the late-sown crops suffered damage. The earlier-sown ones were often sources of infection for nearby, later-planted crops, and the incidence of stem rust in the latter was probably the combined effect of higher temperatures and higher levels of inoculum. It is obvious that consecutive sowings of a susceptible cultivar in the same area pose a great danger to those sown last.

C. CROP DIVERSITY

Authors stating that crop homogeneity only favors the pathogen overlook that large fields sown to a cultivar(s) with a single gene for resistance ineffective against some strains can delay disease onset over a region. In Australia, rust survival during the summer months is often minimal, and this leads to situations where a strain causes a local outbreak, but its progress is checked by large areas sown to a cultivar resistant to it.

On the other hand, cultivation of the same cultivar(s) in different areas facilitates migration of rust strains. For example, if in a region growing resistant cultivars one of the cultivars becomes susceptible to a new strain, but all other cultivars remain resistant, this strain has an advantage over avirulent ones wherever that cultivar is grown. Several popular Australian wheat cultivars (e.g., Gamenya, Condor, Banks) are grown over a wide range of environments, and react differently to stem rust strains.

The prediction of epiphytotics is difficult. The cultivation of Halberd on approximately two-thirds of the South Australia wheat area was a classical example of genetic vulnerability, especially as virulent strains were recorded not only in this state but also in Western Australia, Victoria, and southern New South Wales. Furthermore, most cultivars in the southern wheat belt had no genes for resistance, as stem rust was not considered a serious threat to wheat production. It is possible that due to conditions favoring disease in 1973, aggressive biotypes of *Pgt* developed that were specifically adapted to Halberd. Direct evidence for this hypothesis is lacking. Zadoks (1961) called such rust isolates "field races," as distinct from those differentiated on seedlings in the glasshouse. Johnson and Taylor (1972) showed that a biotype of stripe rust that had a greater capacity to sporulate on the cultivar Joss Cambier was unchanged at the corresponding virulence loci for all known host genes.

D. OVERWINTERING

Because of higher temperatures in Queensland, *Pgt* can produce urediospores throughout the winter. Much of the overwintering is in southern Queensland, including the Darling Downs, but it can also take place further north (Central Highlands). However, since drought conditions in the Central Highlands appear to be the rule rather than the exception, inoculum from this area does not threaten the southern crops of region 1 in most years.

E. DROUGHTS

The Dawson and Callide districts are closer to the coast and less likely to be affected by long droughts. Furthermore, irrigation here creates conditions favorable for pathogen survival occurring mainly on volunteer wheat. Therefore, the resistance of the cultivars in these areas can be a deciding factor in determining which strains survive in a dry year. Furthermore, in wet years when there is potential for severe infection, the inoculum generated could reach the Darling Downs and then northern New South Wales.

The establishment of airborne rust strains over long distances occurs because of the localized drought periods that diminish the earlier strains. The incidence of rust in an area may be extremely low, but conditions favorable to rust development exist elsewhere. Then following the drought, urediospores carried by wind may reintroduce rust to the drought-stricken areas.

F. "*PUCCINIA* PATH"

A very important observation made during the 1973–1974 rust epiphytotic was the west-to-east rust movement: from Ceduna–Streaky Bay to the Eyre Peninsula, then to the main wheat area of South Australia, and, finally, to western Victoria and southern New South Wales. Some inoculum was blown further north and reached northern New South Wales and Queensland (Fig. 2). Thus, in that season, a "*Puccinia* path" existed in Australia.

Following the epiphytotic, it was clear that susceptible wheats (mainly Halberd) grown at Ceduna and on the Eyre Peninsula were a hazard to the main Australian wheat area.

G. RESISTANT CULTIVARS

The use of resistant cultivars has influenced the rust flora in three main ways: (1) by favoring strains more virulent; (2) by allowing virulent mutants to establish themselves; and (3) by reducing the survival chances of some rust strains by using resistant cultivars in vital over-summering areas.

H. SPREAD OF PATHOGEN MUTANTS

Whether a particular mutation will become widespread in Australia depends on the (1) rate of mutation from avirulence to virulence of the pathogen locus involved; (2) difference in the level of resistance of the host to the mutant and original strain; (3) resistance of other cultivars being grown; and (4) chances for survival and distribution of the mutant.

1. Effect of mutation rate. Cultures have different mutation rates to virulence (Luig, 1979). For example, with *Sr26* the mutation rate in the pathogen is so low that it can be assumed that in a heavily infected wheat field, not a single uredium virulent on *Sr26* is present. By contrast, each hectare of wheat reasonably infected with a strain avirulent for *Sr8* will contain many uredia virulent for *Sr8*.

2. Effect of host resistance. The advantages that go with an acquired virulence will depend on the host gene involved and the nature of the mutation. At the extremes, *Sr5* conditions immunity in the field, while another gene, *Sr25,* on its own does not confer a measurable degree of adult plant resistance. The nature of the fungal mutation is also important. When Eureka (*Sr6*) was released in 1938, it was fully resistant to the predominant strain. However, during 1954 to 1960, new strains, although avirulent, showed progressive increase of virulence (Watson and Luig, 1968), and some rust developed.

3. Effect of other cultivars. The importance of the presence of a cultivar(s) with the corresponding gene for resistance is obvious. However, cultivars susceptible to other strains but resistant to the mutant will have an opposite effect.

4. Chances for survival and spread. A new pathotype generally has a better chance for survival if the mutation occurs early in the season, thus allowing it to multiply and spread. The proximity of grass stands (*Hordeum leporinum* Link., *Agropyron scabrum*) and/or volunteer wheat serving as summer/autumn hosts can be important. The latter hosts are frequently found along roads, railway lines, irrigation projects and plant breeding nurseries. Also important in Australia is the area where the mutation occurs. Climatic conditions near ths coast are more favorable for host and pathogen survival. Mutants that occur where prevailing winds can disseminate the new strain over long distances are more likely to spread.

I. INITIAL INOCULUM LEVEL

The initial inoculum (Vanderplank's X_o) is one of the main factors in determining if a stem rust epiphytotic will occur. A dry year or a summer drought drastically reduces the X_o, and winter temperatures are usually too low to allow rapid build-up. Thus, even if conditions favor the pathogen during the critical spring months, no epiphytotic occurs. In an abnormal year, however, when the winter is mild, daytime temperatures rise to 15°C or above. If this is associated with high humidity and/or rain, inoculum will increase to a dangerous level. This happened before the disastrous 1973/1974 rust epiphytotic in region 2.

In some seasons, no rust can be found in certain districts, while in other years, severe epiphytotics develop on the same cultivars in these districts. Obviously genes only delaying the onset of an epiphytotic or conferring a small degree of resistance are of little benefit to a farmer in a severe epiphytotic. However, in years when a mild epiphytotic develops, such genes are beneficial.

In Australia, the infection rate (Vanderplank's r) is very dependent on fluctuations of the weather. In this connection, an important factor is the comparatively low density of stand (dryland seeding rate: 20–45 kg/hectare) in wheat fields, thus providing a lower within-stand humidity and the reduced chance of infection.

J. POSSIBLE EFFECT OF TEMPERATURE

The Australian strain pattern of *Pgt* could be affected by different temperature requirements of the strains involved. The different climates of the areas in which rust survival occurs make this a possible cause for the observed fluctuations. There is a urediospore movement from Queensland to Victoria and South Australia, and vice versa. During the winter and early spring months when the pathogen has to persist and multiply, temperatures in the two southern states are low and would favor biotypes that can develop under such conditions (mean day and night temperature 12° and 7°C, respectively). In contrast, average day temperatures in Queensland during this time of the year are relatively high (mean day and night temperature 17° and 10°C, respectively), and thus temperature optima of the local strains may be different. It is not unreasonable to assume that within a strain types may exist that vary in temperature requirement. Variability of this kind in *Pgt* has been studied little due to the difficulties involved in creating the right environments. For such experiments the fungus has to parasitize adult plants that are grown under field conditions. With certain genes for resistance, however, variation in the fungus regarding temperature requirements (high for some, low for others) to establish a compatible (i.e., susceptible) interaction has been demonstrated (Luig and Rajaram, 1972).

K. Barberries as an Alternate Host

Barberries (*Berberis vulgaris* L.) occur in Tasmania and are frequently infected by *P. graminis* f.sp. *secalis* Eriks. and E. Henn. (stem rust of rye) (*Pgs*) (Watson and Luig, 1958). However, little rye is grown on this island. I believe that the significance of barberries is overstated. In cold climates they are important for pathogen survival, and in the case of a popular cultivar carrying a durable gene for resistance, or a combination of genes, the sexual cycle may produce the matching virulence(s). Several centuries ago, when small-grain cereals comprised heterogeneous land races, probably the main function of the sexual cycle was to provide the pathogen with a multitude of genotypes from which the adapted biotypes were selected. However, with present-day homogeneous monocrops, single rust clones cause epiphytotics.

VI. Stem Rust of Wheat

A detailed account of each rust is beyond the scope of this chapter, but *Pgt* will be discussed in more detail.

A. Races 43, 44, 45, 46, 54, and 55

Investigations in Australia on variability in the pathogen were initiated in 1921 by Waterhouse. Until the 1940s, only the 12 standard differentials of Stakman *et al.* (1962) were used. Originally six races—43, 44, 45, 46, 54, and 55—were found. Because of the variation detected on introduced supplemental differentials possessing genes not present in current cultivars, it must be assumed that the six races did not constitute the full extent of variability that existed at that time.

Except for Acme and Kubanka, races 43, 44, and 54 produced similar infection types on the other differentials (Waterhouse, 1938). Both have *Sr9g*, and Luig (1978) suggested that the corresponding gene(s) for avirulence in the Australian pathogen is mutable. Races 45, 46, and 55 also showed close affinities on the 12 differentials; again, there was variation on Acme and Kubanka, and race 45 produced I.t. 2 on Marquis (*Sr7b*). Waterhouse (1938) tested the Sydney University Wheat Collection with the six races and found that the majority of the resistant wheats fell into two classes, namely, those resistant to 43, 44, and 54 but susceptible to the other three, and vice versa. This confirmed the presence of two race groups and pointed to a dual origin for *Pgt*. The similarities among the races could be explained on the basis of derivation from the same area of origin but at different times.

Several leading Australian cultivars from this period (Nabawa, Yandilla King, Canberra) were resistant to some of the six races; such cultivars probably promoted the spread of the virulent ones. While barberries play no role in the survival of *Pgt* on mainland Australia, the possibility that a new strain arose through the sexual cycle cannot be entirely ruled out. Waterhouse (1934) reported infection of scattered barberries in central New South Wales, and nearby *Agropyron scabrum* and *Bromus racemosus* L. had uredial infection.

B. RACES 126 AND 222

Race 126, which was found in Western Australia in 1925 and rapidly spread through Australia and New Zealand, was a new type. It did not arise through the sexual cycle from the old races, since it possessed a dominant gene for avirulence on plants with *Sr21* (Luig and Watson, 1961). Additionally, 126 was virulent on Reliance (*Sr5*) and all Australian cultivars at the time, and it was more aggressive in the field as well as under controlled conditions (Waterhouse, 1938). Such a strain is unlikely to have arisen by mutation, and again introduction is the hypothesis advanced. The appearance of 126 led to a dramatic decline of the old races; after 1928 only 43 and 45 were detected occasionally. It is thought that all six races have disappeared without leaving derived races behind.

Thus 126 was practically the only race in Australasia from 1929 to 1941. In 1938 Eureka, the first popular stem rust-resistant cultivar, was released introducing a new epidemiological factor. Eureka carries *Sr6* in a very susceptible genetic background ("Vertifolia effect"—see Vanderplank, 1968) and was grown in northwestern New South Wales, an area vital for rust survival at that time. When a mutant virulent for *Sr6* occurred in race 126, Eureka was severely rusted; by 1942 the mutant was widespread. It appears that in the harsh Australian environment extreme susceptibility can lead to epiphytotics under conditions where other "susceptible" cultivars are only slightly rusted.

During the 1940s and 1950s, cultivars with *Sr11* were widely grown in regions 1 and 2. Again mutations for virulence occurred, and in the majority of cases this was accompanied by a simultaneous loss of virulence for Acme and Kubanka, thus changing the race designation from 126 to 222. The cultivar most severely rusted by these mutant strains was Yalta. It is interesting to note that Eureka and Yalta produce semisusceptible infection types with strains of *Pgs* virulent for *Sr6* and *Sr11*, respectively. Luig and Rajaram (1972) suggested that the genes for resistance to *Pgs* act as modifiers of genes for resistance to *Pgt*. If this is so, it could constitute a kind of nonspecific resistance as defined by Vanderplank (1968), and thus play an important role in epidemiology involving wheats with no genes for resistance to *Pgt*.

C. Race 21

In 1954 race 21 (strain 21-0), first isolated from the alpine region in southern New South Wales, became widespread in eastern Australia. Watson (1955), Watson and Luig (1966), and Luig and Watson (1977) described the differences between strains 21-0 and the strains prevalent before 1954. After 1957, the original strain of race 126 and its derivatives were recovered in low frequencies from the eastern wheat belt, and they disappeared about 1962.

It is remarkable that over vast areas one strain can displace another when many widely-grown cultivars are susceptible to both. This suggests intensive competition when, in fact, disease incidence is often very low. Moreover, in 1954, 21-0 lacked the genes for virulence on cultivars with *Sr11* that were planted on the greater part of the area liable to rust (region 1). Nevertheless, 21-0 increased from 11.1% and 9.2% of the isolates collected in Queensland and in northern New South Wales, respectively, in 1954–1955, to 67.9% and 45.6%, respectively, in 1955–1956. In 1957–1958, all 27 isolates from Queensland were identified as 21-0 or 21-2 (virulent for *Sr11*). During the next season, conditions favored rust development in region 1: only 9 of 291 isolates were pre-1954 strains. Thus, for practical purposes, the earlier pathogen types were eliminated in region 1.

Strain 21-0 was soon replaced by 21-2. During the following two decades, mutational changes produced considerable variability in race 21 (Watson and Luig, 1963, 1966; Luig and Watson, 1970, 1977), but strain 21-2 still occurs. Host genes corresponding to those involved in the mutational changes in the pathogen, are utilized in the Australian supplementary set for classification, and can be placed in two groups:

1. Genes present in commercial cultivars		
1[a]	*Sr6*	(Eureka)
2	*Sr11*	(Gabo)
3	*Sr9b*	(Gamenya, Festival, Robin)
4	*Sr36*	(Mengavi, Mendos)
5	*Sr17*	(Spica, Glenwari, Gala)
6	*Sr8*	(Oxley, Condor, Egret)
8	*Sr30*	(Festiguay)
—	*Sr12*	(Windebri, Celebration)
2. Genes for which variation occurred independently of commercial wheat genotypes		
7	*Sr15*	(Thew)
9	Sr_{Agi}	(*Agropyron intermedium* derivative)
10	Sr_{5120}	(Entrelargo de Montijo)
11	Sr_{BB}	(Titan, Barletta Benvenuto)

[a]Numbers indicate position in the strain formula.

Genes for virulence corresponding to host genes in group 1 have accumulated in the fungal population due to selection. The presence of genes for virulence in respect to group 2 probably involves several mechanisms, but much remains obscure.

Since 1957, a strain that became one of the most prevalent among the post-1954 groups was 21-2,3,7. It arose in region 1, where it attacked Gamenya (*Sr9b*), and Festival and Kenora (*Sr9b, Sr15*) in 1961–1962. Virulence for *Sr9b* gave it a distinct advantage over other strains, especially when the widely grown Robin and Raven (*Sr11, Sr9b*) were released in region 2. However, it was the cultivation of Gamenya in most areas of regions 1 and 2 and its popularity in region 3 that ensured the success of 21-2,3,7, and indeed the continued presence of this strain for 25 years. Released in 1958, Gamenya was the fourth ranking cultivar in New South Wales by 1965 (300,000 hectares). In 1966, it ranked second in Australia (800,000 hectares), and in later years it covered about 800,000 hectares in region 3 alone.

The wide distribution of 21-2,3,7 throughout Australasia poses the question as to whether the same rust clone was involved. Perusal of the rust survey data (Luig and Watson, 1970, 1977; N. H. Luig, unpublished) shows that mutants for virulence on Sr_{Agi}, Sr_{51209}, and Sr_{BB} —genes not used in the host population—were not detected among

21-2,3,7 types or were very rare. Thus, it was concluded that with the exception of a small number of mutants, 21-2,3,7 comprised only a single phenotype.

During the early and mid 1960s, Gamenya (*Sr9b*) and Mengavi (*Sr36*) were rusted by two groups of strains, respectively, but the strain combining virulence for both was extremely rare. When Mendos (*Sr11, Sr17, Sr36,* plus *Sr7a*—which is practically ineffective) became widespread in region 1, the same strains combined genes for virulence corresponding with any two of the three main resistance genes in Mendos. Soon, three strains virulent on Mendos were detected, and by 1968 they accounted for 75% of the isolates collected in region 1. The strain most frequently recovered was also virulent on Gamenya and Mengavi.

In 1968, two strains virulent for *Sr30* (Festiguay) were found, and they seemed to be related to the Gamenya-attacking types, but were not derivatives of the strains virulent for Mendos. Soon these strains and their descendants were widespread throughout the eastern wheat belt, but particularly in region 1 where Festiguay was grown.

This led to a rapid decline of the three strains virulent on Mendos. This shift in the rust flora was facilitated by the replacement of Mendos by Gamut (*Sr6, Sr9b, Sr11, SrGt*) in region 1. During the 1973–1974 season, mutations for virulence for *Sr6* took place in the Festiguay-attacking strains, and Gamut also became susceptible. As with Mendos, the pathogen was able to accumulate the necessary genes for virulence in a stepwise process because of the earlier cultivation of wheats with the corresponding single genes for resistance.

In 1960, strain 21-2 was isolated in region 3 from the wheat area (Esperance) closest to South Australia. As 21-2 was the only strain virulent on Gabo (*Sr11*), its spread was facilitated by this cultivar being grown on 48.3% of the Western Australia wheat area. In 1961, 21-2 made up 94.1% of the isolates (Luig and Watson, 1970). Strains of the older race group (126–222) were not isolated in Western Australia after 1962. In 1962 a new strain, 21-1,2 with virulence for *Sr6,* appeared. As Eureka (*Sr6*) and Wongoondy (*Sr6*) were grown extensively in certain rust-liable areas of Western Australia, it was not surprising that 21-1,2 increased at the expense of 21-2. In 1963, 21-1,2 caused substantial crop losses. Luig and Watson (1970) suggested that the Western Australian 21-1,2 had originated as an independent mutation and was not windblown from the eastern wheat belt. They further suggested that under the selection pressure of the 1963 epiphytotic, a type of 21-1,2 evolved that was more aggressive than other strains.

The 21-1,2 attacked Eureka in Queensland and northern New South Wales, but as the area sown to this cultivar declined, so did the strain. By contrast, 21-1,2 remained the predominant strain in Western Australia until 1969 despite a reduction in the area sown to cultivars with *Sr6*. In 1971, it was again predominant in Western Australia. In 1962, 21-1,2 was isolated from South Australia, but remained at low frequencies until 1973 when it was the main strain contributing to the severe epiphytotic that destroyed Halberd (*Sr6, Sr11*) sown on about 60% of the South Australia wheat area.

Race 21 is a very basic type and has occurred in all wheat-growing areas of the world. The two virulences acquired in Australia which resulted in 21-1,2, those for *Sr6* and *Sr11*, made it possible for Western Australia biotype to attack Gabo and later Eureka. Unfortunately, Halberd, grown on two-thirds of the wheat area in South Australia, possessed just these two genes—somewhat a concidence as none of its stated parents possess *Sr6*. Halberd is very susceptible to strains virulent for *Sr6* and *Sr11*, but until 1973–1974, this was masked by the predominance of strains avirulent for *Sr6* in South Australia and Victoria.

D. Races 194 and 326

In 1969, when Halberd was released, two new strains (races 194 and 326) were detected almost simultaneously in South Australia, Victoria, and central New South Wales. Race 326 was distinct from other Australian races by avirulence for *Sr7b* (Marquis). Moreover, both strains possessed the virulence/avirulence combination for *Sr9b/Sr15*, whereas previous Australian strains virulent for *Sr9b* (putative descendants from 21-2,3,7) were virulent for *Sr15*. Later tests showed that the 1969 strains produced distinctly lower infection types on stocks carrying *Sr15* than did preexisting avirulent strains. Luig (1977) found further differences, including the virulence combination *Sr6,11,9b,17,8*, new for this area. De Sousa (1975), Watson (1981), Watson and de Sousa (1983), and Luig (1983) compared the 1969 strains with those from Angola, Zimbabwe, Mozambique, and Malagasy, and concluded that these central African countries had strains similar to those detected in Australia.

The most important virulence possessed by the 1969 strains was that for *Sr6*, because it combined virulence with aggressiveness and good survival ability.

E. POSSIBLE INFLUENCE OF BARLEY

A possible factor in the epidemiology of wheat stem rust that receives little attention is the response of barley to different strains. In 1953, Stakman and Christensen reported that race 59A was more virulent than race 59 on some previously resistant barley cultivars. Unfortunately, this was largely ignored. Little is known about the adult plant reactions of Australian barley cultivars to strains of *Pgt.* N. H. Luig (unpublished) tested several common cultivars with strains representing pre-1954, post-1954, and putative exotic strains of 1969, and found that, generally, the last group was the most virulent. Because of the widespread cultivation of barley, such differential behavior on barley could explain certain shifts in *Pgt.*

VII. The Establishment of Exotic Strains

It is thought that on three occasions, strains of *Pgt* possessing different combinations of virulence and aggressiveness, were introduced to Australia (Luig, 1977). In the first case (1925), race 126 completely replaced the established strains. Likewise, race 21, detected in 1954, supplanted strains 126-5,6,7,11 and its derivatives within a few seasons, despite the lack of genes necessary for survival on cultivars with *Sr6, Sr9b, Sr11,* and *Sr17.* Strains 326-1,2,3,5,6 and 194-1,2,3,5,6, found in 1969, and their derivatives have not completely replaced the derivatives of 21-0; however, during the four seasons 1979/1980 to 1982/1983, they were predominant in the eastern wheat belt. It appears that these two groups of rusts are similar in aggressiveness. The success of the two 1969 strains and of their derivatives is, therefore, due to a combination of their aggressiveness and virulences. No isozyme differences were detected between the two groups by Burdon *et al.* (1982). The authors propose that 21-0, like 194-1,2,3,5,6 and 326--1,2,3,5,6, originated in Africa. Isozyme phenotypes recorded also agreed with the hypothesis that strain 34-2,11 is the product of a hybridization event between races 126 and 21. These events indicate that natural selection overseas can create biotypes better suited to Australian conditions than those selected locally.

Other introductions may have occurred between 1925 and 1980. Recognition of an exotic strain mainly depends on a difference in the virulence combination, and it is impossible to distinguish an exotic strain from a mutant if only one pathogenic difference is involved. For example, as 326-1,2,3,5,6 and 194-1,2,3,5,6—avirulent and virulent for *Sr7b,* respectively—were found simultaneously, it was assumed that both strains were exotic. However, if the detection of 194-1,2,3,5,6 had occurred in a later year, it would have been concluded that 194-1,2,3,5,6 was a mutant from 326-1,2,3,5,6 and not a separate introduction.

VIII. Durable Resistance and Establishment of Virulent Mutants

The most durable gene (Riley, 1973) for stem rust resistance is *Sr26,* present in Eagle, Kite, Jabiru, Avocet, Harrier and Hybrid Titan. Eagle was released in 1971, and by 1974 it was grown on over half a million hectares. The six cultivars now occupy nearly 1 million hectares annually, but no virulence has been detected.

The author proposes that "durable resistance" to *P. graminis tritici* in wheat is the consequence of high specialization of the pathogen. The pathogen is probably so highly specialized that in the absence of a sexual or parasexual cycle, pathotypes cannot undergo substantial changes without loss in aggressiveness.

Mutation for virulence for certain host genes occurs readily in the field, while mutation to avirulence appears to be rare. In Australia, this prevents removal of unnecessary genes for virulence, but the problem is overcome by the occasional introduction of inoculum from overseas.

Unnecessary genes for virulence can also be introduced by exotic strains. If such genes are present in superior pathotypes, they will spread and increase to high frequencies.

Rapid replacement of an allele for avirulence by its allele for virulence in the absence of a host with the corresponding gene for resistance has been observed on several occasions, particularly in the case of *Sr15,* without a satisfactory explanation.

IX. Leaf Rust of Wheat

There has been a tendency to underestimate the importance of wheat leaf rust (*P. recondita* f. sp. *tritici* Rob. ex Desm.) in Australia. In region 1, favorable environmental conditions frequently lead to an abundance of the disease, and susceptible cultivars may be defoliated.

In region 2, cool temperatures during the late winter and early spring normally retard disease development. Furthermore, Halberd, grown extensively during the last 10 years in South Australia—over 50% of the wheat area in some years—possesses a considerable degree of resistance. In most years, prevailing dry conditions in the central wheat areas of region 3 prevent serious outbreaks of leaf rust.

Field surveys to assess variability in *P. recondita tritici* were done by Waterhouse (1952) from 1921 to 1951. Until 1946, the eight standard differentials and Thew (*Lr20*) were used to identify the numerous isolates, but only two biotypes, distinguishable on Thew, were isolated.

In 1945, when Gabo (*Lr23*) covered a small area in northern New South Wales and Queensland, a new virulent strain was detected that rapidly spread. Later Waterhouse noticed variation for Webster (*Lr2a*) in the Gabo-attacking strain. All of Waterhouse's standard races were further separable on Thew.

The extensive cultivation of wheats with Hope-type resistance (*Lr14a*), namely, Glenwari, Spica, Gala, and Lawrence, in the 1950s resulted in rust types virulent for *Lr14a*. In 1958, Gamenya and Mengavi, both with *Lr3*, were released, and again a shift in the rust flora occurred. At this time, virulence for *Lr15* [Kenya 112-E19-J(L)W1483] was also detected (Watson and Luig, 1961).

Further variability resulted from the cultivation of 10 wheats between 1960 and 1980: Festiguay *Lr2a* (released 1963), Mendos *Lr14a* (1964), Timgalen *Lr3, Lr27* (1967), Gatcher *Lr27* (1968), Songlen and Timson *Lr17* (1975), Warimba *Lr3* (1976), Shortim *Lr1, Lr27,* Cook *Lr3* (1977) and Lance *Lr20* (1978). By contrast, the widely grown Egret (1973) and Banks (1980) (probably both with *Lr13*) remain resistant to all Australian isolates.

Survey data indicate north-to-south movement in regions 1 and 2. The rust flora in Western Australia is different from that of the other states, although the data indicate that strains from the east occasionally enter Western Australia and displace older strains. Since 1964, Gamenya (*Lr3, Lr23*) was widely grown in Western Australia—usually more than 50% of the wheat area—and that gave strains with virulence for *Lr3* and *Lr23* a distinct advantage.

Waterhouse (1952) reported the presence of three new races at Lincoln, New Zealand: 34 and 53 in 1942 and 98 in 1948. Watson (1962) also detected a new race (15) in New Zealand, in 1952. Races 34 and 15 are characterized by virulence for *Lr3,* and 98 produced I.t. X on wheats with this gene. At that time *Lr3* was absent from New Zealand and Australian cultivars, and therefore no explanation can be offered for the occurrence of this virulence prior to the release of Gamenya and Mengavi in Australia. Most of the present leaf rust strains of New

Zealand are avirulent for *Lr23* and *Lr14a,* and thus are distinct from those of mainland Australia; however, in this respect they resemble the strains with which Waterhouse worked.

Rees *et al.* (1979) studied the progress and effects of two epiphytotics in 45 cultivars and found considerable differences. Fast rusting was conspicuous in Gabo and its derivatives, Gala, Koda, Mengavi, Gamenya, Mendos, and Gamut.

X. Stripe Rust of Wheat

Stripe rust (*P. striiformis* West.) was first detected in October 1979 in Victoria; it is not known how the rust came to Australia. It was previously thought that stripe rust, if introduced, would not become a serious wheat disease in Australia because of the high summer temperatures (Waterhouse, 1936).

During 1979 stripe rust spread to several regions of Victoria and also to South Australia and southern and northern New South Wales. This rapid spread was undoubtedly facilitated by the extreme susceptibility of some Victorian cultivars, unselected for resistance. In the summer of 1980 the wheat belt of the three states was very dry, but the pathogen survived, and in that season attacked the susceptible cultivars more severely than leaf or stem rust. In 1980, stripe rust was also found in New Zealand (Wellings and McIntosh, 1982). It will be interesting to see if the pathogen can cross the Nullabor Plain (Fig. 2) to Western Australia, or become established in Queensland where temperatures are higher. Stripe rust has only been severe in regions 2 and 4.

Winter temperatures in most of the Australian wheat belt are usually too low for rapid development of leaf and stem rust but favor stripe rust: e.g., night temperatures of 10°C are optimal for infection by stripe rust, while the higher daytime temperatures (15°–18°C) promote sporulation. However, high temperatures in late spring induce resistance in many wheat genotypes.

Many samples of stripe rust from different regions have been processed and tested on the internationally accepted differentials, but only one strain, 104 E137, was identified during the seasons 1979 and 1980 (Wellings and McIntosh, 1982). This strain occurs in Europe, suggesting an introduction from there. A closely related rust, *P. striiformis* var. *dactylidis* Guy. & Mass., stripe rust of cocksfoot, has been known in New Zealand since 1975, but was not found in Australia until 1979 (Wellings and McIntosh, 1982). In California, the wheat-attacking form of *P. striiformis* was found on *Hordeum leporinum* and *Phalaris paradoxa* L., and the two grasses seemed to play an important role in the epiphytotics (Line, 1976). These grasses are common in southern New South Wales and Victoria.

XI. Stem Rust of Rye

Despite an increase in demand for rye bread in Australia, resulting in a much larger area sown to this cereal, stem rust has not been a major problem. This is probably due to the small and separated areas sown to rye. However, persistent attacks by *Pgs* occur where rye is planted near the coast, either commercially or for soil stabilization purposes. In Western Australia, *Pgs* survives on *Agropyron distichum* Beauv. growing in sand dunes along the coast, and when temperatures rise in spring, infection spreads to nearby rye crops. Prior to 1957, individual plants in rye fields often had stem rust, but the causal organism was *Pgt*. In 1957, *Pgs* was found in rye on the central coast of New South Wales. Earlier (1947 and 1951), this *forma specialis* was isolated from a grass (*Deyeuxia monticola* Vickery) growing at high altitudes in the southern alpine region of New South Wales (Waterhouse, 1957).

The source of the original inoculum of *Pgs* in Australia is unknown. It may have been introduced, or alternatively it could have evolved through sexual or somatic hybridization between other *formae speciales*. Watson and Luig (1958) reported that throughout Tasmania, *A. repens* L. infected with *Pgs* contributed most of the teliospore material for infection of barberries. Other grasses infected by *Pgs* and growing in the vicinity of barberries were *A. scabrum* and *Hordeum leporinum*. On seedlings of Little Club wheat, at least four different infection types were recorded when testing the *Pgs* cultures.

Tan *et al.* (1975) studied variability in the Australian *Pgs* flora using self-fertile lines of *Secale cereale* possessing single genes for resistance in an *S. vavilovii* background. The lines proved very efficient in differentiating between cultures collected from well separated locations. As the main cultivar, Black Winter, is fully susceptible to *Pgs*, the reason for such diversity of strains, without any known selection, is obscure.

XII. Leaf Rust of Rye

Puccinia recondita Rob. ex Desm. f. sp. *secalis* (rye leaf rust) frequently attacks crops of rye, mainly in coastal districts (Waterhouse,

1952). In instances where only a few plants are attacked, *P. recondita* f. sp. *tritici* may be the causal agent.

Some plants of Black Winter rye are resistant to both pathogens, and on inbreeding such plants produce mainly resistant offspring (Waterhouse, 1952; Tan, 1973). Wheats are immune or resistant to *P. recondita secalis.*

Several decades ago the area sown to rye was very small, and common grasses were not identified as hosts for *P. recondita* f. sp. *secalis.* Thus, the origin and persistence of this rust is unexplained.

XIII. Stem Rust of Barley

Barley is an important crop, although in the main cereal-growing states, New South Wales, Western Australia, and Victoria, only about 3%, 5%, and 5%, respectively, of the cereal area is sown to it. Like wheat, barley is sown in the late autumn to early winter. Stem rust on barley is caused by *Pgt, Pgs,* and by their hybrids (Luig and Watson, 1972; Burdon *et al.,* 1981). While the first two have a specific cereal host, the hybrids are avirulent on commercial wheat, rye, and oats, but are adapted to *A. distichum* or *A. scabrum.* However, since the commercial host in each case is barley, these rusts are referred to as *P. graminis* f. sp. *hordei* (*Pgh*).

Recently, most rust samples collected from barley in Queensland comprised *Pgh,* with some crops heavily rusted. The remaining samples were mostly *Pgt,* with a few samples of *Pgs.* Studies in the greenhouse using adult plants of several common barley cultivars showed that strains of *Pgh* were more pathogenic on barley than race 21 of *Pgt,* or two strains of *Pgs.* This suggests that the new rust—detected in 1963—is well adapted on barley and probably sufficiently virulent to cause measurable reductions in yield.

Since 1965 some samples from rye were *Pgh.* Although some strains result in a semisusceptible to fully susceptible reaction on most seedlings of Black Winter, there is no indication that adaptation on rye was a factor in the evolution of this new *forma specialis.*

XIV. Leaf Rust of Barley

In all states, crops of barley, especially those in coastal areas, have some leaf rust. The causative agent is *P. hordei* Otth., which is non-pathogenic on the other small-grain cereals. Relatively few races of *P. hordei* have been reported in Australasia, and until 1952 all isolates resembled UN race 16 of Levine and Cherewick (1952). In 1952 a new strain (UN race 14) was detected that was virulent on barleys with *Pa* (McWhirter, 1955). Differences on other testers of Levine and Cherewick, were also noticed. Although only a few collections of barley leaf rust are analyzed annually, it appears that UN race 14 predominated during the last two decades. In 1978/1979, an unexpectedly severe epiphytotic in the Darling Downs of Queensland demonstrated that it can be very destructive when conditions favor it.

Preliminary experiments indicated that UN race 14 was more aggressive than UN 16 (Luig, 1957). This and the several differences in virulence make it unlikely that UN 14 arose as a mutant from UN 16; hence it is probably an introduction.

Investigation of the inheritance of resistance to UN race 14 showed a complex situation at low temperatures, when in some cases interaction of three genes for resistance appeared to be involved. At high temperatures, mostly monofactorial segregation ratios were obtained with both UN races (Luig, 1957).

XV. Stem Rust of Oats

The rust pathogens of oats, *P. graminis* f. sp. *avenae* (*Pga*) Eriks. and E. Henn. (stem rust) and *P. coronata* f. sp. *avenae* Corda (crown rust), are widespread in Australasia. Oats are grown in a wider range of environments than other cereals. This is mainly due to the multipurpose end uses of the crop: grain for human and livestock consumption, forage, green fodder, and hay. In Queensland, oats are grown mainly for green fodder; however, the main oat growing in eastern Australia occurs in Region 2. In Western Australia, large areas are also sown for grain and green fodder.

Oat stem rust is a disease of inland areas, where oats are sometimes severely rusted. The pathogen oversummers on wild oats (*A. fatua* L. and *A. ludoviciana* Durieu) and on volunteer plants but is also found on grasses that normally do not have other forms of *P. graminis* (Waterhouse, 1952; Luig and Watson, 1977). These grasses are *Lamarckia aurea* (L.) Moench., *Vulpia bromoides* L., *Amphibromus neesii* Steud., *Dactylis glomerata*, and *Phalaris* sp. On *Hordeum leporinum* the pathogen must compete against *Pgt*, *Pgs*, and *Pgh*. Sand oats (*Avena strigosa* Schreb.) are grown commercially along coastal New South Wales and Queensland, the main cultivar being Saia.

Waterhouse (1952) studied 910 samples collected throughout Australia and 30 from New Zealand. These cultures were differentiated on hosts with *Pg1*, *Pg2*, and *Pg3*.

Luig and Baker (1973) reported that races possessing virulence for *Pg1*, *Pg2*, Pg3, and *Pg4* were quite widespread during the seasons 1970/1971 and 1971/1972. Virulence for Saia was found mainly in Queensland and northern New South Wales. Recent studies (Oates and Brouwer, 1979; Oates, 1980) confirmed the high degree of variability. The extent of variability in the pathogen was unexpected, as the sexual stage of *P. graminis avenae* rarely occurs, and most Australian oat cultivars lack resistance to *P. graminis avenae*.

XVI. Crown Rust of Oats

Crown rust is widespread in Australia, and oats in coastal areas in eastern Australia are prone to severe attacks. Some crops are ruined before grazing is possible; others lodge or ripen prematurely. *Puccinia coronata* survives on volunteer commercial oats and wild oats that are infected throughout the year.

Surveys covering the period of 1935–1951 were summarized by Waterhouse (1952), who described 13 races. Race 6 accounted for 50% of isolates. Additional surveys were done by Baker and Upadhyaya (1955) in 1952, 1953, and 1954, and by Oates (1980) in the late 1970s. These investigations revealed considerable variability. Oates found that each of the 10 standard differentials was susceptible to more than one strain, and he also recorded variation with respect to several single-gene lines.

XVII. Summary

The genetic aspects of epidemiology dealt herein are mainly of *P. graminis tritici*. Data from the other cereal rusts support the conclusion but are inadequate.

Survey data suggest that rust clones—i.e., strains that cannot be subdivided on additional supplementary differentials—are responsible for the epiphytotics that occur in Australia. Stabilizing selection does not play a major part in pathogen survival.

The evidence from 60 years of rust investigations shows that natural selection among the many rust strains and postulated mutants does not result in the establishment of types that are best suited to the Australasian environment and changing host genotypes. In three instances, putative introductions have replaced the older strains because of their superior aggressiveness.

References

Baker, E. P., and Upadhyaya, Y. M. (1955). Physiologic specialization in crown rust of oats. *Proc. Linn. Soc. N.S.W.* **80,** 240–257.

Brown, J. F., and Shipton, W. A. (1964). Relationship of penetration to infection type when seedling wheat leaves are inoculated with *Puccinia graminis tritici. Phytopathology* **54,** 89–91.

Burdon, J. J., Marshall, D. R., and Luig, N. H. (1981). Isozyme analysis indicates that a virulent cereal rust pathogen is a somatic hybrid. *Nature (London)* **293,** 565–566.

Burdon, J. J., Marshall, D. R., Luig, N. H., and Gow, D. J. S. (1982). Isozyme studies on the origin and evolution of *Puccinia graminis* f.sp. *tritici* in Australia. *Aust. J. Biol. Sci.* **35,** 231–238.

Close, R. C. (1967). Stem rust of wheat in Canterbury. *N. Z. Wheat Rev.* No. 10, pp. 47–52.

de Sousa, C. N. A. (1975). A comparison of strains of *Puccinia graminis* f.sp. *tritici* from Angola and Rhodesia with their counterparts in Australia. M.Sc.Ag. Thesis, University of Sydney, Sydney, Australia.

Johnson, R., and Taylor, A. J. (1972). Isolates of *Puccinia striiformis* collected in England from the wheat varieties Maris Beacon and Joss Cambier. *Nature (London)* **238,** 105–106.

Levine, M. N., and Cherewick, W. J. (1952). Studies on dwarf leaf rust of barley. *U.S. Dep. Agric., Tech. Bull.* **1056,** 1–17.

Line, R. F. (1976). Factors contributing to an epidemic of stripe rust on wheat in the Sacramento Valley of California in 1974. *Plant Dis. Rep.* **60,** 312–316.

Luig, N. H. (1957). Inheritance studies of barley in relation to disease resistance. M.Sc.Ag. Thesis, University of Sydney, Sydney, Australia.

Luig, N. H. (1977). Exotic species in Australia their establishment and success. *Proc. Aust. Ecol. Soc.* **10,** 89–96.

Luig, N. H. (1979). Mutation studies in *Puccinia graminis tritici. Proc. Int. Wheat Genet. Symp., 5th, 1978,* pp. 533–539.

Luig, N. H. (1983). "A Survey of Virulence Genes in Wheat Stem Rust, *Puccinia graminis* f.sp. *tritici*" Parey, Berlin.

Luig, N. H., and Baker, E. P. (1973). Variability in oat stem rust in eastern Australia. *Proc. Linn. Soc. N.S.W.* **98,** 53–61.

Luig, N. H., and Rajaram, S. (1972). The effect of temperature and genetic background on host gene expression and interaction to *Puccinia graminis tritici. Phytopathology* **62,** 1171–1174.

Luig, N. H., and Watson, I. A. (1961). A study of inheritance of pathogenicity in *Puccinia graminis* var. *tritici. Proc. Linn. Soc. N.S.W.* **86,** 217–229.

Luig, N. H., and Watson, I. A. (1970). The effect of complex genetic resistance in wheat on the variability of *Puccinia graminis* f.sp. *tritici. Proc. Linn. Soc. N.S.W.* **95,** 22–45.

Luig, N. H., and Watson, I. A. (1972). The role of wild and cultivated grasses in the hybridization of formae speciales of *Puccinia graminis*. *Aust. J. Biol. Sci.* **25,** 335–342.

Luig, N. H., and Watson, I. A. (1977). The role of barley, rye and grasses in the 1973–74 wheat stem rust epiphytotic in southern and eastern Australia. *Proc. Linn. Soc. N.S.W.* **101,** 65–76.

McEwan, J. M. (1966). The source of stem-rust infecting New Zealand wheat crops. *N. Z. J. Agric. Res.* **9,** 536–541.

McWhirter, K. S. (1955). Studies of the genetics of resistance to rust (*Puccinia hordei* Otth.) and mildew (*Erysiphe graminis hordei* Marchal) in barley. Thesis (Hons.), University of Sydney, Sydney, Australia.

Nix, H. A. (1975). The Australian climate and its effects on grain yield and quality. *In* "Australian Field Crops" (A. Lazenby and E. M. Matheson, eds.), Vol. 1, pp. 183–226. Angus & Robertson, Sydney, Australia.

Oates, J. D. (1980). "Oat Rust Survey 1979–80" (mimeo.). University of Sydney, Sydney, Australia.

Oates, J. D., and Brouwer, J. B. (1979). "Oat Rust Survey 1978–79" (mimeo.). University of Sydney, Sydney, Australia.

Person, C. (1959). Gene-for-gene relationships in host: Parasite systems. *Can. J. Bot.* **37,** 1101–1130.

Rees, R. G. (1972a). Urediospore movement and observations on the epidemiology of wheat rusts in north-eastern Australia. *Aust. J. Agric. Res.* **23,** 215–223.

Rees, R. G. (1972b). *Agropyron scabrum* and its role in the epidemiology of *Puccinia graminis* in north-eastern Australia. *Aust. J. Agric. Res.* **23,** 789–798.

Rees, R. G.., Thompson, J. P., and Goward, E. A. (1979). Slow rusting and tolerance to rusts in wheat. II. The progress and effects of epidemics in *Puccinia recondita tritici* in selected wheat cultivars. *Aust. J. Agric. Res.* **30,** 421–432.

Riley, R. (1973). Genetic changes in hosts and the significance of disease. *Ann. Appl. Biol.* **75,** 128–132.

Stakman, E. C., and Christensen, J. J. (1953). Problems of variability in fungi. *USDA Yearb.—Plant Dis.*, pp. 35–62.

Stakman, E. C., Stewart, D. M., and Loegering, W. Q. (1962). Identification of physiologic races of *Puccinia graminis* var. *tritici. U.S., Agric. Res. Serv., ARS* **E617,** 1–53.

Tan, B. H. (1973). The genetics of resistance to *Puccinia graminis* Pers. in the genus *Secale.* Ph.D. Thesis, University of Sydney, Sydney, Australia.

Tan, B. H., Watson, I. A., and Luig, N. H. (1975). A study of physiologic specialization of rye stem rust in Australia. *Aust. J. Biol. Sci.* **28,** 539–543.

Vanderplank, J. E. (1968). "Disease Resistance in Plants." Academic Press. New York.

Waterhouse, W. L. (1934). Australian rust studies. IV. Natural infection of barberries by black stem rust in Australia. *Proc. Linn. Soc. N.S.W.* **59,** 16–18.

Waterhouse, W. L. (1936). Some observations on cereal rust problems in Australia. *Proc. Linn. Soc. N.S.W.* **61,** 5–38.

Waterhouse, W. L. (1938). Some aspects of problems of breeding for rust resistance in cereals. *J. R. Soc. N.S.W.* **72,** 7–54.

Waterhouse, W. L. (1952). Australian rust studies. IX. Physiologic race determinations and surveys of cereal rusts. *Proc. Linn. Soc. N.S.W.* **77,** 209–258.

Waterhouse, W. L. (1957). Australian rust studies. XV. The occurrence in Australia of stem rust of rye, *Puccinia graminis secalis* E. & H. *Proc. Linn. Soc. N.S.W.* **82,** 145–146.

Watson, I. A. (1955). The occurrence of three new wheat stem rusts in Australia. *Proc. Linn. Soc. N.S.W.* **80,** 186–190.

Watson, I. A. (1958). The present status of breeding disease resistant wheats in Australia (Farrer Oration). *Agric. Gaz. N.S.W.* **69,** 630–631.

Watson, I. A. (1962). Wheat leaf rust in Australia—strain variations for the period 1951–1961. *Comm. Cereal Rust Surv.* No. 1, pp. 1–24 (mimeo.).

Watson, I. A. (1981). Wheat and its rust parasites in Australia. *In* "Wheat science—Today and Tomorrow" (L. T. Evans and W. J. Peacock, eds.), pp. 129–147. Cambridge Univ. Press, London and New York.

Watson, I. A., and Cass Smith, W. P. (1962). Movement of wheat rusts in Australia. *J. Aust. Inst. Agric. Sci.* **28,** 279–287.

Watson, I. A., and de Sousa, C. N. A. (1983). Long distance transport of spores of *Puccinia graminis tritici* in the southern hemisphere. *Proc. Linn. Soc. N.S.W.*, 311–321.

Watson, I. A., and Luig, N. H. (1958). Widespread natural infection of barberry by *Puccinia graminis* in Tasmania. *Proc. Linn. Soc. N.S.W.* **83,** 181–186.

Watson, I. A., and Luig, N. H. (1961). Leaf rust on wheat in Australia: A systematic scheme for the classification of strains. *Proc. Linn. Soc. N.S.W.* **86,** 241–250.

Watson, I. A., and Luig, N. H. (1963). The classification of *Puccinia graminis* var. *tritici* in relation to breeding resistant varieties. *Proc. Linn. Soc. N.S.W.* **88,** 235–258.

Watson, I. A., and Luig, N. H. (1966). *Sr15*—A new gene for use in the classification of *Puccinia graminis* var. *tritici. Euphytica* **15,** 239–250.

Watson, I. A., and Luig, N. H. (1968). Progressive increase in virulence in *Puccinia graminis* f.sp. *tritici. Phytopathology* **58,** 70–73.

Wellings, C. R., and McIntosh, R. A. (1982). Stripe rust—A new challenge to the wheat industry. *Agric. Gaz. N.S.W.* **92**(3), 2–4.

Zadoks, J. C. (1961). Yellow rust on wheat, studies in epidemiology and physiologic specialization. *Tijdschr. Plantenziekten* **67,** 69–256.

11

Epidemiology in Europe

J. C. Zadoks
*J. J. Bouwman**
Department of Phytopathology, Agricultural University,
Wageningen, The Netherlands

*Present address: Agricultural College, Groenezoom 400, 3315 LA Dordrecht, The Netherlands

THE CEREAL RUSTS,
Volume II

ISBN 0-12-148402-5

I. Introduction

The scale of the subject of this chapter is large. In the time dimension, monocyclic processes are taken for granted, but attention is given to polyetic processes. Cause-and-effect relations explaining an epidemic are painted with a coarse brush only. In the distance dimension the emphasis is on long distances, to be expressed in hundreds of kilometers. There is also the dimension of complexity: host and pathogen interact, they form a pathosystem that reacts to climate and weather. Higher orders of complexity arise when humans, interfering with a pathosystem, become its prisoners as much as its masters.

"Matters of scale" have been discussed before (Zadoks and Schein, 1979). At higher scale levels, information is scanty, data are inaccurate, inferences may be less justified, and conclusions may be invalid. Humans are involved, first as the sum total of a large group of individuals—farmers responding to the cereal-rust pathosystems in their fields—and, second, as the institution makers—institutions creating policies to suit their purposes—in many combinations and permutations. Science must tackle the problem of high-order complexity and try to unravel explicit or implicit policies of the past, hoping that the insight acquired can help to improve the policies of the future.

Geographically, the area of interest is Europe, with North Africa and the Near East at the fringe. Temporally, the emphasis is on the last half century, with a historical introduction. The subject matter is largely wheat with its three rusts, as the other cereals have little to contribute to the problem in focus here. That problem is, by and large, the extensity and intensity of epidemics (Zadoks and Kampmeijer, 1977).

II. Historical Note

Greek and Roman classical authors were familiar with the cereal rusts and had a notion about ecological conditions favoring rust attack. Theophrastus (371–286 BC) made pertinent remarks on cultivar susceptibility, the situation of the field, and the effect of weather. His discussions on the influence of morphology on rust infection seem irrelevant, but his observation that grain grown in low-lying damp places is more frequented by rust than that in high and windy places is still correct. The theme was repeated by the knowledgeable Roman author Pliny the Second. He overemphasized the effect of the moon, but he knew that weather conditions conducive to heavy dew were conducive to rust, which is still true (Zadoks, 1982).

The classic wisdom was repeated by Fontana [(1767)1932] and Targioni Tozzetti [(1767)1952], both discussing the 1766 stem rust epidemic on wheat in Italy. They explicitly stated that all of Italy was affected, the first mention of a really large-scale rust epidemic. Tozzetti, also an amateur meteorologist, described how the weather caused late fall sowing and poor emergence, how a cold winter led to a tender and late crop, and how conducive the spring weather was for stem rust. His analysis of the summer preceding the epidemic, a cool wet summer that did not allow summer plowing and that must have been good for oversummering of the rust, is a far forerunner of Zadoks' (1961) theorem on the importance of back-tracking an epidemic through the preceding winter and summer to the previous crop.

Published attention on the alternate hosts began in 1660 with the enactment of the Barberry Laws by the City of Rouen in northern France, decreeing that all barberries be eradicated. No information exists on their effectiveness. In the 1780s, Marshall in England performed the first experiments, but it was a laborious way over Schøler in Denmark (1818 ex Plowright, 1889; Hermansen, 1968) to De Bary (1865), who finally proved that *Aecidium berberidis* and *Puccinia graminis* were the same species.

Famous epidemiological studies were done by Eriksson and Henning (1896) on wheat rusts, especially stripe rust, *P. striiformis*. In Europe, they remained unsurpassed until about 1960, when novel approaches to stem rust epidemiology were developed in the United Kingdom and new data on stripe rust epidemiology were produced in the Netherlands.

Avoidance of damage was mentioned by Pliny. When rust attacks wheat, premature harvesting may give solace. In 1767, both Fontana and Targioni Tozzetti mentioned this practice; the latter stated that harvesting in the milky ripe stage leads to some loss as the grains are shrivelled and light, but letting the stem rust go may lead to one-tenth of a normal yield or less. Premature harvesting was advocated throughout the nineteenth century. Cord drawing was first mentioned by Olivier in the early seventeenth century. It was based on the idea that dew had something to do with rust. If then the dew drops could be shaken off the plants around sunrise by means of a few people dragging a cord over the crop, the rust could be prevented (Fig. 1). Tozzetti suggested a variation, whipping the wheat plants lightly with a bunch of willow twigs, but sadly added that this method is of little avail on large estates.

Fig. 1. Cord drawing as a means of control against premature death of wheat due to sunburn and rust. Romantic picture from Heuzé (1872), suggesting not only a rope but also a kind of net drawn over the wheat crop in early morning to shake off the dew droplets.

III. The Ingredients of an Epidemic

The basic ingredients of a rust epidemic are a suitable host, a suitable pathogen, and suitable weather. When any ingredient is lacking or is not suitable, there is no epidemic. The picture is, however, not a black-or-white one. Vanderplank's (1963) equivalence theorem is applicable to large-scale epidemics in the gray zone of the picture: unusually suitable weather may lead to an epidemic on moderately susceptible cultivars; highly susceptible cultivars may be subject to an epidemic under not so suitable conditions.

A. THE PATHODEME

The geographic distribution of the economically important host is well known (Broekhuizen, 1969); see also Table I. From west to east, the climate of Europe gradually changes from temperate atlantic to continental. Variations occur when (1) higher altitudes lead to conditions similar to the temperate atlantic type during the cereal growing season, and (2) the nearness of large water masses has a similar effect. The first occurs in the Swiss plain at about 500 m altitude, where the growing conditions are very similar to those in the Netherlands. The second case is exemplified by the coastal area on the Black Sea in Romania and Bulgaria, which is so mild and humid that rusts can thrive the year around. In both cases the agrometeorological situation is conducive to stripe rust epidemics. Conversely, mountain chains may cause rain shadows and induce continental climates as in the Castillian plains of Spain. The south to north changes in Europe are more abrupt than the west to east changes. The Alpine ranges cut

Table I

Approximate Wheat Hectareages of European Countries[a]

Country	× 1000 Hectares	Country	× 1000 Hectares
Albania	200	Italy	3500
Austria	300	Malta	1
Belgium + Luxemburg	200	Netherlands	100
Bulgaria	800	Norway	20
Czechoslovakia	1300	Poland	1800
Denmark	100	Portugal	400
Finland	100	Romania	2300
France	4200	Spain	2700
Germany (GDR)	700	Sweden	300
Germany (FRG)	1600	Switzerland	100
Greece	1000	United Kingdom	1300
Hungary	1300	U.S.S.R.[b]	63,000
Ireland	50	Yugoslavia	1700

[a]Source: FAO Production Yearbook 1978.
[b]Including non-European part.

across Europe roughly along a west to east line, following the Pyrenees, the Alps, and the Carpathians. Along the Mediterranean Sea, wheat, barley, and rye are largely grown as short-cycle winter crops without need of vernalization. North of the Alpine ranges, these crops usually are fall-sown crops with a vernalization requirement. The exception is malting barley, which is usually a spring-sown crop. In the far north, the winter season is so long and cold that only spring-sown crops can be grown, which, profiting from the long daylength, complete their life cycle within a short period.

Dates of sowing and harvesting and the duration of the "crop-free period" or "crop gap"—the period between harvest and next emergence—are particularly relevant to epidemiology. Historical data cannot be understood without recognizing the changes over the last 80 years. For western Europe, the "wheat gap" was mapped by Zadoks (1961). This map should be supplemented by the summer rainfall in that period, which is a measure of the chances that a "green bridge" between the end of the preceding and the beginning of the following season is established by volunteer plants. Historical changes in the importance of the green bridge are considerable. In the southern Sweden of Eriksson and Henning (1896), it was not at all unusual to sow next year's crop 1 month before the harvest of the current year's crop. The duration of the crop-free period was zero to even −30 days. This situation was true for some higher alpine valleys in Switzerland until the 1950s. Long crop-free periods were legally enforced in Denmark, where the cultivation of winter barley is forbidden, as it is in some areas of the Netherlands. From the 1940s to the 1970s it was recommended that wheat be sown relatively late, in part to avoid eyespot and take-all diseases; Zadoks (1961) argued that it worked also against stripe rust. At present, following the move for intensive wheat cultivation, there is a tendency to sow earlier; it will have epidemiologic consequences (Fig. 2).

Harvest in the Netherlands has been about mid-August. As this is the month with the highest precipitation, breeders selected for earliness, thus prolonging the wheat gap. Present developments point to the inverse. Intensive wheat cultivation requires high levels of nitrogen throughout the growing season. In the Netherlands, one or two top dressings are applied at stages DC30 (80 kg nitrogen, N) and 38 (40 kg N); sometimes a third top dressing is given at DC55 (30 kg N). In Great Britain and Northern Ireland the number of split nitrogen applications is up to five. The result is a new phenomenon: the leaf area duration increases (Ellen and Spiertz, 1980), and the wheat leaves remain green up to ripening. In fact, heads may ripen before the flag leaf dies, an

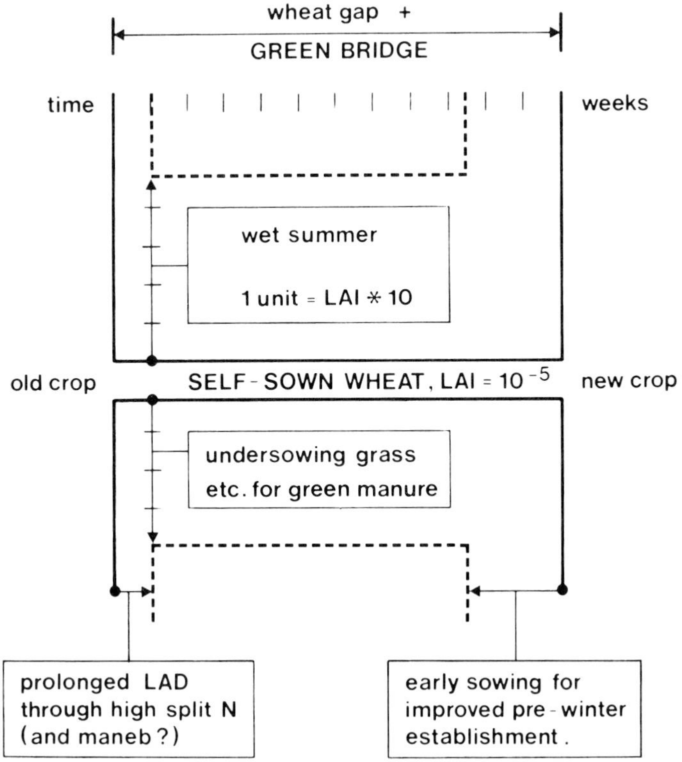

Fig. 2. Schematic representation of the wheat gap and the green bridge. Arrows indicate how various effects tend to shorten and widen the green bridge, thus facilitating the oversummering of foliar pathogens.

unusual sight in Europe. This situation favors rust epidemics, especially of leaf rust, unless the foliage is protected by resistance and/or by chemicals (Darwinkel, 1980).

The green bridge of volunteer wheat between two successive seasons was quite narrow in the Netherlands, especially before World War II, because of repeated and painstakingly neat summer plowing with light horse-drawn plows, mainly for weed control. Presently, wheat is undersown with other crops, leguminous plants or grasses for seed, or—a recent development—for the improvement of soil structure and as green manure. In 1980, for example, 84% of cereals (mainly wheat) in the Dutch Zuyder Zee Polders was undersown with grasses (data by courtesy of J. Meems). The undersowing prevents summer plowing. The wheat grain lost during harvest finds a good seedbed with favorable germination conditions, and the green bridge consisting of volunteers is wide (Fig. 2). Eventually, inoculum of stripe rust, leaf rust, mildew, and other foliar pathogens passes the bridge from previous to following season in large quantities. The epidemiological implications are considerable, but they have not yet been well studied (Broennimann, 1966).

Agronomic practices change gradually and, seen from an epidemiological view, in a pulsating pattern with a wavelength of decades. Patterns as distinguished in the Netherlands affecting the green bridge must occur elsewhere in Europe, but they are not found in the scientific literature. Some East European countries had a successful agrarian reform after World War II, i.e., Hungary and Bulgaria. There, most arable land was put under the jurisdiction of large and scientifically managed Agro-Industrial Complexes, which have decisively changed wheat disease epidemiology. Some of the changes are:

1. Wheat fields are large.
2. Wheat is cultivated in favorable ecological zones only.
3. Old cultivars (long straw, long vegetation period, adequate resistances) were replaced by modern cultivars.
4. High chemical inputs are applied (N, P, K, biocides).
5. New cultivars are relatively susceptible to disease, but ripen early, creating a crop-free period of 2 or more months (Bulgaria).
6. Through mechanization, there is more synchronization of sowing and harvesting.

Most of these changes took place in the period 1955–1975; their consequences are still difficult to fathom. In combination with barberry control, they probably led to a reduction of the importance of *P. graminis* and *P. striiformis* and a relatively high importance of *P. recondita.*

In dry areas, irrigation favors cereal rusts in two ways. First, irrigation, in combination with high N, causes lush growth which makes the wheat relatively susceptible. Second, the wet soil is conducive to regular and prolonged dew formation. Stripe rust was prominent on irrigated wheat in the very dry Ebro valley (J. C. Zadoks, personal observation, 1958). On Cyprus, wheat usually escapes the rust, but on irrigated wheat all three rusts can be present with leaf rust up to 90% and stem rust up to 30% severity (Della, 1978).

Besides the grain crops, cereal rusts have accessory hosts, the grasses. The degree to which grasses effectively play a role in cereal rust epidemiology is not well established; it probably varies between areas and rusts. Zadoks (1959, 1961) reasoned that stripe rust is so specialized in northwestern Europe that grasses no longer play a role in its epidemiology; he also indicated areas where grasses were thought to be important in bridging a wheat gap. This view is still valid, but whether it is also true for southeastern Europe and the Mediterranean area has not yet been established. For stem rust on wheat and crown rust of oats, grasses are important in the circum-Mediterranean area. Malençon (1963) distinguished definite cerealicolous and graminicolous phases in the epidemiology of stem rust of wheat in Morocco. For crown rust of oats, grasses have been mentioned as important accessory hosts and eventual green bridges in Spain (Salazar and Martinez, 1973). Quantitative information on the graminicolous phase of cereal rust epidemics is completely absent.

The genetic make-up of the cereal host has to be considered. Robinson (1976) defined a pathodeme as a population of "a host species in which all individuals have a stated pathosystem character (resistance) in common." Parlevliet (1975) stated that current barley cultivars have no vertical resistance genes against barley leaf rust. If this is true, the total winter plus spring barley population is one pathodeme, notwithstanding that within the pathodeme cultivars differ in partial resistance (Parlevliet, 1976). These differences are relevant with respect to the intensities of various epidemics, but they are not or not yet relevant to pathotypes. Again, Europe is in a state of transition. In several instances, as in the 1932 epidemic of stem rust on wheat, the whole European wheat population can be regarded as one pathodeme, but in recent times gene-specific vertical pathodemes have been introduced in southeastern Europe (Bartoš, 1975a, 1980; Ionescu-Cojocaru, 1979).

B. THE PATHOTYPE

The distribution maps of the Commonwealth Mycological Institute for *P. recondita* (map 226) and *P. striiformis* (map 97) show that these rusts exist throughout Europe. Such maps illustrate extensity without intensity, and extensity only at the specific level. Intensity maps are rare, and if they exist (Fig. 3), it is not always known whether the intensity is based on prevalence, severity, or damage (sensu Zadoks and Schein, 1979).

For more detailed epidemiological studies, specific information on those rust types that behave as epidemiologically recognizable entities is crucial. Data from identification studies of races or of virulence genes can be helpful. They have been published for decades, but the reliability of the older data is questionable. Even though race identifi-

Fig. 3. Mean annual crop losses of wheat due to stripe rust (*Puccinia striiformis*). Curves are isodam curves connecting points of equal damage, expressed in per mill. Data refer to the period 1961–1970, but they are still relevant. (From Rijsdijk and Zadoks, 1976.)

cations are locally reproducible, they were not necessarily comparable between locations. In the late 1950s, it was impossible to reproduce race identifications of wheat stripe rust between Wageningen, the Netherlands, and Brunswick, Germany, both at 52° latitude and only 330 km apart (Zadoks, 1961). In the 1970s, the problem has been largely overcome. In addition, the problem of the identification of so-called field races (Zadoks, 1961, 1966b), difficult to identify in the seedling stage in a greenhouse, has been solved to a large extent.

Race identification of wheat leaf rust has been messy for a long time. Hassebrauk (1959), having spent a lifetime on it, arrived at the point of giving up. J. C. Zadoks (unpublished) tested Dutch isolates on the standard differentials, using seed stocks from different origins, including Manhattan, Kansas. The different seed stocks gave different results. Again, the international situation has much improved, so that comparison of results from different stations can be made with some confidence. As in stripe rust, field races of wheat leaf rust exist (Zadoks, 1966b; also unpublished), but little is known about them.

In large-scale epidemiological studies, physiological races in the classical sense are not necessarily the entities to focus on. Using various devices, such as the avirulence/virulence formula (Loegering and Browder, 1971), genotypes of the rust can be identified in detail. The number of cases with a high degree of detail is still low, so that the detail does not help the epidemiologist. In the following, an attempt is made to use the notion of pathotype as did Robinson (1976). He defined a pathotype as "a population of a parasite species in which all individuals have a stated pathosystem character (pathogenicity or parasitic ability) in common."

C. WEATHER AND CLIMATE

Weather is the actual state of atmospheric conditions as to radiation, temperature, precipitation, wind, etc., at a specific place and time. The ever-varying weather profoundly affects the course of a rust epidemic. The analytical knowledge on responses of rust to environmental stimuli in the monocyclic process has been developed so far that the course of an epidemic can be roughly explained using daily weather data (Zadoks and Rijsdijk, 1971). Nevertheless, the vicissitudes of the weather are such that the epidemiologist prefers to look at periods longer than 1 day to explain large-scale phenomena.

A specific weather type, with steady weather lasting several days, can often be related to a particular circulation pattern. If so, the distinction of weather types is useful, as they pertain to larger areas and longer periods of time, say thousands of square kilometers and weeks. They may even have predictive value when monocyclic phenomena are related to them (Hogg *et al.*, 1969; Zadoks, 1968).

Climate involves atmospheric conditions over a long period. Thirty years' averages of synoptic weather data are often used to characterize climate. Good agroclimatic maps of Europe are available (Thran and Broekhuizen, 1965), but they have not yet been used in epidemiology. Climatic data have been widely used in epidemiology, but there are some risks, including erroneous choices, oscillations, and persistence. (1) Current opinion holds that the Mediterranean area has little stripe rust because of its hot and dry climate. However, (a) during the wheat growing period the weather is usually neither hot nor dry, and (b) climatic data are averages, and growing seasons differ widely, some being cool, moist, and conducive to stripe rust. Outbreaks of stripe rust in the western Mediterranean area in 1977 and 1978 demonstrate the potential (Section VII,C.). Generally, rust can appear at any place where the climate allows for wheat cultivation. (2) Oscillations in climate exist, but they are difficult to identify. They can be confounded by (3) the phenomenon of persistence. Weather is not randomly distributed over successive days; on the contrary, for a number of days one weather type prevails followed by a period of another weather type. A weather type persists for some period. In the same way, hot and dry years and cool and wet years need not be randomly distributed. For various types of years there is a tendency to occur in clusters, nonrandom—persistence again.

Another aspect of atmospheric conditions is their suitability to carry inoculum over large distances. Some cereal rust epidemics are airborne. Inoculum in a source area has to be uplifted into the air, transported over a distance, and deposited in a target area. (1) Free and forced convection bring urediospores from the crop space into the surface air layer. Surface winds may carry inoculum over hundreds of kilometers. Spores can easily be trapped in standard spore traps. (2) Free convection and turbulence can lift a spore cloud up to an altitude of some 3 km, where wind speed is high and spore clouds can travel far and fast. By turbulent deposition the spore cloud erodes at the bottom and finally the spore cloud is detached from the ground (Hirst and Hurst, 1967). Spores must be deposited from the high-drifting cloud by rain, as can be demonstrated by indoor experiments (S. Nagarajan, unpublished) and outdoor rain gauges (Rowell and Romig, 1966). (3) A low-level spore cloud can be caught in a frontal system, uplifted in an occluded air mass, and rapidly transported over hundreds of kilometers.

If an air mass, arriving at a certain point in time and space, is characterized by its spore load, the origin of the spores can be approximated by back-tracking (Hogg, 1961). The result is a geostrophic trajectory of an air mass or travel route, which provides useful information only when rust severity in the areas under the trajectory is known. Weather types, circulation patterns, and air-mass trajectories are related subjects. In Europe, these subjects have not been studied adequately in an integrated manner, taking into account the actual field data on crop, rust, and weather in source and target areas. Specifics are needed on the extent of the source, the amount of inoculum available, its survival during travel, its deposition, the extent and the condition of the target, the weather in the target area, the pathodeme–pathotype compatibility, and so on.

Generally, cyclonic and anticyclonic weather patterns in Europe alternate so frequently that in any season rust can migrate from one place to another. This also applies to the Mediterranean area, where the Mediterranean Sea is no real barrier to rust dispersal according to the substantial circumstantional evidence, especially that from international wheat resistance-testing nurseries.

IV. Rust Dispersal Studies

Rust dispersal takes place at short, medium, and long ranges. Short range means within a crop, medium range between crops within a region, and long range from one region or nation to the other. Here, the interest is in the long-range dispersal. Though dispersal over long distances is a well-established fact, sound evidence is rare in Europe. Before examining the evidence itself, the merits of various types of evidence are to be discussed.

A. TYPES OF EVIDENCE

Ideally, the conclusion that rust has travelled over long distances is based on the following criteria (after Zadoks, 1968):

1. Crop phenology in the source area.
2. Rust phenology in the source area.
3. Weather conditions at the source area.
4. Air trajectories from source to target.
5. Spore content of the air between source and target.
6. Spore trapping data in the target area.
7. Weather conditions at the target area.
8. Crop phenology in the target area.
9. Rust phenology data in the target area.
10. Matching of pathotypes in source and target areas.

There is no study from Europe in which all these items were duely considered; the inferences on long distance dispersal were based on limited data. From a strictly scientific viewpoint, this situation is unsatisfactory. But the evidence sometimes is so dramatic—e.g., for the 1932 epidemic of stem rust on wheat in eastern Europe (Zadoks, 1965)—that the researcher has to set aside scientific scruples.

B. EVIDENCE FROM PATHOTYPES

New pathotypes—that is, new combinations of identifiable virulence genes—may arise monotopically or polytopically. A monotopic origin implies that the new pathotype came into being only once, a rare event indeed, and then spread. A polytopic origin implies that the pathotype has come into being several times, in several places, the events being mutually independent but governed by similar or convergent selection pressures. The problems are twofold. One is the word new, which only means hitherto undetected; detection capability is poor indeed. The other is in the precision of identification. How can one be sure that two different isolates are genetically identical? Brushing doubts aside, there are two types of evidence, single-pathotype evidence and multiple-pathotype evidence.

C. SINGLE-PATHOTYPE EVIDENCE

Single-pathotype evidence is worth considering in special cases. Race 77 of wheat leaf rust is virulent to all standard differentials. Detected at an early date, its occurrence throughout the years was erratic. It appeared massively in eastern Europe in the 1960s once the resistance gene *Lr3* became widely used, mainly through Bezostaya 1 (e.g., Ionescu-Cojacaru and Negulescu, 1974). The introduction of gene *Lr26* (on the rye translocation, to be discussed further) through Aurora, Kavkaz, Clement, and other cultivars provides an even clearer case. The 77 RT pathotype moved west! Did the new pathotype originate monotopically or polytopically, on Aurora and Kavkaz in the East, and on Clement in the West? Briefly, in one of the best defined cases of European rust history the evidence based on criterion 10 is inconclusive.

Zadoks' (1961) conclusions on long distance dispersal of stripe rust on wheat are safer. Based mainly on criterion 10, they are strongly supported by other observations according to criteria 1, 2, 8, and 9. In hindsight, the inference on the direction of dispersal in Spain may be wrong.

D. MULTIPLE-PATHOTYPE EVIDENCE: NURSERY DATA

Inferences on dispersal, or on the lack of dispersal, may be drawn from multiple-pathotype evidence, that is, evidence based on most or all pathotypes of an area. There are two sources of information: (1)

infection spectra from disease testing nurseries, and (2) frequency distributions of races.

Infection spectra from disease testing nurseries can be suggestive of common inoculum. In the 1950s, the infection spectra of wheat nurseries in southern France and Algeria were said to be so similar that they indicated exchange of leaf rust across the Mediterranean, over 600 km (P. J. R. Chevallier, personal communication). Although the inference is probably correct, the evidence is weak. Using modern technology, beginning with multilocational testing of specially designed nurseries, such as the International Wheat Rust Nurseries of the U.S. Department of Agriculture, followed by standardized observations, and perfected by computerized data processing (Kampmeijer, 1981), the evidence from disease testing nurseries can be sharpened. It becomes convincing when supported by race identification, as has been done with the International Yellow Rust Trials, which ran in Europe from 1956 up to this date, under various names.

E. MULTIPLE-PATHOTYPE EVIDENCE: RACE FREQUENCY DATA

Race frequency distributions must be handled with care. They are informative if (1) annual sampling is done in a standardized way, (2) the sampling is absolutely random, (3) the number of samples is large (hundreds to thousands per country), and (4) the handling of the samples is impeccable. None of the European sampling schemes satisfies these requirements.

The only case of random sampling (stratified sample computer-drawn from all fields stored in a data bank) on record comes from Great Britain (James, 1969), and the samples were not used for race identification in any systematic way. Roadside sampling, as described by Zadoks (1966a), is next best, but may be biased as to cultivars grown and chemical treatments applied. Records on methods are usually deficient as to the operational definition of a sample, the number of isolates per sample, their transportation, handling, storage, and so on. The differential hosts used at different sites were not always from the same seed source, and the conditions of testing may vary between seasons and places.

Nevertheless, inferences can be made. Obviously, the racial spectrum of wheat stem rust from a barberry infested site is not identical to that from an agricultural area at large (Table II,a). The differences are significant. At the barberry sites, races 14 and 21, thought to be long-distance races, are underrepresented. Contrarily, there need not be a significant difference between the race spectra of a country over two successive years (Table II,b and c), The example demonstrates that the acceptance of the H_0 hypothesis (i.e., there is no difference between the two race spectra) depends rather on the threshold value of P that is chosen. Two neighboring countries within one agroclimatic zone, such as Spain–Portugal (Table II,d) may already differ considerably, even if three races are chosen that are present in both racial spectra. Of course, the observed difference could be due to differences in methods and materials between the two stations. When two race spectra from different epidemiological zones are compared, such as Spain and Hungary, the test statistic (in this case χ^2) becomes excessively high, indicating very significant differences between the infection spectra (Table II,e).

The results suggest that selected information can be handled. Caution is needed to avoid misinterpretation. Nonparametric tests, such as the chi-square test used here, can guide the interpretation. The result from Table II,e, supposedly means that the two zones Spain and Hungary are epidemiologically different. There is no clue as to what the nature of the difference is. It can be due to differences between pathodemes, pathotypes, climatological conditions, or a combination of these factors. The large differences found here can hardly be due to differences in materials and methods between testing stations. Because at the time of testing specific resistances were rare, and since both areas are large and have a continental climate, the result points toward nondispersal due to geographical separation. This may also be true with wheat leaf rust (Salazar and Brañas, 1973a,b; Negulescu and Cojocaru-Ionescu, 1973). Geographical distance is, however, not an absolute barrier, and in due time a pathotype from southeastern Europe may appear in southwestern Europe, or vice versa, by convergent evolution or by physical transportation.

To what extent the analysis of race frequency distributions can become an analytical tool in epidemiology remains to be seen. A critical reserve is advocated here.

F. SPORE TRAPPING AND AIR TRAJECTORIES

Although distant spore dispersal was suggested by Klebahn in 1904, rust spore trapping work in Europe has hardly been done in a systematic way, when all relevant criteria are taken into account. Little or no rain sampling (Rowell and Romig, 1966) has been done.

The best evidence is provided by a British and an Iberian group.

Santiago (1955, 1962) and Salazar and Brañas (1963) related stem rust trapping work to wind direction and to crop and rust phenology in the target area, thus satisfying in part criteria 6, 7, 8, and 9. They concluded that stem rust came from northwestern Africa. They seem to have avoided the pitfall of sampling local inoculum instead of inoculum coming in over a long distance, but absolute certainty cannot be given. The British group (Hirst *et al.*, 1967a,b; Hogg, 1961) definitely avoided that pitfall, and completed the evidence by back-tracking the trajectories of air masses with and without rust and by intercepting rust spores at high altitudes by means of aircraft-carried spore traps, thus satisfying criteria 4 and 5.

Interesting but fragmentary information was provided by Hermansen (DK) and co-workers, trapping urediospores at places where no cereals were grown (Faroë Islands) (Hermansen and Wiberg, 1972) and at isolated points on the windward side of land masses with cereal cultivation. They concluded that barley leaf rust arriving at the infertile wind-beaten Faroë Islands must have come from western Europe (criteria 6, 7, 8, 9) and that barley leaf rust races new to Denmark have come from Great Britain and Northern Ireland or the Netherlands (criteria 6 through 10). Kølpin Ravn (1906, ex Hermansen, 1973) found stem rust on islands in the Kattegat (between Denmark and Sweden), where no barberry grew (distance from mainland ~50 km, criterion 2). Hermansen *et al.* (1965, 1967) caught urediospores in the air above Denmark at 200–1000 m altitude, but they were not sure that these were cereal rust spores. Buus-Johansen *et al.* (1973) exposed detached barley leaves to air flow at altitudes up to 1500 m and caught *P. hordei* and *P. striiformis* (criterion 5). Hermansen *et al.* (1976) drew trajectories for spore catches of *P. hordei* and *P. recondita* at the North Sea coast in Denmark and pointed to southern Great Britain and the northern Netherlands as possible sources (criteria 4 and 6).

On the basis of extensive field studies (criterion 10), Hermansen (1968, 1973) comes to the interesting conclusion that in Denmark the initial inoculum of *P. graminis, P. coronata, P. hordei,* and *P. striiformis* of barley is almost entirely of exodemic origin.

G. Wind-Pattern Studies

Several studies apply criteria 8 and 9 in combination with general circulation patterns or wind-direction frequency studies. The conclusions of such studies are suggestive, but the inferences made may be incorrect. Savulescu (1953, Fig. 48, p. 149) suggested that rusts were

Table II

Analyses of Race Spectra of Stem Rust by Means of the Chi-Square Test[a]

a. Data from Spain, 1971[b]

Races	2	14	21	24	75	102	111	133	166	186	Σ
B (close to barberry)	0	10	10	3	1	0	0	2	0	3	29
$\bar{B}$ (away from barberry)	1	2	0	0	0	5	4	0	4	2	18
$N = 47$	$df = 9$				$\chi^2 = 35$				$P < 0.001$		

b. Data from Spain, 1968 and 1969[b]

Races	14	17	21	23	24	75	133	176	186	283	Σ
1968	3	1	12	1	0	2	5	1	1	1	27
1969	2	2	11	0	2	0	6	0	0	0	23
$N = 50$	$df = 9$				$\chi^2 = 8.4$				$0.30 < P < 0.50$		

c. Data from Spain, 1969 and 1971[b]

Races	14	17	21	—	24	75	133	—	186	—	Σ
1969	2	2	11	—	2	0	6	—	0	—	23
1971	10	0	10	—	3	1	2	—	3	—	29
$N = 52$	$df = 6$				$\chi^2 = 13$				$0.025 < P < 0.050$		

d. Data from two neighbor countries, 1961

Races		14	21	186	Σ
Spain[c]		95	158	354	607
Portugal[d]		7	31	20	58
$N = 665$	$df = 2$	$\chi^2 = 20$		$P < 0.001$	

e. Data from 1961

Races	10	11	14	15	16	17	21	24	34	40	75	122	133	186	207	279	Σ
Spain[c]	10	2	95	2	2	2	158	14	0	0	2	50	64	354	39	4	798
Hungary[e]	0	0	233	0	0	17	685	0	6	59	0	0	0	0	0	0	1000
$N = 1798$				$df = 15$						$\chi^2 = 997$							$P << 0.001$

[a] N = total number of samples considered per test. df = degrees of freedom. χ^2 = chi square, a test statistic. P = probability of false rejection of H_0 hypothesis.
[b] From Salazar and Brañas (1973a).
[c] From Salazar and Brañas (1968).
[d] From Santiago and Luna Pais (1968).
[e] From Bocsa (1966).

blown into Romania by winds from various directions. Abdel-Hak *et al.* (1974) suggested that leaf rust came into Egypt from the north side of the Mediterranean basin. These are working hypotheses rather than established facts. One such working hypothesis was that wheat stripe rust in the Netherlands was annually blown in from more southerly countries; Zadoks (1961) showed this hypothesis to be incorrect applying criteria 1, 2, 3 and 6, 7, 9, plus 10.

There is plenty of evidence that the air circulation patterns in the Mediterranean basin permit long-distance exchange of particulate matter. The arrival of Sahara dust in Europe and the Alps is a popular piece of evidence. The point here is not that long-distance dispersal is impossible but that in specific cases the scientific proof of long-distance dispersal is incomplete or inadequate.

V. Overseasoning of Rusts

Apart from resistance, the two major barriers to rust build-up are the crop gap in summer and the winter. These barriers can be circumvented or overcome in different ways, e.g., accessory hosts and alternate hosts. The epidemiologist must always distinguish the incidental and the functional aspects of research findings. If once in a while a wheat race of a rust is isolated from a grass, this may be an incidental result without epidemiological significance. The finding may, however, indicate that this rust resides regularly on grasses and that grasses have a function, maybe even an indispensable function, in the rust's annual cycle.

A. OVERSUMMERING OF *P. STRIIFORMIS* ON WHEAT

The oversummering of *P. striiformis* on wheat was well documented by Zadoks (1961). It passes along the narrow green bridge of self-sown, volunteer wheat in most of northwestern Europe. The length and width of the green bridge vary according to agronomic practices (see Section III,A) and weather. Low summer temperatures and/or heavy summer rainfall shorten the bridge in the Netherlands as they prolong the vegetation period and delay the harvest. As grasses are not accessory hosts in the area, rainfall is crucial. Thunderstorm frequency largely determines the amount of self-sown wheat, and therewith the width of the green bridge. Corbaz (1966) in western Switzerland related the oversummering of *P. striiformis* to the number of rain days during the summer. In the Netherlands, stripe rust practically disappeared during the long, hot, rain-free summer of 1959. Observations from an experimental plot indicated that there was sufficient dew for infection. The green bridge did not function because the grains shattered during harvest did not germinate due to lack of water; they were not plowed into the ground as the dry soil was too hard to plow. These conditions, exceptional in northwestern Europe, must be representative of Mediterranean lowland conditions without irrigation.

Records of positive stripe rust observations made during the crop-free period are rare. Most publications contain only casual remarks or circumstantial evidence. A quantitative pattern of oversummering for larger areas can hardly be made. It seems that the sea polders in the Netherlands are prominent oversummering areas, with inoculum dosages of up to 100 m^2 of sporulating leaf area (for the whole country) at the narrowest point of the green bridge. Next most important are the Belgian sea polders and some Swiss alpine valleys suitable for wheat cultivation. Ranking third are those areas where the summer is relatively mild and moist, due to the proximity of the sea, (Great Britain, Norway, France, Germany, Denmark, Romania, Southern Sweden) or to high altitudes (various mountain areas) in Switzerland (Zadoks, 1961; Corbaz, 1966) but also in Belgium, Czechoslovakia, Italy, Romania, and Yugoslavia. Note that at high altitudes the green bridge is always short. Although in the third category favorable oversummering conditions exist, it should be remembered that this is but one element in the genesis of an epidemic (see Section VII,C).

B. OVERWINTERING OF *P. STRIIFORMIS* ON WHEAT

Overwintering presupposes autumn infection of fall-sown winter wheat. Inoculum is either exodemic or from a local source. When oversummering occurs, local sources are always available. During late summer and early autumn, two contrasting population developments take place, at least in northwestern Europe and in high-altitude locations elsewhere. The rust population increases rapidly on volunteer wheat during the "wheat gap"; however, the amount of volunteer wheat decreases more and more rapidly, not due to poor growth of individual plants, but because of fall plowing. Fall infection of recently emerged winter wheat depends on the whims of the weather.

Overwintering of *P. striiformis* on wheat takes place as urediomycelium in living wheat plants (Zadoks, 1961). Sporulating lesions are killed by frost below about −4°C. Latent lesions survive low temperatures as long as the hosting leaves survive. Under snow cover, the wheat and the stripe rust in it are well protected against frost. In the winter of 1955/1956, a severe frost occurred suddenly in northwestern Europe, and most of the wheat in the Netherlands, Belgium, and Northern France was winterkilled. Stripe rust practically disappeared from the northwestern European scene. The overwintering of stripe rust has puzzled people because so little rust is seen during the winter. To explain overwintering, Eriksson (1910) devised his mycoplasma theory, stating that the cytoplasms of host and parasite merged to the extent that any microscopically visible distinction between the two cytoplasms disappeared. Apparently, Eriksson could not accept his own evidence on long latent periods. Zadoks (1961) observed latency periods of up to 118 days. There is indirect evidence of latency periods up to 150 days (under snow cover).

Stripe rust can sporulate and reinfect in the winter during mild spells of weather, when daytime temperatures are about 5°C. Insight in the population dynamics of rust is needed to understand that a sizeable rust population can survive and reproduce during an ordinary northwestern European winter unseen. Simulation models demonstrate the point (Zadoks, 1971; Zadoks and Rijsdijk, 1971, 1972). In the Netherlands, one lesion of stripe rust per hectare surviving the winter is enough to cause a severe epidemic if the spring season is favorable for rust development; that one lesion may easily escape attention of even the most attentive researcher. [For comparison: in potato late blight, *Phytophthora infestans,* one lesion per 100 hectares is enough to generate a severe epidemic (van der Zaag, 1956)].

Whether the inoculum surviving the winter is a threat depends, again, on the weather conditions and the cultivars used. The 1977 epidemic in Czechoslovakia was due to a mild winter and to susceptible (Yugoslavian) cultivars (Slovenčíková and Bareš, 1978). Prolonged periods of heavy rain are not particularly conducive to stripe rust, as the spores are washed from the plants and the low-placed sporulating leaves become covered by mud, dying prematurely. Prolonged spells of dry easterly winds, which can vary from 1 to 6 weeks, completely inhibite dew formation and thereby reinfection. Note that in stripe rust up to 50% of epidemic growth is due to lesion growth and not to reinfection. Regulatory action to ban winter barley has been taken in Denmark (Stapel and Hermansen, 1968) and in parts of the Netherlands to protect the economically more important spring barley from stripe rust, leaf rust, and/or mildew. In view of changing conditions (yield, price), these bans are under reconsideration presently. There is reason to believe that they have been effective, however.

C. OVERSEASONING OF *P. RECONDITA* ON WHEAT

Field observations (J. C. Zadoks, unpublished) in the Netherlands on volunteer wheat, late tillers, and isolated wheat plants (road sides, barns) demonstrate that *P. recondita* follows the pattern of *P. striiformis.* Quantitative data are lacking. There is no indication nor experimental evidence for or against the role of grasses. Greenhouse experiments (Hassebrauk, 1933) using seedlings of grasses, artifical infection, and highly unnatural environmental conditions (low light intensity and high humidity) are irrelevant to the oversummering problem.

Unpublished experiments show that leaf rust urediospores can survive for several days on the leaf and that they need only a short dew period to germinate and penetrate. Secondary pustules permit leaf rust lesions to survive and sporulate, although at low intensity, for over 2 months (Mehta and Zadoks, 1970). Therefore, leaf rust seems to be well equipped to oversummer. It seems no wonder that leaf rust is a threat in northwestern Europe and is a permanent problem in southern and southeastern Europe, where summer conditions are adverse to stripe rust.

The function of grasses in southern and southeastern Europe as accessory hosts is not well established. Typical wheat leaf rust races have been isolated from grasses, but these records are too rare to serve as evidence for a green grass bridge. It is more probable that where the summer is too dry and too hot for wheat volunteers to exist, the grasses wither and also cannot serve as a host for the rust. Irrigated areas may form the exception, but no observations are available.

Evidence on overwintering of *P. recondita* on wheat is scarce and imprecise, and mainly circumstantial.

D. OVERSEASONING OF OTHER CEREAL RUSTS

Stripe rust of barley follows the same pattern as that of wheat, but late tillers, found in those parts of the fields where ripening is delayed due to excess water or overdose of nitrogen, are relatively more important (J. C. Zadoks, unpublished).

The overseasoning of *P. hordei* on barley is similar to that of stripe rust of wheat (J. C. Zadoks, unpublished). Oversummering is not a problem where volunteer barley and late tillers are available. Migration of the rust between winter and spring barley has been well established in the Netherlands and in Denmark (Hermansen, 1964). Tan (1976) in Germany isolated four races and 15 variants overwintering in a single field of winter barley. Reinfections during winter were found by Simkin and Wheeler (1974). Parlevliet and van Ommeren (1976, 1981) saw that *P. hordei* on volunteer plants declined until March and then increased through reinfection, and that rates of decline and subsequent increase were cultivar specific and unrelated. There was an interaction between the rate of replacement of leaves and the rate of reproduction of the rust. Rates differed according to soil type. Interference by mildew explained differences between cultivars to a large extent.

Table III

Some Accessory Hosts (Grass Hosts) of *Puccinia graminis* f. sp. *tritici*[a]

Species	Country	Reference
Aegilops cylindrica Host.	Greece	Skorda, 1966b
Aegilops kotschyi Boiss.	Israel	Gerechter-Amitai and Wahl, 1966
Aegilops ovata L.	Greece	Skorda, 1966b
Agropyron spp. L.	Greece	Skorda, 1966b
Agropyron panormitanum Parl.	Morocco	Arthaud *et al.*, 1964
Anthoxanthum odoratum L.	Italy	Basile, 1964
Bromus mollis L.	Italy	Basile, 1964
Bromus sterilis L.	Greece	Skorda, 1966b
Calamagrostis spp. Adans	Italy	Basile, 1964
Cynosorus echinatus L.	Italy	Basile, 1964
Dactylis glomerata L.	Italy	Basile, 1964
Dasypyrum hordeaceum P. Candary	Morocco	Arthaud *et al.*, 1964
Gaudinia fragilis (L) P.B.	Italy	Basile, 1964
Hordeum bulbosum L.	Israel	Gerechter-Amitai and Wahl, 1966
Hordeum murinum L.	Israel	Gerechter-Amitai and Wahl, 1966
	Morocco	Arthaud *et al.*, 1964
	Italy	Basile, 1964
	Greece	Skorda, 1966b
Hordeum spontaneum C. Koch	Israel	Gerechter-Amitai and Wahl, 1966
Koeleria phleoides (Vill.) Pers.	Italy	Basile, 1964
Lolium spp. L.	Greece	Skorda, 1966b
Lolium perenne L.	Italy	Basile, 1964
Poa annua L.	Italy	Basile, 1964
Triticum dicoccoides Koern.	Israel	Gerechter-Amitai and Wahl, 1966
Triticum villosum Bieb.	Italy	Basile, 1964
Vulpia ciliata Lk.	Italy	Basile, 1964

[a]This list is by no means complete.

In northern Europe, *P. graminis* on wheat and rye and *P. coronota* on oats follow the oversummering pattern of leaf rust on wheat, but usually they arrive late and build up during summer on late tillers and early volunteers. In the milder parts of southern Europe and northern Africa, oversummering may be possible in the hills, but also in the agricultural areas along roads and waters, and in gardens and marshes (Morocco; Malençon, 1961, 1963). J. C. Santiago (personal communication) raised the issue of fall migration by rusts from north to south, as demonstrated in North America. There is no factual evidence for such fall migration in Europe, but in view of the earlier remarks on circulation patterns in Europe, it may happen incidentally. On a regional scale, stem rust, oversummering in the high valleys of the Pyrenees, might migrate to the south of the Iberian Peninsula, but nothing has been substantiated. As said before, oversummering is possible only when a suitable host is present. Skorda (1974) made the significant remark that mid-autumn breeding plots in Greece were infected prior to wheat emergence in commercial fields.

There is evidence that grasses are functional accessory hosts for wheat stem rust (Table III). Malençon (1963) speaks of a graminicolous phase of the rust (in contrast to its cerealicolous phase) along roadsides in Morocco and, during the summer period, in the Atlas mountains. In the Mediterranean area, races of wheat stem rust have regularly been isolated from grasses, but this fact does not prove the existence of a green grass bridge; it only suggests a possibility. Quantitative research in the field is needed. Where barberries are functional alternate hosts, there may not be a need for a green grass bridge.

VI. Alternate Hosts

The literature is imprecise in indicating the relative importance of alternate hosts of cereal rusts in Europe. The fact that they are found to be infected by cereal rusts and occasionally do produce races virulent to cereals proves a potential, but not necessarily a real, danger to cereals. The question of whether alternate hosts are functional in the annual resurgence of cereal rusts cannot be answered categorically.

The alternate host of *P. striiformis* has not yet been identified. That of *P. hordei* seems to be relatively unimportant. The epidemiological significance of *Rhamnus* spp. for oat crown rust is questionable, in Europe at least (Salazar and Martinez, 1973). Only the alternate hosts of *P. graminis* and *P. recondita* will be discussed from an epidemiological viewpoint.

A. STEM RUST

Historical evidence about the role of barberry is overwhelming, thus only few authors can be quoted. Several barberry species are involved (Table IV). Eriksson and Henning (1896) decisively demonstrated the importance of barberries as a source of the then devastating oat stem rust epidemics in Sweden, now a problem of the past. Rye is no longer an important crop, but stem rust epidemics on rye have been a plague to many smallholders toiling over their poor soils. There are no recent records of large sweeping stem rust epidemics on rye, but there are many indications about the local or regional importance near the formerly ubiquitous *B. vulgaris.* In the southern part of the Netherlands, barberry hedges used to surround the churchyards. When Oort (1941) conducted his investigations, he advised a village priest to replace the barberry hedge by something else. The answer he received went some-

Table IV

Alternate Hosts of *Puccinia graminis* f. sp. *tritici*

Species	Country	Reference
Berberis spp. L.	Central Europe	Hassebrauk, 1967
	Ireland	Prendergast, 1966
	Italy	Basile, 1962
	Wales	Broadbent, 1921
Berberis aetnensis Presl.	Italy	Basile, 1964
Berberis amurensis Rupr.	Bulgaria	K'rzhin, 1977
Berberis bidentata Lehm.	Bulgaria	K'rzhin, 1977
Berberis cretica L.	Greece	Skorda, 1964
Berberis ilicifolia Hort. ex C. Koch	Belgium	K'rzhin, 1977
Berberis lycium Hort. ex C. Koch	Belgium	K'rzhin, 1977
Berberis vulgaris L.	Belgium	K'rzhin, 1977
	Spain	Salazar *et al.*, 1968
	Yugoslavia	Špehar, 1962
	Italy	Basile, 1956, 1964

Berberis ilicifolia Hort. ex C. Koch = *B. aetnensis* Presl.
Berberis lycium Hort. ex C. Koch = *B. vulgaris* L.
Berberis cretica L. = *B. vulgaris* L.

thing like this: "I need the barberry to keep the kids out and what the farmers lose in this life they will gain in the next." Fortunately, the problem has solved itself as rye has been replaced by maize.

For wheat stem rust, the situation is confusing. Where barberry grows in the wild or as hedges, stem rust is found. Often, the rust is largely specialized on grasses (Morocco, Atlas Mountains, Spain). Nevertheless, wheat races are found on and around such barberries (Basile, 1968; Massenot, 1961, 1978; Salazar and Brañas, 1973a; Špehar, 1962). As these wheat rust races are often different from the current races found in commercial fields, such barberries can be regarded as epidemiologically nonfunctional, except perhaps as a source of new genotypes (Salazar and Brañas, 1968; Santiago, 1962; Špehar, 1968; Špehar *et al.*, 1976; Skorda, 1966a,b). There are some records of locally or regionally important barberry-based epidemics, as in Bavaria (Germany; Hinke, 1955), France (Massenot, 1961, 1978), Ireland (Prendergast, 1966), Spain (Salazar *et al.*, 1968), and Yugoslavia (J. C. Zadoks, personal observation; Špehar, 1962). In Alpine and subalpine flood plains, barberry (*B. vulgaris*) is often widespread. Whenever wheat cultivation touches these wild lands, stem rust becomes prominent. This was manifested in Switzerland during World War II, when wartime shortages necessitated wheat growing in areas only marginally suitable for cultivation. In Bavaria (Germany; Hinke, 1955), a Barberry Eradication Regulation was enacted with the goal of creating a barberry-free buffer zone of 500 m between the agricultural areas and the barberry-infested wildlands.

It is appropriate to terminate this section by recounting Hermansen's 1968 narrative of a large but involuntary epidemiologic experiment of around 1800. In several countries of Europe, agrarian reform led to the replacement of commons by enclosures. Where ditches or stone banks could not serve as field separations, hedges had to be planted. In Denmark, it was strongly advocated that *B. vulgaris* be used as the hedge shrub. This recommendation was generally followed. Where the barberry was planted, stem rust epidemics appeared within few years. Officialdom ignored the relation between barberry and rust, but private citizens began their own investigations. One schoolmaster, who had begun a business in selling barberry planting material to supplement his income, ended up proving experimentally, in a way that is still acceptable, that the cereal rust epidemics started from barberry. This praiseworthy man, although ignored and forgotten by science, had an impact on stem rust control. When, nearly one century later, the Danish barberry eradication law was enacted (1904), there was little barberry left to eradicate.

B. LEAF RUST

Mention of the aecial stage of wheat leaf rust is scattered throughout the literature, mainly from southern and eastern Europe. Aecia are found regularly on *Thalictrum* spp., but often the check is omitted to see whether the aeciospores infect wheat. Apparently, the aecial stage is functional at times for graminicolous strains of *P. recondita,* but is it functional for wheat-invading strains?

From northwestern Europe, no evidence about the epidemiological functionality of the alternate host for wheat leaf rust is available. The rust thrives without a sexual stage. In the Italian Alps, away from wheat fields, Sibilia *et al.* (1963) and Basile (1971) found *Thalictrum* spp. and surrounding grasses infected by leaf rust, and isolated race 77. Here, the alternate host could have been functional but was not. Tsikaridze *et al.* (1976) in Georgia (U.S.S.R.) found wheat races, among them race 77, first on *Thalictrum foetidum* and later on nearby wheat. Here, the alternate host seems to have been functional in establishing an epidemic on wheat. Such records are rare (Table V).

Of great interest is the finding by d'Oliveira, amply reported by d'Oliveira and Samborski (1966), that there are two groups of wheat leaf rust, one that produces aecia on *Thalictrum* spp. (family *Ranunculaceae*) and one with aecia on *Anchusa* spp., *Lycopsis* spp., and *Echinospermum* spp. (family *Boraginaceae*). Isolates from both groups may belong to the same race, cannot infect *Isopyrum* (*Leptopyrum*) *fumarioides* (*Ranunculaceae*), and cannot cross-hybridize.

Table V

Alternate Hosts of *Puccinia recondita* f. sp. *tritici*

Species	Country	Reference
Anchusa spp. L.	Portugal	d'Oliveira and Samborski, 1966
Clematis mandschurica Rupr.	U.S.S.R.	Azbukina, 1980
Echinospermum spp. Lehm.	Portugal	d'Oliveira and Samborski, 1966
Lycopsis spp. L.	Portugal	d'Oliveira and Samborski, 1966
Thalictrum spp. L.	Portugal	d'Oliveira and Samborski, 1966
Thalictrum flavum L.	Portugal	d'Oliveira, 1941
Thalictrum foetidum L.	U.S.S.R.	Tsikaridze *et al.*, 1976
Thalictrum laserpitiifolium Willd.	Portugal	d'Oliveira, 1941
Thalictrum minus L.	Portugal	d'Oliveira, 1941
Thalictrum speciosissimum O.	Portugal	d'Oliveira and Samborski, 1966

Circumstantial evidence suggests that all these alternate hosts can be functional in Portugal.

Apparently, the epidemiological importance of the alternate host of *P. recondita* on wheat varies according to location. The alternate host may give rise to new races, occasionally. *Anchusa officinalis* normally serves as an alternate host for the rye-inhabiting form of *P. recondita,* though its epidemiological importance is—again—unknown, but at least in one area it also supports the wheat-inhabiting form of *P. recondita.*

VII. Case Studies

A. LEAF RUST OF WHEAT

Georg Riebesel (1892–1965) worked as a breeder in Salzmuende near Halle (German Democratic Republic). From 1926 onward, he studied wheat × rye hybrids. In the years following World War II he selected three major lines: R 47/51, R 51/52, and S 14/44. In 1952 he moved to Aachen (Federal Republic of Germany); taking seeds with him. The line Riebesel 47/51 became a resistance donor to the cultivars Clement (The Netherlands, 1972) and Nautica (The Netherlands, 1976); Riebesel 51/52 was introduced in Germany under the name Weique. The line Salzmuende 14/44 was a resistance donor to Salzmuender Bartweizen (German Democratic Republic, 1958), Kavkaz (U.S.S.R., ~1972) and Aurora (U.S.S.R., ~1972). There is ample evidence that the originally high level of resistance against all three rusts in the resistance donors and their offspring is based on genes in a partial or complete rye *1R*–wheat *1B* chromosome translocation (Zeller, 1972; Zeller and Sastrosumarja, 1972; Bartoš *et al.*, 1973; Bartoš, 1974).

At least 60 European cultivars contain the *Lr3* gene derived from the cultivar Mediterranean. In other words, there has been an immense *Lr3* pathodeme. It remained resistant until "race 77" reappeared, first in eastern Europe. "Race 77" was widely distributed in 1960. The pathotype moved west (Denmark, 1961; Spain, Federal Republic of Germany, Italy, 1962; Czechoslovakia, 1967) but it did not become frequent in the West. Evidently, the pathotype caused a polyetic epidemic over a large area, but, unfortunately, there are no convincing records on the extent of damage.

The Riebesel resistance, present in at least 10 European cultivars, was good during the period of spread of "race 77." Bartoš (1980) found the subtraces UN 10-14 SaBa (Czechoslovakia, 1968) and UN 13-77

SaBa (Czechoslovakia, 1970), which were virulent to the Riebesel resistance, and by 1972 UN 13-77 SaBa was already the predominant 77-pathotype in Czechoslovakia. In 1973, a subrace of race 77 overcoming the Riebesel resistance became widespread in eastern Europe. This "Riebesel pathotype" has been described in different ways:

Czechoslovakia	1970	Race 77-SaBa (SaBa from Salzmuender Bartweizen)	Bartoš, 1975a,b
Romania	1973	Race 77-73 (73 from year of detection 1973)	Negulescu and Ionescu-Cojocaru, 1974
U.S.S.R.	1973	Race 77/VSK (73) (VSK from cultivars infected: Skorospelka 35, Kavkaz)	Voronkova and Sidorina, 1974

It is not known whether the "Riebesel pathotype" originated monotopically or polytopically. The evidence suggests that the various designations (77-SaBa, 77-73, and 77/VSK) stand for one and the same genotype, which became the dominant 77-subtype in 1973. The new pathotype apparently caused much damage or at least much concern. Later it also appeared in the west (Fig. 4) on Clement, registered in at least six countries of the European Community. In some areas of eastern Europe, where the summer is too hot for the rust to overseason (e.g., Bulgaria), the now susceptible cultivars Kavkaz and Aurora are still grown.

B. STEM RUST OF WHEAT

The hypothesis that stem rust of wheat migrates annually from south to north along two major tracts (Zadoks, 1965; Hogg *et al.*, 1969) has never been seriously contested. The West European Tract follows the Atlantic Coast from Morocco to Great Britain and the Netherlands. The East European Tract might have its origin in Greece or even further south. It splits into two branches, one going northwest along the Danube and the other north along the Karpathians and then fanning out northwest over Poland and Scandinavia and northeast over Ukrainia.

The West European Tract is well documented, but little new evidence became available after its postulation in 1965. There is certainly a "green wave" from south to north, dates of flowering and harvesting in the south being earlier than in the north, as in North America.

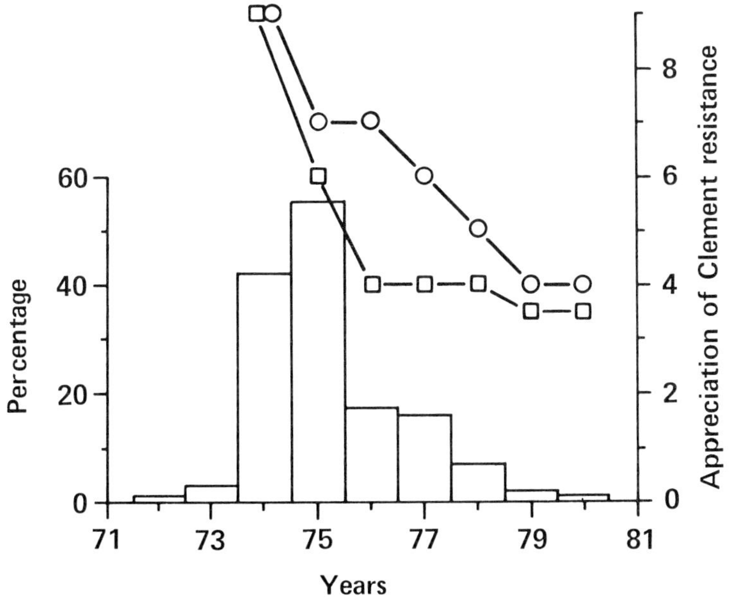

Fig. 4. Prevalence of Clement in the Netherlands and appearance of compatible rust races. Clement derived its resistance from the rye translocation. The appearance of compatible rust races is shown as a decline of the resistance as recorded in the Rassenlijst, the Dutch list of recommended cultivars. Abcissa, years. Vertical: left-hand ordinate and columns, percentage of total wheat hectareage occupied by Clement; right-hand ordinate and lines, appreciation of Clement resistance by Rassenlijst on scale of 0–10 (good). Circles, *Puccinia recondita;* Squares, *Puccinia striiformis.*

Irregularities, however, overshadow the general pattern, which does not apply to sowing dates. The idea of a stem rust epidemic riding the green wave is certainly an oversimplification. It could happen, however, because most of the wheat cultivars along the tract, certainly in its northern part, are susceptible to stem rust. It does happen once in a while, as may be inferred from Massenot's (1978) 1977 data on France. It may explain the single, strong, and early focus found by W. Hoogkamer in 1981 in the southwest Netherlands. Massenot also indicated different sources of inoculum: (1) the Moroccan source discussed earlier, (2) barberry sources in the northern half of Spain, and (3) a few local barberry sources as in Burgundy. Different sources produce epidemics of different types with different racial spectra.

The East European Tract has been postulated on the basis of evidence mainly from the 1932 epidemic. The severe 1951 epidemic in Scandinavia (Denmark, Sweden, Finland), initially difficult to explain, possibly confirms the existence of an East European Tract. Hermansen

(1968) attributed the great impact of the epidemic in Denmark to the lateness of the crops in 1951. He reported references to late crops with stem rust in Bulgaria, therewith indicating a source that was maybe not particularly strong but large. Donchev (1979) states that stem rust in Bulgaria neither oversummers nor overwinters and that barberry, although present, is far from the wheat fields and thus nonfunctional. He supposes that spring infections are exodemic and come from the south, mentioning Greece, Italy, and North Africa. In view of Skorda's (1974) statements, Greece may be a source as well as Turkey. North Africa could be reduced to include only Egypt, where stem rust has its peak appearance in about April. Zambettakis (1966) captured urediospores in Greece and found that stem rust appeared early on the isle of Kreta and in Attica and the Peloponnesus. He states that on Kreta and in Turkey the uredial stage overwinters on grasses, and that urediospores can infect Balkan countries in April. According to him, the Balkan races are similar to those in and possibly come from the Near East, but they appear 1 month later (Zambettakis, 1974).

Bartoš (1980) somewhat confirmed the northwestern branch of the East European Tract, stating that in Czechoslovakia stem rust is exodemic. Whereas races 21 and 14 had been dominant for many years, virulence to *Sr5* was predominant in 1972, and this was attributed to the extensive growing of Bezostaya 1 and related cultivars in the source area, the Lower Danube Plain (Bulgaria, Romania).

C. STRIPE RUST OF WHEAT

Stripe rust of wheat is generally thought to be the rust of the Atlantic zone of Europe, but once in a while the rust seemingly changes its behavior. This was the case in some recent outbreaks in the western Mediterranean basin. Serious stripe rust attacks occurred in 1977 in Italy, Tunisia, and eastern Algeria, and in 1978 in Spain. What has happened?

Applying the idea that the rust can be anywhere where the wheat is susceptible and the weather is favorable, provided that the rust gets enough time to spread and develop, weather data have been subjected to a primary analysis. For a number of stations (Table VI), monthly weather data were compared to 30-years means. Data were compacted to 3-month periods, called ASO (August + September + October), NDJ (November + December + January), and FMA (February + March + April). The summer months were disregarded. A clear pattern emerged when the data from the wheat growing seasons 1974–1979 were compared to the long-term average (Fig. 5).

Table VI

Synoptic Weather Stations Studied and Following the Pattern Indicated in the Text

Country	Town	Latitude	Longitude	Altitude (m)
France	Nîmes	43.52° N	04.24° E	62
	Toulouse	43.38° N	01.22° E	153
Italy	Pescara	42.26° N	14.12° E	16
	Roma	41.48° N	12.14° E	3
Spain	Madrid	40.24° N	03.41° W	657
	Sevilla	37.22° N	06.00° W	13
Tunisia	Tunis	36.50° N	10.14° E	5

Epidemics in Italy (Vallega and Zitelli, 1979), Tunisia (Ghodbane, 1977), and eastern Algeria (J. C. Zadoks, personal information) followed the following set of conditions:

1. ASO 1976: relatively low temperature, relatively high precipitation.
2. NDJ 1976/1977: relatively high temperature, relatively high precipitation.
3. FMA 1977: relatively high temperature.

Conditions for (1) could ensure oversummering and fall multiplication, for (2) continued and rapid multiplication during early winter, and for (3) fast multiplication at late winter or early spring time. In Spain, the same set of conditions prevailed, but there is no epidemic on record (C. H. van Silfhout, personal communication). A possible cause might be the hot summer of 1976.

In Italy the wheat cultivar rusted was mainly Irnerio, known to be susceptible to stripe rust. In Tunisia, commonly cultivated semidwarfs such as Soltane and Siete Cerros were severely damaged. In eastern Algeria, "Mexican" types had been grown.

A follow-up in 1978 from Italy showed no stripe rust epidemic, although in nurseries there was severe infection, as in 1977 (spring conditions were also similar). In Tunisia, practically no stripe rust could be found in the main wheat growing area (J. C. Zadoks, personal observation), as a new and resistant Tunisian cultivar was grown. In eastern Algeria, severe stripe rust was found (J. C. Zadoks, personal observation) on isolated fields of "Mexican" types. In southern Spain, however, a severe epidemic occurred on Siete Cerros and Mahissa I, then widely grown, and some other cultivars (Nagarajan *et al.*, 1984).

One hypothesis is that stripe rust had already settled all over Spain

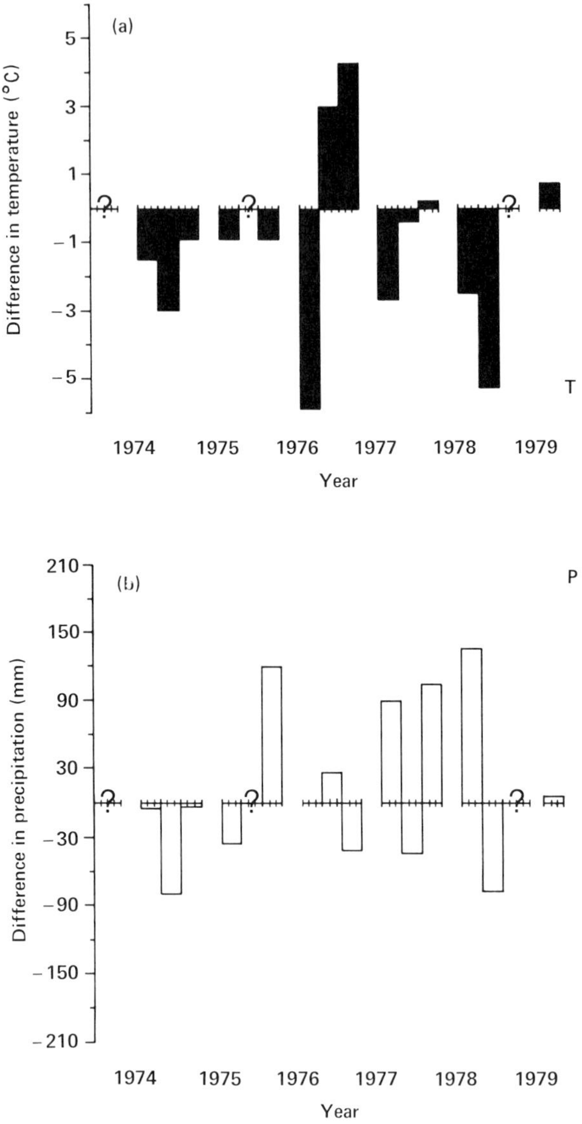

Fig. 5. Temperature and precipitation at Rome (Italy) in the years 1974/1979, shown as deviates from the 30-year mean values. (a) The daily mean temperatures in °C were averaged per month, and the difference between the monthly average observed and the 30-year mean value of the monthly average was determined. For each group of 3 months, these differences were added and shown as a column in the histogram. (b) As in (a) but using daily precipitation (mm) instead of daily mean temperature. ? = No information available. Data by courtesy of the Royal Netherlands Meteorological Institute, De Bilt.

in 1977, although at low severities. The summer of 1977 was relatively cool, thus favoring local oversummering at scattered locations. The winter of 1977/1978 was relatively warm and wet: conditions permitting rapid multiplication of the rust. The spring of 1978 was relatively warm and wet, at least in the Madrid area. It is well known in Spain that wet years are years with high yield potential and high rust potential (J. Salazar, personal communication).

The correlation between Spain and eastern Algeria in 1978 inasfar as "Mexican" wheats were grown, and the contrast between Spain and Tunisia are striking. The stripe rust epidemics in the western Mediterranean basin were not necessarily due to new rust races, as conventional races of the "French" type were identified (40 E 136 and 41 E 136; R. W. Stubbs, personal communication) along with North African types (38 E 16) of relatively recent detection. Siete Cerros, the main wheat responsible for the 1978 epidemic in Spain, has a general susceptibility to stripe rust according to field test in the Netherlands with four Dutch field races (Nagarajan *et al.*, 1984).

VIII. Final Remarks

Our picture of the large-scale epidemiology of wheat rusts in Europe is far from complete. Rarely have the methodological requirements of Section IV, A been satisfied in such a way that sound proof could be given for any statement. Most evidence was circumstantial, and not too good. This is particularly true for the retrospective epidemiological studies reported here. Nevertheless, these retrospective studies reinterpreting data from the literature have contributed most to recent knowledge. For future progress, more complete documentation also of haphazard or stray observations is needed. Real progress, however, can only be made by well-designed international recording schemes and experiments. Let us hope that the future will bring an economic tide more favorable to such concerted efforts.

References

Abdel-Hak, T. M., El Shehedi, A. A., and Nazim, M. (1974). The source of inoculum of wheat leaf rust in relation to wind directions in Egypt. *Egypt. J. Phytopathol.* **6,** 17–25.

Arthaud, J., Guyot, L., and Malençon, G. (1966). Comparative biometric studies of the formae speciales of black rust (*Puccinia graminis* Pers.) living on the wild grasses of the Moroccan Atlas. *Proc. Cereal Rusts Conf., 1964,* pp. 204–206.

Azbukina, Z. (1980). Economical importance of aecial hosts of rust fungi of cereals in the Soviet Far East. *Proc. Eur. Mediterr. Cereal Rusts Conf., 5th, 1980,* pp. 199–201.

Bartoš, P. (1974). On additional genes for stem- and leaf rust resistance in some European wheat cultivars possessing resistance derived from rye. *Cereal Rusts Bull.* **2,** 10–11.

Bartoš, P. (1975a). On the presence of the gene *Sr5* in some European cultivars. *Cereal Rusts Bull.* **3,** 27–28.

Bartoš, P. (1975b). Physiologic specialization of wheat leaf rust (*Puccinia recondita*) in Czechoslovakia in the years 1970–1975. *Ved. Pr. Vysk. Ustavu Rastl. Vyroby Praha-Ruszyne* **20,** 7–12.

Bartoš, P. (1980). Physiologic specialization of stem rust and leaf rust of wheat in Czechoslovakia in the last 17 years. *Cereal Rusts Bull.* **8,** 9–11.

Bartoš, P., Valkoun, J., Košner, J., and Slovenčíková, V. (1973). On the genetics of rust resistance of the wheat cultivar Kavkaz. *Cereal Rusts Bull.* **1,** 27.

Basile, R. (1956). Razze fisiologiche di *Puccinia graminis tritici* Erikss. et Henn. isolate da ecidioconidi di *Berberis vulgaris* raccolto durante l'estate 1956, in zone alpine. *Boll. Stn. Patol. Veg.* **14,** 183–188.

Basile, R. (1962). Alcune razze fisiologiche di *Puccinia graminis* var. *tritici* identificate da ecidioconidi di *Berberis* raccolto in Italia nel 1961. *Boll. Stn. Patol. Veg.* **19,** 1–6.

Basile, R. (1964). Razze fisiologiche di ruggini identificate da ecidioconidi prelevati da *Berberis aetnensis* Presl. e *Berberis vulgaris* e da uredoconidi di *Puccinia graminis* var. *tritici* Erikss. et Henn. e di *P. recondita* Rob. provenienti da Graminacee spontanee raccolto in Italia durante gli anni 1962 e 1963. *Phytopathol. Mediterr.* **3,** 79–85.

Basile, R. (1968). Prevalence and distribution of the most important physiologic races of *Puccinia graminis* var. *tritici* in the various regions of Italy during the period 1953–1965. *Proc. Cereal Rusts Conf., 1966,* pp. 104–107.

Basile, R. (1971). Prevalenza e distribuzione delle più importanti razze fisiologiche di *Puccinia recondita* var. *tritici* nelle diverse regioni Italiane negli anni 1953–1965. *Ann. Ist. Sper. Patol. Veg.* **11,** 39–68.

Bocsa, E. (1966). Physiologic specialization of brown and black rust of wheat in Hungary. *Proc. Cereal Rusts Conf., 1964,* pp. 321–323.

Broadbent, W. H. (1921). Report on barberry and the black rust of wheat survey in SW. Wales. *J. Minist. Agric. (G.B.)* **28,** 117–123.

Broekhuizen, S. (1969). "Agro-ecological Atlas of Cereal Growing in Europe. II. Atlas of the Cereal-growing Areas in Europe." Pudoc, Wageningen/Elsevier, Amsterdam.

Broennimann, A. (1966). Yellow rust problems in Switzerland. *Proc. Cereal Rusts Conf., 1964,* pp. 77–81.

Buus-Johansen, H., Hermansen, J. E., and Westenbaek-Hansen, H. (1973). Trapping of powdery mildew conidia and urediospores at Gedser Reef Lightship and live spores in the upper air. *Årsskr.–Vet.- & Landbohoejsk. (Copenhagen),* pp. 86–94.

Corbaz, R. (1966). Notes sur la rouille jaune du froment en Suisse romande [*Puccinia glumarum* (Schmidt) Eriksson et Henning]. *Phytopathol. Z.* **56,** 40–53.

Darwinkel, A. (1980). Grain production of winter wheat in relation to nitrogen and diseases. I. Relationship between nitrogen dressing and yellow rust infection. *Z. Acker- Pflanzen-bau* **149,** 299–308.

de Bary, A. (1865). Neue Untersuchungen ueber die Uredineen, insbesondere die Entwicklung der *Puccinia graminis* und den Zusammenhang derselben mit *Aecidium berberidis. Monatsber. K. Preuss. Akad. Wiss.,* pp. 15–50.

Della, A. (1978). Cereal rusts in Cyprus in 1976/77. *Cereal Rusts Bull.* **6,** 21.
d'Oliviera, B. (1941). Notes on the production of the aecidial stage of some cereal rusts in Portugal. *Rev. Appl. Mycol.* **20,** 353–354.
d'Oliviera, B., and Samborski, D. J. (1966). Aecial stage of *Puccinia recondita* on *Ranunculaceae* and *Boraginaceae* in Portugal. *Proc. Cereal Rusts Conf., 1964,* pp. 133–150.
Donchev, N. (1979). The leaf rust problem on wheat in Bulgaria in 1977–78. *Cereal Rusts Bull.* **7,** 16–20.
Ellen, J., and Spiertz, J. H. J. (1980). Effects of rate and timing of nitrogen dressings on grain yield formation of winter wheat (*T. aestivum* L.). *Fert. Res.* **1,** 177–190.
Eriksson, J. (1910). Ueber die Mykoplasmatheorie, ihre Geschichte und ihren Tagesstand. *Biol. Centralbl.* **30,** 618–623.
Eriksson, J., and Henning, E. (1896). "Die Getreideroste, Ihre Geschichte und Natur sowie Massregeln gegen dieselben." Norstedt, Stockholm.
Fontana, F. (1932). Observations on the rust of grain. *Phytopathol. Classics* **2,** 1–40.
Food and Agriculture Organization (FAO) (1978). "Production Yearbook," Vol. 32. Food Agric. Organ. U.N., Rome.
Gerechter-Amitai, Z. K., and Wahl, I. (1966). Wheat stem rust on wild grasses in Israel. Role of wild grasses in the development of the parasite and in breeding for resistance. *Proc. Cereal Rusts Conf., 1964,* pp. 207–217.
Ghodbane, A. (1977). Stripe rust on wheat. *FAO Plant Prot. Bull.* **25,** 212.
Hassebrauk, K. (1933). Gräserinfektionen mit Getreiderosten. *Arb. Biol. Reichsanst. Land- Forstwirtsch., Berlin-Dahlem* **20,** 165–182.
Hassebrauk, K. (1959). Zur physiologischen Spezialisierung des Weizenbraunrostes (*Puccinia recondita* Rob. = *P. triticina* Erikss.) in Deutschland und andere europaeischen Staaten im Jahre 1958. Kritische Bemerkungen zur Methodik der Rassenbestimmung. *Nachrichtenbl. Dtsch. Pflanzenschutzdienstes (Stuttgart)* **11,** 43–45.
Hassebrauk, K. (1967). Zur Epidemiologie des Schwarzrostes in Mitteleuropa. *Phytopathol. Z.* **60,** 169–176.
Hermansen, J. E. (1964). Notes on the appearance of rusts and mildew on barley in Denmark during the years 1961–1963. *Acta Agric. Scand.* **14,** 33–51.
Hermansen, J. E. (1968). Studies on the spread and survival of cereal rust and mildew disease in Denmark. *Friesia* **8,** 161–359.
Hermansen, J. E. (1973). Aspects of the history of aerobiology in Denmark. *Bull. Ecol. Res. Comm., NFR* **18,** 11–16.
Hermansen, J. E., and Wiberg, A. (1972). On the appearance of *Erysiphe graminis* f. sp. *hordei* and *Puccinia hordei* in the Faroës and the possible primary sources of inoculum. *Friesia* **10,** 30–40.
Hermansen, J. E., Buus-Johansen, H., Westenbaek-Hansen, H., and Carstensen, P. (1965). Notes on the trapping of powdery mildew conidia and urediospores by aircraft in Denmark in 1964. *R. Vet. Agric. Univ., Yearb., Copenhagen,* pp. 121–129.
Hermansen, J. E., Buus-Johansen, H., and Westenbaek-Hansen, H. (1967). A method of trapping live powdery mildew conidia and urediospores in the upper air. *R. Vet. Agric. Univ., Yearb., Copenhagen,* pp. 77–81.
Hermansen, J. E., Torp, U., and Prahm, L. P. (1976). Evidence of long-distance dipersal of live spores of *Puccinia hordei* and *P. recondita* f. sp. *tritici. Cereal Rusts Bull.* **4,** 31–35.
Heuzé, G. (1872). "Les plantes alimentaires," Vol. 1. Librairie Agricole de la Maison Rustique, Paris.
Hinke, F. (1955). Bekaempfungsmassnahmen gegen die Berberitze zur Verhuetung von Schwarzrostschaeden an Getreide. *Pflanzenschutz* **12,** 1–4.

Hirst, J. M., and Hurst, G. W. (1967). Long-distance spore transport. *In* "Airborne Microbes" (P. H. Gregory and J. L. Monteith, eds.), pp. 307–344. Cambridge Univ. Press, London and New York.

Hirst, J. M., Stedman, O. J., and Hogg, W. H. (1967a). Long distance spore transport: Methods of measurement, vertical spore profiles and the detection of immigrant spores. *J. Gen. Microbiol.* **48,** 329–355.

Hirst, J. M., Stedman, O. J., and Hurst, G. W. (1967b). Long distance spore transport: Vertical sections of spore clouds over the sea. *J. Gen. Microbiol.* **48,** 357–377.

Hogg, W. H. (1961). Meteorology in relation to recent black rust epidemics. *Trans. Br. Mycol. Soc.* **44,** 137–138.

Hogg, W. H., Hounam, C. E., Mallik, A. K., and Zadoks, J. C. (1969). Meteorological factors affecting the epidemiology of wheat rusts. *WMO, Tech. Note* **99,** 1–143.

Ionescu-Cojocaru, M. (1979). Leaf and stem rust of wheat in Romania in 1978. *Cereal Rusts Bull.* **7,** 21–22.

Ionescu-Cojocaru, M., and Negulescu, F. (1974). The complementary effect of at least two genes for resistance to *Puccinia recondita* f. sp. *tritici* operating in the winter wheat cultivars Aurora and Kaukaz. *Cereal Rusts Bull.* **2,** 16–18.

James, W. C. (1969). A survey of foliar diseases of spring barley in England and Wales in 1967. *Ann. Appl. Biol.* **63,** 253–263.

Kampmeijer, P. (1981). Differ: A procedure to find new differential varieties in large cultivar-isolate reaction matrices. *Cereal Rusts Bull.* **9,** 9–13.

Klebahn, H. (1904). "Die wirtswechselnden Rostpilze: Versuch einer Gesamtdarstellung Ihrer biologischen Verhaeltnisse." Borntraeger, Berlin.

K'rzhin, Kh. (1977). New races of *Puccinia graminis* f. sp. *tritici* Erikss. et Henn. isolated from aecidia in Bulgaria. *Rastenievud. Nauki* **14,** 118–123.

Loegering, W. Q., and Browder, L. E. (1971). A system of nomenclature for physiologic races of *Puccinia recondita tritici. Plant Dis. Rep.* **55,** 718–722.

Malençon, G. (1961). Epidémiologie de la rouille noire des céréales au Maroc. *C.R. Seances Acad. Agric. Fr.* **47,** 587–594.

Malençon, G. (1963). Les données écologiques du développement estival da la rouille noire (*Puccinia graminis* sens. lat.) dans le Moyen Atlas. *C.R. Séances Acad. Agric. Fr.* **49,** 279–291.

Massenot, M. (1961). Epidémiologie de la rouille noire des céréales en France. *C.R. Séances Acad. Agric. Fr.* **47,** 594–600.

Massenot, M. (1978). Changes in the race composition of *Puccinia graminis* f. sp. *tritici* in France, in 1977. *Cereal Rusts Bull.* **6,** 14.

Mehta, Y. R., and Zadoks, J. C. (1970). Uredospore production and sporulation period of *Puccinia recondita* f. sp. *tritici* on primary leaves of wheat. *Neth. J. Plant. Pathol.* **76,** 267–276.

Nagarajan, S., Kranz, J., Saari, E. E., Seiboldt, G., Stubbs, R. W., and Zadoks, J. C. (1984). An analysis of the 1978 epidemic of yellow rust on wheat in Andalucia, Spain. *Phytopathol. Z.* **91,** 159–170.

Negulescu, F., and Ionescu-Cojocaru, M. (1973). Physiologic races of *Puccinia recondita tritici* and *Puccinia graminis tritici* occurring on wheat in Romania during 1968–1970. *Cereal Rusts Bull.* **1,** 35–37.

Negulescu, F., and Ionescu-Cojocaru, M. (1974). The outbreak of a new form of race 77 of *Puccinia recondita* f. sp. *tritici* on wheat cultivar Aurora in Romania in 1973. *Cereal Rusts Bull.* **2,** 19–22.

Oort, A. J. P. (1941). Is de berberis een gevaar voor de graancultuur? *Neth. J. Plant Pathol.* **47,** 112–119.

Parlevliet, J. E. (1975). Partial resistance of barley to leafrust, *Puccinia hordei.* I. Effect of cultivar and development stage on latent period. *Euphytica* **24,** 21–27.

Parlevliet, J. E. (1976). Partial resistance of barley to leaf rust, *Puccinia hordei.* III. The inheritance of the host plant effect on latent period in four cultivars. *Euphytica* **25,** 241–248.

Parlevliet, J. E., and van Ommeren, A. (1976). Overwintering of *Puccinia hordei* in the Netherlands. *Cereal Rusts Bull.* **4,** 1–4.

Parlevliet, J. E., and van Ommeren, A. (1981). The development of the leaf rust, *Puccinia hordei,* population during winter, spring and early summer in 11 winter barley cultivars. *Neth. J. Plant Pathol.* **87,** 131–137.

Plowright, C. B. (1889). "A Monograph of the British Uredineae and Ustilagineae." Kegan Paul, Trench & Co., London.

Prendergast, A. G. (1966). Black stem rust in Ireland in 1963. *Proc. Cereal Rusts Conf., 1964,* pp. 183–186.

Rijsdijk, F. H., and Zadoks, J. C. (1976). Assessment of risks and losses due to cereal rusts in Europe. *Proc.—Eur. Mediterr. Cereal Rusts Conf., 4th, 1976,* pp. 60–62.

Robinson, R. A. (1976). "Plant Pathosystems." Springer-Verlag, Berlin and New York.

Rowell, J. B., and Romig, R. W. (1966). Detection of urediospores of wheat rusts in spring rains. *Phytopathology* **56,** 807–811.

Salazar, J., and Brañas, M. (1963). Five years' results of black rust spore counting. *Epidemiol. Biometeorol. Fungal Dis. Plants, Proc. NATO Adv. Study Inst., 1963,* 1 p.

Salazar, J., and Brañas, M. (1968). Physiologic specialization of *Puccinia graminis* f. sp. *tritici* in Spain (years 1961–1967). *Proc. Cereal Rusts Conf., 1966,* pp. 99–101.

Salazar, J., and Brañas, M. (1973a). Physiologic races of wheat black rust (*Puccinia graminis* Pers. var. *tritici* Eriks. et Henn.) detected in Spain in the years 1968–1971. *Cereal Rusts Bull.* **1,** 21–23.

Salazar, J., and Brañas, M. (1973b). Physiologic races of *Puccinia recondita* f. sp. *tritici* detected in Spain, years 1968–1971. *Cereal Rusts Bull.* **1,** 28–29.

Salazar, J., and Martinez, M. (1973). Physiologic races of crown rust (*Puccinia coronata* Cda. f. sp. *avenae* Erikss) detected in Spain during the period 1969–1971. *Cereal Rusts Bull.* **1,** 19–20.

Salazar, J., Alonso, M., and Brañas, M. (1968). Epidemiological observations on wheat stem rust. *Proc. Cereal Rusts Conf., 1966,* pp. 34–35.

Santiago, J. C. (1955). Epidemiology of wheat stem rust in Portugal. *Agron. Lusit.* **17,** 275–295.

Santiago, J. C. (1962). Estudos de epidemiologia e de especialização fisiológica da ferrugem negra do trigo. *Melhoramento* **15,** 7–89.

Santiago, J. C., and Luna Pais, A. (1968). Evolution of physiologic specialization of the wheat stem rust fungus in Europe during the period 1958–1963. *Proc. Cereal Rusts Conf,, 1966,* pp. 303–317.

Savulescu, T. (1953). "Monografie uredinalelor din Republica Populara Romana." Ed. Acad. Rep. Dep. Romȃne.

Sibilia, C., Basile, R., and Boskovic, M. (1963). Ricerche su *Thalictrum* sp. condotte nei pascoli alpini nel 1962 e 1963. *Boll. Stn. Patol. Veg.* **21,** 93–97.

Simkin, M. B., and Wheeler, B. E. J. (1974). Overwintering of *Puccinia hordei* in England. *Cereal Rusts Bull.* **2,** 2–4.

Skorda, E. A. (1966a). Physiologic specialization of wheat stem rust in Greece. *Proc. Cereal Rusts Conf., 1964,* pp. 177–182.

Skorda, E. A. (1966b). Studies on the physiologic races of the wheat stem rust (*Puccinia graminis tritici*) in Greece during the eight year 1955–1962. *Ann. Inst. Phytopathol. Benaki* [N.S.] **7,** 157–176.

Skorda, E. A. (1974). Physiologic races of *Puccinia graminis* var. *tritici* in Greece during the period 1963–1969. *Cereal Rusts Bull.* **2,** 7–9.

Slovenčiková, V., and Bareš, I. (1978). Epidemic of stripe rust in Czechoslovakia in 1977. *Cereal Rusts Bull.* **6,** 15–18.

Špehar, V. (1962). The influence of *Berberis vulgaris* and spontaneous gramineae on the occurrence and intensity of the attacks of black rust of wheat (*P. graminis tritici*) in Croatia. *Boll. Stn. Patol. Veg.* **20,** 133–138.

Špehar, V. (1968). Physiologic races of stem rust in Western Yugoslavia and their importance as early foci of new races for Europe. *Proc. Cereal Rusts Conf., 1966,* pp. 65–68.

Špehar, V., Vlaković, V., and Ković, B. (1976). The role of *Berberis* spp. on appearance of virulent physiological races of *Puccinia graminis* f. sp. *tritici*. *Proc. Cereal Rusts Conf., 1976,* pp. 63–67.

Stapel, C., and Hermansen, J. E. (1968). Growing of winter barley forbidden in Denmark. *Tidskr. Landoekonomi* **155,** 218–230.

Tan, B. H. (1976). Recovery and identification of physiologic races of *Puccinia hordei* from winter barley. *Cereal Rusts Bull.* **4,** 36–39.

Targioni Tozzetti, G. (1952). True nature, causes and sad effects of the rust, the bunt, the smut, and other maladies of wheat and of oats in the field. *Phytopathol. Classics* No. 9, pp. 1–159.

Thran, P., and Broekhuizen, S. (1965). "Agro-ecological Atlas of Cereal Growing in Europe. I. Agro-climatic Atlas of Europe." Pudoc, Wageningen/Elsevier, Amsterdam.

Tsikaridze, O. N., Tsereteli, G. H., and Gogava, T. I. (1976). Role of *Thalictrum foetidum* in the renewal of *Puccinia recondita* infection in spring. *Soobshch. Akad. Nauk. Gruz. SSR* **81,** 713–715.

Vallega, V., and Zitelli, G. (1979). Epidemics of yellow rust on wheat in Italy. *Cereal Rusts Bull.* **6,** 17–22.

Vanderplank, J. E. (1963). "Plant Diseases: Epidemics and Control." Academic Press, New York.

van der Zaag, D. E. (1956). Overwintering and epidemiology of *Phytophthora infestans,* and some new possibilities of control. *Neth. J. Plant Pathol.* **62,** 89–156.

Voronkova, A. A., and Sidorina, L. I. (1974). Genetic characteristics of race 77, which induced a severe epiphytic of leaf rust in the rust-resistant wheat varieties Avrora and Kavkaz. *Sov. Genet. (Engl. Transl.)* **10,** 1195–1199.

Zadoks, J. C. (1959). Cereal and grass rusts of the Netherlands. *Robigo* **8,** 1–3.

Zadoks, J. C. (1961). Yellow rust on wheat. Studies in epidemiology and physiologic specialization. *Neth. J. Plant Pathol.,* **67,** 69–256.

Zadoks, J. C. (1965). Epidemiology of wheat rusts in Europe. *FAO Plant Prot. Bull.* **13,** 97–108.

Zadoks, J. C. (1966a). On the dangers of artificial infection with yellow rust to the barley crop in the Netherlands; a quantitative approach. *Neth. J. Plant Pathol.* **72,** 12–19.

Zadoks, J. C. (1966b). Field races of brown rust of wheat. *Proc. Cereal Rusts Conf., 1964,* pp. 92–93.

Zadoks, J. C. (1968). Meteorological factors involved in the dispersal of cereal rusts. *In* "Proceedings of the Regional Training Seminar on Agrometeorology" (A. J. Borghorst, ed.), pp. 179–194. Wageningen, The Netherlands.

Zadoks, J. C. (1971). Systems analysis and the dynamics of epidemics. *Phytopathology* **61,** 600–610.
Zadoks, J. C. (1982). Cereal rusts, dogs, and stars in antiquity. *Garcia de Orta, Sér. Est. Agron., Lisboa* **9,** 13–20.
Zadoks, J. C., and Kampmeijer, P. (1977). The role of crop populations and their deployment, illustrated by means of a simulator, EPIMUL 76. *Ann. N.Y. Acad. Sci.* **287,** 164–190.
Zadoks, J. C., and Rijsdijk, F. H. (1971). A calculated guess of future rust development in cereal crops. *Proc. Indian Natl. Sci. Acad., Part B* **37,** 440–443.
Zadoks, J. C., and Rijsdijk, F. H. (1972). Epidemiology and forecasting of cereal rusts, studied by means of a computer simulator named EPISIM 1972. *Proc. Cereal Rusts Conf., 1972,* pp. 293–296.
Zadoks, J. C., and Schein, R. D. (1979). "Epidemiology and Plant Disease Management." Oxford Univ. Press, London and New York.
Zadoks, J. C., Chang, T. T., and Konzak, C. F. (1974). A decimal code for the growth stages of cereals. *Weed Res.* **14,** 415–521.
Zambettakis, C. (1966). La propagation des spores dans l'air, utilisation d'un capteur-mesures. *Sci. Nat.* **75,** 2–20.
Zambettakis, C. (1974). Influence des facteurs géographiques sur l'épidémiologie des certaines maladies fongiques des plantes. *C. R. Soc. Biogéogr. Séance* **451,** 35–48.
Zeller, F. J. (1972). Cytologischer Nachweis einer Chromosomensubstitution in dem Weizenstamm Salzmünde 14/44 (*T. aestivum*). *Z. Pflanzenzuecht.* **67,** 90–94.
Zeller, F. J., and Sastrosumarja, S. (1972). Zur Cytologie der Weizensorte Weique (*T. aestivum* L.). *Z. Pflanzenzuecht.* **68,** 312–321.

12

Epidemiology in the Indian Subcontinent

S. Nagarajan

Regional Station, Indian Agricultural Research Institute, Simla, India

L. M. Joshi

Division of Mycology and Plant Pathology, Indian Agricultural Research Institute, New Delhi, India

THE CEREAL RUSTS,
Volume II

ISBN 0-12-148402-5

I. Introduction

Archaeological excavations at Mergarh, Baluchistan, of the Indus Basin have revealed several kinds of grain, i.e., *Hordeum distichum, H. vulgare, H. vulgare* var. *nudum, Triticum monococcum, T. dicoccum, T. durum,* and *T. aestivum,* from the period 6000 B.C. (Jarrige and Meadow, 1980). Rig Veda, the oldest Aryan writings of ancient India, which are about 4000 to 5000 years old, dealt with wheat culture (Kohli, 1968), showing that wheat cultivation in the Indian subcontinent occurred very early.

In the Indian subcontinent, wheat (*Triticum aestivum* and *T. durum*) is one of the major winter cereals, and annual production is around 48–50 million metric tons (Table I). Wheat yields are affected by all the three rusts, stem rust (*Puccinia graminis* Pers. f. sp. *tritici* Ericks and Henn), leaf rust (*P. recondita* Rob. ex. Desm. f. sp. *tritici*), and stripe rust (*P. striiformis* Westend), in addition to various other diseases (Joshi *et al.*, 1975b).

Table I

Area and Production of Wheat in the South Asian Nations[a]

Country	Area in 1000 hectares	Production in 1000 metric tonnes
Afghanistan	2,400	2,200
Bangladesh	265	494
Bhutan	65	64
India	22,200	34,982
Nepal	356	415
Pakistan	6,696	9,944
Total	31,982	48,099

[a]From Anonymous (1980).

II. Nature and Recurrence of Wheat Rusts

A. Orography of the Subcontinent

The Indian subcontinent (Fig. 1) is separated from the southern parts of Asia by the massive Himalaya and the Hindukush mountains. The great snow-capped Himalayan arc, with many tall peaks under perpetual snow, stretches over 2500 km, from the gorges of the Indus to Brahmaputra. The Himalaya mountains can be divided into four major areas: (1) the Assam or eastern Himalayas; (2) the central or Nepal Himalayas; (3) the Kumaon or western Himalayas; and (4) the northwest or Punjab Himalayas.

The mean annual 10°C isotherm lies at an elevation of 3960 m in the eastern Himalayas and descends to 3,050 m in northwestern parts (Mani, 1968). Thus, the northwestern Himalayas are cooler than in the east, which is reflected in the survival and recurrence of the cereal rusts. Apart from this, there is a vast difference in temperature at a given height between northern and southern slopes of the mountain. The enormous valleys of the Himalayan chain, running in all directions, permit oversummering of the cereal rusts, when the plains of India pass through the scorching summer. In Nepal valleys, farmers are contented if wheat yields pay the cost of cultivation of rice, which is their main cereal crop. This is reflected in the poor cultivation practices followed in approximately 60% of the area sown to wheat in Nepal (Kayastha, 1970).

South of the Himalayan arc is the Great Indo-Gangetic Plain, where

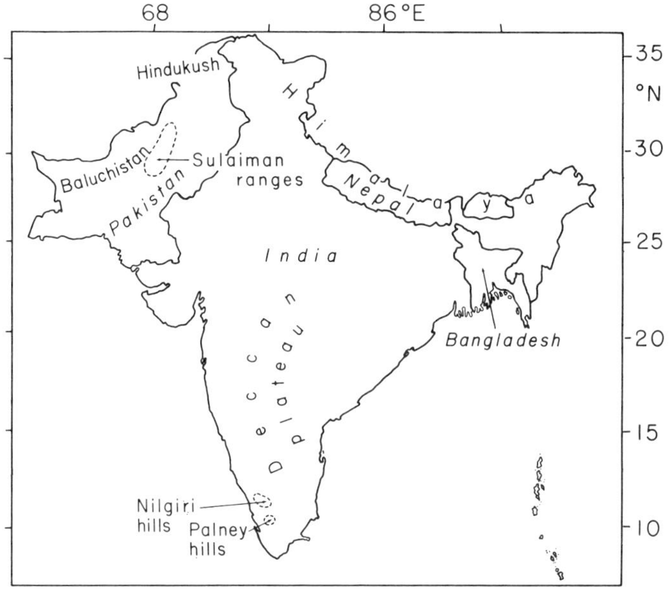

Fig. 1. The Indian subcontinent. (The territorial boundaries of nations shown in the figures are neither accurate nor authentic.)

wheat is grown extensively during winter. This alluvial belt, stretching from around the Indus River to the delta of the Ganges and Brahmaputra, is a continuum. Further down in the Deccan Plateau, the highlands of central and peninsular India, rain-fed bread and durum wheats are grown in the deep black soils. In the sandy Sindh and the arid zone, wherever there is assured irrigation, wheat is grown. In the extreme south, around 10–12°N, are the Nilgiri and Palney hills (> 2000 m), and where wheat is cultivated throughout the year.

B. INDIAN FAMINES

The first stem rust epidemic was recorded in 1786 A.D. in central India, and since then, many epidemics have occurred. Famine and crop disaster in the overpopulated, food deficit nations were predicted to occur during the 1970s (Paddock and Paddock, 1967). Genetic vulnerability of Indian wheats leading to crop failure due to rusts, similar to the one that occurred in the United States during 1953 and 1954, was predicted by Paddock (1970). This fear is disputed based on 200 years of historical records. It has been observed that the delicately balanced surplus/deficit wheat production of the subcontinent may be upset if a pandemic occurs. However, the food problem only becomes acute if a rust epidemic is preceded or followed by a poor rice crop, due to monsoon failure. Nagarajan and Joshi (1975) could not substantiate the fear that high yielding dwarf varieties would lead to famine conditions as a result of crop failure due to wheat rusts. In an analysis on the variability of annual wheat yields from 1909 to 1975, Waggoner (1979) observed that bumper crops and failures will continue and international trade can continue relieving famine.

C. BASIC INVESTIGATIONS

"Tamra rog," meaning copper diseases, is the Sanskrit name for the wheat rusts, referring to the yellowish-colored spores that the pathogen produced. In the early days in India, rusts were attributed to the curse of God, and stem and leaf rusts were ascribed to different genders of the same organism. Major Sleeman (1839) linked the flax rust with the rust of wheat and tried sanitation methods by uprooting flax to protect wheat. He observed a relationship between November precipitation and subsequent appearance of stem rust, which was later scientifically validated (Nagarajan, 1973). Sleeman collected the infected debris from the epidemic affected fields of 1828–1829 and they were identified as *Uredo* species. His attempts probably were the first scientific investigations of stem rust of wheat on the Indian subcontinent. Barclay (1887) observed the aecial stages on Barberry and related it with *P. graminis.* Butler and Hayman (1906) gave a climatological explanation for the recurrence and development of the rusts in the Indo-Gangetic Plain. Mehta (1925) initiated a systematic study on the wheat rusts and subsequently demonstrated that the aecia on *Berberis* spp. were not those of *P. graminis* f. sp. *tritici* (Mehta, 1940). In the Himalayas the types of aecia were found to be those of *P. graminis agropyri* (Prasada, 1947), *P. brachypodii* (Misra, 1965; Payak, 1965), and P. *poae-nemoralis* (Joshi and Payak, 1963).

A large number of grasses were tested for susceptibility (Prasada, 1948), and only *Agropyron longearistatum, A. semicostatum, A. repens, Bromus patulus, Brachypodium sylvaticum* were moderately infected by stem rust. In the Nilgiri hills, natural infection on *B. sylvaticum* was observed (Pathak *et al.,* 1979), while there were other grasses that were moderately resistant (Bahadur *et al.,* 1973). Therefore, grasses are potential reservoirs of inoculum in the hills (Joshi and Merchanda, 1963). In the Indo-Gangetic Plain, wheat rust urediospores do not survive the hot and dry summer months (Mehta, 1940). The rust oversummers in the hills on self-sown or volunteer plants or on the out-of-season crop. Mehta (1952) recommended cultivation of resistant cultivars in the hills and removal of out-of-season wheat in all the hilly tracts to minimize the epidemic threat in the plains. As follow-up, wheat cultivation in the hills where rust oversummers was suspended, but the exercise produced conflicting claims (Ramakrishnan, 1950; Misra, 1953). It is now clear that legislation did not reduce the disease, as the pathogen still survived on grasses.

Mehta (1952), despite a number of misfits in the trajectories, concluded that the stem rust spreads from the Himalaya as well as from the Nilgiri and Palney Hills in South India (Fig. 2,a–d). He thought that stem rust survived and spread from along the Himalayas to the Indo-Gangetic Plains. He indicated that Central Nepal was a more important source than other parts of the Himalaya. Mehta (1940) observed that *Thalictrum* spp. were not functional in the recurrence of leaf rust. Leaf rust urediospores spread annually by air from oversummering foci in North and South India, while stripe rust spread mainly from the northwestern Himalayas. By virtue of South India being warmer, stripe rust (a cool-season rust) never spreads from the Nilgiri and Palney Hills to the adjoining plains.

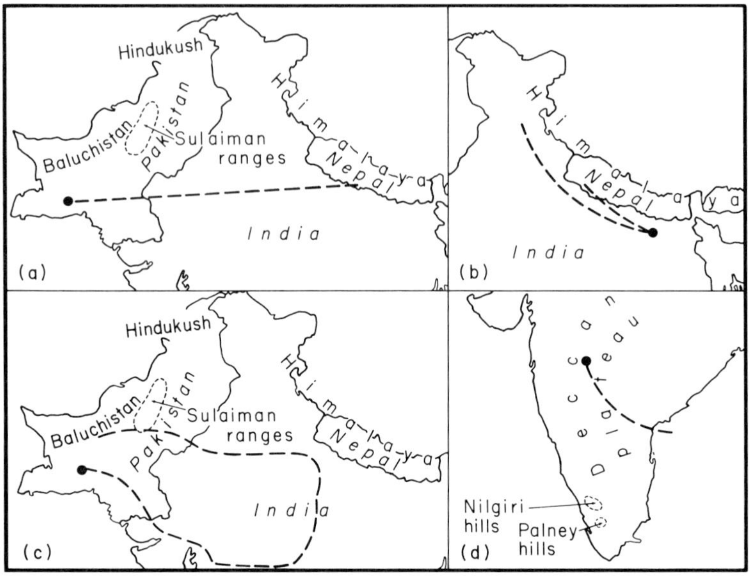

Fig. 2. Trajectories for different heights (Mehta, 1952) showing Himalaya, Baluchistan, and Southern Hills as active stem rust sources. (a) February 23, 1937: source, Nepal; deposition, Sindh. (b) Janauary 23, 1937: source, Nepal and West Himalaya; deposition, North Bihar. (c) February 11, 1935: source, Baluchistan; deposition, Sakrand. (d) December 24, 1935: source, southern hills; deposition Parbhani.

D. SITUATION IN OTHER COUNTRIES

In Afghanistan, all the wheat rusts occur (Hassan, 1965), but no information on the nature and recurrence of the wheat rusts is available. Epidemiological studies on the stem rust in what now is Pakistan and Bangladesh were conducted by Mehta (1940, 1952). He observed the pathogen to oversummer in the Sulaiman ranges, Hindukush mountains, and Baluchistan area and to spread to the wheat-growing plains of the Indus Basin. Generally, stem rust appears earliest along the coastal regions of Pakistan, in Sind. Severity of the disease is higher here as the weather conditions are more favorable for the multiplication of the disease (Hassan and Kirmani, 1963). Observations of Mehta (1940) on the role of *Berberis* spp., and *Thalictrum* spp. also apply to Pakistan and Bangladesh. Leaf rust oversummers in the hilly tracts of Pakistan and descends into the plain to cause infection. It occurs in all the provinces and may at times reach epidemic proportions. The time of arrival of leaf rust in Bangladesh varies between years; as a result, early sowings of even susceptible cultivars can result in disease escape (Joarder *et al.*, 1981). Due to warm weather, stripe rust is not of economic consequence in Bangladesh. Stripe rust develops only between December to April in the Kathmandu Valley of Nepal, while during other months no disease is seen. As the temperature is low from November to March, virtually no multiplication of stem rust occurs during this period. On the contrary, leaf rust that has a lower temperature adaptation survives better even during February (C. B. Karki, unpublished).

III. The Stem Rust Puzzle

A. DISEASE MONITORING

When the Mexican wheats were introduced in the mid 1960s, a need for disease monitoring system was felt. Since 1967, systematic surveys have been conducted and the information released periodically in the form of a "Wheat Diseases Newsletter." A team of plant pathologists drove the national highway and every 25 ± 5 km checked fields, noting the disease prevalence, severity, and other relevant information.

In addition, trap nurseries consisting of different cultivars and lines were planted at many locations to provide disease information from remote areas. Samples collected by the mobile survey team and from trap nurseries were sent to the Regional Station, Indian Agriculture Research Institute (IARI), Simla, for physiologic race identification. Based on this information, it was possible to reexamine theories on the nature and recurrence of wheat rusts. Joshi *et al.* (1971) concluded that the Nilgiri and Palney Hills in South India were the primary stem rust foci for the winter grown wheat crop in the plains, while the cold temperatures in the Himalayas curtailed the disease to an insignificant level. As wheat is grown throughout the year in the Nilgiri and Palney Hills, there are wheat and other grass hosts available for the pathogen to survive on nearly all year. Aerobiological spore trapping studies, backed by field surveys of the Indo-Gangetic Plains, Himalayas, and central Nepal, showed that stem rust is not active in the Himalayan valleys between November and March (Nagarajan and Joshi, 1977).

B. EARLIER ASSUMPTIONS

In 1952, Mehta found that "irrespective of the source of wind that brought the inoculum, whenever heavy infection (70% or above) was observed within a week from the reported appearance; date of initial outbreak of stem rust at the location was put back by 3–4 weeks for purposes of drawing trajectories." Similarly, Mehta used an allowance of 10–15 days for cases when there was 20–25% crop infection. The inoculum source was inferred from the trajectories. This led him to conclude that both South and North Indian hills served as a source of rust inoculum. To result in a 70% severity in 3–4 weeks, a heavy initial inoculum input is needed; however, this was not noted by Mehta (1952). Artificial epiphytotics created at Delhi showed that stem rust needs 55–60 days to reach epidemic levels (> 65% severity) when the initial disease level was one pustule per 238 tillers (Nagarajan, 1973). Disease gradient studies conducted at Karnal, North India, showed that stem rust spreads very slowly there due to the cool weather, while leaf rust spreads at an intermediate rate, and stripe rust at the fastest rate (Fig. 3) (Joshi and Palmer, 1973). Hence, it is evident now that the time allowance made by Mehta (1952) in drawing trajectories for northern India was too short.

C. LONG-DISTANCE DISPERSAL AND DEPOSITION OF PRIMARY INOCULUM

1. Deposition by Rain Scrubbing

Each year stem rust moves more than a thousand kilometers between its source and the furthest fields. Such a spread cannot occur at ground level, mainly because of the frictional layer of the atmosphere and the rugged topography. In the United States, primary inoculum of stem rust arrives with rain (Rowell and Romig, 1966; Roelfs *et al.*, 1970). Following these findings, extensive rain sampling in the Indian subcontinent revealed that stem rust urediospores are rain-deposited (Nagarajan, 1973). Tropical cyclones of late October and November, and the associated rainfall, enable efficient transportation and deposition of urediospores from Nilgiri and Palney hills to central India in less than 68 hr (Nagarajan *et al.*, 1976). On those occasions when it took more than 120 hr for an air mass to cover this distance no infection occurred, which probably was due to loss of spore viability.

2. Role of Weather Factors

After a detailed study of the weather conditions that influence spore transport and deposition, Nagarajan and Singh (1975) postulated the "Indian Stem Rust Rules" (ISR). These were a set of upper-air synoptic

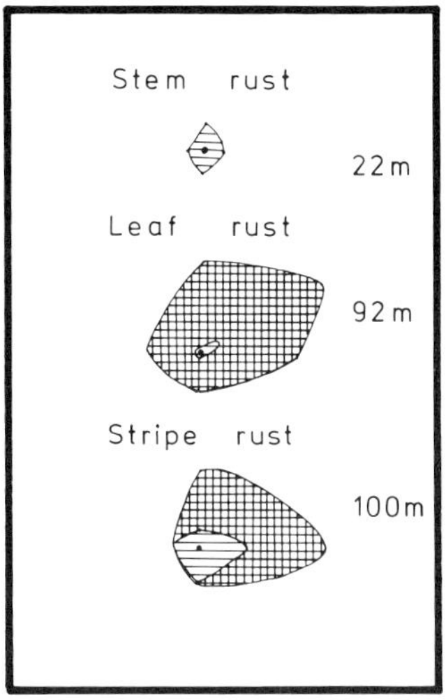

Fig. 3. Dispersal gradient of stem, leaf, and stripe rust from point source mapped after 75 days at Karnal, northwestern India (Joshi and Palmer, 1973). Dark, square and lined area denote moderately heavy, light and traces of infection, respectively. Maximum distance of spread from source given in meters.

situations that enabled northerly spread of urediospores (Fig. 4). The 700-millibar synoptic weather situations that constitute the ISR are:

1. Storm/depression occurs in the Bay of Bengal or in the Arabian Sea between 65 and 81° east, 10–15° north, and ends over central India.
2. A persistent high pressure cell exists over south–central India (not far from Nilgiri Hills).
3. A deep trough extends into South India, caused by the onward movement of western disturbances.

If one or a combination of these situations occur, then disease may appear in central India. These situations result in precipitation, which scrubs the inoculum from upper air. In such a case, stem rust first appears exactly below the locations where the cyclone dissipates. If there are a number of favorable cyclones, then primary infections appear over a large area, enabling a regional epidemic to develop; otherwise, isolated local epidemics occur.

While discussing the 1829 epidemic of Madhya Pradesh, Sleeman (1839) had written that a heavy rain started on November 20, 1828, when a greater part of the land was already sown. He observed that the crop was completely destroyed by stem rust by the end of March 1829.

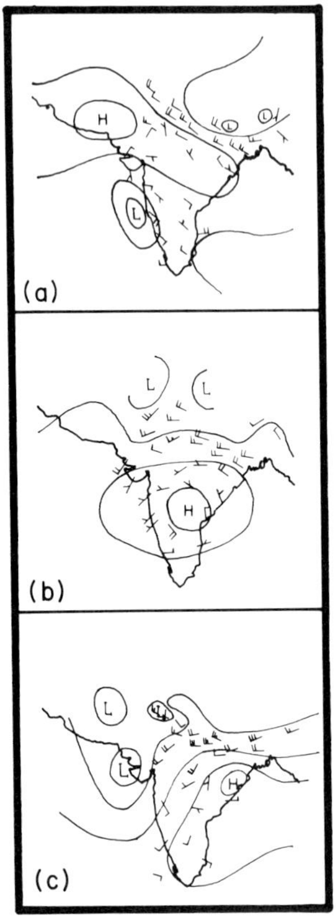

Fig. 4. The Indian stem rust rules (ISR) (Nagarajan and Singh, 1975). A, B, Synoptic situations correspond to (a) ISR-1, (b) ISR-2, and (c) ISR-3 (Section III,C,2) respectively.

The heavy rain that occurred during November 1828 was probably due to a tropical cyclone, ISR-1. An analysis of the weather data from Madhya Pradesh showed an association between greater precipitation and higher minimum temperature in the epidemic years (Ekbote and Misra, 1945). The conducive ground level conditions created by the weather anomaly of November 1946 were considered to be the reason for the 1947 epidemic (Mallik, 1958). It is now known that these weather conditions also enabled transportation and deposition of primary inoculum, in addition to creating favorable conditions for rust infection. Years with November precipitation have nights warmer than normal by 5–9°C; as a result, rapid stem rust multiplication occurs (Nagarajan, 1973). From central India, rust spreads across the country along with western disturbances and associated weather. In

Table II

Data for the Stem Rust Prediction Attempts in India and Results Recorded from 1972 through 1978[a]

Rain sample				Weather				
Location	Date	Stem rust urediospore count	Probable hours of transport	During transport	After deposition	Crop stage (days after seeding)	Forecast details	Remarks
Powerkheda	28.10.72	30	80	Cyclone over Somalia, East Africa, and extreme South India. Anticyclone over Central India.	Cool and cloudy	20–30	Rust may appear late November.	Leaf rust widespread, stem rust isolated. Weather curtailed disease development.
Jabalpur	25.11.72	5220	64	Cyclone in Bay of Bengal dissipated over Jabalpur.	Cloudy rainy weather for 4 days.	40	Rust expected early January around Jabalpur.	In late January severely infected fields observed.
Powerkheda	25.11.72	80	64					
Nagpur	25.11.72	10	44					
Pusa	25.11.72	Trace	?					
Kholapur	20.11.74	30	36	Moist southerly winds from Arabian Sea.	Favorable conditions for infection.	25	Rust predicted by mid December.	Stem rust reported on December 12.
Dapoli	22.11.76	10	?	Tropical cyclone at 700 mb high winds.	Plentiful moisture and cool weather for many days.	30	Rust expected early January.	
Kholapur	22.11.76	10	24					
Hoshangabad	7. 8.76	120	56					
Indore	8. 8.76	30						
Jabalpur	20. 8.76	15						
Jabalpur	21.11.77	30	?	Tropical cyclone at 700 mb, high winds.	Moist and cool for many days.	30–40	Rust expected by late December.	Rust appeared in the predicted area on tall native wheat in late December.
Niphad	23.11.77	1370	?					
Powerkheda	23.11.77	80	48					
Poona	23.11.77	10	38					

[a]In 1973, 1975, and 1978, no cyclone occurred in November and stem rust did not appear in central India before heading of wheat in late February.

years when ISR-1 is not satisfied, central India remains free from epidemics of stem and leaf rust (Table II), but a few rust uredia may appear late in the season if weather conditions satisfy only ISR-2 and/or ISR-3. In these situations, rust occurs late in the crop season, and stem severity remains negligible.

D. THE NEW CONCEPT

Years of study had shown that stem rust spreads primarily from the Nilgiri and Palney Hills, and that the Himalaya area has hardly any effect. Stem rust urediospores that survive in the South Indian Hills are carried by the tropical cyclone and are deposited along with rain over central India. Under favorable conditions, infections are established on the 1-month-old crop, and the subsequent development depends on the weather conditions of January and February. Spread of stem rust throughout India from these initial central Indian pockets results from certain weather situations called the western disturbances (Fig. 5). The occurrence of tropical cyclones can be monitored from weather satellites, and accordingly a disease appearance prediction system has been developed and validated. As stem rust spread is unidirectional, it has opened new possibilities for the scientific management of the disease.

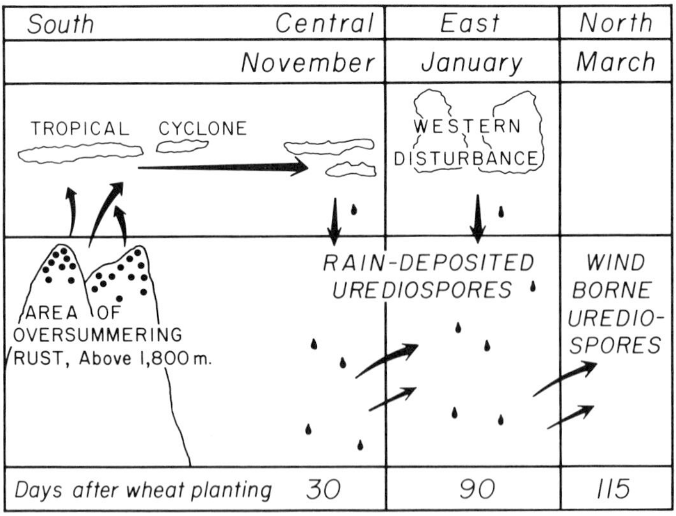

Fig. 5. An empirical model for the spread of stem rust for a normal sown wheat crop (Nagarajan, 1977).

IV. Leaf and Stripe Rusts in the Indo-Gangetic Plain

A. TIME AND PLACE OF APPEARANCE OF LEAF RUST

An analysis of 14 years of survey data covering the Indo-Gangetic Plains showed that leaf rust first appears along the Himalayan foothills. In most cases leaf rust becomes established in central and eastern Uttar Pradesh (UP) and Bihar between late December and mid January. These foci occur to a distance of up to 270 km from the nearest foothills. Due to katabatic winds, primary infections appear to a maximum distance of 60 km along the Himalayan foothills in Jammu and Kashmir, Punjab, Haryana, and western UP (Joshi *et al.*, 1977). By mid February, foci in the warmer eastern tract expand and become many times larger than the isolated pockets along the foothills. During early February leaf rust appears uniformly all over western UP and parts of Haryana, showing the advance of infection from central and eastern UP, westward. By the end of February the less active infections along the foothills of Punjab, Haryana, and western UP start spreading and, in a fortnight, become mixed with the actively spreading eastern source. Thereafter, the rust from the two sources is indistinguishable. Based on this geophytopathological information, an empirical model on the spread of leaf rust over northern India has been proposed (Fig. 6) (Nagarajan, 1977).

B. CLIMATIC EXPLANATION

By mid January, well established infections occur throughout the warmer, eastern part of India. Due to differential radiation, warmer pockets with adiabatic wind occur, and these probably carry the urediospores to mid to low altitude levels. Such spores are transported over hundreds of kilometers to parts of northwestern India along with weather systems called "western disturbances." These are mid to low altitude, low-pressure systems that are formed near the Caspian Sea area, and move towards India every 3–4 days during nonmonsoon months (Pisharoty and Desai, 1956). When these systems linger over the Indo–Pakistan border, they cause rain or snow during the winter months. Urediospores that move westward from the infection centers in eastern UP and Bihar are deposited along with the accompanying rain over parts of Punjab, Haryana, and western UP in addition to the already existing disease. Many such disturbances occur enabling urediospore transport and deposition from January to March, and ini-

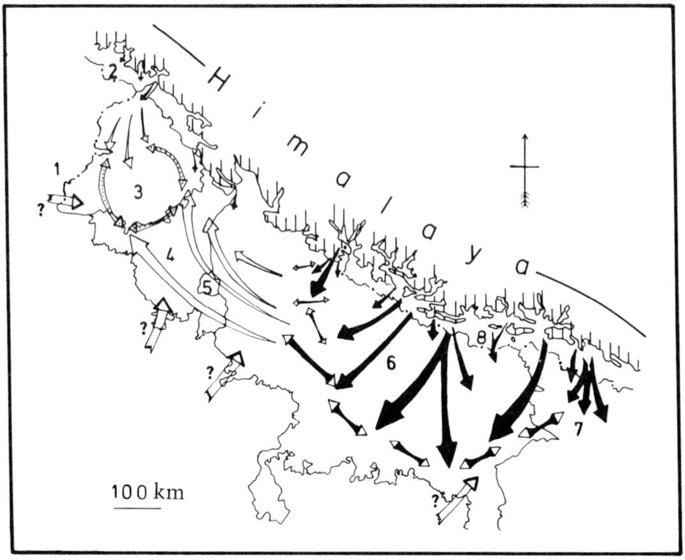

Fig. 6. Empirical model on the spread of leaf rust over the Indo-Gangetic Plain, for a normally sown crop (Nagarajan, 1977). Black arrows denote movement from infection centers up to January 15, white-headed arrows for early February, white arrows for mid-February, and dotted arrows for mid March. Questioned arrows show stray exchange late in the season. Numbers 1–8 stand for Pakistan, Jammu and Kashmir, Punjab, Haryana, Delhi, Uttar Pradesh, Bihar, and Nepal, respectively. Himalaya 1 km height is denoted with vertical lines.

tial infections appear in these areas by late February. Thus, with frequent passages of western disturbances between January and March, large depositions of inoculum from the east occur and continued periods of leaf wetness are assured. Climatic data reveal the number of rainy days is more important in determining the build-up of rusts over northwestern India than the amount of rainfall (Nagarajan and Joshi, 1978a; Butler and Hayman, 1906).

C. NATURE AND RECURRENCE OF STRIPE RUST

Stripe rust oversummers throughout the Himalaya, Hindukush, and Sulaiman mountain ranges and in the northwestern Frontier Province, but the largest amounts are in the mountains and valleys of Indus and its tributaries (Mehta, 1952; Joshi *et al.*, 1977). Stripe rust occurs along the foothills of Himalaya in Jammu and Kashmir, Punjab, Haryana, Himachal Pradesh, and western UP during early January. In India, primary infections occur near Jammu, Gurdaspur, Pathankot, Ropar, and

Jagadhari, where the Tawi, Ravi, Beas, Satluj, and Jamuna rivers enter the plains, respectively. As the river openings are along straight steep valleys, katabatic wind flow carries the urediospores to the foothills (Nagarajan, 1977). Cool night temperatures with dew are common along the foothills and assure successful infection. From such secondary foci, disease spreads gradually during February into the adjoining plains. Reports from Pakistan also indicate that primary foci of stripe rust appear along the foothills by January, and that its spread and development is confined to the cooler parts of western Punjab, and the northwestern Frontier Area (Hassan, 1978). Years with stripe rust epidemics have been characterized by a greater number of rainy days due to the frequent passage of western disturbances, The rains extend the cool winter and delay crop maturity by a week to 10 days and, as a consequence, greater terminal severity is reached. Passage of the western disturbances that favor stripe and leaf rust epidemics over northwestern parts of the Indian subcontinent follows certain geographical tracts (Nagarajan, 1980). So, by studying these systems, predictions of epidemics can be made, assuming the occurrence of a susceptible host and virulent pathogen.

V. Leaf and Stripe Rusts in South India

Leaf and stripe rust also survive in the Nilgiri and Palney Hills throughout the year. Leaf rust effectively spreads to central India by wind and then is deposited by rain. Conditions that favor transportation and deposition of leaf rust urediospores are identical with those for stem rust. Following the ISR and the stem rust prediction method (Nagarajan and Singh, 1975, 1976) appearance of leaf rust can also be predicted. Currently, however, the amount of disease likely to develop subsequently cannot be predicted. Stripe rust of wheat and barley occurs in the Nilgiri and Palney Hills but never effectively spreads to the warmer, central and peninsular India; thus stripe rust is not important in this area, except in the South Indian hills.

VI. Pathogen Variability

A. Wheat Rust Flora

Epidemiologically, South Asia is a single zone and occasionally there is a movement of inoculum from Turkey to India (Nagarajan, 1971;

Kirmani, 1980). The evolutionary center of wheat and its near relatives is located in eastern Turkey, the Caucasus mountain area, Georgia, and Armenia of the U.S.S.R. There is a great degree of variability in the pathogen population here. A gradual eastward spread of new virulences of stripe rust from Turkey to India is associated with the "western disturbances" in weather (Nagarajan, 1971). Frequent occurrence of these weather systems during the wheat-growing months creates favorable conditions for the severe development of leaf and stripe rusts (Nagarajan, 1980). When the host has common susceptibility, and these weather systems occur, then races and biotypes spread across countries in waves or jumps. Kirmani (1980) evaluated 38 wheat cultivars of Pakistan, to 18 differing stripe rust cultures collected from Turkey, Afghanistan, Iran, and India (Fig. 7). He observed the Turkish races to be more diverse and virulent and rendered a majority of the lines susceptible; Afganistan races were intermediate in virulence, and the Indian races were the least virulent.

Iran has about 7–8 million hectares of wheat, which is mostly grown

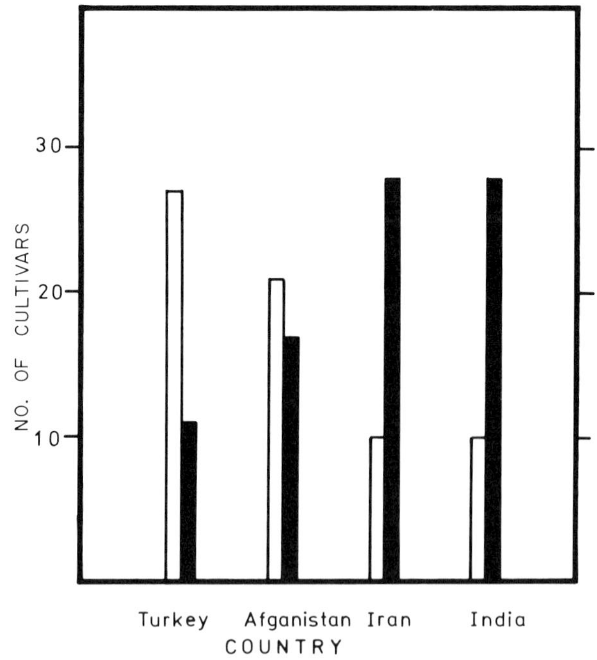

Fig. 7. Average response of 38 wheat cultivars to 18 cultures of stripe rust in the subcontinent and neighboring countries (Kirmani, 1980). Black and white bars represent resistant and susceptible cultivars, respectively.

Table III

Race Flora of the Wheat Rusts in the Asian Subcontinent and Adjoining Countries

Rust disease	Races found in				
	Iran	Afghanistan	Pakistan	India	Bangladesh
Stem	11, 14, 15, 17, 19, 116	11, 15, 17, 40	9, 11, 15, 15B, 17, 21, 24, 34, 40, 42, 117	14, 15, 15B, 17, 21, 21A-1, 24, 34, 34A, 40, 40A, 42, 42B, 42B-1, 72, 75, 117, 117A-1, 184, 194, 222, 226, 294	11, 15, 17, 21, 34, 40
Leaf	57, 114, 122, 143, 167	20, 77, 184	12, 20, 57, 77, 144, 150, 184	10, 11, 12, 12A, 12B, 16, 17, 20, 26, 61, 63, 70, 77, 77A, 77A-1, 104, 104A, 104B, 106, 107, 108, 131, 162, 162A, 162B	15, 20, 57, 184, 77
Stripe	14, 14A, 18, 19, 19A, 20, 20A, 25, 25A	—	—	13, 14, 14A, 19, 20, 20A, 24, 31, 38, 38A, 57, A, D, E, F, G, H, I	

in the north from the Caspian Sea area to the Afghanistan–Pakistan border. The central highlands of Iran are a dry, arid zone where wheat cultivation is sparse. A large number of races and biotypes has been recorded (Table III) in Iran in recent analysis of the wheat rusts (Bamdadian, 1973, 1980; Adbel-Hak, 1963). Afghanistan, believed to be one of the secondary evolutionary centers of wheat, is likely to have diverse race flora. A few samples collected from Nangarhar area of Afghanistan (Hassan, 1965) reveal similarity in the race flora to that of Pakistan. Prior to the partition of British India as Pakistan, Bangladesh, and India, Mehta (1952) had identified stem rust races 15, 21, 40, and 42, leaf rust races 10, 20, 63, 106, 107, and 108, and stripe rust races A, 19, and 31 from the area that presently forms Pakistan. Subsequently, Hassan (1965), Hassan *et al.* (1965, 1967, 1977), and Kirmani (1980) have reported many new races and biotypes (Table III).

In 1923, two wheat stem rust samples from Pusa, India, were identified by M. N. Levine, University of Minnesota, to be race 15 and 27 (Mehta, 1940). In India, systematic race analysis work was initiated in 1931 (Mehta, 1940), and has continued since (Singh *et al.*, 1979).

Hassan *et al.* (1965, 1967) believed that the stem and leaf rust race flora of Pakistan and Bangladesh (then East Pakistan) were similar. There is a paucity of information from Nepal, Bhutan, and Bangladesh, making it difficult to generalize, but it is believed that at least many of the leaf and stripe rust races are common throughout South Asia.

B. VARIABILITY AND MONITORING METHODS

In developing countries, breeding for disease resistance is the only option available to minimize loss due to rusts. To achieve this, the virulence identification system that was originally followed was modified so that intelligent exploitation of host resistance genes is possible. For stripe rust Pakistan has opted to follow Johnson *et al.* (1972) without change (Hassan *et al.*, 1977; Kirmani, 1980), and the avirulence/virulence formula is being used for leaf rust analysis (Hussain *et al.*, 1980). Analytical procedures have been updated, and so far 19 leaf rust virulence groups have been identified. In India, the utility of *Lr* genes was first evaluated (Sawhney *et al.*, 1977), and subsequently various identification procedures have been suggested. Leaf rust races in India are now identified based on the response of 13 hosts with different *Lr* genes (Nagarajan *et al.*, 1981).

Mutation or parasexuality is probably the cause of new races in the subcontinent (Sharma and Prasada, 1970). One of the factors that influence pathogen dynamics is host resistance. In the last 15 years stem rust virulence patterns have undergone a dramatic change, mainly due to the changes in cultivars grown. It has been observed that under asexual reproduction, the chance of emergence of related biotypes is greater than for an entirely new race (Watson, 1970). Similar observations have been made in India (P. Bahadur, S. Nagarajan, and S. K.

Table IV

New Stem Rust Races or Biotypes Identified in India between 1965–1980[a]

Race/ biotype	Year	Cultivar	Location
40	1975–1976	HD 4513	Wellington
40A-1	1974	*Triticum durum*	Wellington
11A	1974	*T. durum*	Wellington
117A-1	1976–1977	NI 5439	Dharwad
24A	1980–1981	MP-195, A-9-30	Powerkheda

[a] P. Bahadur, S. Nagarajan, and S. K. Nayar (personal communication).

Nayar, personal communication). The number of races and biotypes recorded between 1965 and 1980 is given in Table IV, which validates this point. The Himalaya region does not favor the evolution of new races or biotypes of *P. graminis* f. sp. *tritici,* while frequent variation occurs in the pathogen in the Nilgiri and Palney Hills. Evolution of races in the southern hills and the adjoining area is approximately at a rate of one in every 3–4 years. Although the role of mutation and parasexuality should not be underestimated, a catalytic role is played by host resistance, which induces an increase in the frequency of races that overcome it. Thus, races of low frequency that were not found earlier may suddenly increase, influenced by the host susceptibility (Mac Key, 1980).

VII. Disease Management Approaches

A. Boom and Bust Cycle

Biffen demonstrated that resistance to stripe rust could be governed by a dominant allele. Such a resistance was soon widely exploited (Vasudeva *et al.*, 1962) until its consequences were highlighted by Vanderplank (1963). He characterized this sort of resistance as "vertical." Cultivars with vertical resistance and desirable agronomic traits rapidly attract the attention of the growers and the area planted increases. The changed situation provides a selection pressure on the pathogen for races that can infect the cultivar. Such races increase faster than the others under these circumstances, and when favorable environmental conditions prevail, epidemics occur. When the hitherto resistant cultivar becomes susceptible, the process of incorporating resistance is intensified. The vicious "boom and bust cycle" is typical of overexploitation of vertical resistance. The boom is the sudden increase in the area of a resistant cultivar, while the bust refers to the process of a virulent pathogen appearing and creating an epidemic. The 1949, 1972, and 1973 wheat rust epidemics in India (Joshi *et al.*, 1975a) and the 1978 stripe rust epidemics of Pakistan (Kidwai, 1979) are examples of the bust part of the cycle.

Crop vulnerability is "a narrow genetic base of a crop exposed to a broad and unknown genetic basis of the pathogen" (National Academy of Sciences, 1972). The boom and bust cycle and genetic vulnerability are interlinked. By applying the "Peter Principle" (Peter and Hull, 1969), two basic principles can be developed to explain the vulnerability and the recurrent bust cycles: (1) Every agronomically superior cultivar increases in its hectarage, making it more vulnerable. (2) The vulnerability can be reduced by making available new host cultivars that have additional or different sources of resistance.

These two principles hold wherever vertical resistance is used. However, there are many cultural methods available to delay the recurrent bust cycles.

1. Mixed Cropping

Farming in developing countries is more a means of subsistence, and therefore, mixed-crop farming is a common practice. Raising many crops in the same field buffers against drought, pests, diseases, and other natural calamities and yet gives adequate food to sustain life. Wheat, barley, and bengal gram (chickpea) are grown in stands in the same field, and this practice can be exploited to minimize disease epidemics. In a mixed crop there is species diversity within a field, and as a result the rate of disease spread is delayed, consequently reducing the terminal disease severity. If high–yielding strains of all the crops are available, then the very practice of mixed farming can be exploited to minimize disease losses.

2. Gene Deployment

"A centrally planned, properly executed, strategic use of useful vertical resistance genes over a large geographical tract to minimize the risk of epidemics or pandemics of plant diseases" is gene deployment (Browning *et al.,* 1969). They are of the opinion that the *Puccinia* path of the United States can be divided into zones wherein different resistant genes can be deployed. This genetical diversity along the *Puccinia* path minimizes the spread of effective inoculum between zones. When the stem rust epidemiology was investigated existence of a *Puccinia* path in the Indian subcontinent was evident (Nagarajan and Joshi, 1980). This path is also divisible into subzones depending on the time and mode of arrival of the primary inoculum (Fig. 8a). Therefore, there is a possibility of gene deployment in central and peninsular India against stem rust of wheat. Similar possibilities exist in the Indo-Gangetic Plain for managing leaf rust also (Fig, 8b) (Nagarajan *et al.,* 1980a). A quick change in cultivars averted or delayed the possible epidemic of 1976 in India, while in Pakistan where cultivars were not changed, a severe leaf rust epidemic occurred (Hassan, 1978). In a broader sense, it appears that leaf rust epidemics can be checked by frequent cultivar changes at both the inoculum source and the remote

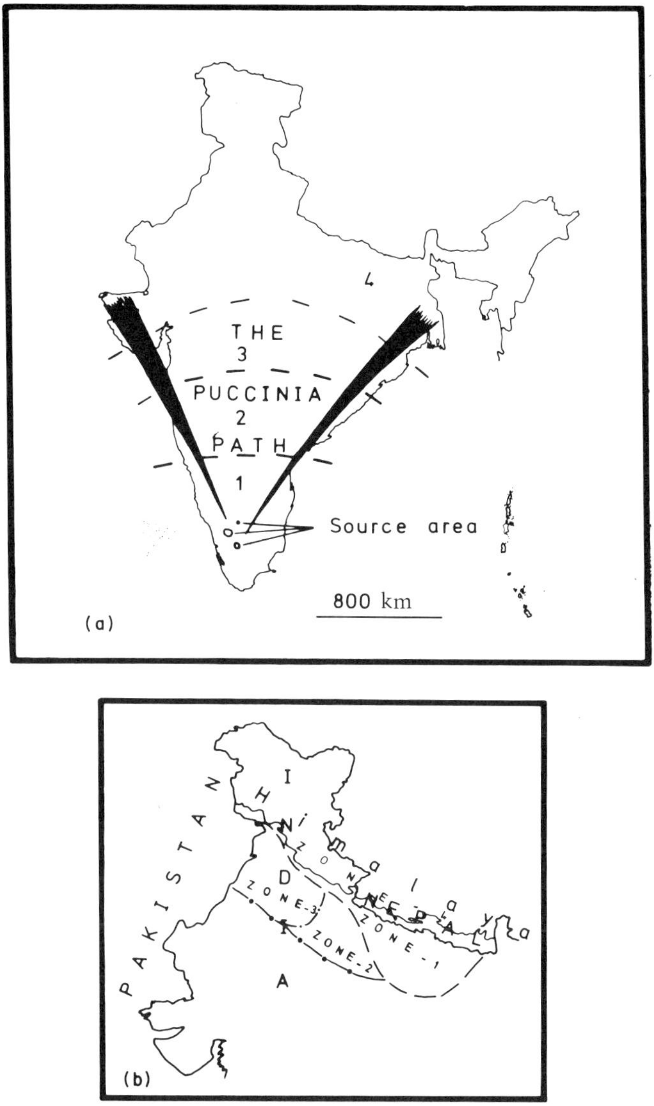

Fig. 8. (a) The predominantly unidirectional spread of stem rust and the existence of the *Puccinia* path has opened up the possibilities of gene deployment against the disease (Nagarajan and Joshi, 1980). Division into zones is based on the nature of arrival of the primary inoculum. (b) Indo-Gangetic Plain for leaf rust can be divided into four major zones for purposes of gene deployment. Zone 4 is the main leaf rust summering area; it subsequently spreads from Zone 1 to Zone 3.

wheat areas and this would be most effective if the sources of resistance were diverse (Nagarajan *et al.*, 1978).

3. *Multilines and Other Means*

The multiline approach popularized by Borlaug (1953) aims at creating within-crop diversity. By using lines that are morphologically similar but differ genetically for resistance it is possible to create within-crop diversity. Multilines, therefore, are an assembly of lines that have different resistant genes while possessing similar agronomic traits such as maturity, height, grain color, etc. The genetical diversity created by multiline reduces the initial level of infection (X_0 level), though it does not alter the apparent rate of infection. Gill *et al.* (1981) showed that it effectively delays the buildup and spread of stripe rust. Multilines are grown only on a small area in India, so their effect cannot yet be determined.

Investigations on slow rusting by Kulkarni and Chopra (1980) showed that the gene action was additive. These authors believed that with suitable modifications in the selection procedure improvement for this trait is feasible. Various components that influence the slow-rusting behavior of leaf rust and their nature of inheritance were investigated. They found that latent period and initial number of uredia per unit area had negligible influence on the slow-rusting behavior. The wheat cultivars Agra Local, K.68, Lal Bahadur, and Kalyansona have a susceptible reaction to race 122 of *P. graminis* f. sp. *tritici*, but greatly differ in the way disease increases (i.e., in infection ratio) and in amount and duration of sporulation (Pande *et al.*, 1979).

B. PREDICTION SYSTEMS

1. *Disease Appearance*

Developing a prediction system is a prerequisite for an efficient management of the wheat rusts and there are two phases: (1) predicting the appearance, and (2) the course of disease development. Pathogens that spread long distances, such as the wheat rusts, need to recur annually to create an epidemic, but they do not arrive annually at the same time and in the same amounts. So predicting the initial disease appearance is the first step in developing a forecasting system.

In seeking to explain the long-distance transport of rust spores from southern India to the north and their arrival in rain, Nagarajan (1973) examined the synoptic conditions associated with stem rust infections in central India. A set of synoptic rules, called the Indian stem rust rules (Section III,C,2) were formulated. Urediospores were shown to be transported around the 700-millibar level as the majority of the air trajectories at that level could be traced back to the source area, the South Indian hills. Urediospores deposited after more than 120 hr of travel in upper air failed to cause infection, even with favorable conditions at the ground level. Infection also failed to occur if the post-deposition conditions were too dry. Using this information, a method to predict stem rust appearance was developed (Nagarajan and Singh, 1976). Probable area and approximate time of disease appearance has been accurately predicted since 1972 (Fig. 9). The occurrence of stem rust was subsequently monitored by field surveys (Table II).

2. *Disease Build-up*

Once the disease appears, subsequent development is dependent on the prevailing local weather conditions. Based on field epiphytotic studies, Nagarajan and Joshi (1978b) developed a linear model for predicting stem rust severity for 7 days in advance:

$$Y = -29.3733 + 1.820\, X_1 + 1.7735\, X_2 + 0.2516\, X_3$$

where Y = expected disease severity after 1 week, X_1 = mean disease severity for the past week, X_2 = mean weekly minimum temperature (°C) expected for the next week, X_3 = mean maximum relative humidity expected in the next week and -29.3733 is a constant.

This regression model was tested by means of a multilocation test (Karki *et al.*, 1979). If the disease appearance and severity prediction approaches were coupled with a simulator model, it should be possible to make predictions in a more systematic way.

3. *Bioclimatic Model for Leaf Rust*

Development of leaf rust over the Indo-Gangetic Plain is related to the number of rainy days. Western disturbances and the associated

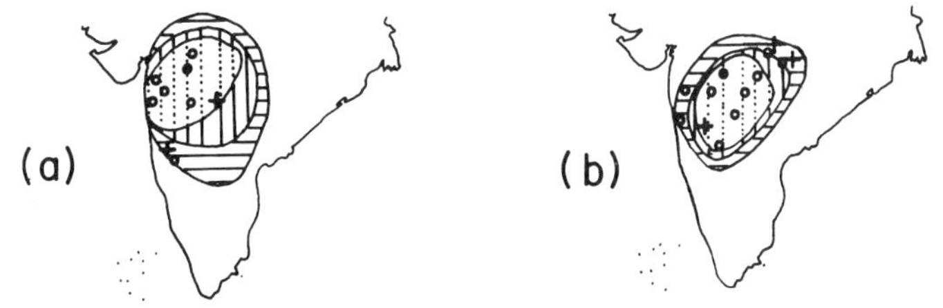

Fig. 9. Early December predictions on stem and leaf rust appearance, with different grades of emphasis, (a) 1976/1977 and (b) 1977/1978 crop years. Places where stem (+) and leaf (○) rust appeared 30–40 days later are marked (Nagarajan and Joshi, 1980).

rainfall enable bulk deposition of inoculum of diverse pathogen virulences over northwest India from the eastern side. It also provides moisture for pathogen infection, enhancing the possibilities of epidemics. Thus, a bioclimatic model for northwestern India was developed (Nagarajan *et al.*, 1980b). The following criteria need to be satisfied if an epidemic is to occur.

1. During the period of January 15 to January 20, a wheat disease survey in UP and North Bihar should detect infections in at least five or six sites, separated by a minimum of 25 ± 5 km.
2. The number of rainy days from January to mid April over northwestern India should be at least double the normal of 12 days.
3. Over northwestern India the weekly mean maximum temperature during March to mid April should be ± 1°C of normal (26°C).

Details of the survey route were given by Joshi *et al.* (1971, 1977). It is assumed that all the cultivars grown over the Indo-Gangetic Plain are susceptible, although in reality they are not. Therefore, these rules must be interpreted with caution while predicting epidemics. When the first and second factors are satisfied, then a severe epidemic tends to occur, and if only partly fulfilled, local outbreaks will probably occur. When this is coupled with a hot, dry period during early April, there is an additional loss created by the interaction of rust and atmospheric drought (Nagarajan *et al.*, 1980b). If the mean maximum temperature is low from January to mid April, stripe rust tends to become serious in northwestern India.

C. DISEASE MONITORING THROUGH REMOTE SENSING

1. *Use of Weather Satellites*

The detection, identification, and evolving strategies for disease control are important to sustain increased food production. The conventional field survey (Joshi *et al.*, 1971, 1977) is laborious, time-consuming, and covers only 25% of the rural areas. Developing remote-sensing techniques such as aerial survey from aircraft or satellite would be a quick means of crop health reconnaissance. Through weather satellites, it is now possible to track storms, measure cloud cover, monitor sea temperature, and even estimate rainfall. The tropical cyclone (ISR-1) can be monitored by the weather satellite, as the bright cloud mass on the satellite photographs (Fig. 10) corresponds to clouds that have already penetrated significantly into the troposphere. Such clouds

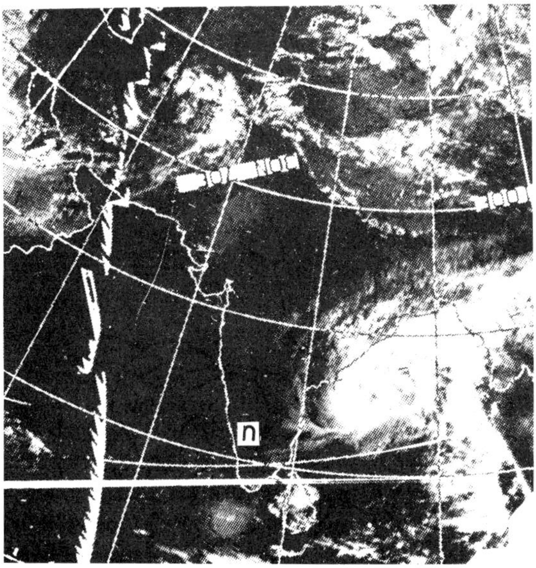

Fig. 10. Weather-satellite imagery of the subcontinent for November 5, 1969. The white mass of cloud due to tropical cyclone satisfies ISR-1 (Section III,C,2).

are likely to contain areas of precipitation, and therefore it is possible to postulate the deposition of stem rust inoculum. The tracks of cyclones mapped by the weather satellite NOAA-5 and the paths of the urediospore trajectory are identical (Nagarajan and Singh, 1973; Nagarajan and Joshi, 1980), and therefore one can be substituted for the other. Rust infections were found exactly below where the cyclone dissipated, which was detected in the infrared imagery of the weather satellite. It is possible to monitor the subsequent weather conditions that influence rapid multiplication of the diseases from the weather satellite. So Nagarajan and co-workers (personal communication) observed that the cumulative, percentage value of cloud cover over the Indo–Pakistan area can be used as an index to monitor ground-level conditions favorable for stripe rust. The stripe rust epidemic year of 1978 had significantly higher amounts of cloud cover during the wheat growing period, compared to the disease-free 1977. The leaf rust epidemic year 1976 had an intermediate value (Fig. 11) and was favored by the relatively warm but wet weather conditions.

Continued cold weather and precipitation extends crop maturity and enhances the development of stripe rust, while intermediate situations favor the development of leaf rust. During relatively dry years, rust development seldom occurs. The daily maximum temperature of Ludhiana, India, shows that 1978 was cooler by about 3.5°C/day than 1977. Thus, a relationship may exist between percentage area under cloud cover and maximum temperature at ground level. If there are reasons to predict an unhealthy crop based on weather satellite imageries and other information, confirmation can be made through the earth resource satellite.

2. *Landsat and Its Usage*

Large-area crop inventory experiments (LACIE) of the United States is the prime experiment on the Landsat II and III to assess wheat yields of major wheat-growing nations. The same satellite can be used to differentiate healthy from diseased crops. In an exploratory study, Landsat frames of MSS Channel-6 (0.7–0.8 μm) and MSS Channel-7 (0.8–1.1 μm) were obtained for the years 1977 and 1978 covering the subcontinent. These images were analyzed to test if there was any signature difference in reflectance between the healthy and the diseased areas. Preliminary investigations showed that the diseased crop can be differentiated from that of the healthy, in MSS Channel-6 (Nagarajan, 1980). In Channel-6 there was a contrasting difference between the Indian and the Pakistan side during the epidemic year of 1978 while the disease-free years of 1977 showed overlapping (Fig. 12). The differences were less obvious in MSS Channel-7. As the subcontinent is one agroclimatic zone, the observed differences in the reflec-

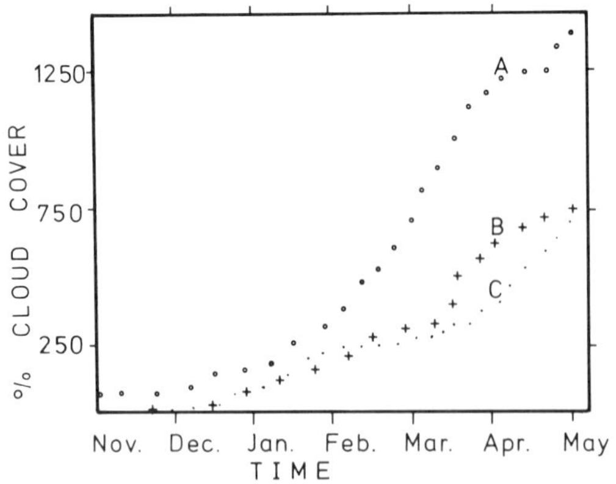

Fig. 11. Cumulative percentage area under cloud cover over northwestern parts of the Asian subcontinent for some discrete days, as monitored by US-DMSP satellite (Nagarajan, 1980). Note that 1978, the stripe rust year (A) had values that contrasted to those in 1977 the no disease year (C), while the leaf rust year (B) was intermediate.

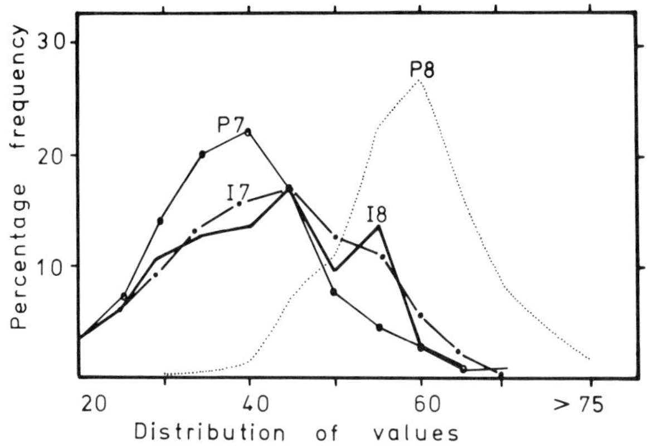

Fig. 12. Frequency distribution of the reflectance values covering northwestern India (I) and Pakistan (P) for the years 1977 (7) and 1978 (8) from Landsat-II negatives, MSS channel 6 (0.7–0.8 μm). Note that the Indian side had similar values in the 2 years, while Pakistan differed due to a 1978 stripe rust epidemic (Nagarajan, 1980).

tance through Landsat were possibly due to the disease. By using both weather and Landsat, it will be possible to monitor wheat-crop health. As part of the food resources management system, developing a technology to monitor wheat production from space would be of great utility.

VIII. Food Resources Management

During the last 15 years, South Asian countries have emerged towards self-sufficiency in food, mainly due to the increased production in wheat, rice, and other crops. Wheat contributes substantially to the food buffer, a protection against the recurrent monsoon failure. Food shortages in India in the past were mainly due to the failure of the monsoon-dependent rice crop, and when the subsequent wheat crop also suffered, it aggravated the food scarcity. Therefore, an appropriate prediction system for wheat diseases to warn of an epidemic needed to be developed. In the event of the inevitable rust epidemic, forecast by mid March, alternate strategies to maintain the same level of food grain production are as follows:

1. As wheat arrives at the markets in June, the State Corporations should attempt to purchase the available surplus wheat. This would

provide a food buffer, to feed the people through the public distribution system.

2. Available energy for irrigation and harvest should be diverted to those states where the disease severity is predicted to be less so that a marginal increase in production can be attained.

3. Advance planning can be made for a bumper rice crop by reallocation of inputs.

These and other managerial steps may help achieve the same national food grain production even during epidemic years. The disease forecasting approach would then help to achieve the same level of food grain rather than protecting the already diseased crop.

IX. Possible Future Trends

N. E. Borlaug in 1953 said, "We believe our ability to foresee the future in this uncertain world is severely limited."

It is evident that to visualize how things will take shape in subsequent years is a tough job. This makes it all the more tempting to think how things will be. By the year 2000 A.D. the population of India will be about 1 billion. This means to provide the same level of wheat consumption as today, India will need 56–58 million tons of grain or an increase of 20 million tons from the present level of production. The impetus on the high-yielding cultivars must continue and the emphasis on disease resistance may be even higher. Reducing the genetic vulnerability of the crop to rust epidemics will be the prime job of the plant pathologists. Appropriate management of the host resistance genes through deployment, pyramiding genes, etc. will require greater attention. The systems approach to manage the biological yield constraints, such as the EPIPRE (Zadoks *et al.*, 1980), will be tested, even in the developing countries. Monitoring crop health and management of the food resources is likely to get greater emphasis as the pressure on land increases and the international grain trade shrinks.

Acknowledgment

We thank David J. Royle, Long Ashton, for his critical comments.

References

Abdel-Hak, T. (1963). Geographic distribution of physiologic races of wheat rusts in the Near East. Paper presented at the 1st FAO Near East Wheat and Barley Training Center, Cairo, Egypt.

Anonymous (1980). "FAO Production Yearbook," Vol. 34, pp. 96–97. FAO, Rome.

Bahadur, P., Singh, S., Goel, L. B., Sharma, S. K., Sinha, V. C., Ahmad, R. U., and Singh, B. P. (1973). Impact of grass introduction on cereal rusts in India. *Indian J. Agric. Sci.* **43,** 287–290.

Bamdadian, A. (1973). Physiologic races of *Puccinia recondita* in Iran (1968–1972). *Cereal Rusts Bull.* **1,** 45–47.

Bamdadian, A. (1980). Epidemiology of yellow rust of wheat (*Puccinia striiformis.*) in Iran. *Proc.—Eur. Mediterr. Cereal Rusts Conf., 5th, 1980,* pp. 219–222.

Barclay, A. (1887). A descriptive list of uredinales occurring in the neighborhood of Simla (Western Himalaya). *J. Asiat. Soc. Bengal.* **56,** Part II, 350–375.

Borlaug, N. E. (1953). New approach to the breeding of wheat varieties resistant to *Puccinia graminis tritici. Phytopathology* **43,** 467(abstr).

Browning, J. A., Simons, M. D., Frey, K. J., and Murphy, H. C. (1969). Regional deployment for conservation of oat crown rust resistant genes. *Spec. Rep.—Iowa Agric. Home Econ. Exp. Stn.* **64,** 49–56.

Butler, E. J., and Hayman, J. M. (1906). Indian wheat rusts with a note on the relation of weather to rust on cereals. *Mem. Dep. Agric. India, Bot. Serv.* **1,** 1–57.

Ekbote, R. B., and Misra, S. D. (1945). Attacks of black stem rust of wheat in Nerbudda Valley. *Indian J. Genet. Plant Breed.* **5,** 122.

Gill, K. S., Nanda, G. S., Aujla, S. S., Singh, G., and Sharma, Y. R. (1981). Studies on multilines in wheat. II. Progression of yellow rust in mixtures of isogenic lines in varying proportions. *Indian J. Genet. Plant. Breed.* **41,** 124–129.

Hassan, S. F. (1965). Some physiologic races of leaf and stem rust of wheat in Afganistan in 1963–64. *West Pak. J. Agric. Res.* **3,** 233–234.

Hassan, S. F. (1978). Rust problems in Pakistan. *In* "Wheat Research and Production in Pakistan" (M. Tahir, ed.), Vol. I, pp. 90–93. PARC, Islamabad, Pakistan.

Hassan, S. F., and Kirmani, M. A. S. (1963). Occurrence of physiologic race 17 of *Puccinia graminis tritici* in Pakistan. *Agric. Pak.* **14,** 395–396.

Hassan, S. F., Kirmani, M. A. S., and Hussain, M. (1965). Physiologic races of stem rust of wheat in Pakistan during 1961–64. *West Pak. J. Agric. Res.* **3,** 17–20.

Hassan, S. F., Kirmani, M. A. S., and Hussain, M. (1967). Some new records of physiologic races of *Puccinia recondita* Rob. ex. Desm. f. sp. *tritici* in Pakistan. *West Pak. J. Agric. Res.* **5,** 179–183.

Hassan, S. F., Hussain, M., and Rizvi, S. A. (1977). Investigations on rusts of wheat in Pakistan. *Cereal Rusts Bull.* **5,** 4–10.

Hussain, M., Hassan, S. F., and Kirmani, M. A. S. (1980). Virulences in *Puccinia recondita.* Rob. ex. Desm. f. sp. *tritici* in Pakistan during 1978 and 1979. *Proc.—Eur. Mediterr. Cereal Rusts Conf., 5th, 1980,* pp. 179–184.

Jarrige, J.-F., and Meadow, R. H. (1980). The antecedents of civilization in the Indus Valley. *Sci. Am.* **243,** 122–133.

Joarder, O. I., Islam, R., Rahman, S., and Eunus, A. M. (1981). Effect of seeding date on yield and other agronomic traits of some wheat varieties grown on irrigated lands in Bangladesh. *Indian J. Agric. Sci.* **51,** 489–493.

Johnson, R., Stubbs, R. W., Fuchs, E., and Chamberlain, N. H. (1972). Nomenclature for physiologic races of *Puccinia striiformis* infecting wheat. *Trans. Br. Mycol. Soc.* **58,** 475–480.

Joshi, L. M., and Merchanda, W. C. (1963). *Bromus japonicus* Thunb., susceptible to wheat rusts under natural conditions. *Indian Phytopathol.* **16,** 312–313.

Joshi, L. M., and Palmer, L. T. (1973). Epidemiology of stem, leaf, and stripe rusts of wheat in northern India. *Plant Dis. Rep.* **57,** 8–12.

Joshi, L. M., and Payak, M. M. (1963). A *Berberris aecidium* in Lahaul Valley, Western Himalayas. *Mycologia* **55,** 247–250.

Joshi, L. M., Saari, E. E., and Gera, S. D. (1971). Epidemiology of wheat rusts in India. *INSA Bull.* **37B,** 449–453.

Joshi, L. M., Srivastava, K. D., and Ramanujam, K. (1975a). An analysis of brown rust epidemics of 1971–72 and 1972–73. *Indian Phytopathol.* **28,** 138 (abstr.).

Joshi, L. M., Srivastava, K. D., Singh, D. V., Goel, L. B., and Nagarajan, S. (1975b). "Annotated Compendium of Wheat Diseases in India." Indian Counc. Agric. Res., New Delhi.

Joshi, L. M., Nagarajan, S., and Srivastava, K. D. (1977). Epidemiology of brown and yellow rusts of wheat in north India. I. Time and place of appearance. *Phytopathol. Z.* **90,** 116–122.

Karki, C. B., Pande, S., Thombre, S. B., Joshi, L. M., and Nagarajan, S. (1979). Evaluation of a linear model to predict stem rust severity. *Cereal Rusts Bull.* **7,** 3–7.

Kayastha, B. N. (1970). Wheat and barley improvement in Nepal. *Proc. FAO Rockefeller Found. Wheat Semin. 3rd, 1970,* pp. 85–95.

Kidwai, A. (1979). Pakistan reorganises agricultural research after harvest disaster. *Nature (London)* **277,** 169.

Kirmani, M. A. S. (1980). Report on "Comparative evaluation of wheat varieties from Pakistan against stripe rust in seedling stage." IPO, Wageningen, The Netherlands (unpublished report).

Kohli, S. P. (1968). "Wheat Varieties in India," Tech. Bull. No. 18. Indian Counc. Agric. Res., New Delhi.

Kulkarni, R. N., and Chopra, V. L. (1980). Slow-rusting resistance: Its components, nature and inheritance. *Z. Pflanzenkr. Pflanzenschutz.* **87,** 562–573.

Mac Key, J. (1980). Genetics of race-specific phytoparasitism on plants. *Proc. Int. Congr. Genet., 14th, 1978,* pp. 363–381.

Mallik, A. K. (1958). An examination of the crop yields at crop-weather stations with special reference to rainfall. *Indian J. Meteorol. Geophys.* **9,** 1–8.

Mani, M. S. (1968). "Ecology and Biography of the High Altitude Insects." Junk, The Hague.

Mehta, K. C. (1925). Yellow and brown rusts of wheat. A preliminary note on their annual recurrence. *Proc. Indian Sci. Congr., 12th, 1925,* p. 191.

Mehta, K. C. (1940). "Further Studies on Cereal Rusts in India," Sci. Monogr. No. 14. Imp. Counc. Agric. Res., Delhi.

Mehta, K. C. (1952). "Further Studies on Cereal Rusts in India," Part II, Sci. Monogr. No. 18. Indian Counc. Agric. Res., Delhi.

Misra, A. P. (1953). Influence of the suspension of summer cultivation of wheat and barley on the incidence of black rust. *Indian J. Agric. Sci.* **23,** 47–54.

Misra, D. P. (1965). Rusts connected with barberry in central Nepal. *Indian Phytopathol.* **18,** 66–70.

Nagarajan, S. (1971). Spread of yellow rust of wheat from Turkey to India—a hypothesis. *Epidemiol. Plant Dis., Proc. Nato Adv. Study Inst., 1971,* Vol. 1, p. 79.

Nagarajan, S. (1973). Studies on the urediospore transport of *Puccinia graminis tritici* and epidemiology of stem rust of wheat in India. Ph.D. Thesis, University of Delhi.

Nagarajan, S. (1977). Meteorology and forecasting of epidemics of diseases. *Symp. Basic Sci. Agric., 1977,* pp. 136–143.

Nagarajan, S. (1980). "Epidemiological Investigations on the Outbreak of Wheat Rusts During 1978 in Pakistan and Spain, with an Emphasis on Forecasting and Management," Techn. Rep. Alexander von Humboldt Foundation, Bonn, West Germany.

Nagarajan, S., and Joshi, L. M. (1975). A historical account of wheat rust epidemics in India, and their significance. *Cereal Rusts Bull.* **3,** 29–33.

Nagarajan, S., and Joshi, L. M. (1977). Sources of primary inoculum of wheat stem rust in India. *Plant Dis. Rep.* **61,** 454–457.

Nagarajan, S., and Joshi, L. M. (1978a). Epidemiology of brown and yellow rusts of wheat over north India. II. Associated meteorological conditions. *Plant Dis. Rep.* **62,** 186–188.

Nagarajan, S., and Joshi, L. M. (1978b). A linear model for a seven-day forecast of stem rust severity, *Indian Phytopathol.* **31,** 504–506.

Nagarajan, S., and Joshi, L. M. (1980). Further investigations on predicting wheat rusts appearance in central and peninsular India. *Phytopathol. Z.* **98,** 84–90.

Nagarajan, S., and Singh, H. (1973). Satellite television cloud photography as a possible tool to forecast plant disease spread. *Curr. Sci.* **42,** 273–274.

Nagarajan, S., and Singh, H. (1975). The Indian stem rust rules—A concept on the spread of wheat stem rust. *Plant Dis. Rep.* **59,** 133–136.

Nagarajan, S., and Singh, H. (1976). Preliminary studies on forecasting wheat stem rust appearance. *Agric. Meteorol.* **17,** 281–289.

Nagarajan, S., Singh, H., Joshi, L. M., and Saari, E. E. (1976). Meteorological conditions associated with long distance dissemination and deposition of *Puccinia graminis tritici* uredospores in India. *Phytopathology* **66,** 198–203.

Nagarajan, S., Joshi, L. M., Srivastava, K. D., and Singh, D. V. (1978). Epidemiology of brown and yellow rusts of wheat in north India. III. Impact of varietal change. *Plant Dis. Rep.* **62,** 694–698.

Nagarajan, S., Joshi, L. M., Srivastava, K. D., and Singh, D. V. (1980a). Epidemiology of brown and yellow rusts of wheat in north India. IV. Disease management recommendations. *Cereal Rusts Bull.* **7,** 15–20.

Nagarajan, S., Sinha, S. K., Joshi, L. M., and Saari, E. E. (1980b). Interaction between kernel filling in wheat and leaf rust infection in North India. *Pflanzenschutz* **87,** 221–226.

Nagarajan, S., Nayar, S. K., and Bahadur, P. (1981). "The Proposed Brown Rust of Wheat (*Puccinia recondita* f. sp. *tritici*) Virulence Monitoring System," Res. Bull. No. 1. IARI, R.S. Simla.

National Academy of Sciences. (1972). "Genetic Vulnerability of Major Crops." NAS, Washington, D.C.

Paddock, P., and Paddock, W. (1967). "Famine 1975. America's Decision: Who Will Survive?" Little, Brown, Boston, Massachusetts.

Paddock, W. C. (1970). How green is the green revolution? *BioScience* **20,** 897–902.

Pande, S., Joshi, L. M., and Nagarajan, S. (1979). Quantifying infection and sporulation as possible parameters for measuring host resistance. *Proc. Int. Wheat Genet. Symp., 5th, 1978,* Vol. II, pp. 1087–1097.

Pathak, K. D., Joshi, L. M., and Chinnamani, S. (1979). Natural occurrence of *Puccinia graminis tritici* on Brachypodium sylvaticum during off season in Nilgiri hills. *Indian Phytopathol.* **32,** 308–309.

Payak, M. M. (1965). *Berberis* as the aecial host of *Puccinia brachypodii* in Simla hills (India). *Phytopathol. Z.* **52,** 49–54.
Peter, L. F., and Hull, R. (1969). "The Peter Principle." William Morrow Co., Inc., New York.
Pisharoty, P. R., and Desai, B. N. (1956). "Western disturbances" and Indian weather. *Indian J. Meteorol. Geophys.* **7,** 333–338.
Prasada, R. (1947). Discovery of the uredo-stage connected with the aecidia so commonly found on species of *Berberis* in the Simla Hills. *Indian J. Agric. Sci.* **17,** 137–151.
Prasada, R. (1948). Studies on rusts of some of the wild grasses in the neighborhood of Simla. *Indian J. Agric. Sci.* **18,** 165–176.
Ramakrishnan, T. S. (1950). Wheat rusts from Madras. *Sci. Cult.* **15,** 362–363.
Roelfs, A. P., Rowell, J. B., and Romig, R. W. (1970). Sampler for monitoring cereal rust uredospores in rain. *Phytopathology* **60,** 187–188.
Rowell, J. B., and Romig, R. W. (1966). Detection of uredospores of wheat rusts in spring rains. *Phytopathology* **56,** 807–811.
Sawhney, R. N., Nayar, S. K., Singh, S. D., and Chopra, V. L. (1977). Virulence pattern of the Indian leaf rust races on lines and varieties of wheat with known Lr genes. *SABRO J.* **9,** 13–20.
Sharma, S. K., and Prasada, R. (1970). Somatic recombination in the leaf rust of wheat caused by *Puccinia recondita* Rob. ex. Desm. *Phytopathol. Z.* **67,** 240–244.
Singh, S., Goel, L. B., Nayar, S. K., Sharma, S. K., and Chatterjee, S. C. (1979). Prevalence and distribution of physiologic races of wheat rusts in India during 1976–78 crop seasons. *Indian Phytopathol.* **32,** 417–420.
Sleeman, W. H. (1839). Extracts from Major Sleeman's diary. *Trans. Agric. Hortic. Soc. India* **6,** 77–79.
Vanderplank, J. E. (1963). "Plant Diseases: Epidemics and Control." Academic Press, New York.
Vasudeva, R. S., Prasada, R., Lele, V. C., Joshi, L. M., and Pal, B. P. (1962). "Rust resistance in Varieties of Wheat and Barley in India," Res. Ser. No. 32. Indian Counc. Agric. Res., New Delhi.
Waggoner, P. E. (1979). Variability of annual wheat yields since 1909 and among nations. *Agric. Meteorol.* **20,** 41–45.
Watson, I. A. (1970). Changes in virulence and population shifts in plant pathogens. *Annu. Rev. Phytopathol.* **8,** 208–230.
Zadoks, J. C., Rijsdaijk, F. H., and Rabbinge, R. (1980). EPIPRE, a systems approach to supervised control of pests and diseases of wheat in the Netherlands. *In* "The Management of Pest and Disease Systems." IIASA, Vienna, Austria.

13

Epidemiology in North America

A. P. Roelfs

Cereal Rust Laboratory, Agricultural Research Service, U.S. Department of Agriculture, University of Minnesota, St. Paul, Minnesota

THE CEREAL RUSTS,
Volume II

ISBN 0-12-148402-5

I. Introduction and History

The classical studies of epidemics of plant disease include many examples involving the cereal rusts. Among the first plant disease epidemics that were studied as they developed on a regional basis were the wheat stem rust epidemics that occurred in the United States in 1923 and 1925 (Stakman and Harrar, 1957). Earlier epidemics had been studied after the disease had become severe over a large area or, in some cases, years later. The studies by Stakman and his co-workers became the basis for the area of phytopathology that became known as epidemiology.

Chester (1946) in his monograph provides an insight of early botanists and their concern about the rusts. He also reevaluated previous opinions on epidemiology, to which he added his experience, and then laid the basis for the regional development of the leaf and stem rusts of wheat. The rust diseases are particularly apt to cause epidemics, because these fungi have a high capacity to produce spores, the urediospore is well adapted for wind transportation, and there is a vast area of cultivated host plants.

A. Wheat Stem Rust

Records of the earliest epidemics in colonial North America are scant and most were lost. However, at least in localized areas, wheat stem rust (caused by *Puccinia graminis* Pers. f. sp. *tritici*) must have been a problem because barberry eradication laws were passed in Massachusetts, Rhode Island, and Connecticut in 1754, 1766, and 1776, respectively. The north central region of the United States suffered a severe wheat stem rust epidemic in 1878 (Hamilton, 1939). This epidemic affected southern Minnesota, northern Iowa, and Wisconsin. As more of the northern Great Plains was planted to wheat, the frequency of reported regional epidemics increased. Bolley (1891) raised the question concerning the source of inoculum that generated stem rust epidemics. He noted that epidemics could arise from local (aeciospores) or from general (urediospores) inoculum sources. Regional epidemics occurred in 1904 (Carleton, 1905), 1916, 1923, 1925, 1935, 1937, 1953, and 1954 (Stakman and Harrar, 1957). The epidemic of 1904 was more extensive than the one of 1878 and included Minnesota, North Dakota, South Dakota, and to a lesser degree extended northward into Canada and south and eastward through Iowa, Nebraska, Kansas, and Wisconsin. The next three epidemics of 1916, 1923, and 1925 were similar in size to the 1904 epidemic. The 1916 epidemic caused a series of actions that resulted in programs for barberry eradication, pathogen quarantine, breeding for disease resistance, determination of pathogen variation (Chapter 1), and disease epidemiology. Again in 1935, 1937, 1953, and 1954, epidemics affected approximately the same area as the 1904 epidemic did and occurred when a shift in the principal pathogen race resulted in the majority of the wheat cultivars being susceptible to that race. Large regional epidemics, such as those that extended as far south as Kansas, probably resulted from urediospore movement from a southern overwintering source. Winds in this area move generally from the south in the spring and summer. Infected barberry has seldom been found as far south as Kansas. The regional epidemic of 1878 could have been an exception, because it was limited to northern areas; however, the epidemics resulting from documented aeciospore spreads were always much more local in nature. The extent of the eradication of the barberry by 1930 makes it obvious that the epidemics of 1935, 1937, 1953, and 1954 were not the direct result of aeciospore spreads. The epidemics from aeciospore spreads generally occurred annually but resulted in serious disease losses in an area of less than 2.9 km^2 downwind of the infected bushes. Epidemics resulting from aeciospore spreads were studied by Beeson (1923), Melhus *et al.*, (1920), Stakman *et al.* (1927), Durrell and Lungren (1927), and Walker (1927).

Early stem rust epidemics in Canada occurred in 1904, 1909, 1916, 1919, 1923, 1925, and 1927 (Bailey, 1928). The epidemic of 1916 was the most severe, causing a loss of 200 million dollars. Although the epidemics of 1904, 1909, 1919, 1925, and 1927 were less severe than in 1916, Canada still lost an average of 25 million dollars per year from 1918 through 1927. During the next 30 years, epidemics were less frequent, occurring in 1935, 1937, and 1938 when race 56 became established in North America. The last series of epidemics occurred in 1953, 1954, and 1955 with the establishment of race 15-B (Peturson, 1958). Craigie (1945) thought that the epidemics of 1925, 1927, 1930, 1935, 1937, and 1938 were not directly related to aeciospore spreads from barberry.

The difference in years that epidemics were reported in the adjacent areas of the United States and Canada can be related to differences in (1) crop maturity at disease onset, (2) host resistance used, and (3) environmental conditions. In addition, epidemic years were designated by differences from the normal amount of disease that occurred in the area.

According to Borlaug (1954), severe epidemics of stem rust occurred in Mexico in 1938 and 1939. Another occurred in central Mexico in

1948 (Rupert, 1951). In most years, however, epidemics were avoided in spite of many susceptible host cultivars and a year-round source of inoculum. Borlaug (1954) stated that the Mexican farmer had "learned to manipulate dates of planting and to use irrigation water properly to minimize the losses from rust." Late-planted wheat maturing in June or wheat planted during the summer was often seriously damaged by stem rust, but this practice was seldom used on a regional basis.

B. WHEAT LEAF RUST

The severity of the wheat stem rust problem in North America overshadowed the effects of wheat leaf rust caused by *Puccinia recondita* Rob. ex. Desm. f. sp. *tritici* (Chester, 1946) in the early part of the twentieth century. However, a major regional epidemic in 1938 throughout much of central United States on winter wheat emphasized its importance. Leaf rust primarily affects the leaf blades, whereas stem rust also attacks the stem; thus there is less time for a leaf rust epidemic development during the grain-filling period. Yield losses are usually less from leaf rust than from stem and stripe rust. Growers have often overlooked leaf rust losses because the disease is usually more severe in warm years with adequate rainfall conditions, which also favor wheat growth. Thus, even with heavy leaf rust infections, yields often can be higher than the long-term average. However, when leaf rust is controlled by chemicals or other means in these years, yields are much higher than the average. Chester (1946) stated that losses due to leaf rust were often underestimated. The severe losses caused by leaf rust have generally been in winter wheat areas where the disease overwintered. Few cases of severe regional leaf rust epidemics on spring wheat have been observed. However, losses can be significant. Losses of 1–5% commonly occur on spring wheat (Roelfs, 1978). Epidemics of statewide importance in Georgia occurred in 1935, 1937, and 1972; in Illinois in 1922, 1932, 1935, 1938, 1939, 1940, 1945, 1946, 1948, 1949, and 1950; in Indiana in 1927, 1931, 1932, 1935, 1937, and 1945; in Iowa in 1927 and 1938; in Kansas in 1938, 1949, 1957, and 1974; in Missouri in 1938, 1945, and 1957; in Nebraska in 1938; in Oklahoma in 1921, 1938, 1945, 1973, 1974, and 1975; in Pennsylvania in 1929, 1931, 1932, 1934, 1935, 1936, 1937, 1938, 1939, 1945, 1946, 1949, and 1950; in South Carolina in 1972, 1973, and 1974; in South Dakota in 1965; and in Texas in 1949 and 1974 (Roelfs, 1978). The leaf rust epidemics that occurred in 1974 in California, Kansas, Oklahoma, Texas, and South Carolina, developed from three separate sources of inoculum. The frequent epidemics in Illinois, Indiana, and Pennsylvania were curbed from 1950 to 1981 by the use of resistant cultivars; however, in 1982 leaf rust was again epidemic in the eastern United States. The epidemics that occurred in Georgia and South Carolina resulted in part from the cultivation of the very susceptible but high-yielding cultivar Blueboy.

Severe leaf rust epidemics have occurred infrequently on a regional basis where spring wheat is the principal crop. However, losses from 1 to 10% occur frequently enough that leaf rust resistance is an important objective in wheat breeding in Canada (Chapter 15). Current cultivars in the rust liable area possess effective multigene resistance (Chapter 2).

Leaf rust in Mexico is often an important disease; however, records of epidemics are not readily available. The crop is planted late enough in the fall to avoid heavy seedling infections in the fall, but planted early enough to mature before the rainy season, and losses are generally light. Failure to do this results in an epidemic such as occurred in 1977 (Dubin and Torres, 1981). Evidence shows little direct relationship of leaf rust epidemics in Mexico to those in the rest of North America.

C. WHEAT STRIPE RUST

In the United States and Canada, the same *forma specialis* of *Puccinia striiformis* West. attacks wheat and barley. Stripe rust of wheat is adapted to cooler temperatures than are the other two rusts in the United States. It has been a major disease of fall-planted wheat only and to nearby spring wheat but only in the area west of the Rocky Mountains where summer temperatures are cool, as in parts of Idaho, Oregon, Washington, and western Montana, as well as in California during the winter. Most epidemics of stripe rust have resulted from early fall infections that survived the winter and then spread in the early spring. Although not totally clear, evidence indicates the source of inoculum for initial infection is from nearby sources of volunteer wheat, wheat planted for a cover crop in orchards, or from wild grasses (Shaner and Powelson, 1973; Tollenaar and Houston, 1967; Burleigh and Hendrix, 1970). A series of stripe rust epidemics occurred between 1915 and 1935 and again from 1959 through 1961. Otherwise the disease was present in minor amounts until 1976 (Roelfs, 1978). Since 1976, stripe rust has been an important disease in this area.

Stripe rust has not caused serious epidemics in the major wheat-growing areas east of the Rocky Mountains. It was first reported in central Texas in 1941 and again in 1953 and 1956 (Futrell, 1957). Then in 1957, it occurred from Texas to South Dakota and Wyoming on the north. This epidemic was followed by a wider distribution in 1958, when it was found throughout the Great Plains. Since then, stripe rust has frequently been found in the south central areas of the United States with little loss except to individual growers in Texas. Stripe rust was again present from Texas to North Dakota in 1981, but in only trace amounts.

Stripe rust has been relatively unimportant in Canada, because most of the crop is planted in the spring. Stripe rust in Canada is most commonly found in Alberta (Sanford and Broadfoot, 1932), which is adjacent to western Montana.

Although stripe rust is common in Mexico and can become very severe on highly susceptible cultivars, there seems to be little information available on serious epidemics. It has caused losses locally in the high valleys, but no epidemic years were reported by Campos (1960).

D. OTHER RUSTS

Oat stem rust (caused by *P. graminis* Pers. f. sp. *avenae*), like wheat stem rust, has been the most severe in the north central states and adjacent Canada (Chapter 4). Fewer epidemics of oat stem rust have occurred than of wheat stem rust (Roelfs, 1978), which in part could be due to a much smaller area planted to oats (approximately 20% that of wheat) and because oat foliage is more frequently winterkilled to the crown than wheat in southern Oklahoma and north central Texas. Currently, the most common race, NA 27, is virulent on most of the cultivars. Occurrences of epidemics in the north central states are highly correlated with the date of disease onset (Roelfs and Long, 1980). Oats is a minor crop in Mexico; however, oats and wild oats (*Avena fatua* L.) are commonly rusted.

Crown rust (caused by *Puccinia coronata* Cda.) is the most serious rust disease of oats in North America (Chapter 5). Epidemics are common in the southern states, where the disease overwinters, and again on late seeded oats in Minnesota, the eastern Dakotas, and Manitoba (Roelfs, 1978). Crown rust is common in Mexico. In the southern states of the United States and Mexico, the loss of oat forage is probably as important as or more important than losses in grain production.

Stem rust (caused by *P. graminis* Pers.) epidemics on barley develop more slowly than on wheat for a number of reasons that vary in importance with year and location. Barley as a host has a shorter vegetative period, grows at lower temperatures, is more likely to winter kill to the crown, and is grown on smaller areas in Mexico and the southern states than is wheat. In addition, barley has a level of resistance to *P. graminis* both as seedling and as adult plant (Steffenson *et al.*, 1982a,b). Epidemics of stem rust on barley have occurred most frequently in Iowa, Minnesota, North Dakota, Illinois, South Dakota, and Wisconsin (Roelfs, 1978).

Leaf rust of barley (caused by *Puccinia hordei* Otth.) has been even less important than stem rust, mainly due to the small area planted to winter barley, the only known host on which the disease can overwinter, and because of the earliness of the crop, which limits the time for disease development (Roelfs, 1978).

The rusts of rye (caused by *P. graminis* Pers. f. sp. *secalis* and *P. recondita* Rob. ex Desm. f. sp. *secalis*) are not a major problem in North America. The crop is very winter-hardy, and stem and leaf rust diseases can survive the winter into southern Canada (Johnson and Green, 1952). Rye is cross-pollinated, resulting in some diversity for resistance, and rye is grown on small widely scattered areas. Stem rust has never caused major losses in the United States, and leaf rust has caused only local losses (Roelfs, 1978). Rye leaves normally senesce at an earlier growth stage than in other cereals; therefore, leaf rust must be severe early in the season to have a major effect on yield. Most of the inoculum is endogenous to the area where the disease develops.

Rusts of corn have been minor diseases in the United States, regardless of the millions of hectares of corn grown (Chapter 7). This may be partly due to host resistance, but also to the lack of an overwintering host (Kingsland, 1975) such as occurs with the winter planted cereals. When present the disease has been the most common on late-planted fields (Wallin, 1951; Futrell, 1975) and on sweet corn.

Rust of sugarcane is relatively new to North America and is discussed in Chapter 8 with rusts of sugarcane.

II. Wheat Production and Rust Epidemics

The cereal rust epidemiology studies in North America have mainly involved the wheat rusts. Those at least nationwide in scope were of wheat stem rust in Canada (Craigie, 1945; Peturson, 1958), Mexico (Stakman *et al.*, 1940; Borlaug, 1954), and the United States (Stakman, 1934; Stakman and Harrar, 1957). Epidemics of leaf rust of wheat in the

United States were reviewed by Chester (1946). Stripe rust epidemics were studied in California (Tollenaar and Houston, 1967), Oregon (Shaner and Powelson, 1973), Washington (Burleigh and Hendrix, 1970), Montana (Sharp and Hehn, 1963), and Canada (Sanford and Broadfoot, 1932). A review article by Rajaram and Campos (1974) reviewed the epidemiology of the three rusts in the Western Hemisphere. Due to the abundance of information on the epidemiology of the wheat rusts, the rest of this chapter will be concentrated on them, but will draw on information from epidemiological studies of the other cereal rust diseases when it is advantageous to do so. Additional information on epidemiology of the other diseases will be found in the specific chapters on that disease in this volume.

Wheat is the principal cultivated crop in much of North America, especially in the drier areas (Fig. 1). The approximately 40 million hectares of wheat grown annually in North America is comprised of about 400 different cultivars; however, only about 20 of these are grown on over 400 thousand hectares each (Table I).

A. WHEAT AND RUST IN MEXICO

Wheat is intensively cultivated in isolated areas across Mexico where irrigation water is available. Wheat is generally seeded in November and early December; however, volunteer plants can exist in the summer crops. The period November through April is generally the dry season, which creates the need for irrigation. Wheat is generally harvested in April and May and is followed by a summer crop of corn. Rust is usually light through the growing season, but leaf and stem rust can become severe near maturity with the onset of the summer rains. The major wheat-producing areas are the El Bajio, mainly in the state of Guanajuato (Gj) (Fig. 1), and the Pacific coastal plain, mostly in the states of Sonora (So) and Sinaloa (Si). The cultivars grown are selections made and released by INIA (Instituto Nacional de Investigaciones Agricolas) from the wheat program of Centro Internacional de Mejoramiento de Maíz y Trigo (CIMMYT). Stripe, leaf, and stem rusts occur but the cultivars are generally resistant and epidemics have been few in the past 30 years. The most recent leaf epidemic in the state of Sonora resulted in part from a virulent race oversummering on volunteer wheat plants and then moving to early-planted wheat (Dubin and Torres, 1981). The data are meager, but it appears that the rusts normally survive throughout the year in the intermountain highlands and, with favorable conditions, can move northeast into the southern

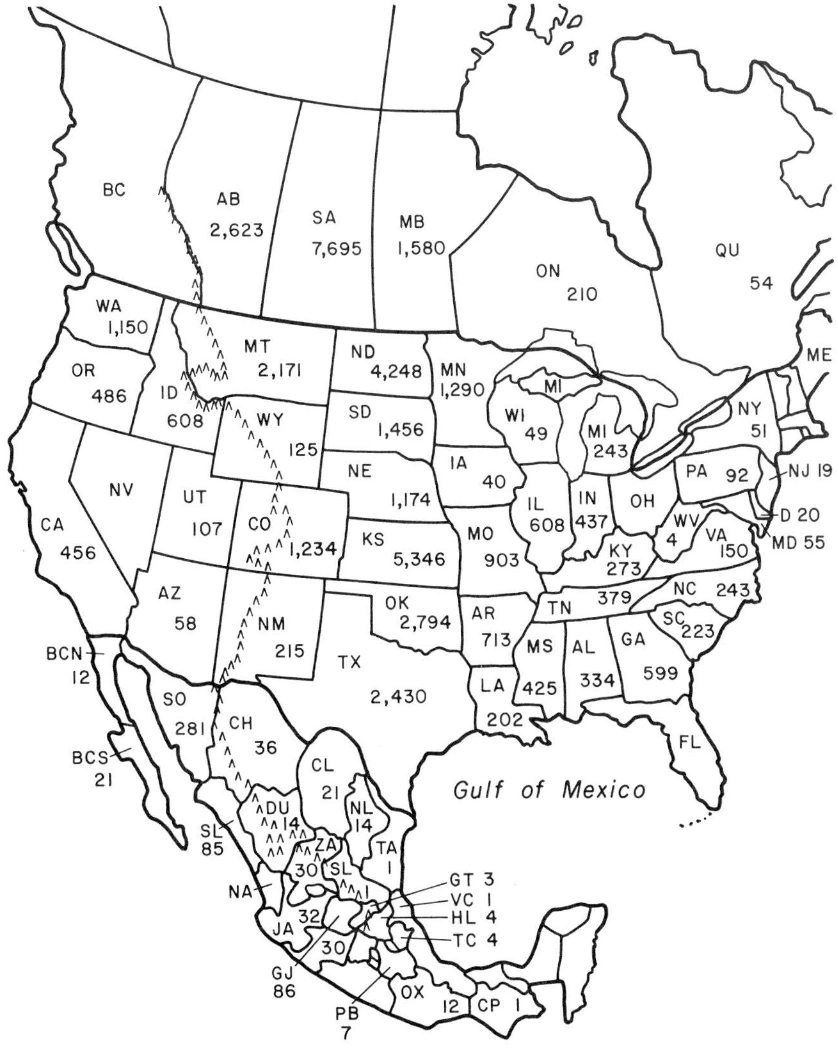

Fig. 1. Distribution of wheat in North America. Values are in 1000 hectares using the period 1978–1981 as a guide. Figures for Mexico were obtained from various officials of Instituto Nacional Investigaciones Agricolas, for the United States from Briggle *et al.* (1982), and for Canada from the Canadian Cooperative Wheat Producers.

Table I

The Principal Wheat Cultivars in North America and Their Rust Resistance as Known[a]

Cultivar	Hectares	*Sr* gene(s)	*Lr* gene(s)	Market Class
Neepawa	6,884,160	*5,+*	*13*	HRS[b]
Centurk	1,813,296	*5,6,8,9a,17*	—[c]	HRW
Tam W101	1,527,973	—	—	HRW
Scout	1,414,769	*2,17*	—	HRW
Olaf	1,195,444	*8,9b,12,Wld-1*	*2a,10,+*	HRS
Eagle	1,133,600	*2*	—	HRW
Scout 66	993,941	*2,17*	—	HRW
Arthur 71	916,561	*5,36*	*9*	SRW
Sage	890,808	*17,24*	*24*	HRW
Butte	745,760	*6,8,9g,Wld-1*	*10*	HRS
Era	731,075	*5,6,12,17*	*10,13,+*	HRS
Arthur	724,075	*5,36*	—	SRW
Wakooma	704,160	?	?	Dur[b]
Sinton	663,880	?	*10,13,+*	HRS[b]
Waldron	657,446	*5,11,Wld-1,+*[d]	*2a,10,+*	HRS
Bento	655,920	?	*1,2a,12,13*	HRS[b]
Ward	605,619	*9e,+*	?	Dur
Abe	546,144	*5,36*	*9*	SRS
Triumph	529,664	*Tmp*	—	HRW
Larned	418,812	—	—	HRW
Triumph 64	418,004	*Tmp*	—	HRW
New cultivars with high potential use				
Newton	1,997,460[e]	—	*1,11,+*[f]	HRW
Tam W105		—	—	HRW
Columbus		?	*13,16*	HRS

[a]Data are 1982 for Canada and 1979 for the United States; no cultivar from Mexico is grown on 400,000 hectares.
[b]Canadian cultivars.
[c]Normally susceptible in the field.
[d]Segregates for *Sr5* and *Sr11*.
[e]1980.
[f]Segregates for *Lr11*.

United States and northwestward into the major production areas in Sinaloa and Sonora. Barley production in Mexico is limited now, although earlier it was an important source of stem rust inoculum (Borlaug, 1954). Wheat leaf rust appears to survive throughout the year in at least trace amounts in all but the production areas of the northwest. There seems to be relatively little direct movement of rust from

Mexico into the main production areas of the United States in most years. The pathogen races in Mexico and south Texas are similar, but generally are different from those in the central and north central areas of the United States (Roelfs *et al.*, 1982). Although early evidence indicated an annual interchange of inoculum from north to south across North America in the fall (Stakman and Harrar, 1957), since 1970 with less rust in the United States and Canada this has not been true. Race 15 has predominated in areas north of Mexico but has not been found in Mexico. Leaf rust virulence for either *Lr9* or *Lr24* is common in south Texas and the very small wheat-producing area along the Texas border in Mexico, but neither of these virulences has been found in the central or northwest areas of Mexico. With wheat rusts oversummering at least as far south as Kansas in the summer (Burleigh *et al.*, 1969b), inoculum from the far north probably causes few if any infections in the southern United States and Mexico.

B. WHEAT AND RUST IN THE UNITED STATES

Wheat is grown in 42 of the states in the United States. Approximately 80% of the wheat is planted in the fall. The remaining 20% is planted in the spring to bread and durum wheats (Table I). Fall seeding starts in late August on the Canadian border and ends in mid December along the coastal area of the Gulf of Mexico, and southern Texas. Spring planting starts in April as soon as the ground can be worked and continues into early May. Harvesting of the crop starts in the south in May and continues northward until it reaches the Canadian border in early to mid August. About 370 different cultivars of wheat are grown annually (Briggle *et al.*, 1982), but generally a few cultivars make up most of the area. Because of the differences in grain characteristics of the five major classes of wheat, diversity is maintained on a national, if not a regional, basis (Table I). Cultivars are developed by both private and public breeding programs. The 17 most popular cultivars grown in the United States in 1979, were developed in eight different breeding programs. In the higher rainfall area (over 900 mm) in the eastern third of the United States, soft red winter wheat (SRW) is grown. In the southern two-thirds of the Great Plains (Texas, Oklahoma, Kansas, Nebraska, Colorado), hard red winter wheat (HRW) is grown with a narrow band extending through western South Dakota, southern North Dakota, into north central Montana. The hard red spring (HRS) and durum (Dur) wheats are grown in the Dakotas, Minnesota, and northeastern Montana. In the Pacific northwest, Michigan, and New York, white winter wheats (WW) are the most important. Most of the wheat in the United States (90%) is grown under dry land conditions, except in the drier western portions of Kansas, Oklahoma, Texas, and the eastern portions of Colorado and Wyoming, where a portion of the crop is irrigated. There are also important areas of irrigation in California and Washington (Reitz, 1976). There are many areas where planting practices vary from this outline, but they are small and too numerous to be considered on a continental basis.

Leaf rust is a major disease of wheat throughout most of the United States, with the greatest losses historically occurring along a band from Pennsylvania through Kansas and Oklahoma (Roelfs, 1978). Leaf rust can overwinter on wheat wherever the fall-infected leaves survive through the winter. Although it probably is rare, leaf rust has overwintered on wheat as far north as southern Wisconsin (Christman, 1904). Rarely does overwintering occur north of 41°N latitude but it usually survives as far north as 40°N latitude. Leaf rust can also overwinter throughout most of the Pacific northwest.

Stem rust is a major disease of spring and durum wheats in the north central states and extends southward into Nebraska (Roelfs, 1978). It originally survived the winter in the northern states as teliospores, and then by subsequent infection of the barberry in the spring it continued its cycle. However, an extensive barberry eradication program had eliminated most of the barberry growing near grain fields by the mid 1920s (Stakman and Harrar, 1957). Currently, inoculum for this area is principally from overwintering rust south of 34°N (Stakman, 1923), but rust can overwinter farther north (Paxton, 1921). In the southern half of the country, stem rust seldom causes any loss (Roelfs, 1978), because the winter wheat heads soon after the last freeze and leaves the pathogen little time to increase to epidemic levels. However, adequate inoculum is produced so that it can be blown northward in sufficient amounts to initiate epidemics. The date of the first detected infection in the northern areas is closely related to yield loss, with greater losses occurring in years when the disease arrives earlier (Hamilton and Stakman, 1967; Roelfs and Long, 1980) than usual.

Stripe rust has been the major rust of wheat west of the Rocky Mountains. Epidemics have occurred in cycles (Section I,C). These cycles may well be related to variation in temperatures between years (Coakley and Line, 1981a,b). Stripe rust also occurs in the central third of the United States. The inoculum may be from the highlands of central Mexico (Guanajuato, Gj; Fig. 1), arriving in December and January and infecting susceptible wheat in Texas, Louisiana, and Arkansas. No study of the epidemiology of this population has been attempted.

C. Wheat and Rust in Canada

Wheat is an important crop in Manitoba, Saskatchewan, and Alberta. Nearly two-thirds of the area is of a single cultivar, and due to quality standards, most cultivars are rather closely related. The major wheat crop is hard red spring (87%), with the remainder mostly durum wheat along with small amounts of winter wheat in Alberta, Ontario, and the Maritime provinces. Little if any rust overwinters on wheat in Canada, except possibly for stripe rust in Alberta or leaf rust in Ontario. Most of the inoculum comes from the United States.

III. Sources of Inoculum

Although the hosts are different, most of the cereal rust fungi have two basic sources of inoculum for the cereal host, urediospores and aeciospores. The exceptions are stripe rust, for which there is no known alternate host, and sugarcane rust, which is autoecious. However, the principal source of inoculum for all of the cereal rusts in North America currently is from urediospores.

In epidemiological studies, inoculum is often spoken of as exogenous, from outside the area under consideration, and as endogenous, from inside the area under consideration. These terms are not always clearly distinguishable, because what is an endogenous source of inoculum for field A can also be an exogenous inoculum source for a nearby field B. Thus, when it is said that the state of Texas serves as a source of exogenous inoculum for the northern states, it must be understood that some fields in Texas probably also have exogenous sources. Once the initial infection occurs in any field or on any plant, generally the endogenous inoculum produced there is the primary source of inoculum for further disease development.

The initial source of annual infection of wheat in North America is almost exclusively by urediospores. These spores may have been produced on volunteer plants nearby or on the wheat crop to the south. Other uredial hosts probably have a minor role, except perhaps *Hordeum jubatum* L. (for wheat and rye stem rust) in the north central states. However, *Avena fatua* L. may play an important role in oat stem and crown rust epidemics in some areas. Although urediospores can, if unweathered, survive on crop debris (Orr and Tippetts, 1971) for a short period, they probably have little potential for infecting the next crop.

A. Aeciospores

Aeciospores are released singly or in clumps, and only in approximately 100% humidity. However, the numbers of spores produced are very large. An average-sized barberry bush in southern Minnesota was the source of an estimated 64 billion spores in a few days (Stakman *et al.*, 1927).

The alternate hosts in North America were an important source of inoculum for *P. graminis* and *P. coronata*. Barberry, *Berberis vulgaris* L., once was a major inoculum source of wheat, oat, and rye stem rust in the north central states of the United States and Canada, and can still be a source of early inoculum for oat stem rust from Pennsylvania through southern Ontario, and of wheat stem rust in eastern Washington and northern Idaho. Most of the aeciospores currently isolated in the eastern United States are of *P. graminis* f. sp. *secalis,* which spreads to noncereal grasses. The barberry eradication programs of the United States and Canadian governments and their subdivisions have removed nearly all barberry from close proximity to small grain cereal fields. However, a few barberry bushes remain in most of the north central states of the United States and eastern Canada.

The importance of aeciospores as inoculum was in their genetic diversity (Roelfs and Groth, 1980; Groth and Roelfs, 1982). Aeciospores also served as a source of inoculum early in the season (Roelfs, 1982). Epidemics that were initiated by aeciospore infections were generally local in effect (Stakman *et al.*, 1927). Most aeciospores travel less than 100 m; however, a few spores can travel much longer distances. Lambert (1929) found effective spreads were limited to 3.5 km. Urediospores resulting from the initial aeciospore infection generally formed a fan-shaped infected area with a length of up to 15–30 km and a width of one-fourth the length. Hutton (1927) indicates a spread of 160 km was known. Aeciospore-initiated epidemics were usually north of the 40° parallel (Stakman, 1923). Viability of teliospores were generally reduced when they were produced in hot dry conditions further south. However, aeciospores were observed farther south at high elevations (2000 m) (Stakman, 1923), and epidemics resulting from aeciospores were documented in Colorado by Durrell and Lungren (1927) south of the 40° parallel. The Colorado locations varied from 1200 to 1500 m elevation compared with 300 m in eastern Nebraska at 40° at the southern edge of barberry-initiated epidemics. Local epidemics that were the result of initial inoculum from barberry were documented in Illinois and Indiana (Beeson, 1923), Iowa (Melhus *et al.*, 1920), Minnesota (Stakman *et al.*, 1927), North Dakota, South Dakota, and Wisconsin (Walker, 1927).

Rhamus cathartica L. was associated with epidemics of crown rust in the northern United States and eastern Canada. Inoculum from *Rhamus* has never resulted in as serious regional losses as has the inoculum from barberry. This difference is due in part to the oat crop having lesser value, the smaller area of oats grown in a region, and because crown rust attacks only the leaves and leaf sheaths. Thus, there is less tissue available that senesces earlier than true stem spike tissue, which also serves as a host for the stem rusts. Current evidence, including race stability, pathogen uniformity, and few local early epidemics, would indicate that the sexual cycle is essentially a local problem occurring principally in limited areas of Iowa, Minnesota, Wisconsin, and Ontario.

B. UREDIOSPORES

Urediospores are the principal source of inoculum of all the cereal rusts in the United States. A single uredium can produce over 1000 urediospores per day for a period of several weeks (Section VI). These spores are relatively long-lived (Bromfield, 1967; Peltier, 1925; Reed and Holmes, 1913; Thiel, 1938; Johnson and Green, 1952; Mohamed, 1960), survive in the field away from the host for periods of several weeks (Stakman *et al.*, 1923; Orr and Tippetts, 1971), and can be carried long distances (Christensen, 1942) before they are generally scrubbed from the air by rain (Rowell and Romig, 1966).

IV. Exogenous Inoculum

Exogenous urediospores seldom, if ever, arrive in such amounts as to directly cause an epidemic, but increase in inoculum from the resulting infections often can initiate an epidemic. Exogenous inoculum is mostly ineffective. The loss in numbers and viability of urediospores during transport is tremendous. Even upon deposition, most viable urediospores are deposited on nonvegetative material, nonhost plants, or resistant cultivars, or under environmental conditions that are unfavorable for germination, penetration, and establishment.

A. Spore Trapping

Exogenous inoculum can be detected by spore trapping or by direct observation of disease occurrence. Spore trapping has been done in North America for over 60 years. Initially, spores were trapped by impaction on microscope slides that had been thinly coated with petroleum jelly. Although large numbers of spores are trapped, impaction traps are very sensitive to local inoculum (Roelfs *et al.*, 1968). Volumetric traps have also been used in North America, but on a more limited scale. Although spores are trapped, correlations are poor with initial arrival of inoculum (exogenous). This is probably due to several factors: trapping of nonviable urediospores, viable spores but conditions unfavorable for infection, failure to properly distinguish between urediospores of closely related species or *formae speciales*, resistance of host plants, and an inability to sample enough air volume. The most effective exogenous inoculum is deposited during rain storms, when volumetric samplers currently used are rather ineffective. Thus, in the United States, rain samples are the best device for the detection of exogenous inoculum (Rowell and Romig, 1966; Roelfs *et al.*, 1970). To be most effective, the inoculum source must be far enough away so that the spore clouds are relatively randomly dispersed. Urediospores of a similar morphology cannot already be present in the area of the trap, and the rainfall pattern needs to be such that sampler locations can be maintained with assurance that at least one will be in the area of each storm.

B. Monitoring Disease Spread

Because of the problem of recognizing urediospore morphological differences between some species, *formae speciales*, and races, it is necessary to do field surveys of the diseases to detect inoculum movement. These surveys have been conducted over most of North America for over 60 years.

The following techniques have been developed by the Cereal Rust Laboratory, partly because of cost and partly because they are effective in detecting rust, especially when it is present in trace amounts. Survey routes are chosen that cross the major small-grain cereal-producing areas when the majority of the crop is from late boot to early dough growth stages. The routes chosen are normally all-weather roads, and surveys are made on approximately the same date annually, adjusting only plus or minus 7 days for annual differences in the crop growth stage. Once in the area, the surveyor stops at the first small-grain cereal field after the odometer reading is divisible evenly by 10. Subsequently the surveyor stops every 20 miles (32 km) or at the first field after that distance. The surveyor normally spends approximately 15 min in each field or plot (or at a stop in case of several adjoining fields). Notes are taken on environmental conditions, host condition, host maturity, cultivar grown, rust severity and prevalence, loss potential, and other diseases. Collections of urediospores are made for physiological race determinations. If rust is common, the surveyor examines a strip approximately 1 m wide along a loop extending 100 m into the field and reports an average, except when the loop is obviously atypical to the field. When rust is scarce, the surveyor uses previous experience to determine the area of the field that is most likely to be rusted. However, data recorded is for the field mean. Experimental, demonstration, and trap plots are examined to ensure that the resistance of the commercial cultivars are not excluding rust development. For example, in the hard red spring wheat areas of the Northern Great Plains, in recent years stem rust is seldom found in commercial fields; however, susceptible cultivars in experimental plots are often severely rusted. Thus, the surveyor rapidly determines the potential for rust in an area and the virulence for the commonly occurring cultivars in the experimental plots. If no rust is found on susceptible cultivars, no extra effort is required in the commercial fields; however, if susceptible cultivars in research or demonstration plots are rusted, the surveyor usually makes a special effort to check in nearby fields or check additional fields to verify the extent of the disease occurrence. When fields are examined more frequently than every 32 km, the additional stops are recorded as special stops. Occasionally a change of route will be necessitated by local conditions affecting the host, disease, or road. This can only be done when the surveyor is well versed in both agriculture and roads in the area.

C. CHARACTERISTICS OF SPREAD FROM EXOGENOUS INOCULUM

Disease resulting from exogenous inoculum characteristically has the oldest infections at a standard plant height (Roelfs and Rowell, 1973). The initial infections from exogenous inoculum frequently are at the top of the canopy when the inoculum arrives. The secondary infections tend to be at the same height or just above or below older infections. The oldest infections are generally distributed at random through the area (Rowell and Roelfs, 1971), except when the environment is so marginal for rust infections that it occurs only in certain areas of the field or in certain fields. However, the distribution over a larger area will still be random. These infections are often still high in the canopy when sporulation occurs, which results in the spores spreading very rapidly horizontally, and often vertically, up from the field.

V. Endogenous Inoculum

In a regional approach to epidemiology, the term endogenous inoculum is normally used to refer to inoculum produced within a given field or local area, although in experimental epidemiology the term may be used for very much smaller units. Endogenous inoculum is the immediate cause of nearly all cereal rust epidemics, even though the initial infections may have been caused by exogenous inoculum.

A. Spore Trapping

Spore traps can be used as an estimate of the disease severity. Our experience indicates that impaction-type traps are the most effective, and one may choose from the volumetric type (rotobar) (Asai, 1960) or wind-impaction types such as a rod or slide (Roelfs, 1970; Bromfield *et al.*, 1959). Volumetric samplers can be used, but they may be more likely to be influenced by exogenous inoculum than are impaction traps. Although it is not always possible or even desirable to do so, spores usually should be collected over areas with a radius of 10 m (Roelfs, 1972) if spore concentrations are to be representative of what happens in large fields. Traps exposed outside of the crop are affected by distance from the source and wind direction (Roelfs, 1972), causing erratic data and low spore numbers (Roelfs *et al.*, 1972; Roelfs, 1972). Neither sedimentation nor rain samples are recommended for studying endogenous inoculum. Spore trapping can be a very valuable tool in studying epidemiology in plots. However, until an electronic counter is developed that rapidly distinguishes particles by color, shape, and size, it is currently too expensive to adequately study epidemics over millions of hectares by spore trapping.

B. MONITORING DISEASE SPREAD

The frequency or intensity of disease caused by endogenous inoculum decreases as distance from the source increases. The best-documented natural spreads of this type in the United States are those of *P. graminis* from infected barberry bushes (Section III,A) and from urediospore spreads by Celik (1974), Elliot (1960), Emge and Shrum (1976), Kingsolver (1980), and Underwood *et al.* (1959). Spread from endogenous inoculum is found around volunteer plants in newly seeded winter wheats and when the disease spreads from a winter to a nearby spring wheat, and most commonly in the spring from an overwintering focus. The spread usually is fan-shaped with its origin at the upwind source. This spread pattern can be modified by areas of nonhost plants or local environmental conditions or both. The downwind spread follows the direction the wind is blowing at the time most of the spores are released and when conditions were favorable for infection after their dispersal. Thus, downwind may not be the mean direction of air flow during any specific time period.

Disease resulting from endogenous inoculum sources generally occurs relatively early in the season, often within a few weeks of emergence, or after a period of very unfavorable environmental conditions for the disease, such as winter for a fall-planted crop. Generally the host is growing rapidly so that infection at the top of the canopy may be under the canopy when sporulation occurs. Thus, many of the spores are trapped within the canopy, which results in a heavy infection low on the plant that can be overlooked by inexperienced observers. These foci of severely rusted plants grow gradually in size but seldom reach more than a meter in diameter before reaching the top of the canopy. Once the disease reaches the canopy top, it rapidly spreads horizontally across the field. These foci may have severities of 10–100%, and the ground beneath them may take on the color of urediospores before the disease has spread horizontally more than a meter. Because these foci generally result from a single infection that survives the unfavorable environmental conditions of winter, they seldom occur in a random manner, but occur in specific sites, e.g., sheltered, most fertile, or wetter parts of the field.

C. CHARACTERISTICS OF SPREAD FROM ENDOGENOUS INOCULUM

The most noticeable difference between infections from endogenous and exogenous inoculum is that the oldest infections from endogenous sources are generally low in the canopy. Foci are common, and horizontal spread is limited until the disease reaches the top of the canopy. Foci are nonrandom in distribution, and different foci are often the result of a different pathogen race. Foci are likely to occur in some predictable even though erratic pattern. For example, they may occur near old volunteer plants, on plants protected by snow cover along a tree row (Eversmeyer and Skidmore, 1974) or snow fence, etc. Normally on a single cultivar the most severe rust infection occurs in the earliest plantings.

VI. Urediospore Movement

Urediospores of the cereal rusts are passively released, transported by air currents, and deposited under the forces of gravity, impaction, and rain. Urediospores vary in size and shape depending on the fungus, but the size usually ranges from about 20 to 30 μm (Savile, Vol. 1). Urediospores have a terminal velocity of 0.97 to 1.24 cm/sec (Ukkelberg, 1933). Thus, in still air, a spore would fall 0.67 m/min, 40 m/hr, or approximately 0.9 km/day. At this rate of descent a spore at 1500 m (about half the normal maximum height of heavy spore concentrations) in a 13.3 m/sec (30 mph) wind would travel 1760 km before landing. Of course, with this wind speed the air is composed of many eddies, updrafts, etc., which tend to extend the time a spore is airborne. Urediospores generally move in greatest numbers when the plants are rapidly growing (warm to hot temperatures), foliage remains dry (low relative humidity), and near mid-day (usually the windiest and driest period of the day).

The shape of a urediospore probably is important to its response to transport and deposition, but much more information is required about this relationship. In studying different impaction traps under field conditions, Roelfs *et al.* (1968) found that urediospores of leaf and stem rust of wheat vary in their frequency of impaction on both vertically exposed microscope slides (2.54 × 7.62 cm) and 5-mm-diameter rods. On the rods, the urediospores of leaf rust tended to be on the mid arc facing the wind, while the majority of stem rust urediospores were more on the sides. These differences in impaction may relate to differences in the size and shape of urediospores.

Large numbers of urediospores are produced from a uredium, and this production continues for several weeks, resulting in the production of 100,000 urediospores for wheat stem rust (Katsuya and Green,

1967; Stakman, 1934) and over 28,000 for wheat leaf rust (Chester, 1946). The urediospores are released on a diurnal cycle peaking near mid-day (Asai, 1960). Most are trapped by the canopy and are deposited by sedimentation or impaction. Sedimentation is deposition due to the force of gravity while impaction results from urediospores being driven onto plant tissue by air movement within the canopy. Thus, of the millions of urediospores produced per acre, only a small fraction escape the canopy.

A. TRANSPORTATION

As the urediospore escapes the canopy, it is influenced by eddies of many types. Again, most of the urediospores are probably deposited within a 100 m of the source (Roelfs, 1972). Only 10% of the urediospores in a horizontal plane above the canopy are in the same plane 100 m downwind. However, because of the large numbers of spores produced, significant numbers of urediospores are transported vertically. These urediospores transported vertically generally decrease in number with elevation but are relatively numerous up to 3300 m above rusted fields (Stakman *et al.*, 1923). Claims of movements of viable urediospores have been reported for distances of over 8000 km. However, some of the distances are questionable because ground surveys were inadequate to establish that there were no spores produced between the stated source and the point of investigation. This is important because even 0.4 hectare of wheat with less than a 10% severity can produce a trillion urediospores (1.02×10^{12}) (Rowell and Roelfs, 1971). Furthermore, urediospores trapped of the various rusts are often difficult or impossible to distinguish microscopically with certainty on an individual spore basis. From most traps, it has also been impossible to determine if the urediospore was alive when it was trapped. Perhaps some of the new trapping methods will solve this last problem (Schwarzbach, 1979). The longest documented single movement by urediospores of *P. graminis* in North America was at least 680 km. Infections of wheat stem rust were found on volunteer wheat plants at two locations at Churchill, Manitoba, which is separated from the nearest wheat-producing area near Winnipeg by forest, lakes, and tundra (Newton, 1938). The world record may well be the transport of spores from Africa to Australia (see Chapter 10). However, Stakman (1934) found that few spores were in the air more than 400 km from a source area.

Vertical distribution of urediospores of wheat stem and wheat leaf rust was studied 1–6 m above the canopy by Eversmeyer *et al.* (1973). They found a decreasing number of urediospores of *P. recondita* as height increased, but an increasing number of *P. graminis.* This difference was attributed to differences in amounts of exogenous inoculum. Additional studies are needed on the vertical spore concentrations over isolated infected sources.

B. DEPOSITION

Urediospore deposition is caused by three different mechanisms; (1) sedimentation, (2) impaction, and (3) rain scrubbing. Transportation of spores within the crop canopy usually ends by sedimentation onto leaves, stems, or the ground by the force of gravity. The longer the spore remains airborne within the canopy, the more likely it is to be impacted by small eddies on a nearby stem or leaf. Those urediospores that escape the canopy by moving out vertically are carried by stronger wind eddies, updrafts, and currents. Although most are still deposited in the first 100 m from the source (Roelfs, 1972), the farther a spore travels, the less likely it is to be deposited by sedimentation or impaction. Therefore, the importance of deposition by rain scrubbing (Rowell and Romig, 1966) increases. Samples of rain water analyzed for urediospore content were successfully used to predict disease onset in the north central area of the United States (Roelfs *et al.*, 1970).

VII. Factors Affecting Epidemic Development

Epidemics of the cereal rusts are affected by four independent yet related factors: host, pathogen, environment, and time. The relationship between these factors and the disease is such that an independent discussion of each is not realistic, but is so complex that understanding and studying all the interactions together is difficult.

A. HOST RESISTANCE—PATHOGEN VIRULENCE

As pointed out in Chapters 5 and 6 in Volume 1 of this series, a host is susceptible if it does not possess a gene for resistance, or if the pathogen culture has virulence matching the host gene(s) for resistance. Besides this race-specific resistance, most cultivars have additional non-race-specific resistance [at least not known to be race-specific (Chapter 16)]. Both types of resistance exist in nearly all cultivars (Table I) grown in areas of North America where a rust disease has been historically important. In 1972, Blueboy, a soft red winter wheat cultivar that had the ineffective *Lr1* and *Lr10* resistance, was grown in Georgia, and losses of 50% were incurred (Roelfs, 1978), even though leaf rust does not usually cause serious losses in wheat in Georgia. In 1974, the soft red winter wheat cultivar McNair 701 was introduced in Florida, Georgia, and South Carolina, where stem rust seldom occurs; however, the pathogen rapidly increased and severe losses were incurred (Roelfs, 1978). In both of these cases, other susceptible cultivars (no detectable race-specific resistance to the pathogen genotypes present) suffered much less loss. Thus, in addition to knowing the virulence of the pathogen population (Vol. 1, Chapter 5) and resistance of the host, it becomes necessary to study nonspecific type of resistance of the host.

Additionally, in asexual-reproducing organisms like most of the cereal rusts, important differences may exist between cultures for characteristics other than virulence. Aggressiveness measured as reproduction potential was studied by Katsuya and Green (1967). Many other uncharacterized differences exist between cultures. However, if these differences are polygenic in nature, they may remain together long enough in asexually reproducing organisms to be important for detailed study.

In oat stem rust, the most common race is NA-27, which is virulent on the *Pg-2* and *Pg-4*, which are the only resistance genes widely used. In commercial oat fields, scattered plants of wild oats (*Avena fatua* L.) exist that have no known resistance. Races Na-27 and NA-16 both occur on the wild oats, with the frequency of NA-16 greater than on commercial oat cultivars, although the numbers of urediospores of race NA-27 deposited on the wild oats must be 15–1000 times greater than that of NA-16. The greater frequency of NA-16 on wild oats apparently indicates more aggressiveness in this race than in race NA-27 on wild oats.

B. Date of Disease Onset

Hamilton and Stakman (1967) found that epidemics of wheat stem rusts in the central third of the United States were related to the date of rust appearance. Roelfs and Long (1980) found in the north central part of the United States that over a 38-year period the date of disease onset explained about 20–40% of the estimated yield losses of oats from stem rust. Date of onset explained a greater portion of the variation than any other factor studied, including temperature and rainfall. The most prevalent race NA-27 is virulent on nearly all the cultivars grown in North America; thus resistance–virulence was not a factor (Roelfs and Long, 1980). The pathogen in this case was windborne from a southerly location, so the relationship between date of disease onset and losses in the north should be expected. The relationship between disease onset and loss could be improved with better data. Disease severity on date of disease onset was not considered and it varied from a single uredium (less than 0.001 severity) to 20% (200 uredia per tiller). Another variable not considered in this study was the host growth stage at time of disease onset. An early infection generally results in a greater disease loss, all other factors being the same. In the north central area studies, the planting data within and between years may vary as much as 21 days. In this study, the yield loss was estimated for the entire state. Had actual yield losses been available for the fields in which the rust was reported, the date of disease onset in oat stem rust might have explained most of the variation in losses due to oat stem rust.

C. Initial Disease Severity

In areas where the disease overwinters, the "critical month" principle (Chester, 1943) is useful in predicting losses. Essentially this theory is based on the premise that if rust is present before winter, some disease will survive the winter. However, in the spring when the host can grow at temperatures too cool for disease increase, or when the old leaves die before the rust spreads from them to the newer leaves, a critical period occurs for the survival of the disease. At the end of this critical period, Chester's critical month, the increase of disease is very constant between years, because environmental conditions are similar and the disease at the end of the critical period is closely correlated with final severity. Although the critical month theory appears to be crude and is quite simple, it has worked relatively well over the years. The environmental conditions that favor wheat during its reproductive process also are favorable for the obligate parasite of wheat. The environmental conditions after the "critical" month are those in the broad area of the fungal growth curve, one standard deviation either side of the optimum for the pathogen when a change of several degrees in temperature has little effect on the pathogen. During and before the critical month (just before the boot stage of the host in Oklahoma),

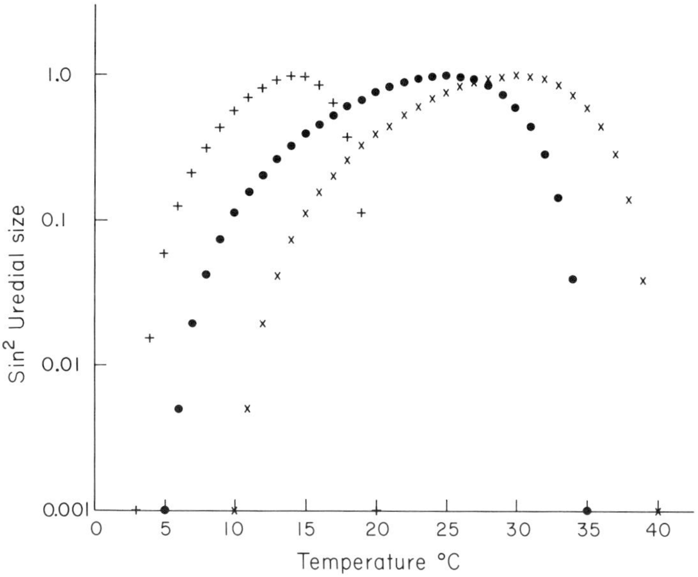

Fig. 2. The effect of temperature on fungal growth for stripe +, leaf ●, and stem X diseases of wheat using Schrodter sin² transformation, as modified by Dirks and Romig (1970).

conditions can be in the tail of the growth curve, which is near the minimum for pathogen growth, where a single degree difference in temperature has a major effect on fungal growth (Fig. 2).

The frequency of initial infection and/or the frequency of infection on a given date are also useful for predicting losses due to disease. The common measurement of disease severity is based on the percentage of the maximum possible infection present. The scale most used worldwide is the modified Cobb Scale (Peterson *et al.*, 1948). Another sacle was designed by James (1971) and is based on percent of the leaf tissue diseased (percentage of the green tissue occupied). Either scale is satisfactory, but the scale used should be indicated. Prevalence is used by most cereal rust workers to indicate the percentage of tillers infected (plants are usually not possible to identify without pulling them). However, in studying the entire epidemic it may be desirable to use a single value for disease severity and prevalence. This value can be calculated determining the number of ureida per tiller that equal 1%. Kingsolver *et al.* (1959) found for stem rust that 10 uredia per tiller equals 1% severity on the modified Cobb Scale. Thus, 1 uredium/tiller = 0.1%, 1 uredium/10 tillers = 0.01%, 1 uredium/100 tillers = 0.001%, etc. For wheat leaf rust, Burleigh *et al.* (1969a) found that 18 uredia/tiller equals 1% and thus 1.8 uredia/tiller = 0.1%, 1.8 uredia/10 tillers = 0.01%, etc. The number of crown rust and rye leaf rust uredia per 1% would be similar to wheat leaf rust and for barley leaf rust the number would be larger. Disease severity is correlated with numbers of spores impacted just above the canopy (Burleigh *et al.*, 1969a), which can also be used to estimate disease losses.

D. ENVIRONMENTAL FACTORS

Many workers have studied the effect of environmental factors on epidemics in North America (Table II). It would appear that nearly every possible combination of temperature—maximum, minimum, winter, summer, $\sin^2$ transformed, weekly means, monthly means, annual means, degree days, etc.; precipitation—days of rain, hours of

Table II

Major Studies of the Effect of Environmental Factors on Cereal Rust Epidemics in North America

Disease	Factor(s) studied	Reference
Wheat stripe rust	Temperature, dew	Burleigh and Hendrix, 1970
	Temperature, precipitation	Line, 1976
	Temperature, precipitation	Coakley, 1979
	Temperature	Coakley and Line, 1981b
	Temperature, precipitation, degree days	Coakley and Line, 1981a
Wheat stem rust	Temperature	Walster, 1921
	Temperature, precipitation	Tehon and Young, 1924
	Temperature	Stakman and Lambert, 1928
	Temperature, wind, precipitation, rainy days	Lambert, 1929
	Temperature, precipitation, rainy days	Peltier, 1933
	Temperature, humidity, precipitation	Atkins, 1936
	Temperature, hours of sun, wind, precipitation	Craigie, 1945
	Temperature, rain	Wallin, 1964a,b
	Temperature	Katsuya and Green, 1967
	Dew	Prabhu and Wallin, 1970
	Temperature, precipitation, dew	Burleigh *et al.*, 1972
Oat stem rust	Temperature, precipitation	Roelfs and Long, 1980
Wheat leaf rust	Temperature, precipitation	Tehon, 1927
	Temperature, precipitation, dew	Eversmeyer and Burleigh, 1970

rain, amount of rain, monthly means, etc.; relative humidity; and hours of free moisture have generally failed to provide a means of explaining a major proportion of the variation when host resistance–pathogen virulence is not a factor. Perhaps the recent work on stripe rust of wheat by Coakley (1979) and Coakley and Line (1981a,b) will prove to be an exception to these failures. Their findings of differences in winter and spring temperatures are not that far from Chester's (1943) "critical month." In most of the other studies, the disease studied was wheat stem rust in the prairies of the central third of the United States. Here, wide variations of temperatures can occur within a kilometer or even a few meters at night, and precipitation can vary even more (e.g., 1 cm in a single storm or by 6–10 cm within a growing season). Temperature and hours of free moisture vary with height within the canopy. Thus, even if temperature and precipitation play a very important part in epidemic development as most epidemiologists believe, the variation between points within the canopy and over an area no bigger than a field is such that some points are very favorable for, and others very unfavorable for, disease development at any given time. Although additional measurements may solve this problem in small uniform plots, there are severe financial and labor restrictions in conducting such measurements in commercial fields.

VIII. The Future

Attempting to study disease epidemiology on a national basis when over 30 million hectares are grown presents a mammoth problem. Methods are available for determining the race-specific resistances in the 300 cultivars grown; however, because the change over of cultivars grown annually is near 10%, continued work is required. The effect of resistance not known to be race-specific resistance is poorly characterized currently in most cultivars. Virulence of the pathogen populations of stem rust and crown rust in the United States and Canada are well characterized, as is wheat leaf rust in Canada. The other diseases are not regularly studied or published data are lacking. Only hints exist on aggressiveness differences in any of the pathogen populations, and the stability of the pathogen both genetically and environmentally remains to be investigated. Little methodology is available for this latter type of study.

The effects of the race-specific genes for resistance on epidemic development under field conditions needs to be studied throughout the plant growth period. Studies of effects of multilines, mixtures of host genotypes in a field, as well as mixtures of different host cultivars or crops in adjacent fields over large areas need to be conducted.

Studies of disease in commercial fields are now limited by our inability to cover large areas and still detect small foci or low severities over local areas. Perhaps spore trapping would solve some of these problems if the counting could be done by mechanical means. Another potential method for measuring disease severity in the future may be by remote sensing (Colwell, 1956; Kanemasu *et al.*, 1974) as that technology develops.

References

Asai, G. N. (1960). Intra- and inter-regional movement of uredospores of black stem rust in the Upper Mississippi River Valley. *Phytopathology* **50,** 535–541.

Atkins, I. M. (1936). Ecological factors in north Texas related to the 1935 stem rust epidemic. *Plant Dis. Rep., Suppl.* **93,** 31–41.

Bailey, D. L. (1928). Studies on cereal diseases. IV. Stem rust in western Canada. *Can. Dep. Agric. Bull.* [N.S.] **106,** 1–31.

Beeson, K. E. (1923). Common barberry and black stem rust in Indiana. *Purdue Univ., Dep. Agric. Ext., Bull.* **118,** 1–8.

Bolley, H. L. (1891). Wheat rust: Is the infection local or general in origin? *Agric. Sci.* **5,** 259–264.

Borlaug, N. E. (1954). Mexican wheat production and its role in the epidemiology of stem rust in North America. *Phytopathology* **44,** 398–404.

Briggle, L. W., Strauss, S. L., Hamilton, D. E., and Howse, G. H. (1982). Distribution of the varieties and classes of wheat in the United States in 1979. *U.S., Agric. Res. Serv., Stat. Bull.* **676,** 1–107.

Bromfield, K. R. (1967). Some uredospore characteristics of importance in experimental epidemiology. *Plant Dis. Rep.* **51,** 248–252.

Bromfield, K. R., Underwood, J. F., Peet, C. E., Grissinger, E. H., and Kingsolver, C. H. (1959). Epidemiology of stem rust of wheat. IV. The use of rods as spore collecting devices in a study of the dissemination of stem rust of wheat uredospores. *Plant Dis. Rep.* **43,** 1160–1168.

Burleigh, J. R., and Hendrix, J. W. (1970). The winter biology of *Puccinia striiformis* West in the Pacific Northwest. *Tech. Bull.—Wash. Agric. Exp. Stn.* **65,** 1–17.

Burleigh, J. R., Romig, R. W., and Roelfs, A. P. (1969a). Characterization of wheat rust epidemics by numbers of uredia and numbers of urediospores. *Phytopathology* **59,** 1229–1237.

Burleigh, J. R., Schulze, A. A., and Eversmeyer, M. G. (1969b). Some aspects of the summer and winter ecology of wheat rust fungi. *Plant Dis. Rep.* **53,** 648–651.

Burleigh, J. R., Eversmeyer, M. G., and Roelfs, A. P. (1972). Development of linear equations for predicting wheat leaf rust. *Phytopathology* **62,** 947–953.

Campos, T. A. (1960). Importancia de las razas fisiologicas de *Puccinia graminis* var.

tritici Eriks. y Henn. en la producion de variedades de trigo resistentes a la roya. *Esc. Nac. Agric. Bol. Tech.* **2,** 1–102.

Carleton, M. A. 1905. Lessons from the grain-rust epidemic of 1904. *Farmers' Bull.* **219,** 1–24.

Celik, N. (1974). The relation of variety and distance from an inoculum source to severity of wheat leaf rust infection. M.S. Thesis, Oklahoma State University, Norman.

Chester, K. S. (1943). The decisive influence of late winter weather on wheat leaf rust epiphytotics. *Plant Dis. Rep., Suppl.* **143,** 133–144.

Chester, K. S. (1946). "The Nature and Prevention of the Cereal Rusts as Exemplified in the Leaf Rust of Wheat." Chronica Botanica, Waltham, Massachusetts.

Christensen, J. J. (1942). Long distance dissemination of plant pathogens. *In* "Aerobiology" (F. R. Moullon, ed.), Publ. No. 17, pp. 78–87. Am. Assoc. Adv. Sci., Washington, D.C.

Christman, A. H. (1904). Observations on the wintering of grain rusts. *Trans. Wis. Acad. Sci., Arts Lett.* **15,** 98–107.

Coakley, S. M. (1979). Climate variability in the Pacific Northwest and its effect on stripe rust disease of winter wheat. *Clim. Change* **2,** 33–51.

Coakley, S. M., and Line, R. F. (1981a). Climatic variables that control development of stripe rust disease on winter wheat. *Clim. Change* **3,** 303–315.

Coakley, S. M., and Line, R. F. (1981b). Quantitative relationships between climatic variables and stripe rust epidemics on winter wheat. *Phytopathology* **71,** 461–467.

Colwell, R. H. (1956). Determining the prevalence of certain cereal crop diseases by means of aerial photography. *Hilgardia* **26,** 223–286.

Craigie, J. H. (1945). Epidemiology of stem rust in Western Canada. *Sci. Agric. (Ottawa)* **25,** 285–401.

Dirks, V. A., and Romig, R. W. (1970). Linear models applied to variation in numbers of cereal rust urediospores. *Phytopathology* **60,** 246–251.

Dubin, H. J., and Torres, E. (1981). Causes and consequences of the 1976–1977 wheat leaf rust epidemic in northwest Mexico. *Annu. Rev. Phytopathol.* **19,** 41–49.

Durrell, L. W., and Lungren, E. A. (1927). Barberry eradication and sources of black stem-rust in Colorado. *Colo., Agric. Exp. Stn., Bull.* **315,** 1–18.

Elliot, A. M. (1960). Studies on the epidemiology of certain races of *Puccinia graminis tritici.* MS Thesis, University of Minnesota, Minneapolis.

Emge, R. G., and Shrum, R. D. (1976). Epiphytology of *Puccinia-striiformis* at five selected locations in Oregon during 1968 and 1969. *Phytopathology* **66,** 1406–1412.

Eversmeyer, M. G., and Burleigh, J. R. (1970). A method of predicting epidemic development of wheat leaf rust. *Phytopathology* **60,** 805–811.

Eversmeyer, M. G., and Skidmore, E. L. (1974). Wheat leaf and stem rust development near a wind barrier. *Plant Dis. Rep.* **58,** 459–463.

Eversmeyer, M. G., Kramer, C. L., and Burleigh, J. R. (1973). Vertical spore concentrations of three wheat pathogens above a wheat field. *Phytopathology* **63,** 211–218.

Futrell, M. C. (1957). Wheat stripe rust epiphytotic in Texas in 1957. *Plant Dis. Rep.* **41,** 955–957.

Futrell, M. C. (1975). Puccinia polysora epidemics on Maize associated with cropping practice and genetic homogeneity. *Phytopathology* **65,** 1040–1042.

Groth, J. V., and Roelfs, A. P. (1982). Effect of sexual and asexual reproduction on race abundance in cereal rust fungus populations. *Phytopathology* **72,** 1503–1507.

Hamilton, L. M. (1939). Stem rust in the spring wheat area in 1878. *Minn. Hist.* **20,** 156–164.

Hamilton, L. M., and Stakman, E. C. (1967). Time of stem rust appearance on wheat in the western Mississippi basin in relation to the development of epidemics from 1921 to 1962. *Phytopathology* **57,** 609–614.

Hutton, L. D. (1927). Barberry eradication reducing stem rust losses in wide areas. *U.S. Dep. Agric., Yearb.* pp. 114–118.

James, W. C. (1971). An illustrated series of assessment keys for plant diseases, their preparation and usage. *Can. Plant Dis. Surv.* **51,** 39–65.

Johnson, T., and Green, G. J. (1952). Overwintering of urediospores of rye stem rust in Manitoba. *Phytopathology* **42,** 403–404.

Kanemasu, E. T., Schimmelpfennig, H., Chin Choy, E., Eversmeyer, M. G., and Lenhert, D. (1974). ERTS-1 data collection systems used to predict wheat disease severities. *Remote Sens. Environ.* **3,** 93–97.

Katsuya, K., and Green, G. J. (1967). Reproductive potentials of races 15B and 56 of wheat stem rust. *Can. J. Bot.* **45,** 1077–1091.

Kingsland, G. (1975). Overwintering of *Helminthosporium maydis,* causing southern corn leaf blight, and *Puccinia sorghi,* causing corn leaf rust. *Proc. Am. Phytopathol. Soc.* **2,** 136 (abstr.).

Kingsolver, C. H. (1980). Progression of stem rust epidemics. *Prot. Ecol.* **2,** 239–246.

Kingsolver, C. H., Schmitt, C. G., Peet, C. E., and Bromfield, K. R. (1959). Epidemiology of stem rust. II. (Relation of quantity of inoculum and growth stage of wheat and rye at infection to yield reduction by stem rust). *Plant Dis. Rep.* **43,** 855–862.

Lambert, E. B. (1929). The relation of weather to the development of stem rust in the Mississippi Valley. *Phytopathology* **19,** 1–71.

Line, R. F. (1976). Factors contributing to an epidemic of stripe rust on wheat in the Sacramento Valley of California in 1974. *Plant Dis. Rep.* **60,** 312–316.

Melhus, I. E., Durrell, L. W., and Kirby, R. S. (1920). Relation of the barberry to stem rust in Iowa. *Iowa State Coll,, Res. Bull.* **57,** 283–325.

Mohamed, H. A. (1960). Survival of stem rust urediospores on dry foliage of wheat. *Phytopathology* **50,** 400–401.

Newton, M. (1938). The cereal rusts in Canada. *Emp. J. Exp. Res.* **6,** 125–140.

Orr, G. F., and Tippetts, W. C. (1971). Deterioration of uredospores of wheat stem rust under natural conditions. *Mycopathol. Mycol. Appl.* **44,** 143–148.

Paxton, G. E. (1921). Observation on the overwintering of the black stem rust of wheat in the vicinity of Madison, Wisconsin, MS Thesis, University of Wisconsin, Madison.

Peltier, G. L. (1925). A study of the environmental conditions influencing the development of stem rust in the absence of an alternate host. IV. Overwintering of urediniospores of *Puccinia graminis tritici. Res. Bull.*—Nebr., Agr. Exp. Stn. **35,** 1–5.

Peltier, G. L. (1933). Relation of weather to the prevalence of wheat stem rust in Nebraska. *J. Agric. Res. (Washington, D.C.)* **46,** 59–73.

Peterson, R. F., Campbell, A. B., and Hannah, A. E. (1948). A diagrammatic scale for estimating rust intensity of leaves and stems of cereals. *Can. J. Res., Sect. C* **26,** 496–500.

Peturson, B. (1958). Wheat rust epidemics in Western Canada in 1953, 1954, and 1955. *Can. J. Plant Sci.* **38,** 16–28.

Prabhu, A. S., and Wallin, J. R. (1970). Relation of weather to development of infection foci on wheat stem rust. *Plant Dis. Rep.* **54,** 959–963.

Rajaram, S., and Campos, A. (1974). Epidemiology of wheat rusts in the western hemisphere. *CIMMYT Res. Bull.* **27,** 1–27.

Reed, H. S., and Holmes, F. S. (1913). A study of the winter resistance of the uredospores of *Puccinia coronata.* Cda. *Va., Agric. Exp. Stn., Annu. Rep., 1911, 1912,* pp. 78–81.

Reitz, L. P. (1976). Wheat in the United States. *U.S., Agric. Res. Serv., Agric. Inf. Bull.* **386,** 1–57.
Roelfs, A. P. (1972). Gradients in horizontal dispersal of cereal rust uredospores. *Phytopathology* **62,** 70–76.
Roelfs, A. P. (1978). Estimated losses caused by rust in small grain cereals in the United States—1918–76. *Misc. Publ.—U.S., Dep. Agric.* **1363,** 1–85.
Roelfs, A. P. (1982). Effects of barberry eradication on stem rust in the United States. *Plant Dis.* **66,** 177–181.
Roelfs, A. P., and Groth, J. V. (1980). A comparison of virulence phenotypes in wheat stem rust populations reproducing sexually and asexually. *Phytopathology* **70,** 855–862.
Roelfs, A. P., and Long, D. L. (1980). Analysis of recent oat stem rust epidemics. *Phytopathology* **70,** 436–440.
Roelfs, A. P., and Rowell, J. B. (1973). Wheat stem rust epidemic potential in 1972. *Plant Dis. Rep.* **57,** 434–436.
Roelfs, A. P., Dirks, V. A., and Romig, R. W. (1968). A comparison of rod and slide samplers used in cereal rust epidemiology. *Phytopathology* **58,** 1150–1154.
Roelfs, A. P., Rowell, J. B., and Romig, R. W. (1970). Sampler for monitoring cereal rust urediospores in rain. *Phytopathology* **60,** 187–188.
Roelfs, A. P., McVey, D. K., Long, D. L., and Rowell, J. B. (1972). Natural rust epidemics in wheat nurseries as affected by inoculum density. *Plant Dis. Rep.* **56,** 410–414.
Roelfs, A. P., Long, D. L., and Casper, D. H. (1982). Races of *Puccinia graminis* f. sp. *tritici* in the United States and Mexico in 1980. *Plant Dis.* **66,** 205–207.
Rowell, J. B., and Roelfs, A. P. (1971). Evidence for an unrecognized source of overwintering wheat stem rust in the United States. *Plant Dis Rep.* **55,** 990–992.
Rowell, J. B., and Romig, R. W. (1966). Detection of urediospores of wheat rusts in spring rains. *Phytopathology* **56,** 807–811.
Rupert, J. A. (1951). "Rust Resistance in the Mexican Wheat Program," Foll. Tec. No. 7. Oficina de estudies especiales secretaria de Agricultura y Ganaderia, Mexico D. F.
Sanford, G. B., and Broadfoot, W. C. (1932). Epidemiology of stripe rust in western Canada. *Sci. Agric.* **13,** 77–96.
Schwarzbach, E. (1979). A high throughput jet trap for collecting mildew spores on living leaves. *Phytopathol. Z.* **94,** 165–171.
Shaner, G., and Powelson, R. L. (1973). The oversummering and dispersal of inoculum of *Puccinia striiformis* in Oregon. *Phytopathology* **63,** 13–17.
Sharp, E. L., and Hehn, E. R. (1963). Overwintering of stripe rust in winter wheat in Montana. *Phytopathology* **53,** 1239–1240.
Stakman, E. C. (1923). The wheat rust problem in the United States. *Proc. Pan-Pac. Sci. Congr. (Aust.)* **1,** 88–96.
Stakman, E. C. (1934). Epidemiology of cereal rusts. *Proc. Pac. Sci. Congr., 5th, 1933,* Vol. 4, pp. 3177–3184.
Stakman, E. C., and Harrar, J. G. (1957). "Principles of Plant Pathology." Ronald Press, New York.
Stakman, E. C., and Lambert, E. B. (1928). The relation of temperature during the growing season in the spring wheat area of the United States to the occurrence of stem rust epidemics. *Phytopathology* **18,** 369–374.
Stakman, E. C., Henry, A. W., Curran, G. C., and Christopher, W. N. (1923). Spores in the upper air. *J. Agric. Res. (Washington, D.C.)* **24,** 599–606.
Stakman, E. C., Melander, L. W., and Fletcher, D. G. (1927). Barberry eradication pays. *Stn. Bull.—Minn., Agric. Exp. Stn.* **55,** 1–24.

Stakman, E. C., Popham, W. L., and Cassell, R. C. (1940). Observations on stem rust epidemiology in Mexico. *Am. J. Bot.* **27,** 90–99.
Steffenson, B. J., Wilcoxson, R. D., and Roelfs, A. P. (1982a). Field reaction of selected barleys to *Puccinia graminis. Phytopathology* **72,** 1002 (abstr.).
Steffenson, B. J., Wilcoxson, R. D., and Roelfs, A. P. (1982b). Reactions of barley seedlings to stem rust, *Puccinia graminis. Phytopathology* **72,** 1140 (abstr.).
Tehon, L. R. (1927). Epidemic diseases of grain crops in Illinois, 1922–1926. The measurement of their prevalence and destructiveness and in interpretation of weather relations based on wheat leaf rust data. *Bull.—Ill. Nat. Hist. Surv.* **17,** 1–96.
Tehon, L. R., and Young, P. A. (1924). Notes on the climatic conditions influencing the 1923 epidemic of stem rust on wheat in Illinois. *Phytopathology* **14,** 94–100.
Thiel, A. F. (1938). The overwintering of urediniospores of *Puccinia graminis tritici* in North Carolina. *Elisha Mitchell Sci. Soc.* **54,** 247–255.
Tollenaar, H., and Houston, B. R. (1967). A study of the epidemiology of stripe rust, *Puccinia striiformis* West., in California. *Can. J. Bot.* **45,** 291–307.
Ukkelberg, H. G. (1933). The rate of fall of spores in relation to the epidemiology of black stem rust. *Bull. Torrey Bot. Club* **60,** 211–228.
Underwood, J. F., Kingsolver, C. H., Peet, C. E., and Bromfield, K. R. (1959). Epidemiology of stem rust of wheat. III. Measurements of increase and spread. *Plant Dis. Rep.* **43,** 1154–1159.
Walker, W. A. (1927). Black stem rust and the barberry eradication campaign in Wisconsin. *Wis., Agric. Exp. Stn., Bull.* **84,** 1–20.
Wallin, J. R. (1951). An epiphytotic of corn rust in the North Central region of the United States. *Plant Dis. Rep.* **35,** 207–211.
Wallin, J. R. (1964a). Texas, Oklahoma, and Kansas winter temperatures and rainfall, and summer occurrence of *Puccinia graminis tritici* in Kansas, Dakotas, Nebraska, and Minnesota. *Int. J. Biometeorol.* **7,** 241–244.
Wallin, J. R. (1964b). Summer weather conditions and wheat stem rust in the Dakotas, Nebraska, and Minnesota. *Int. J. Biometeorol.* **8,** 39–45.
Walster, H. L. (1921). Rust and the weather. *Science* **53,** 346.

14

Disease Modeling and Simulation

P. S. Teng
*Kira L. Bowen**
Department of Plant Pathology, University of Minnesota, St. Paul, Minnesota

I. Introduction

A cereal rust epidemic occurs when a set of conditions relating to the disease tetrahedron of pathogen, host, environment, and humans is fulfilled (Zadoks and Schein, 1979). Humans, because of their ability to

*Present address: Department of Plant Pathology, University of Illinois, Urbana, Illinois 61801

THE CEREAL RUSTS,
Volume II

ISBN 0-12-148402-5

exert drastic changes in their natural environment, have, through agriculture, destabilized many of the coexistential relationships apparent in wild pathosystems. Humans must often be viewed as the vertical apex in the disease tetrahedron, dictating the mode and intensity of interaction between the three horizontal apices of pathogen, host, and environment. Here we are concerned with using a relatively modern tool, computer modeling, to help us better understand how such delicate relationships have been disrupted, and in understanding, to direct our efforts at the development of rational strategies for managing the impact of rust epidemics on crop yield.

A. USE OF MODELS AND SIMULATION IN THE STUDY AND MANAGEMENT OF PLANT DISEASES

Modeling is not a new tool in scientific inquiry, although it has become a powerful and sometimes specialized activity with recent developments in computer technology. Computer modeling is a way of putting the "systems approach" into action (Teng and Zadoks, 1980). The basis of this approach is that any complex system cannot be properly understood by ad hoc studies on its components. An integrated approach is needed since the "whole is more than the sum of its parts." A plant disease epidemic is a complex system of many interacting components. Epidemiology may be considered "the study of the dynamic interactions between pathogen and host, and the disease that results, as influenced by man and environment" (Kranz, 1974). The complexity of a disease epidemic and its dynamic interaction with the environment renders epidemiology ideal for a systems approach and for computer studies (Teng, 1983). Also, by its very nature, epidemiology requires mathematics, since it is concerned with quantitative questions.

A disease epidemic comprises two biological subsystems—the host and the pathogen. These two subsystems are the interacting components that give rise to disease. Environmental factors affect the individual and interactive behavior of each subsystem. Furthermore, as humans create agricultural systems, they influence the development of the epidemic by their cultural practices. Each biological subsystem in an epidemic is a hierarchy of other subsystems, each of which has spatial and temporal properties. For example, a field epidemic of wheat leaf rust caused by *Puccinia recondita* f. sp. *tritici* is a polycycle at one level of the hierarchy, a monocycle at a lower level, and an infection process at the next. If each component at any level of the hierarchy is affected by its own set of environmental factors, then our minds are limited in keeping track of the many environmental effects on the dynamics of a rust epidemic. A computer model can alleviate the limitations of the human mind, and even enhance our ability to integrate multifactor effects on rust epidemics. The model removes much of the laborious work required in interpreting epidemiological phenomena, allowing concentration on concepts and responses. The systems approach facilitates construction of a model that captures all current knowledge on epidemiology of a particular rust disease. The computer provides a means of executing that model, and the researcher is given the liberty of experimenting with input–output patterns using the model. A large part of any research involves data collection and analysis. Often, there is insufficient time left for interpretation and conceptualization. Systems modeling, by providing a logical procedure for incorporating large amounts of data into a coherent and holistic entity, should allow time for the important scientific need to THINK!

Later in this chapter, we distinguish between system (simulation) and nonsystem (mathematical function) models. System modeling of plant disease epidemics using computers is a relatively recent event in plant pathology. The first series of models were developed during the late 1960s by two groups, one in the Netherlands headed by J. C. Zadoks of The Agricultural University, Wageningen, and the other in the United States headed by P. E. Waggoner and J. G. Horsfall of the Connecticut Agricultural Experiment Station (Zadoks, 1971; Waggoner and Horsfall, 1969; Waggoner *et al.*, 1972). The Dutch group has used cereal leaf rusts for much of their modeling, while the Connecticut group has used other diseases. In the United States, a flexible disease epidemic simulator was reported by Shrum (1975) for wheat stripe rust. Epidemics of barley leaf rust caused by *P. hordei* have also been simulated in New Zealand (Teng *et al.*, 1980).

With the first decade of experience in computer simulation of rust epidemics now behind us, it is not surprising to see some proposals that this tool be made more acceptable to those who are less quantitatively inclined (Teng and Zadoks, 1980). Indeed, advances in computer software are making available "user friendly" languages that require only an understanding of the principles of modeling, since the computer handles the mechanics (Teng, 1983).

Epidemiology is the encompassing study of cereal rust pathology, and the systems approach is the means to achieve a semblance of holism in research.

B. THE SYSTEMS APPROACH AND MODELING

The systems approach is a philosophy of holism. The methodological arm of this philosophy has been called systems research. Because this methodology is relatively new, and there is no core of established terms, it is important that terms be defined.

A term commonly encountered in modeling is systems analysis (Kranz and Hau, 1980). This is the aspect of systems research that includes defining the epidemic system to be modeled, identifying its subsystems, their interrelationships, and the environmental variables that influence these subsystems. Systems analysis is a process of disassembly, and usually leads to the formulation of an *initial system model.* This initial system model may be a conceptual model that represents the researcher's perception of the epidemic. A systems analysis of any pathogen–host interaction has the tangible benefit of identifying missing data in any part of that disease cycle. This process can be used to direct research, especially if the conceptual model is translated into a simplified computer model (Teng *et al.*, 1977). Regardless of the resulting model's value for forecasting, the systems analysis part of a study may be potentially useful in shaping our ideas on epidemiology (Teng and Zadoks, 1980).

The model resulting from systems analysis attempts to simulate the behavior of the real epidemic. The terms "simulate" and "model" are closely associated with the systems approach. Simulation is the process of designing a model to represent a system, and using the model as an experimental tool to understand and manage the system. The word "system" is used to mean any aggregation of biological processes or subprocesses that contribute to a common function, e.g., dissemmination in a monocycle is comprised of the processes of spore liberation, dispersal, and deposition. A "model" is an abstract representation of an entity in a form different from the original. We have used this term in a broad sense to reflect our view that it is not possible to give a definitive meaning to the word. Historically, Churchman *et al.* (1957) distinguished three types of models: iconic models, which visually represent aspects of a system (e.g., a photograph); analogue, which represent parts of a system using other properties; and symbolic, which represent properties of a system using symbols like numbers. Computer models fall into the last group. The dynamics of interaction in a rust epidemic are captured and duplicated by symbols acting within the computer.

The process of developing a system model to simulate rust epidemics may be divided into distinctive yet interrelated steps (Teng and Zadoks, 1980). The first step is to clearly define the objectives for developing the model, and to determine the necessary detail. A model that is to be used for research into fundamental epidemiological processes will require more detail than a model for predicting rust progress in a given environment.

An explanatory model is one that attempts to account for biological relationships in its structure while an empirical model (e.g., a regression function) does not claim to incorporate any cause-effect relationships. Although the distinction at times is hazy, specifying whether an explanatory or empirical model is to be developed will help in the modeling process.

Step two in system modeling involves defining the scope of the disease system and its hierarchical structure. Because most natural systems can increase in complexity both up and down a hierarchy, it is essential that the boundaries of the modeled system be well defined. What is within the boundary is then the system proper, while what is outside is the system environment. Driving variables—e.g., temperature and dew—originate from the system environment and influence the individual and interactive behavior of the subsystems in the system proper. Some variables can also originate as a by-product of the system proper and leave the system, e.g., spores. The identification of the hierarchical structure leads to the distinction between the subsystems to be modeled (e.g., spore germination) and their associated driving variables (e.g., temperature), based on current knowledge of the system. The structure of the system may later be modified, and additional variables may be specified as information becomes available. This organizational process identifies grey areas or gaps in our current knowledge of rust epidemiology and has often shown that too much research is concentrated on processes like germination, while others, like spore dispersal, have been ignored.

Step three is the formulation of an initial system model. An initial system model may be a set of blocks and arrows (flow charts) showing the interrelationships between different processes in a monocycle (Teng *et al.*, 1977). It may also be a computer program that translates these blocks and arrows into state variables and rate equations (Fig. 1). This program can be used to aid in ranking the processes within a monocycle for their influences on the disease progress curve (polycycle) generated through many monocylces (Teng and Zadoks, 1980). Response patterns produced by an initial system model have been used to test hypotheses regarding the role of monocyclic processes like latent period and spore production in governing the rate of disease increase (Zadoks, 1971; Teng *et al.*, 1977).

Step four is to manipulate available data on rust biology into a form

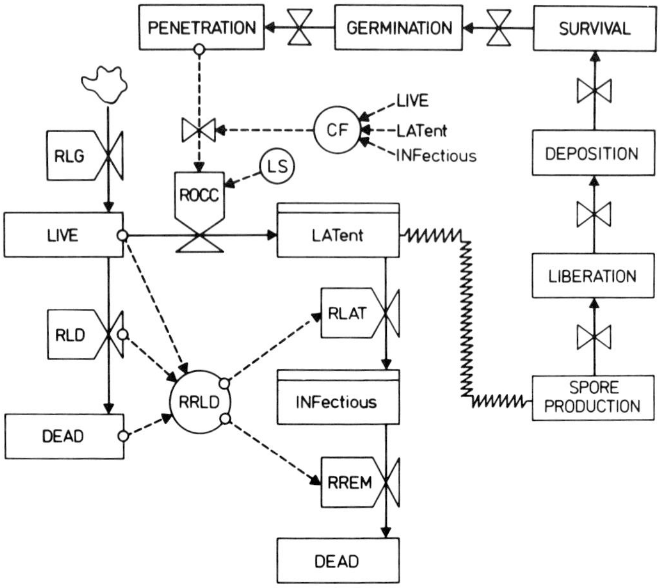

Fig. 1. Relational diagram of a detailed system model for a leaf rust. After Teng and Zadoks (1980).

suitable for use in a detailed simulation model, or, when data are missing, to obtain data experimentally. The information required to model rust polycycles comes from data on the effects of specific factors on processes within a monocycle (e.g., the effect of temperature on spore production per uredium). This is a single-stimulus single-response function, where the stimulus is temperature and the response is spore production. More common is the situation with multiple stimuli and a single response—e.g., the effects of temperature, age of uredium, and density of uredia per unit area leaf surface on spore production (Teng *et al.*, 1980). This latter relationship requires a factorial experiment, conducted with controlled environment, for elucidation. In Section II we discuss how the different processes within a rust monocycle can be quantified.

Step five is the determination of how to translate data into a computer program that represents the structure of the model system. Teng and Zadoks (1980) have noted that many techniques are available for this translation process, from high-level computer languages like FORTRAN to special simulation languages like CSMP. The choice of program structure and language is determined largely by what the computer language has been developed for and the computing skills of the modeler.

Assembly of the detailed system model and its translation for computer execution are demanding. However, validation of the model is more exacting, as it requires exhaustive testing (Teng, 1981a). Only after a model has been rigorously evaluated can one be confident of its use for predictions. It is meaningless to apply any model results to disease management until the validity of the model has been established. There is a growing feeling among modelers that the validation step in modeling has received insufficient attention.

C. RUST BIOLOGY RELEVANT FOR MODELING

Current simulation models have addressed only one stage—the uredial stage. The reason is that this repeating stage (of the polycycle) results in field epidemics, many of which cause economic loss. A monocycle is a single infection cycle, initiated when a spore is deposited on a host surface, and ending with the death of the uredium. A polycycle is the spatial and temporal overlap of many monocycles in the same season. Rust epidemics that span several seasons have been called polyetics by Zadoks (1972). Polyetics can include all the stages in a rust life cycle and may occur on several host species. Polyetics consist of a level of complexity beyond the cropping system of the polycycle, and we know of no attempt to model them.

The monocycle is the central theme in a modeling effort on rust epidemics (Teng *et al.*, 1980), but the model should also account for the overlap of monocycles in the field. This overlap can result when new generations of uredia are produced before a parent generation dies. In modeling the rust monocycle, it is necessary to be clear about component processes, and to have precise working definitions of these processes. Modeling forces a sharpening of our definitions of the component processes in monocycles (see Fig. 2). For example, when Teng (1978) started work with barley leaf rust, it was difficult to use any published data because of imprecise definitions in the literature. Incubation period, latent period, and generation time were treated synonymously by workers (Teng and Close, 1978). Yet, when distinctive processes are identified by an author, these terms are often found to denote different time spans (Fig. 2). This inconsistency in terminology on the basic rust infection cycle has made quantitative comparisons between different sources of information difficult. In general, modeling has played an important role in encouraging a critical examination of field and laboratory measurements.

The processes in a monocycle that are generally specified for modeling are spore production, liberation, dispersal, deposition, germination, penetration, latent growth, uredium eruption, uredium growth, and uredium death. Some of these processes have been aggregated (e.g., dissemination consisting of liberation, dispersal, and deposition) or omitted (e.g., uredium growth), depending on the rust and data availability (Shrum, 1975; Teng *et al.*, 1980). Each of these processes may be further divided into subprocesses, e.g., germination may be divided into water imbibition, germ-tube appearance, germ-tube elongation. In

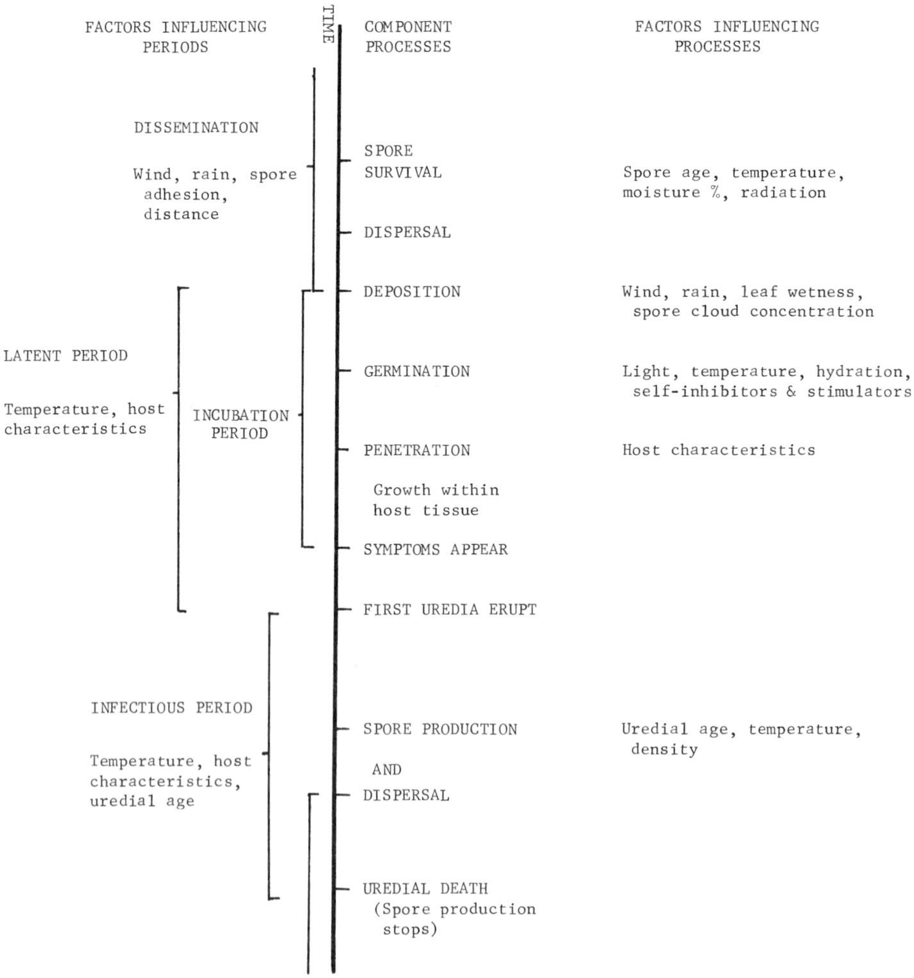

Fig. 2. Diagram of a rust infection cycle.

progressing from monocycle to process to subprocess, the time interval for the activity generally decreases. In modeling, integration time becomes smaller. Each process or subprocess is described, at any time, by the value of a state variable (Teng and Zadoks, 1980). The transition from one to another, as from deposition to germination, is done by transferring the value of the deposition state variable to the germination state variable. The change in state between processes can be visualized as being governed by a series of valves (transition rates), one between each of two state variables (Fig. 1). From a modeling perspective, therefore, a monocycle may be visualized as a linked series of state variables, from spore production to uredium death, where there is commonly a decrease in the value for each state variable, after the first, within the same monocycle. The amount of decrease at each valve is described by a rate equation that dictates how fast one state changes into the next, as affected by specific environmental or host factors.

The biological nature of field epidemics requires that a model account for the daily change in environmental conditions that may favor a particular process in the monocycle. Usually, every day of the cropping season does not have conditions producing leaf wetness necessary for spore germination. Thus, if germination is considered the starting point in a monocycle, then new monocycles are initiated in an erratic manner. Furthermore, the length of each monocycle depends on prevailing conditions during that part of the season. With foliar diseases, the change in amount of leaf area available for infection during the season must also be considered. Two approaches have been used to represent the host entity in modeling: leaf area index (Zadoks, 1971; Shrum, 1975) and leaf area per tiller (Teng *et al.*, 1980). Simulations of an epidemic occurring in a field may be as actual numbers of lesions per unit area of that field, or the percentage foliar severity per tiller. Although disease is usually modeled over time, simulations also attempt (implicitly or explicitly) to convey some spatial property to the modeled amount of disease. With a stochastic simulation, model output is a mean amount of disease with a variance value and thus resembles a sample mean from a population with a particular distribution. This distribution may be over time or space. EPIMUL (Kampmeijer and Zadoks, 1977) is a truely spatial model and capable of simulating epidemics over time as well as spatially over the modeled area.

Environmental factors are usually the major variables that drive the rust epidemic system; host and pathogen physiology, whenever they affect the epidemic, must also be recognized. One point of influence is in the measurable expression of pathogen virulence/host resistance, e.g., the infection ratio (Mehta and Zadoks, 1970) or receptivitiy

(Rowell, 1981). These terms describe the ratio of number of erupted uredia divided by number of deposited spores per unit leaf area. A high infection ratio, e.g., 0.2, indicates that the race/cultivar combination has potential for high rate of disease increase (Teng *et al.*, 1977). System models of rust epidemics can easily account for changes in this ratio with plant age, which occurs with adult plant resistance.

A model must rationally represent biology. Thus, if basic data on rust biology is absent, the task of assembling a realistic model is difficult. Furthermore, errors made in hypothese concerning epidemiological processes are translated into the model. While the modeling process can guide our understanding of biology, it is also important that effort be put into conceptualization of any epidemic (Teng and Gaunt, 1980).

II. Modeling the Rust Monocycle

Those processes specified for mathematical modeling of the rust monocycle are spore dissemination, spore survival, spore germination, latent period, and spore production. Models may deal with a single component or with several components in one mathematical equation. The mathematical models used may be single-variable functions or may be a series of equations whose use depends on which conditions are met. Equations vary from a simple linear form to a complex quadratic.

The components outlined in Fig. 2 are discussed in this section in relation to the environmental factors that "drive" each component. How the individual components may be quantified and modeled and how these component models may be incorporated into the complete rust system model are discussed.

A. Spore Dissemination

Three components of the monocycle make up dissemination. These are liberation, spore dispersal, and deposition. Liberation of urediospores is the actual release of spores by physical factors such as wind and is dependent on wind speed and spore adhesion (Gregory, 1973). The movement of spores from sporophores to where they are deposited is dispersal (Zadoks and Schein, 1979). Deposition is affected by wind and rain, and is the termination of flight due to landing (Zadoks and

Schein, 1979). In modeling, dissemination by mechanical means is not often considered.

Dispersal has been measured by many researchers for different kinds of fungal spores and particles (Gregory, 1973). The observation of disease spread from one single spot in a field to a larger area of coverage over time may provide an estimate of dispersal. The mathematical description of cloud diffusion has also been used to describe dispersal (Schrodter, 1960; Gregory, 1973). Spore traps have often been placed in different parts of a field in which a small group of plants have been inoculated (Eversmeyer and Kramer, 1980; Roelfs, 1972). The number of spores trapped at given distances from the "source" provides data for evaluating dispersion. Deposition has been studied by Teng (1978) and Chamberlain (1967), among others, by observing the rate of fall of spores in a settling tower. This is a simplified and ideal condition. Detailed descriptions on how to handle these processes are summarized by Gregory (1973).

Teng (1978) modeled liberation and deposition as stochastic events. Kampmeijer and Zadoks (1977) incorporated the three processes of dissemination into spore dispersal. Spores produced at a known source diffuse over the field in a two-dimensional distribution. Waggoner and Horsfall (1969) dealt with *Alternaria solani* spores in a manner which could be applied to rust fungi. The number of spores liberated is determined as a proportion of spores produced. They derived this proportion from a function of wind speed for determining shear stress. The proportion of spores thus liberated that land on host material is the ratio of attained leaf area index (LAI) to maximum attainable LAI. Shrum (1975) used a gamma distribution for downwind and a normal distribution for crosswind determination of how far and in what direction a liberated spore will travel. The mean of the gamma function was made equivalent to Schrodter's (1960) "spore landing mean" by calculating a wind vector from the wind speeds and directions which act on the spores. The standard deviation for the crosswind normal was conditional on how far downwind the spore had traveled. The percentage of disseminated spores that landed on host material was a function of the leaf area index.

More sophisticated empirical models can be developed from published data. Liberation of rust urediospores has been related to wind speed, diurnal periodicity (Hirst, 1953; Asai, 1960; Kramer and Pady, 1966; Pady *et al.*, 1965), and relative humidity or actual leaf wetness (Asai, 1960). Wind-tunnel experiments on *P. hordei* urediospores suggested two linear relationships exist relating liberation to wind speed—one for speeds below 4 km/hr and one for speeds above that

(Teng, 1978). Complex mathematical models have also been developed for dispersal. Gregory (1973) presents some of these, one of which is:

$$Q(x) = Q(0) \exp \frac{(-2px^{(1-m/2)})}{\pi C(1-m/2)}$$

where p = deposition coefficient, m = turbulence, C = coefficient of diffusion, $Q(0)$ = number of spores released at source, and x = distance from source. Other models of particle movement may be found in studies of air pollution and cloud diffusion. Roelfs (1972) presented a fairly simple equation for dispersal gradients of leaf and stem rusts derived from his studies on horizontal dispersal. Description of barley leaf rust spores over time in a settling tower graphed as an asymmetrical logistic curve (Teng, 1978) from which a regression equation can be developed. Gregory (1973) presented equations relating deposition to spore concentrations, distance from source, wind, and size of rain droplets.

B. SPORE SURVIVAL

In the rust monocycle, survival has been considered the phase between liberation and deposition (Teng *et al.*, 1977; Teng and Zadoks, 1980). Measurement of survival may differ from one study to another. Three distinct "survival" factors were presented by Shrum (1975). He distinguished loss of fecundity for spores produced in aging lesions, loss of viability during flight, and loss of viability after deposition. A careful definition of survival is needed so that proper measurements and applications are made. Survival has been measured as the proportion of spores remaining viable after being subjected to a set of conditions or as the percent of spores which germinate or the percent that produce lesions after inoculation.

Shrum's first survival factor in EPIDEMIC, loss of viability before liberation, is affected by temperature and lesion age. As temperature increases and lesions mature, urediospores being produced decrease in viability (Shrum, 1975). The second factor, loss of viability during dissemination, and the third, loss after deposition, are usually treated similarly and are often the only survival phenomena measured. Spores exposed to high temperatures, high light intensities (Hwang, 1942; Smith, 1966; Bromfield, 1967), and high moisture content (Bromfield, 1967) decrease in viability. Generally, the percentage of nonviable spores of *P. graminis, P. recondita, P. coronata, P. hordei,* and *P. striiformis* follow a sigmoid curve when graphed against length of exposure to these conditions. In Shrum's (1975) model, viability of wheat stripe rust spores decreases rapidly at first then more slowly until viability reaches zero at 20 days after liberation. Teng (1978) found that the loss of viability of *P. hordei* spores also follows a roughly sigmoid curve with increased exposure to high light intensities (sunny days).

C. SPORE GERMINATION

Germination is the event of the rust monocycle during which a spore in a resting state is transformed to an active state. Several people have distinguished developmental phases within the germination process (Hess *et al.*, 1975; Allen, 1965). Each of these phases may be modeled separately, although they usually are not. The process of germination is commonly considered complete when the germ-tube length is greater than the spore diameter (Zadoks and Groenewegen, 1967).

Both endogenous and exogenous factors affect spore germination. Exogenous factors include light, temperature, hydration of the spore, and conditions prevailing when the spore was produced. Endogenous factors include self-inhibitors and self-stimulators. Data must be collected on the effects of these factors in order to develop a model for this phase. A common way of obtaining data for exogenous-factor effects is to measure the proportion of spores that germinate after they have been lightly dusted on water agar plates or slides and exposed to varying factor levels.

Empirical data dealing with endogenous factors are scarce. Yarwood (1956) found that germination of *Uromyces phaseoli* spores decreased with increasing spores per unit area. Kuhl *et al.* (1971) observed that the germination of *Puccinia* spores was promoted when in low densities. Teng (1978) found no differences in germination due to spore concentrations ranging from 5×1000 to $10 \times 10{,}000$/ml. The effects of other environmental factors on germination are discussed in Volume 1 of this series.

Temperature and light effects can be modeled through linear and quadratic equations if interactions are not considered. Shrum (1975) used a linear regression equation coupled with a temperature summation procedure for modeling the relationship between germination and temperature. The accumulation of hour-degrees toward germination occurred only when free moisture was available, and was never initiated except during hours of darkness. In modeling the germination of *P. hordei* urediospores, Teng (1978), like Shrum (1975), treated moisture as a binary (yes or no) activity—only when free moisture was available could germination occur. Also, *P. hordei* germination occurred only in the dark. The data on time–temperature relationships for germination that Teng (1978) collected was fitted to a quadratic equation using regression:

$$\ln \mathrm{GP} = 1.22182 + 0.82268\,H + 0.07899\,T - 0.04630\,H^2 + 0.00011\,T^2 - 0.01177\,HT$$

where GP = germination percentage, H = germination period in hours, and T = germination temperature (°C). Although germination has generally been modeled as a distinct process, the final germination percentage might be more aptly treated as a stochastic variable (Teng, 1978) following a normal distribution over time (Zadoks and Groenewegen, 1967).

D. Latent Period

Vanderplank (1963) defined latent period as the time between spore deposition and uredium eruption. This includes the processes of deposition, germination, fungal penetration, and growth within plant tissue until uredia appear. The first two processes have been discussed and are relatively easy to describe. The latter two processes are harder to study because these processes cannot be seen without destructive sampling, which interrupts the integrity of the parasitic relationship.

More studies on penetration and postpenetration events are needed. Rowell (1981) presented empirical data for the events between penetration and colonization by *P. graminis;* the different values observed with different host cultivar/pathogen race combinations show the complexity involved.

In his model for wheat stripe rust, Shrum (1975) included a parameter for penetration but gave it no value because of the lack of information. Shrum's parameter for latent period varies with temperature. The simulator sums degree-hours from −2°C to 28°C and considers the latent period complete when the total reaches 5568.

Teng (1978) found that the latent period of *P. hordei* increased in a diurnal temperature regime (mean day 32°C, mean night 13°C), while constant temperatures increasing from 10° to 24°C decreased the latent period from 13.5 to 6.4 days. The data gathered by Teng (1978) for *P. hordei* on barley was subjected to regression analysis. The resulting function for predicting latent period (LP) in his model was:

$$\mathrm{LP} = 19.6926 - 0.08492\,T^2 + 0.002534\,T^3$$

where T = ambient temperature (°C). Simkin and Wheeler (1974) also found a decrease in length of latent period with increasing temperatures from 2° to 30°C with *P. hordei.* This same effect of temperature on latent period was observed with *P. graminis* (Melander, 1935), *P. polysora* (Cammack, 1961), and *P. recondita* (Eyal and Peterson, 1967). However, temperatures between 20° and 35°C seemed to have no effect on latent periods of *P. coronata* (Kochman and Brown, 1975).

E. SPORE PRODUCTION

Spore production is the phase of disease development after lesions erupt from plant tissue and while infective propagules are produced. Estimates of spore production are usually made by collecting spores at frequent intervals from the time the uredium erupts until no new spores are detectable. Experimental removal of spores may be through the use of a vacuum collector, or by brushing or gently tapping them from plant tissue. The collected spores are then weighed or counted.

Spore production may be modeled as a constant number of spores per infected plant for a given period of time (the infectious period). This spore production figure is determined experimentally by taking the average spore production from a number of infected plants, over the entire infectious period, using a wide range of conditions and with replication. This constant can then be modified to fit the particular rust being modeled and also the disease severity of the infected plants. For example, spore production on plants with a 1% level of disease may be some number x, and this could be modified for 10% disease simply by multiplying by 10.

In their model to simulate epidemics of *P. striiformis* in a multiline, Kampmeijer and Zadoks (1977) modeled spore production in a simple manner. Disease increase on plants was based on the percentage of plant tissue covered with lesions (disease severity, x) and a "daily multiplication factor." In EPIMUL, they do not consider any environmental effects that are known to influence this component. The treatment of spore production as a constant number for an infectious period is simplistic. Light and temperature differences, as well as the pustule density (i.e., disease severity), have been shown to affect spore production (Smith, 1966; Eyal and Peterson, 1967; Mehta and Zadoks, 1970; Prabhu and Wallin, 1971). Shrum (1975) has included relative humidity in calculations of spore production for his wheat stripe rust simulation model.

The effect of environmental variables on a process may be modeled by a series of modifying coefficients. For example, Smith (1966) found that spore production increased fourfold when temperature rises from 9° to 18°C, then doubles again with an increase in temperature to 23°C. Thus, starting with a base figure for spore production, y, when the temperature T is 9, then at $T = 18$, spore production equals $4y$, and at $T = 23$, $2(4y)$. An equation predicting spore production may also be developed from this same data with the one independent variable being temperature. Shrum's (1975) simulation model started with the maximum number of spores possible for the lesion area, then reduced this for lesion age. A series of gamma functions, from regression of data from Schmitt *et al.* (1964), further modified the maximum spore production number to account for current temperatures and relative humidities.

Unlike Shrum, who developed his functions from published data, Teng (1978) based his functions on data collected from his greenhouse experiments on the *P. hordei*–barley system. Effect of temperature and uredinia age on spore production were described by Teng as a response surface (three-dimensional graph of interactions). Functions, with two independent variables, were derived by regression analysis of data from the response surface. These functions were used to predict spore production in BARSIM.

Spore production is often modeled together with infectious period, where the age of a lesion affects the number of spores produced by that lesion until it dies (Teng *et al.*, 1980). On any day of an epidemic after the first, there will be an age distribution of lesions, initially all from a single monocycle and later from overlapping monocycles. The infectious period, and different-aged lesions having different spore productivity, may be modeled using the "boxcar train concept" (Teng, 1978). This allows representation of many overlapping infectious periods, each period being a train, and each day (dt) of the simulation being one boxcar. The contents (lesions) in a boxcar with the same position in a train will therefore have the same age, and the total number of lesions of a certain age can be accumulated daily.

A single meteorological variable may affect several component processes of the monocycle directly and each in a different way, or indirectly through the effect it has on the host. Host characteristics by themselves also affect some of the disease processes discussed above. Many of these rust monocycle interactions can be modeled as part of a system model or individually as nonsystem models. The selection of events to model and the level of complexity of the model will depend on the modeler's abilities and the intended final use of the model.

III. Modeling the Rust Polycycle

A. NONSYSTEM MODELS

Nonsystem models are generally single equations that do not deal with interrelationships between the variables of a model or combine aspects of time and space. A nonsystem model may be holistic but does not deal with the hierarchical nature of the process being represented. These models may represent a level in the hierarchy of a system and can be used for building system models. Many nonsystem models are used for studying the effect of a single variable on a process.

The variables and the form of a single equation model are frequently chosen intuitively and then refined through statistical analysis. For example, Burleigh *et al.* (1972a,b) developed a linear equation for predicting wheat leaf rust severities and crop losses from that disease. Twelve biological and meteorological variables were measured for possible inclusion in their model, but only combinations of six variables that explained the most variance in predicting disease were used. These were free moisture, rainfall, disease severity observed at an earlier date, the plant growth stage, a fungal growth value, and spore numbers. Burleigh *et al.* (1972a) used stepwise multiple regression to formulate the equation. They repeated the method in developing an equation for predicting crop loss from disease severity at eight stages in the wheat's development. This crop loss model is holistic because it considers aspects of the entire rust epidemic from planting through harvest.

Many nonsystem models that predict rust epidemics involve the use of biological growth models. A general model for "growth" of a disease epidemic is Vanderplank's (1963) exponential model: $dy/dt = ry$. This states that the change in disease with time equals the disease proportion times the rate of increase. This is strictly a temporal model, dealing only with disease increase over time. This equation has two other forms where the only random variable is the disease proportion—the monomolecular and logistic equations (Madden, 1980). The rate of disease increase differs depending on the environment and biological factors. When the environment is favorable, r may be very high. A rough estimate of r from any particular place and growing season can be calculated by rearranging the exponential growth model and correcting for finite resources such as food and time:

$$r = [1/(t2-t1)]\,[\ln(x2/1-x2) - \ln(x1/1-x1)]$$

where $t1$ and $t2$ are two points in time with different disease severities ($x1$ and $x2$) (Vanderplank, 1963).

Disease spread can also be predicted through nonsystem models. Spatial models for the distribution of inoculum have evolved from the models of Schrodter (1960), Gregory (1973), and Lambert *et al.* (1980). Common spatial models predict *y*, amount of disease, at distance *x* from a source. Parameters in spatial models describe the infection gradient and source productivity. Roelfs (1972) developed a nonsystem model for the downwind movement of *P. graminis* and *P. recondita* urediospores from a source. The form of his equation was $\log y = \log a + bx$, where *y* is the number of spores/cm^2, *b* is a measure of the gradient, and *x* is the number of spores/cm^2 at the source.

Nonsystem models have been used with much success over the years. They are easier to work with than system models since they are usually static, and because the interrelationships between variables are ignored. Where such interrelationships of time, space, and environment have significant impact on epidemic development, a system model is preferred. System modeling also allows the inclusion of large numbers of parameters which are necessary for predicting epidemics.

B. SYSTEM MODELS

A system model involves the integration of parts making up a process. System models may deal with the integration of time and space or may only represent the temporal aspects of an epidemic. A prediction of crop losses from an initial disease severity based on phytopathological, meteorological, and physiological concepts can be a system model without ever considering disease spread. In the same way, a system model may predict disease spread without considering disease increase.

In assembling a system model, intuition and experience are also necessary for choosing the variables and form of relationships. The approach used for developing a system model is to first outline an "initial system model" (as in Section I of this chapter). Once this is done, information can be fitted to the model. This may involve the integration of existing nonsystem models (single equations) that serve as components of the system. For example, Burleigh's two models could be integrated into a simple system model with which crop losses can be predicted from six variable inputs (Burleigh *et al.*, 1972a,b). Modifications are often required to make nonsystem models useful within the dynamic system model. Certainly beginning with available information is less time-consuming than starting with nothing. The variables in the equation, for example, would now vary according to the values of other variables, i.e., when certain meteorological components are outside stated limits, fungal growth may cease.

System models have a distinct advantage over most nonsystem models since they are dynamic. That is, they can vary over time. For example, a system for predicting disease increase using Vanderplank's exponential model may be developed that allows that rate of the epidemic to vary as conditions dictate. Disease dispersal gradients can also be changed with wind direction in a dynamic spatial model.

C. MODEL STRUCTURE FOR COMPUTER SIMULATION

The structures of detailed system models have some general characteristics in common, regardless of the method of programming or the computer language used. These characteristics include ways of keeping time (the model clockwork), of keeping track of delays, and of accepting input data and presenting model results (Shrum, 1975; Teng, 1978). The computer language used for the model would determine the particular structure of these characteristics. If the simulation model has an integration time of one day, then, within each day, all or none of the processes in a monocycle may be successful. Figure 1 shows the structure of a monocycle, depicted as a relational diagram, and used as a basis for modeling field epidemics of wheat leaf rust (Teng and Zadoks, 1980).

One important "structure element" in any system model is the "delay." Two major delays in a rust monocycle are the latent period and the infectious period. The capability to program delays is essential for any computer simulation of a natural system. In BARSIM-I, delays were simulated using the "boxcar train concept" (Teng, 1978). This enables the model to keep track of different numbers of successful penetrations on each day and of their different lengths of latency. If all successful penetrations in one day resulted in erupted uredia the next day, then no delays would be needed and the "bookkeeping" in the model would be much simpler.

As an example of model structure, BARSIM-I was divided into five sections (Teng *et al.*, 1980). Section I contained computer code that accepted input data on environment and crop and modified these data for use in the model. This is referred to as a "front end program" by computer specialists. In Section II of BARSIM-I, a whole tiller simulation is performed, in which the monocycle (Fig. 2) is the repeating entity between days. Sections III and IV, respectively, simulate leaf rust development on leaves 1 and 2 down from the spike. Section V of the system model uses disease values from Sections II–IV to estimate yield reduction on that tiller and to output the results in a comprehensible form.

D. EVALUATION

Evaluation of models consists of both verification and validation (Teng, 1981a). This is true for any model. The model's structure should be checked to ensure that it behaves as intended. This is verification and may consist simply of making sure the variables of a nonsystem model are necessary to and appropriate toward the outcome. System models are verified in much the same way. Verification of a model is an effort to prevent errors, and includes checking for incorrect data entry and maintaining documentation for parts of the model. The model is then debugged or tested with selected data to identify and correct programming errors after they have occurred.

Validation is the process of comparing model results to comparable real world data. Burleigh *et al.* (1972a,b) validated their models by direct comparison with actual field data. They found that the model for predicting disease severity (Burleigh *et al.*, 1972a) could account for 67–71% of the variation observed in the field plots from which input data was collected. Considering the range of environments and general nature of their prediction equation, an error of up to 30% may be reasonable.

Shrum (1975) compared model epidemics to field epidemics and found that EPIDEMIC accurately predicted disease progress trends, although differences occurred between daily field estimates and the modeled values. Teng (1978) found the system model BARSIM also successfully predicted barley leaf rust epidemics in fields. The model's accuracy was checked both through a parametric test and a nonparametric statistical test (the Smirnov test). BARSIM was successful in predicting epidemics using these objective tests (Teng, 1978).

E. USING MODELS AS EXPERIMENTAL TOOLS

Models that perform with sufficient accuracy for the purposes for which they were developed are useful as experimental tools. Often it is possible to obtain information, using a model, that is not obtainable through actual experimentation. Models are useful when variable values are outside the range of natural occurrences, when the real world of the model is not in existence (i.e., a Minnesota wheat field in January), or when actual experimentation is too costly, time-consuming, or ineffective (Dent and Blackie, 1979).

Nonsystem models such as that of Burleigh *et al.* (1972a,b), for predicting disease and crop losses are useful in identifying potential epidemics early enough for effective fungicide applications, thus aiding in management decisions. Spatial distribution models (Lambert *et al.*, 1980) are useful for demonstrating "differences associated with dispersal variables such as varieties, planting density and other factors." According to Madden (1980), temporal models are desirable for "(1) evaluating control strategies; (2) predicting future levels of disease; and (3) verifying plant disease simulators and forecasters."

Nonsystem models may be overly simple, but they permit examination of the model's consequences through mathematical reasoning (Madden, 1980). The development of single-equation models is much less complex and less time-consuming than developing a system model. Generally the results of a nonsystem model are easier to understand, as they involve fewer calculations and inputs. Nevertheless, nonsystem models only summarize complex growth processes (Madden, 1980).

System models have the complexity to describe biological phenomena in more detail than nonsystem models. They provide the ability to integrate spatial and temporal aspects of epidemics and are more flexible than single equation models. Shrum (1975) developed EPIDEMIC as an extremely flexible model for predicting a large class of diseases by making certain "modules" modifiable so they can fit other diseases.

System models are powerful tools in epidemic analysis. They can be used for evaluating cause and effect relationships at the population level by regulating certain model conditions (Shrum, 1975). System models also allow examination of various effects on the outcome (Teng, 1978). Like nonsystem models, system models can assist in management decisions and allocation of resources. They may be used for helping bridge the communication gap between farmers and researchers, and for directing research to areas of systems for which knowledge is lacking (Dent and Blackie, 1979).

F. MODELING CROP LOSS DUE TO RUST

Much is now known about the quantification of disease loss, the physiological basis for loss, and the estimation of regional losses (Teng and Gaunt, 1980). Modeling crop losses in cereals due to disease, including the rusts, was reviewed by Teng (1981c, 1985), Teng and Gaunt (1980), and James and Teng (1979). Thus, only the salient points related to modeling rust losses are briefly presented.

This chapter stresses the hierarchical nature of rust epidemics. In attempting to quantify the effect of rust on cereal yield, different levels of complexity are again encountered. Any modeling effort is influenced by the level at which quantification is desired. Furthermore, it is common that the level being modeled dictates the type of model that may be developed and the feasibility of generating experimental data. Crop loss models in cereals have been divided into three categories (Teng, 1981c): empirical, conceptual, and explanatory. The most common of these are the empirical models, developed using field data from experiments in plots or data from many single plants.

Empirical models may be developed using statistical curve-fitting methods, such as multiple least squares regression, and evaluated on their precision using established statistical criteria like r^2 (Teng, 1985). These models fit the range of cropping and disease situations represented by the data, and generally are specific to a region. Empirical models on rust loss have yield loss as the dependent variable and one or more representations of rust intensity as the independent variable(s). Several types of empirical models may be distinguished. The critical or single point model defines the relationship between yield loss and rust intensity at one assessment time during crop growth, i.e.:

$$\%\,\mathrm{YL} = f(X1)$$

where %YL is percentage yield loss, and $X1$ is rust intensity at a growth stage denoted by 1. Examples of critical-point models are those for estimating losses due to wheat stripe rust (Doling and Doodson, 1968; Mundy, 1973; King, 1976), wheat stem rust (Romig and Calpouzos, 1970), and barley leaf rust (King and Polley, 1976; Teng *et al.*, 1979).

Multiple point models estimate loss from two or more sequential assessments of rust intensity, and their form may be represented as

$$\%\,\mathrm{YL} = f(X1, X2\ldots),$$

where X is rust intensity at various growth stages. Examples of multiple-point models are those for wheat leaf rust (Burleigh *et al.*, 1972a,b) and barley leaf rust (Teng *et al.*, 1979).

Area under the disease progress curve (AUDPC) models estimate loss from the area under a disease progress curve defined between two times or growth stages, i.e.,

$$\% \text{YL} = f(\text{AUDPC})$$

Examples of this type of model are those for barley leaf rust (Teng *et al.*, 1979) and for the combined effects of wheat stem and leaf rust (Buchenau, 1975).

Each of the three types of empirical models has its use. Critical-point models are the simplest; although they do not account for variable infection rate of a rust epidemic, they are useful for loss estimation over large areas. In general, critical-point models developed using rust intensities at a late growth stage are more applicable in the field than those for an early growth stage. This is because a critical-point model assumes that rust intensity before or after the time of assessment will be comparable to those encountered in the original experiments for developing the model (Teng, 1981b). Early work by Kirby and Archer (1927) on wheat stem rust showed that yield loss could be predicted with increasing accuracy as the number of sequential assessments was also increased, i.e., as a model changes from critical to multiple point. Multiple point models are more accurate for rust epidemics with variable infection rates, or of long duration. The AUDPC models may be considered intermediate between the critical- and multiple-point models. While explaining the effect of different rust intensities on yield during a similar period, they cannot distinguish epidemics that have the same area under the curve but subtend different periods of growth (James, 1974). Multiple-point models have been favored for use in rust management because of their increased predictive ability under uncertain conditions (Teng *et al.*, 1980).

A modification of the multiple point concept was used by Calpouzos *et al.* (1976) for developing a model for wheat stem rust. Yield loss in the model was estimated from two epidemic parameters, the onset time and the epidemic slope. The model made use of sequential assessments through the slope parameter and recognized the importance of growth stage through the onset parameter. Teng and Montgomery (1981) have suggested that a logical development in empirical modeling would be response-surface models with the form

$$\% \text{YL} = f(X, \text{growth stage})$$

where yield loss is predicted from rust intensity and growth stage (cf. critical or multiple point, where yield loss is predicted from rust intensity at one or more stages). This type of model may be visualized as a three-dimensional response surface, with yield loss on the y axis, rust intensity on the x axis, and growth stage on the z axis (see Teng and Gaunt, 1980).

Empirical models developed using multiple regression are inherently linear (Teng, 1985). Further, regression does not confer a cause–effect relationship to any model. In modeling rust loss, there are indications that at lower rust intensities there may not be a linear relationship between rust and loss (Burleigh *et al.*, 1972a,b; Teng *et al.*, 1979). Similarly, at high rust intensities, a nonlinear relationship may also exist. Tammes (1961) proposed a generalized, conceptual representation of how crop yield may respond nonlinearly to stress, and recently Madden *et al.* (1981) have developed a flexible Weibull model for characterizing this relationship. This model has not been tested using rust data.

The physiology of yield in health cereals has been intensively studied, and there are studies in the physiology of yield loss due to disease (Gaunt, 1978, 1980). Using published information, Teng and Gaunt (1980) suggested that with most cereal disease it is possible to postulate the effect of the disease on yield components when the duration and onset of epidemics are known. They proposed a conceptual model to represent the impact of disease intensity at various growth stages on yield reduction. A conceptual model focuses attention on data collection for modeling and may be used to select treatment points in multiple-treatment experiments (Teng, 1985). The model provides a framework for evaluating "causal" or "noncausal" effects of disease on loss, as indicated by empirical regression models. For example, in reexamining data on barley leaf rust (Teng *et al.*, 1979), the use of early growth stages was unsound because no high rust intensities were encountered during those stages in the field (Teng and Gaunt, 1980). The conceptual model may be visualized as a three-dimensional response surface of yield loss to disease intensity and crop growth stage. For many purposes, empirical models may not need a sound physiological rationale, but certainly comparison of any multiple-point model with a conceptual model will enable deletion of some noncausal functions from the multiple-point model and lead to increased precision.

The effect of rust on yield can be explained at the plant physiological process level through interrelationships between fungal and host metabolism. Until recently, it was not feasible to consider modeling losses at this level of the hierarchy since one requirement is the availability of crop physiological models. Several such models have been developed (Loomis *et al.*, 1979; Loomis and Adams, 1980), and they may be "coupled" with disease or insect models to give an integrated model that represents a disease or insect loss system (Teng, 1981c; Ruesink, 1981). Published details on coupled rust–cereal models is still unavailable. In theory, explanatory disease loss models at the physiological level function independently of locality influences but would require substantial environmental monitoring. They are potentially the most applicable category of crop loss models.

G. PREDICTING RUST EPIDEMICS

The prediction of rust intensity is of interest for crop loss assessment and disease management. Early work is documented in Chester (1946) and involves the identification of important environmental variables that favor rust development. More recent work appears to have emphasized the development of system and nonsystem models for predicting rust intensity at a single site, or its spread from another site.

With wheat stem rust in the United States, Eversmeyer *et al.* (1973) used multiple linear regression to define models that predicted rust severity 7, 14, 21, and 30 days in the future, from a number of variables: rust severity on day of prediction, weekly and cumulative number of urediospores trapped, cultivar, growth stage, maximum and minimum temperature, a fungal temperature growth function, and a weekly infection function. They found that stem rust could not be accurately predicted using only environmental variables, and that a disease intensity assessment significantly improved predictions. Vanderplank (1963) used stem rust data to demonstrate his "compound interest" equation, and Rowell and Roelfs (1976) have noted that, in general, if three parameters are known—present rust intensity, infection rate, and growth stage—then it should be possible to predict epidemic progress. This was confirmed in a study by Calpouzos *et al.* (1976) using an extensive data base consisting of 469 epidemics. The idea was also used by Buchenau (1970) to develop a simple, grower-oriented method for projecting rust development. Buchenau's (1970) method, although simple, was an elegant integrator of epidemiological knowledge on stem rust, identifying leaf wetness as the single most important factor in rust development. The above models all predict rust severity in a proportion term (e.g., percent). Leonard (1969) developed a mathematical equation for estimating the number of sporulating and nonsporulating lesions from the initial number of lesions, number of rust generations, and the number of secondary lesions per lesion of the previous generation. This appears to be an improvement of Vanderplank's (1963) compound interest equation as it includes time delays.

Significant progress has been made by Nagarajan and co-workers to predict, spatially and temporally, stem rust increase in the Indian subcontinent. These workers developed a simple, linear regression model that predicted, with reasonable accuracy, stem rust severity over a 7-day period:

$$y = -29.3733 + 1.820\,X1 + 1.7735\,X2 + 0.2516\,X3,$$

where y is observed mean disease severity, $X1$ is mean disease severity 7 days before prediction day, $X2$ is mean weekly minimum temperature, and $X3$ is mean weekly maximum relative humidity for the prediction period (Nagarajan and Joshi, 1978; Karki *et al.*, 1979). Using spore trapping and meteorological information, they developed a means of predicting the movement of stem rust northward in the Indian peninsula, a situation resembling the "Puccinia pathway" in North America (Nagarajan *et al.*, 1977; Nagarajan and Joshi, 1980).

The development of wheat leaf rust epidemics has also been successfully predicted using multiple linear regression by Eversmeyer and Burleigh (1970) and Burleigh *et al.* (1972b). As with stem rust, they found that environmental variables were not sufficient to account for the variation in rust severity, and that biological variables such as rust severity on the day of prediction and crop growth stage were needed in the regression models. A different approach was used by Coakley and Line (1981) to predict epidemics of wheat stripe rust. They estimated rust severity from cumulative values of degree-days using a base of 7°C, where daily degree-day equals daily average temperature minus 7°C, giving either positive or negative degree-days. Wheat stripe rust has also been predicted by Oort (1968) in numbers of lesions, from an equation using latent period, infectious period, number of new infections per day, and a calculated, simple positive root.

The examples just discussed have dealt with nonsystem models. System simulation models of rust epidemics have inherently better predictive ability than the nonsystem models because, as was discussed earlier, system models can account for the effects of individual environmental variables on one or more processes in the rust infection cycle. However, system models like EPIDEMIC for wheat stripe rust and BARSIM for barley leaf rust require detailed weather data before they are operational. For example, BARSIM-I required the following daily variables: dew, maximum temperature, minimum temperature, minimum leaf temperature, rainfall, and sunshine. In a strict sense, the predictive ability of both system and nonsystem models is retrospective, i.e., they can predict rust intensity on a certain day given conditions that have occured. Nonsystem regression models like those of Burleigh *et al.* (1972b) have limited abilities for prediction. For a system model to have a truely prospective prediction, it would require reliable weather forecasts.

The cereal rusts have a worldwide distribution and impact on cereal productivity, and although there appears to have been much research on predictive models in countries outside North America and Western Europe, much of the literature is not generally available. Wheat stem rust forecasting has been reported in the U.S.S.R. by Minkevich (1976), Minkevich and Zakharova (1977), and Stepanov *et al.* (1977). This last group of workers has published a rather elaborate system model for wheat stem rust, which may be considered a spatial, stochastic model. Wheat stripe rust epidemics have been successfully simulated and predicted in the Peoples Republic of China (Tseng, 1961, and personal communication).

Although it is common to predict just rust intensity, the prediction of other epidemic parameters has also been reported. With wheat stem rust, Leonard (1969) developed the following equation for predicting the rate of disease increase in a multiline population:

$$r(\mathrm{m}) = r(\mathrm{s}) + (n/t) \ln (m)$$

where $r(\mathrm{m})$ is rate of rust increase in a plot of multiline varieties, $r(\mathrm{s})$ is rate of increase in a plot of susceptible variety, m is the proportion of susceptible plants in a host mixture, n is the number of rust generation, and t is the number of days of rust increase. With wheat stripe rust, Tseng (1961) predicted an infection rate that resembled Vanderplank's (1963) apparent infection rate r:

$$r = 1.16X + 0.1Y + 0.95Z - 2.6$$

where X is the number of rainy days in a month, Y is the total rainfall in millimeters in that month, and Z is the average daily temperature for the month.

We have discussed two major ways of predicting rust epidemics, through the use of either single-equation models or system-simulation models. Other predictive systems are feasible but have not been tested on the rusts (Krause and Massie, 1975). These may be viewed as simple integrators of weather, and predict favorable conditions for infection.

IV. Concluding Remarks

A rust epidemic model should not represent the finale of a research project on epidemiology. However, because of the effort needed in developing a usable, valid model of any rust pathosystem, normally, insufficient resources remain to research model implementation and utility. Certainly, as part of the modeling process, setting a well-defined objective for building that model is requisite to the success of the effort. However, with advances in computer technology, the assembly of detailed models may become a relatively easy process, and may lead to more available resources for researching the biology relevant for modeling. This may also mean more time for model evaluation and experimentation, and for research on the modeling process itself as it relates to different end users.

Models are potentially powerful tools in the research process; they expand our inductive and deductive abilities by removing the task of dealing with detail and by stressing clarity in conceptualization. We have discussed the systems analysis aspect of modeling, it should be part of every project on epidemiology. Too often, ad hoc research has resulted in overemphasis on selected components of epidemics, mostly because they are easy to work on, and excluded components that may contribute more to the understanding of disease epidemiology.

It is our view that, as modern epidemiology becomes more quantitative, the systems approach and modeling will become increasingly part of the repertoire of an epidemiologist. Modern epidemiology is quantitative, and requires input from many nontraditional disciplines; this interdisciplinary requirement of rust epidemiology will benefit from a move toward quantification and holism.

References

Allen, P. J. (1965). Metabolic aspects of spore germination in fungi. *Annu. Rev. Phytopathol.* **3,** 313–342.

Asai, G. N. (1960). Intra- and inter-regional movement of uredospores of black stem rust in the upper Mississippi River Valley. *Phytopathology* **50,** 535–541.

Bromfield, K. R. (1967). Some uredospore characteristics of importance in experimental epidemiology. *Plant Dis. Rep.* **51,** 248–252.

Buchenau, G. W. (1970). Forecasting profits from spraying for wheat rusts. *S. D. Farm Home Res.* **21,** 31–34.

Buchenau, G. W. (1975). Relationship between yield loss and area under the wheat stem rust and leaf rust progress curves. *Phytopathology* **62,** 1317–1318.

Burleigh, J. R., Roelfs, A. P., and Eversmeyer, M. G. (1972a). Estimating damage to wheat caused by *Puccinia recondita tritici. Phytopathology* **62,** 944–946.

Burleigh, J. R., Eversmeyer, M. G., and Roelfs, A. P. (1972b). Development of linear equations for predicting wheat leaf rust. *Phytopathology* **62,** 947–953.

Calpouzos, L., Roelfs, A. P., Madson, M. E., Martin, F. B., Welsh, J. R., and Wilcoxson, R. D. (1976). A new model to measure yield losses caused by stem rust in spring wheat. *Minn., Agric. Exp. Stn., Tech. Bull.* **307,** 1–23.

Cammack, R. H. (1961). *Puccinia polysora:* A review of some factors affecting the epiphytotic in West Africa. *Rep. Commonw. Conf. Plant Pathol.* **6,** 134–138.

Chamberlain, N. H. (1967). Deposition of particles to natural surfaces. *Microbes. Symp. Soc. Gen. Microbiol.* **17,** 138–164.

Chester, K. S. (1946). "The Nature and Prevention of the Cereal Rusts as Exemplified in the Leaf Rust of Wheat." Chronica Botanica, Waltham, Massachusetts.

Churchman, C. W., Ackoff, W. L., and Arnoff, E. L. (1957). "Introduction to Operations Research." Wiley, New York.

Coakley, S. M., and Line, R. F. (1981). Quantitative relationships between climatic variables and stripe rust epidemics on winter wheat. *Phytopathology* **71,** 461–467.

Dent, J. B., and Blackie, M. J. (1979). "Information Systems for Agriculture." Appl. Sci. Publ. London.

Doling, D. A., and Doodson, J. K. (1968). The effect of yellow rust on the yield of spring and winter wheat. *Trans. Br. Mycol. Soc.* **51,** 427–434.

Eversmeyer, M. G., and Burleigh, J. R. (1970). A method of predicting epidemic development of wheat leaf rust. *Phytopathology* **60,** 805–811.

Eversmeyer, M. G., and Kramer, C. L. (1980). Horizontal dispersal of urediospores of *Puccinia recondita* f. sp. *tritici* and *P. graminis* f. sp. *tritici* from a source plot of wheat. *Phytopathology* **70,** 683–685.

Eversmeyer, M. G., Burleigh, J. R., and Roelfs, A. P. (1973). Equations for predicting wheat stem rust development. *Phytopathology* **63,** 348–351.

Eyal. Z., and Peterson, J. L. (1967). Uredospore production of five races of *Puccinia recondita* Rob. ex. Desm. as affected by light and temperature. *Can. J. Bot.* **45,** 537–540.

Gaunt, R. E. (1978). Crop physiology: Disease effects and yield loss. *In* "Proceedings of the Australian Plant Pathology Society Workshop on Epidemiology and Crop Loss Assessment" (R. C. Close *et al.*, eds.), pp. 9-1 to 9-12.

Gaunt, R. E. (1980). Physiological basis of yield loss. *Misc. Publ.—Minn., Agric. Exp. Stn.* **7,** 98–111.

Gregory, P. H. (1973). "The Microbiology of the Atmosphere," 2nd ed. Leonard Hill, London.

Hess, H. L., Allen, P. J., Nelson, D., and Lester, H. (1975). Mode of action of methyl cis-ferulate, the self inhibitor of stem rust uredospore germination. *Physiol. Plant Pathol.* **5,** 107–112.

Hirst, J. M. (1953). Changes in atmosphere spore content: Diurnal periodicity and the effects of weather. *p12Trans. Br. Mycol. Soc.* **36,** 375–393.

Hwang, L. (1942). The effect of light and temperature on the viability of certain cereal rusts. *Phytopathology* **32,** 699–711.

James, W. C. (1974). Assessment of plant diseases and losses. *Annu. Rev. Phytopathol.* **12,** 27–48.

James, W. C., and Teng, P. S. (1979). The quantification of production constrains associated with plant diseases. *Appl. Biol.* **4,** 201–267.

Kampmeijer, P., and Zadoks, J. C. (1977). "EPIMUL, a Simulator of Foci and Epidemics in Mixtures of Resistant and Susceptible Plants, Mosaics and Multilines." Centre for Agricultural Publishing and Documentation, Wageningen.

Karki, C. B., Pande, S., Thombre, S. B., Joshi, L. M., and Nagarajan, S. (1979). Evaluation of a linear model to predict stem rust severity. *Cereal Rusts Bull.* **7,** 3–7.

King, J. E. (1976). Relationship between yield loss and severity of yellow rust recorded on a large number of single stems of winter wheat. *Plant Pathol.* **25,** 172–177.

King, J. E., and Polley, R. W. (1976). Observations on the epidemiology and effect on grain yield of brown rust in spring barley. *Plant Pathol.* **25,** 63–73.

Kirby, R. S., and Archer, W. A. (1927). Diseases of cereal and forage crops in the United States in 1926. *Plant Dis. Rep., Suppl.* **53,** 110–208.

Kochman, J. K., and Brown, J. F. (1975). Host and environmental effects on post-penetration development of *Puccinia graminis avenae* and *P. coronata avenae. Ann. Appl. Biol.* **81,** 33–41.

Kramer, C. L., and Pady, S. M. (1966). A new 24-hour spore sampler. *Phytopathology* **56,** 517–520.

Kranz, J. (1974). Comparison of epidemics. *Annu. Rev. Phytopathol.* **12,** 355–374.

Kranz, J., and Hau, B. (1980). Systems analysis in epidemiology. *Annu. Rev. Phytopathol.* **18,** 67–83.

Krause, R. A., and Massie, L. B. (1975). Predictive systems: Modern approaches to disease control. *Annu. Rev. Phytopathol.* **13,** 31–47.

Kuhl, C. L., Maclean, D. J., Scott, K. J., and Williams, P. G. (1971). The axenic culture of Puccinia species from uredospores: Experiments on nutrition and variation. *Can. J. Bot.* **49,** 201–209.

Lambert, D. H., Villareal, R. L., and MacKenzie, D. R. (1980). A general model gradient analysis. *Phytopathol. Z.* **98,** 150–154.

Leonard, K. J. (1969). Factors affecting rates of stem rust increase in mixed plantings of susceptible and resistant oat varieties. *Phytopathology* **59,** 1845–1850.

Loomis, R. S., and Adams, S. S. (1980). The potential of dynamic physiological models for crop loss assessment. *Misc. Publ.—Minn. Agric. Exp. Stn.* **7,** 112–117.

Loomis, R. S., Rabbinge, R., and Ng, E. (1979). Explanatory models in plant physiology. *Annu. Rev. Phytopathol.* **30,** 339–367.

Madden, L. V. (1980). Quantification of disease progression. *Prot. Ecol.* **2,** 159–176.

Madden, L. V., Pennypacker, S. P., Antle, C. E., and Kingsolver, C. H. (1981). A loss model for crops. *Phytopathology* **71,** 685–689.

Mehta, Y. R., and Zadoks, J. C. (1970). Uredospore production and sporulation period of *Puccinia recondita* f. sp. *tritici* on primary leaves of wheat. *Neth. J. Plant Pathol.* **76,** 267–276.

Melander, L. W. (1935). Effect of temperature and light on development of uredial stages of Puccinia graminis. *J. Agric. Res. (Washington, D.C.)* **50,** 861–880.

Minkevich, I. I. (1976). Selection and utilization of disease forecasting models for plant protection. *In* "Modeling for Pest Management" (R. L. Tummala, D. L. Haynes, and B. A. Croft, eds.), pp. 171–175. Michigan State Univ. Press, East Lansing.

Minkevich, I. I., and Zakharova, T. I. (1977). Long-term forecast of brown rust of winter wheat in the Forest-Steppe Zone of the USSR in relation to changes in solar activity. *J. Gen. Biol. (Moscow)* **38,** 372–379 (in Russian).

Mundy, E. J. (1973). The effect of yellow rust and its control on the yield of Joss Cambier Winter Wheat. *Plant Pathol.* **22,** 171–176.

Nagarajan, S., and Joshi, L. M. (1978). A linear model for a seven-day forecast of stem rust severity. *Indian Phytopathol.* **31,** 50–506.

Nagarajan, S., and Joshi, L. M. (1980). Further investigations on predicting wheat rust appearance in central and peninsular India. *Phytopathol. Z.* **98,** 84–90.

Nagarajan, S., Singh, S., Joshi, L. M., and Saari, E. E. (1977). Prediction of *Puccinia graminis* f. sp. *tritici* on wheat in India by trapping the uredospores in rain samples. *Phytoparasitica* **5,** 104–108.

Oort, H. J. P. (1968). A model of the early stage of epidemics. *Neth. J. Plant Pathol.* **74,** 177–180.

Pady, S. M., Kramer, C. L., Pathak, V. K., Morgan, F. L., and Bhatli, M. A. (1965). Periodicity in airborne cereal rust urediospores. *Phytopathology* **55,** 132–134.

Prabhu, A. S., and Wallin, J. R. (1971). Influence of temperature and light on spore production of *Puccinia graminis tritici*. *Phytopathology* **61,** 120–121.

Roelfs, A. P. (1972). Gradients in horizontal dispersal of cereal rust uredospores. *Phytopathology* **62,** 70–76.

Romig, R. W., and Calpouzos, L. (1970). The relationship between stem rust and loss in yield of spring wheat. *Phytopathology* **60,** 1801–1805.

Rowell, J. B. (1981). Relation of post-penetration events in Idaed 59 wheat seedlings to low receptivity to infection by *Puccinia graminis* f. sp. *tritici*. *Phytopathology* **71,** 732–736.

Rowell, J. B., and Roelfs, A. P. (1976). Wheat stem rust. *In* "Modeling for Pest Management" (R. L. Tummala, D. L. Haynes, and B. A. Croft, eds.), pp. 69–79. Michigan State Univ. Press, East Lansing.

Ruesink, W. G. (1981). Environmental inputs for crop loss models. *In* "Plant Protection: Fundamental Aspects" (T. H. Kommedahl, ed.), pp. 131–134. Burgess, Minneapolis, Minnesota.

Schmitt, C. G., Hendrix, J. W., Emge, R. G., and Jones, M. W. (1964). "Stripe Rust, *Puccinis striiformis* West," Fort Detrick Tech. Rep. No. 43. Plant Sci. Lab., Fort Detrick, Frederick, Maryland.

Schrodter, H. (1960). Dispersal by air and water—the flight and landing. *In* "Plant Pathology: An Advanced Treatice" (J. G. Horsfall and A. E. Dimond, eds.), Vol. 3, pp. 169–227. Academic Press, New York.

Shrum, R. (1975). "Simulation of Wheat Stripe Rust (*Puccinia striiformis* West.) Using EPIDEMIC: A Flexible Plant Disease Simulator," Prog. Rep. No. 347. Pennsylvania State University, University Park.

Simkin, M. B., and Wheeler, B. E. J. (1974). The development of Puccinia hordei on barley cv. Zephyr. *Ann. Appl. Biol.* **78,** 225–235.

Smith, R. S. (1966). The liberation of cereal stem rust uredospores under various environmental conditions in a wind tunnel. *Trans. Br. Mycol. Soc.* **49,** 33–41.

Stepanov, K. M., Terekhov, V. I., Sanin, S. S., Afonin, S. P., and Solodukhina, L. D. (1977). A possible model for a seasonal forecast of stem rust of wheat. *Mycol. Phytopathol.* **2,** 155–161 (in Russian).

Tammes, P. M. L. (1961). Studies of yield losses. II. Injury as a limiting factor of yield. *Neth. J. Plant Pathol.* **67,** 257–263.

Teng, P. S. (1978). System modelling in plant disease management. Ph.D. Thesis, University of Canterbury, New Zealand (unpublished).

Teng, P. S. (1981a). Validation of computer models of plant disease epidemics: A review of philosophy and methodology. *Z. Pflanzenkr. Pflanzenschutz* **88,** 49–63.

Teng, P. S. (1981b). Use of regression analysis for developing crop loss models. *In* Crop Loss Assessment Methods" (L. Chiarappa, ed.), Suppl. 3, pp. 51–59. CAB/FAO, Rome.

Teng, P. S. (1981c). Modeling disease-loss systems in cereals. *In* "Plant Protection: Fundamental Aspects" (T. H. Kommedahl, ed.), pp. 122–127. Burgess, Minneapolis, Minnesota.

Teng, P. S. (1985). Assessment of crop losses in cereals. *In* "Reviews of Tropical Plant Pathology" (A. P. Raychauduri, ed.) (in press).

Teng, P. S., and Close, R. C. (1978). Effect of temperature and uredinium density on urediniospore production, latent period, and infectious period of *Puccinia hordei* Otth. *N. Z. J. Agric. Res.* **21,** 287–296.

Teng, P. S., and Gaunt, R. E. (1980). Modelling systems of disease and yield loss in cereals. *Agric. Syst.* **6,** 131–154.

Teng, P. S., and Montgomery P. R. (1981). Response surface models for common rust of corn. *Phytopathology* **71,** 895 (abstr.).

Teng, P. S., and Zadoks, J. C. (1980). "Computer Simulation of Plant Disease Epidemics," pp. 23–31. McGraw-Hill Yearbook of Science and Technology, New York.

Teng, P. S., Blackie, M. J., and Close, R. C. (1977). A simulation analysis of crop yield loss due to rust disease. *Agric. Syst.* **2,** 189–198.

Teng, P. S., Close, R. C., and Blackie, M. J. (1979). Comparison of models for estimating yield loss caused by leaf rust (*Puccinia hordei*) on Zephyr barley in New Zealand. *Phytopathology* **69,** 1239–1244.

Teng, P. S., Blackie, M. J., and Close, R. C. (1980). Simulation of the barley leaf rust epidemic: Structure and validation of BARSIM-I. *Agric. Syst.* **5,** 55–73.

Tseng, S. M. (1961). On the mathematical analysis of the epidemic of wheat stripe rust. I. The rate of disease progress. *Acta Phytophylacica Sin.* **1,** 35–38.

Vanderplank, J. E. (1963). "Plant Diseases: Epidemics and Control." Academic Press, New York.

Waggoner, P. E., and Horsfall, J. G. (1969). EPIDEM. A simulator of plant disease written for a computer. *Bull.—Conn. Agric. Exp. Stn., New Haven* **698,** 1–80.

Waggoner, P. E., Horsfall, J. G., and Lukens, R. J. (1972). EPIMAY. A simulator of southern corn leaf blight. *Bull.—Conn. Agric. Exp. Stn., New Haven* **729,** 1–84.

Yarwood, C. (1956). Simultaneous self-stimulation and self-inhibition of uredospore germination. *Mycologia* **48,** 20–24.

Zadoks, J. C. (1971). Systems analysis and the dynamics of epidemics. *Phytopathology* **61,** 600–610.

Zadoks, J. C. (1972). Methodology for epidemiological research. *Annu. Rev. Phytopathol.* **10,** 253–276.

Zadoks, J. C., and Groenewegen, L. J. M. (1967). On light-sensitivity in germinating uredospores of wheat brown rust. *Neth. J. Plant Pathol.* **73,** 83–102.

Zadoks, J. C., and Schein, R. D. (1979). "Epidemiology and Plant Disease Management." Oxford Univ. Press, London and New York.

PART IV

Control

15

Resistance of the Race-Specific Type

P. L. Dyck
E. R. Kerber
Agriculture Canada Research Station,
Winnipeg, Manitoba, Canada

I. Introduction

The terms race-specific and non-race-specific have received much attention in discussions on breeding for disease resistance (Van-

ISBN 0-12-148402-5

derplank, 1968; Nelson, 1978; Ellingboe, 1975, 1978). Race-specific or vertical resistance implies resistance to some pathogen isolates and not to others and is relatively simply inherited. Nonspecific or horizontal resistance implies resistance to all isolates of the disease organism and is often polygenically determined. With race specificity as the basis for distinction between the two types of resistance, once a pathogen isolate has overcome nonspecific horizontal resistance it must be reclassified as specific or vertical. Specific resistance has also been defined as resistance to infection or hypersensitive resistance, while nonspecific or horizontal resistance permits infection but reduces colonization or spread of the disease. This implies that intermediate levels of resistance must be horizontal; generally, however, such resistance is race-specific. Nelson (1978) has suggested that the two kinds of resistance are conditioned by the same genes. A host gene may give hypersensitive resistance to some isolates and a rate-reducing resistance to others. Ellingboe (1975) has concluded that "nonspecific resistance is that resistance which has not yet been shown to be specific."

Race-specific resistance is conditioned by the interaction of specific genes in the host with those in the pathogen. The genetic principles underlying this interaction were established by Flor (1955), who, working with flax and its rust (*Melampsora lini: Linum usitatissimum* host–parasite system), showed that whether a cultivar is resistant or susceptible to a physiologic race of the pathogen depends on the genotype for resistance or susceptibility of the host and the genotype for virulence or avirulence of the race of the pathogen. A similar system has been shown to exist for most of the cereal crops and their rust pathogens. Such a gene-for-gene system seems a logical consequence of the coevolution of a host and its obligate parasite in nature, and also in "man-guided" evolution, which has seen the pathogen adapt repeatedly to overcome the resistance of new host cultivars. This ability of the pathogen to generate new virulent forms necessitates an ongoing search for new sources and types of resistance that can be utilized in breeding for disease resistance.

The different types of race-specific resistance, their sources, and their use in developing rust resistant cultivars is discussed. The discussion of race-specific resistance includes hypersensitive and intermediate or moderate seedling resistance, mature- or adult-plant resistance, and resistance due to genes with an additive or cumulative effect. There is an increasing consensus that all types of resistance must be utilized in the development of a breeding strategy to produce cultivars with stable rust resistance.

II. History of Race-Specific Resistance

In North America a severe wheat stem rust epidemic in 1904 stimulated the study of the rust disease and methods of control (Stakman, 1955). In 1905, a breeding program to develop rust resistant spring wheats was initiated in the United States. Also in that year, Biffen (1905) first reported on the Mendelian inheritance of stripe rust resistance in wheat. Another severe North American epidemic in 1916 developed early in the season and attacked susceptible cultivars, including the early-maturing cultivar Marquis. This epidemic further stimulated various aspects of rust research and resulted in the initiation of numerous breeding programs to develop resistant cultivars. Ceres, the first cultivar bred for rust resistance, was distributed in North Dakota in 1926 (Stakman, 1955). Because of its rust resistance and good quality, it was soon grown over a large area. By 1928 a race virulent on Ceres was found. This became the most prevalent race and caused severe epidemics in 1934 and in 1937. This probably was the beginning of the so-called "boom and bust" cycle of resistant cultivars succumbing to new, virulent biotypes of the pathogen.

Several new stem-rust-resistant cultivars were released in the 1930s, including Thatcher, which had resistance derived from Iumillo durum wheat. In subsequent years Thatcher was grown over a large spring wheat area and remained resistant until the early 1950s, when it was damaged by race 15B. However, prior to 1950 it was being replaced by other cultivars that had the Hope or H-44 type of resistance to stem rust and resistance to leaf rust. These cultivars were also damaged by race 15B. In 1954 Selkirk, which has gene *Sr6* plus the H-44 type of resistance, was released. The cultivars released since the 15B epidemic, including Manitou, which has the Thatcher type of resistance plus *Sr6*, have provided good resistance to the North American population of stem rust (Green and Campbell, 1979). Since the release in 1930 of stem rust resistant cultivars in Canada, losses due to stem rust have occurred only during the race 15B epidemics of the early 1950s when a partial breakdown of the Hope and H-44 types of resistance occurred.

In Australia (Watson, 1981), from 1920 to 1950, cultivars with single genes for stem rust resistance were released. The single genes used were *Sr6, Sr9b, Sr11, Sr17* or *Sr36(Tt-1)*. These cultivars had apparent good resistance conferred in each by a single gene and were well received by the farmer. However, the rust population adapted rapidly and became virulent on the newly released cultivars. After 1950 cultivars were released with various combinations of these and several other genes. Since then the fungus has not been able to combine the matching virulence genes required to overcome the host gene combinations.

According to Roelfs (1978), the reduced frequency of stem rust epidemics in the United States during the past 25 years resulted from the use of resistant cultivars and the removal of the barberry. In the *CIMMYT Review* (International Maize and Wheat Improvement Center, 1981), the comment is made that "stem rust resistance in CIMMYT bread wheats was stabilized in the late 1950s and has been retained." It appears that the pyramiding of primarily specific genes in breeding for resistance to stem rust of wheat has been reasonably successful in several countries. On the other hand, breeding of cultivars with stable resistance to leaf rust has not been as successful. In Canada the first leaf-rust-resistant cultivars (with Hope resistance-*Lr14a*) released in the late 1930s became susceptible by 1945 (Johnson and Newton, 1946). Selkirk, with genes *Lr10* and *Lr16* (the only effective gene), was released in 1953 and was resistant to the prevalent races until approximately 1962. Manitou, released in 1965, and several related cultivars released since then, all with gene *Lr13* for adult-plant resistance, are now almost fully susceptible. Each of these Canadian cultivars had only one effective gene for resistance to leaf rust at the time of release. To further quote from the *CIMMYT Review* (International Maize and Wheat Improvement Center, 1981), cultivars resistant to leaf rust at the time of release "usually become susceptible after two or three years of commercial productions."

In barley a single dominant gene *T* (the Peatland gene) controls resistance to stem rust in both the seedling and adult-plant stages (Andrews, 1956). Although this is the only important source of resistance in barley and is still highly effective in many countries, some virulent strains of rust have been found (Johnson, 1961).

Major specific genes controlling resistance to stripe rust in wheat have been easily overcome by the pathogen. In the United States breeders are using minor recessive genes (Sharp, 1973), while in the United Kingdom "durable resistance," that is, resistance that has remained effective over time, is being used (Johnson, 1978).

III. Types of Specific Resistance

A. Hypersensitivity

Probably the most common race-specific resistance that has been used in breeding programs is the hypersensitive type. It is characterized by macroscopic lesions at the infection sites. The early collapse and death of the host cells at these sites prevents the further growth of the fungal hyphae. This definition implies "an active resistance mechanism in which the rapid death of the host cells around the point of infection prevents colonization" (Robinson, 1976). However, it has been suggested that the necrotic hypersensitive response does not determine the incompatible reaction but that it is only an incidental stress symptom to the disease (Mayama *et al.*, 1975).

Many of the genes for leaf rust resistance in wheat confer a hypersensitive response, including *Lr1*, the *Lr2* alleles, *Lr3*, *Lr10*, and many of the genes transferred from related species (Browder, 1980). The best known example for wheat stem rust is *Sr6*. Genes for a hypersensitive resistance to crown rust in oats are common, including many of the those found in *Avena sterilis*.

B. Immunity

According to the classification of the types of rust infection (Stakman *et al.*, 1962), an immune response is indicated by the absence of visible lesions on the host plant. It is generally agreed that plants immune to diseases are immune to infection by the pathogen. "An immune plant is a non-host" (Robinson, 1976). Gene *Sr5* has been considered a classical example of a gene for immunity. However, Rohringer *et al.* (1979) showed that when gene *Sr5* is in a Chinese Spring background, macroscopically visible lesions are produced. Furthermore, many of the genes for hypersensitive reaction can produce an immune response to some rust cultures and visible fleck infection types to other cultures, as for example, the *Lr2* alleles (Dyck and Samborski, 1974). Since an immune or hypersensitive reaction depends on host genetic background and/or the rust culture used, the presence or absence of visible lesions does not imply two different types of specific genes. The identified specific genes for immunity may all be hypersensitive.

C. Moderate

Many of the genes for specific rust resistance give a moderate or intermediate level of resistance. In the seedling stage the resulting infection types can range from type 1 to type 3. With this type of resistance, the pathogen penetrates the host and some rust development occurs before an incompatible reaction becomes apparent. Varying amounts of urediospores are produced. These genes do not prevent colonization but reduce the rate of spread of the pathogen.

The seedling resistant reaction associated with wheat leaf rust genes *Lr11, Lr16, Lr17, Lr18,* and *Lr30* is of an intermediate type. With wheat stem rust, genes *Sr8,* the *Sr9* alleles, *Sr22, Sr24,* and *Sr33* are of this type. Usually genes that give moderate resistance in the seedling stage give the same type of resistance in the adult stage.

D. ADULT-PLANT RESISTANCE

Resistance that is first apparent in older plants is termed adult- or mature-plant or postseedling resistance. In early literature it was frequently referred to as field resistance. The onset of adult-plant resistance can vary. According to Anderson (1966), the common wheat cultivars Frontana and Klein Aniversario have a postseedling type of leaf rust resistance that appears as early as the third leaf stage. The adult-plant resistance of the cultivar Exchange is not fully expressed until after the emergence of the flag leaf (Samborski and Ostapyk, 1959). Genes for adult-plant resistance may be effective against a wide spectrum of rust races. In fact, Robinson (1976) states that all adult-plant resistance is of the horizontal type. However, race specificity has been found for several of the adult-plant genes for resistance in wheat including *Lr12* and *Lr13.* Gene *Lr22b,* an adult-plant gene in the cultivar Thatcher, gives resistance to only one known North American race. Some of the genes for adult-plant resistance in common wheat are of interspecific origin, including *Sr2* originally transferred from tetraploid Yaroslav emmer (McFadden, 1930) and *Lr22a* from *Aegilops squarrosa* (Dyck and Kerber, 1970).

Based on studies with stripe rust of wheat, Johnson (1978, 1981a) introduced the term "durable resistance" to refer to resistance that has remained effective in cultivars widely grown for many years. This type of resistance, the durability of which can be judged only in retrospect, can be either complexly or simply inherited. A cultivar with durable resistance to stripe rust is Cappelle-Desprez, which has several specific genes for seedling resistance and several for adult-plant resistance. Several, but not all, of the adult-plant factors appear to be responsible for the durable resistance. These factors appear to be associated with chromosome 5BS-7BS. However, recent evidence (Johnson, 1981b) indicates that the genetic background is important to the expression of durability. Hare and McIntosh (1979) have suggested that the Hope-derived adult-plant gene *Sr2* for stem rust resistance may be durable.

IV. Expression of Specific Resistance

A. TEMPERATURE SENSITIVITY

Temperature affects the expression of many genes for disease resistance. Some genes become ineffective at high temperatures, while others become ineffective under low temperatures. A classical example of high-temperature breakdown is that of *Sr6*, a gene for resistance to stem rust in wheat. Many workers have found *Sr6* to be inactivated by high temperatures. Green *et al.* (1960) tested an isogenic line of *Sr6* in Marquis wheat to a large number of races of stem rust under high and low temperatures and observed that resistance was inhibited at high temperature. Loegering and Harman (1969) showed that this breakdown occurs in the range of 20°–23°C. *Lr20*, a gene for leaf rust resistance, has also been shown to be temperature sensitive (Jones and Deverall, 1977). It is partially effective at 26°C and completely ineffective at 30.5°C. Most of the genes for resistance to stem rust in oats are temperature-sensitive (Martens, Chapter 4). The use of some of the genes sensitive to high temperature may be limited to cultivars grown in temperate climates. Gassner and Straib (1931) showed that some cultivars become less resistant to disease under low temperature. Wheat cultivars Malakof, Norka, and Democrat were resistant at 18.7°C but became susceptible at a lower temperature.

Some of the minor additive genes for resistance to stripe rust from wheat cultivar PI 178383 express the greatest resistance at high temperatures, while other minor genes confer more resistance at a lower temperature (Lewellen and Sharp, 1968). By selecting among infected plant populations grown at different temperatures, Sharp *et al.* (1976) obtained plants that were resistant over a wide range of temperatures. Presumably in this way they were able to combine different genes with additive effects.

Temperature sensitivity may assist in identifying certain genes that are masked when in combination with other genes. Thus, temperature sensitivity may be a tool that can be used in combining or pyramiding several resistance genes into one cultivar.

B. GENE INTERACTIONS

When a cultivar has several genes for resistance to the same disease, it is generally assumed that the genes act independently. A cultivar with two genes, each determining a different level of resistance, usually exhibits the rust reaction phenotype of the most effective gene; the gene conferring the least resistance is masked. The most effective gene is epistatic to those that condition a less resistant reaction. Furthermore, a cultivar with two or more genes will be resistant to all of the rust races to which the genes are effective separately.

However, genes for disease resistance do not invariably act independently. The gene action may be complementary; genes at different loci, or their products, may interact to give higher levels of resistance. Extreme forms of complementary resistance require the presence of two or more genes for the resistance to be expressed. In the oat cultivar Bond, two such complementary genes give resistance to crown rust (Simons *et al.*, 1978). Hosts with either gene alone are susceptible while plants with both genes are resistant to crown rust (Baker, 1966). Martens *et al.* (1981) reported on resistance to stem rust of oats that was conferred by complementary genes involving *Pg-12* and a second gene or a suppressor. This type of complementary action has usually been between recessive genes.

There are numerous examples of genes for disease resistance that interact to give an enhanced level of resistance (Schafer *et al.*, 1963; Knott, 1957; Voronkova, 1980). This complementary interaction, which may be additive, results in a higher level of resistance than that conferred by the genes singly. Dyck (1977) found that PI 58548 has two genes for seedling resistance to leaf rust, one giving a 1+ infection type and the second a 2+. When combined the two genes interact to produce a ;1 infection type. They also interact to produce superior adult-plant or field resistance. More recent studies (Dyck and Samborski, 1982; Samborski and Dyck, 1982) have shown additional interactions between each of two different pairs of genes for seedling resistance, between a pair of adult-plant genes, and between a pair of seedling and adult-plant genes. In general, leaf rust cultures avirulent on both of the combined genes showed the interaction for enhanced resistance, although there were several combinations that showed an interaction with races avirulent on one of the genes and virulent on the other. It should be noted that not all genes that result in intermediate levels of resistance will, when in combinations with other genes, interact to give superior resistance.

Clifford (1975) suggested that specific resistance genes may give a form of residual resistance to pathogen strains possessing the corresponding virulence genes. Such a residual or "ghost" effect has been described for the *Triticum: Erysiphe graminis* system by Martin and Ellingboe (1976) and by Nass *et al.* (1981). Nelson (1981) combined a number of these genes having a residual effect, which he termed "defeated" genes, and obtained effective reduction of disease development. Samborski and Dyck (1982) noted that seedlings of the cultivar Columbus with gene *Lr16* gave an incompatible reaction to several cultures of leaf rust that are virulent on *Lr16*. Presumably, an interaction between *Lr16* and *Lr13*, a gene for adult-plant resistance also present in Columbus, resulted in an incompatible phenotype similar to that produced by *Lr16* with avirulent cultures. It is possible that some of the interactions observed may be similar to the "ghost" effect of ineffective genes. It appears that some resistance genes having very little individual effect can interact with or modify other genes to condition a more genetically complex resistance.

There are also examples of nonallelic additive interactions in stem rust of wheat. Knott (1957) noted that resistance genes *Sr10, Sr11, Sr12*, and particularly *Sr9* were important modifiers of gene *Sr7*. Luig and Rajaram (1972) studied the stem rust reaction of homozygous and heterozygous combinations of *Sr5* and *Sr9b, Sr5* and *Sr13, Sr6* and *Sr8*, and *Sr8* and *Sr9b*. Additive gene interactions were observed especially when *Sr6* was involved. It would appear that some genes are more sensitive to nonallelic interaction than others.

Samborski and Dyck (1982) made a four-way cross between four backcross lines, each with a single gene resulting in a low level of resistance to leaf rust. Highly resistant lines were obtained from selections made in the F_2 and F_3 generations. These highly resistant selections probably had an accumulation of the genes derived from the four parents.

Sharp and co-workers state that resistance to stripe rust is controlled by two different types of resistance genes (reviewed by Robellen and Sharp, 1978). They found that each of the two cultivars PI 178383 and Chinese 166 had a different dominant major gene. Each gene gave a high level of resistance that was largely unaffected by the environment. However, F_2 plants lacking the major genes segregated for additional genes that gave some resistance themselves or acted as modifiers of the heterozygous major genes. Up to three minor genes were accumulated in lines with good levels of resistance. These and other minor genes for resistance to stripe rust are generally sensitive to environmental influence. There is controversy as to whether these minor genes for resistance to stripe rust are true resistance genes or modifiers in a gene complex that interacts to determine the rust reaction of the host plant. We consider them as resistance genes that are important in the development of cultivars with a genetically complex type of disease resistance.

Since combined genes for disease resistance do not necessarily act independently but may interact to give a quantitative improvement in resistance, the following should be noted:

(1) The value of combining a number of genes for resistance into one cultivar is emphasized. Complex resistances can be much more effective than anticipated and may be long-lasting.
(2) Attempts to assign genotypes to cultivars and introductions on the basis of testing them with a series of rust cultures may be misleading. Interactions may mask the presence of individual genes.

C. INHIBITORY EFFECTS

Genes conditioning host resistance can also be inhibited or suppressed by nonallelic genes. The resistance to Canadian leaf rust cultures conferred by gene *Lr23* is suppressed by a gene in Thatcher, but this suppression is only partially effective with Australian cultures (McIntosh and Dyck, 1975). Kerber and Green (1980) observed that Canthatch nullisomic 7D is much more resistant to several cultures of stem rust than normal disomic Canthatch. They concluded that chromosome 7DL carries a gene that inhibits the expression of one or more genes for rust resistance present on other chromosomes of Canthatch.

D. BACKGROUND EFFECTS

The genetic background can affect the expression of specific genes for resistance. Several such genes, particularly those conferring resistance to stem rust and leaf rust of wheat, have been backcrossed into different cultivars. Alleles *Lr2a, Lr2b,* and *Lr2c* were backcrossed into the cultivars Thatcher, Prelude, and Red Bobs; they were most effective in the Thatcher background, intermediate in Prelude, and least effective in Red Bobs (Dyck and Samborski, 1974).

A gene for resistance may be dominant in one genetic background and recessive in another. Consequently, the susceptible parent in a cross can influence the degree of dominance of a gene as has been shown by Anderson (1966) for gene *Lr2.*

The reaction conferred by a gene may be dominant relative to one race of a pathogen and recessive to another (Knott and Anderson, 1956; Lupton and Macer, 1962). It has been suggested that this phenomenon may be due to two closely linked genes, the expression of one being dominant and the other recessive. Hooker and Saxena (1971) tested this with gene *Rp3* in maize. By using a linked marker they were able to screen large testcross populations for crossovers. Since none were detected, the two supposed component genes of *Rp3,* one dominant and one recessive, would have to be less than 0.02 map units apart.

E. ALLELISM

Most genes for disease resistance are inherited independently of each other. When two or more genes are on the same chromosome, they may show varying degrees of linkage. In some cases the genes are either tightly linked or they are alleles, that is, they are at the same locus on a chromosome. Such tight linkage, or multiple allelism may restrict the number of genes that can be combined into one cultivar. In theory, a self-pollinated crop can be homozygous for only one gene at a locus. However, at several loci that were assumed to be multiple alleles for disease resistance, two or more of the alleles were recombined in coupling linkage, and they then behaved as one gene. In oats, stem rust resistance genes *Pg-3* and *Pg-9,* assumed to be alleles, have been combined (Koo *et al.,* 1955). Similarly, in wheat the two alleles at the *Lr14* locus have been combined (Dyck and Samborski, 1970).

Saxena and Hooker (1968) suggest that the *Rp1* locus in maize, which may have as many as 14 different alleles for resistance to *P. sorghi,* consists of a series of tandem duplications of the original gene. These duplications have gradually differentiated to give resistance to different races of the rust. They suggest that the different alleles may consist of one or more combinations of the original gene and/or its modified duplicates. They also suggest the possibility of synthesizing a gene at one of these complex loci (e.g., the *Rp1* complex has a large number of alleles with crossover values ranging from 0.10 to 0.37%) that would confer resistance to many cultures by systematically recombining several of the alleles.

Mayo and Shepherd (1980), using a modified cis–trans test for functional allelism, found that several of the *M* alleles for resistance to flax rust were in fact separate, closely linked loci. They combined two of the *M* genes in the coupling phase where each of the genes functioned independently. Thus, it may be possible to combine three or more of the *M* genes in coupling to construct a complex resistance genotype.

Some alleles at a locus, or closely linked genes, appear to be functionally related as they exhibit a similar phenotype. In wheat, each of the two alleles at the *Lr14* locus for resistance to leaf rust gives a mesothetic infection type but to different races (Dyck and Samborski,

1970). Also in wheat, each of the different alleles at the *Sr9* locus for resistance to stem rust conditions a type 2 infection (Roelfs and McVey, 1979). In oats the alleles or functionally related genes for resistance to stem rust, *Pg-3* and *Pg-9,* also give resistance to crown rust (McKenzie *et al.*, 1968).

Allelism, together with a scarcity of resistance genes, has been a particular problem in the development of stem rust resistant oat cultivars. Until recently it was assumed that there were only seven genes for resistance at three loci. It was suggested that these might involve three chromosomes belonging to a homoeologous series (McKenzie *et al.*, 1970). Two of the alleles at one locus were combined by Koo *et al.* (1955), who suggested that this was a complex locus consisting of pseudoalleles. Several additional genes at different loci have more recently been found (Martens *et al.*, 1980).

V. Sources of Specific–Type Resistance

Due to the continual evolution of rust pathogens to form new and virulent biotypes, there must be a constant search for germ plasm possessing resistance to the various cereal rusts. Three natural sources of the specific type of resistance are available: (1) cultivars that have been produced since the advent of modern plant breeding; (2) land races or primitive cultivars that predate the advanced cultivars; and (3) the relatives—both cultivated and wild—of the crop species under consideration. Although listings of germ-plasm collections maintained at various institutions are available (Ayad *et al.*, 1980; Creech and Reitz, 1971), relatively few details are known of specific accessions in regard to rust resistance, geographic origin and availability of seed. A fourth potential source of disease resistance is through the induction of mutations by various mutagenic agents.

A. WITHIN HOST SPECIES

The most obvious and immediate source of resistance would be local and international collections of old and of contemporary cultivars and breeding stocks of the crop to be improved. This source would be particularly appropriate for new breeding programs. Should suitable resistance not be identified in such material, the search can then be extended to primitive cultivars or so-called land races. This should include a thorough search of comprehensive collections derived from the geographic centers of diversity (Leppik, 1970; Zohary, 1970). Because of their great genetic diversity, primitive cultivars and land races are more likely sources of new resistance genes than material derived from breeding programs. The known genes for rust resistance in most of the cereal crops are now being utilized on an international scale; consequently, few new genes for resistance can be expected from advanced breeding stocks.

B. RELATED CULTIVATED SPECIES AND WILD SPECIES

Plant breeders and pathologists are increasingly aware of the necessity to broaden the genetic pool from which effective sources of resistance can be drawn (Hooker, 1977; Knott, 1979; Krull and Borlaug, 1970; Moseman *et al.*, 1979; Watson, 1970). Ample evidence indicates a significant reservoir of resistance is available among the relatives of cultivated cereal crops (Dinoor, 1977; Gerechter-Amitai and Loegering, 1977; Kerber and Dyck, 1979; Pasquini, 1980; The, 1976). These have usually been the last resort because of problems in transferring such resistance, which is generally of the race-specific type, to commercially acceptable cultivars.

The earliest examples of the transfer of rust resistance to a cereal crop from related species trace to the investigations of American workers in the 1920s and 1930s. Hayes *et al.* (1920) produced the stem-rust-resistant common wheat cultivar Marquillo from a cross between susceptible Marquis common wheat and resistant Iumillo durum wheat. Later, McFadden (1930) crossed the highly disease-resistant tetraploid wheat Yaroslav emmer with Marquis, from which the two selections Hope and H-44 were developed, which at that time were highly resistant to stem rust and leaf rust. These two strains appear in the pedigree of numerous contemporary common wheat cultivars. Following these pioneering successes, this field of investigation was largely neglected.

In the past two to three decades, attention has again been directed toward the exploitation of the relatives of the cereal crops as sources of disease resistance. This renewed interest may be attributed to (1) the acquisition of much significant knowledge on the phylogenetic and cytogenetic relationships between cultivated cereal crops and their related species, (2) the development of techniques such as embryo culturing and the postpollination application of growth regulators to enhance seed development of interspecific crosses, and (3) the development of cytogenetic stocks and methodology that permit the transfer of alien genetic material between normally nonhomologous or between homoeologous chromosomes of the donor and recipient crop species (Feldman, 1979; Knott, 1971; Knott and Dvorak, 1976; Riley and Kimber, 1966; Riley *et al.*, 1968; Sears, 1956, 1972, 1981; Stalker, 1980).

The effective utilization of the cultivated and wild relatives of the cereal crops must begin with the acquisition of a substantial collection of the related species to be surveyed. It should be emphasized that the stocks or accessions of the species to be evaluated should be representative of the geographic regions from which maximum variation for resistance can be expected—the centers of genetic host diversity (Harlan and Zohary, 1966; Rajhathy and Thomas, 1974; Zohary, 1970).

In the following discussion on the transfer of rust resistance to cultivated cereal crops from their relatives, the methodology and examples are taken primarily from wheat (*Triticum*). Most of the significant advances have been made within this genus, although some of the techniques are also applicable to barley, rye, and particularly oats, which, like common wheat and durum wheat, belong to a polyploid complex of species.

The relatives of common wheat ($2n = 42 =$ AABBDD) that have been utilized to a limited extent as sources of disease resistance include the immediate tetraploid ($2n = 28 =$ AABB) and diploid ($2n = 14 =$ AA) progenitors within the wheat genus, the closely related genus *Aegilops, Secale,* and some species of *Agropyron.* The cytogenetic procedure to be used for the transfer to common wheat or durum wheat of genetic material from these related species is primarily dictated by their phylogenetic affinity and genomic constitution (Feldman, 1979; Knott and Dvořak, 1976; Riley and Kimber, 1966; Sears, 1981). The procedure most likely to prove successful depends on whether the chromosomes (genomes) of the donor species possessing the resistance are homologous or nonhomologous with those of the recipient wheat (Feldman, 1979; Knott and Dvořak, 1976; Sears, 1972, 1981).

1. *Transfer of Resistance Involving Homologous Chromosomes*

When the various related species have a genome(s) that is homologous with at least one of the genomes of cultivated common wheat, transfer of resistance is relatively simple. These species include the immediate tetraploid (AABB) and diploid (AA and DD) progenitors of the cultivated wheat. Because normal chromosome pairing and genetic recombination occurs between homologous genomes in hybrids produced from crosses between wheat and these species, the transfer of genes is possible providing crossability and F_1 sterility barriers are not a hindrance.

a. Direct Crosses. Where the parents have a genome in common, the transfer of disease resistance to cultivated wheat from a related species by conventional crossing and selection usually presents little difficulty. Examples of this procedure were the transfer of the stem rust resistance gene *Sr22* from diploid wheat to both tetraploid (Kerber and Dyck, 1973) and hexaploid wheat (The, 1973). Similarly, stem rust resistance was transferred directly from tetraploid to hexaploid wheat (Hayes *et al.*, 1920; Kerber and Dyck, 1973; McFadden, 1930; Knott, 1979). In these hybrids, pairing between the common genome(s) of the donor species and recipient wheat is usually complete or nearly so. Partial or complete sterility of the tetraploid × diploid and hexaploid × diploid hybrids can be overcome by backcrossing to the respective tetraploid and hexaploid parental cultivars.

b. Bridge Crosses. Bridge crosses may be used where the transfer of genetic material, usually between different levels of ploidy, is difficult or impossible by direct hybridization. Kerber and Dyck (1973), for example, transferred the gene *Sr22* first from diploid (AA) to tetraploid (AABB) and then to hexaploid wheat (AABBDD). Although the direct cross between some genotypes of hexaploid and diploid wheats can be made, it is often difficult and the hybrid is sterile.

Another bridging method applicable to the transfer of genetic material to both tetraploid and hexaploid wheats is with natural amphiploids that have the A genome in common with these two wheats or the D genome in common with hexaploid wheat. The amphiploid selected for resistance is crossed to wheat, and the partially fertile hybrid is then backcrossed several times to the wheat cultivar to obtain meiotically stable, fertile plants. Natural amphiploids that could be used in this manner include *T. timopheevii* (AAGG), *Ae. cylindrica* (CCDD), *Ae. ventricosa* (DDMvMv), and others (Feldman, 1979). Similar procedures can be employed to transfer genes to the polyploid wheats from synthetically produced amphiploids having one or more genomes in common with wheat. Kerber and Dyck (1969) and Dyck and Kerber (1970) transferred two genes for leaf rust resistance, *Lr21* and *Lr22a*, and a gene for stem rust resistance, *Sr33* (Kerber and Dyck, 1979), from *Ae. squarrosa* ($2n = 14 =$ DD) to common wheat by first producing synthetic hexaploids ($2n = 42 =$ AABBDD) from the hybrid between tetraploid wheat and the resistant *Ae. squarrosa* strains. The resistant synthetic hexaploid was then crossed and backcrossed several times to a common wheat cultivar to incorporate the resistance genes into a suitable genotypic background.

2. *Transfer of Resistance Involving Nonhomologous Chromosomes*

The transfer of genetic material to cultivated wheat from more distantly related species poses considerable difficulty because their chromosomes have differentiated from those of wheat to the extent that no pairing and genetic recombination between them normally occurs. Nevertheless, cytogenetic procedures are available by which genetic exchange can be induced between wheat chromosomes and those of related species.

a. Transfer of Resistance by Induction of Homoeologous Chromosome Pairing. Numerous species related to wheat have a genome(s) that is homoeologous (genetically and structurally similar) with those of common wheat. The three genomes of wheat themselves are homoeologous, having presumably descended from a common evolutionary ancestor. The genetic control or suppression of homoeologous chromosome pairing is largely due to a gene, *Ph,* on chromosome 5B (Wall *et al.,* 1971b). This gene normally prevents homoeologous pairing not only within wheat but also between homoeologues of wheat and of related or alien species when combined in hybrids. Three cytogenetic procedures have been developed by which the 5B effect can be nullified, thereby inducing pairing and recombination between homoeologues of wheat and related species in hybrid material.

Pairing between an alien chromosome and a homoeologue of wheat can be induced by crossing monosomic 5B of wheat with the species from which resistance is to be transferred. Pairing between the alien and wheat homoeologues will occur in F_1 plants that are deficient for chromosome 5B. However, the hybrids produced are highly sterile and the chance of inducing the desired gene transfer is very low when only a few seeds are produced on 5B-deficient plants. To overcome the high sterility to some extent, nullisomic-5B–tetrasomic-5D can be used in place of monosomic 5B. The use of the 5B-deficient method can be simplified and made more efficient by first producing alien substitution or alien addition lines in which the alien chromosome bears the gene(s) for resistance. The utilization and variations of the chromosome 5B-deficient procedure have been given by Riley and Kimber

(1966) and Sears (1972, 1981). This method was successfully used in the transfer to common wheat of rust resistance from rye (Joshi and Singh, 1979) and leaf rust resistance from *Agropyron elongatum* (Sears, 1972, 1973).

Another approach to the induction of homoeologous pairing is to suppress, rather than delete, the activity of *Ph* on chromosome 5B by adding the genome of certain forms of *Ae. speltoides* or *Ae. mutica.* Riley *et al.* (1968) applied this technique for the transfer to common wheat of stripe rust resistance from *Ae. comosa.* They produced a wheat stock with a disomic addition of the *comosa* chromosome bearing the gene for resistance which was crossed to *Ae. speltoides* to induce pairing between the *comosa* chromosome and its wheat homoeologue. Several backcrosses to common wheat eventually resulted in the cultivar Compair in which chromosome 2D carries a segment from the *comosa* chromosome that conditions resistance. Dvořak (1977) directly transferred genes for leaf rust resistance from *Ae. speltoides* to common wheat by taking advantage of the ability of this diploid to suppress *Ph* and thereby permit homoeologous pairing in the wheat–*speltoides* hybrid.

The most desirable situation for the induction of homoeologous pairing is the use of mutants of the *Ph* locus such as *Ph1a* and *Ph1b* obtained by Wall *et al.* (1971a) and Sears (1977, 1981), respectively. When used in crosses, these mutants have the advantage over the nullisomic 5B procedures by decreasing the amount of aneuploidy in the progeny of the hybrids, increasing fertility, and allowing recombination to occur between 5B and alien homoeologues.

b. Transfer of Resistance by Induced Chromosome Translocations. The use of irradiation to induce translocations for the transfer of disease resistance to a crop species from a related species involving nonhomologous chromosomes has been applied with considerable success since the initial procedure developed by Sears (1956) was employed to transfer leaf rust resistance from *Ae. umbellulata* to wheat. Briefly, Sears produced a line in which an isochromosome of *umbellulata,* bearing the resistance gene, was added to the wheat complement. Irradiation of this line and pollination of the cultivar Chinese Spring with pollen from the irradiated plants resulted in a 42-chromosome derivative in which a segment of chromatin bearing the resistance gene *Lr9* was translocated to chromosome 6B. Since then this procedure or modifications of it (Knott, 1971; Riley and Kimber, 1966; Sears, 1972) have been employed to induce transfers to wheat chromosomes of genetic material from nonhomologous chromosomes of related species, particularly from *Agropyron* and *Secale.* Transfers of resistance to the rusts so induced include, in addition to *Lr9,* leaf rust resistance genes *Lr19, Lr24,* and *Lr25* and the stem rust resistance genes *Sr24, Sr25, Sr26,* and *Sr27* (McIntosh, 1973, 1979).

Now that various cytogenetic procedures have been devised by which the chromosome 5B pairing activity can be nullified, so inducing pairing and genetic recombination between homoeologous chromosomes of wheat and those of related species, it is likely that the use of irradiation will be limited to crosses in which homoeologous association does not occur. Transferring genetic material by induction of homoeologous pairing has the advantage of restricting the size of the alien chromosomal segment and reducing the possibility of introducing undesirable linkages that usually accompany most irradiation-induced translocations.

A review of the literature on the transfer of rust resistance to cereal cultivars from related species reveals several problems and difficulties may be encountered. Some of these with related aspects are as follows.

1. Resistance transferred to crop cultivars from related species may be linked with undesirable agronomic or quality characters (Knott, 1971; Knott and Dvořak, 1976, 1981).
2. Resistance transferred from a lower to a higher level of ploidy is often decreased or "diluted," as expressed by infection type (Dyck and Kerber, 1970; Kerber and Dyck, 1969, 1973, 1979), and in some cases may be completely suppressed (Kerber and Green, 1980; The and Baker, 1975).
3. Resistance of the race-specific type derived from alien species probably will be no more durable than that available within the cultivated crop. Virulent strains of the pathogen have been known to overcome this type of resistance (Knott, 1971; Johnson and Gilmore, 1980; Parlevliet, 1981).
4. Race-specific genes for resistance identified in the relatives of cereal crops are likely to be different from those known in cultivars of the crop (Kerber and Dyck, 1969, 1973, 1979; Knott, 1979).

C. Induced Mutations

The interest shown 15–25 years ago in the use of mutagenic agents for the induction of rust resistance in cereal crops has waned in recent years. This can be attributed in part to the meager improvements obtained for the substantial efforts used to test and screen large populations. This applies particularly to polyploid crops such as wheat and oats in which the observable mutation rate is substantially reduced by the buffering effect of duplicated genetic material. Nevertheless, some successes have been reported. In oats, Simons and Frey (1977) detected mutants having greater tolerance to crown rust as a result of ethyl methanesulfonate treatment. Skorda (1977) obtained wheat mutants induced by irradiation that were resistant or partially resistant to stem rust and stripe rust. Similarly, Borojevic (1979) observed increased resistance to leaf rust among mutant lines derived from irradiation of susceptible wheat cultivars. It is noteworthy that in wheat none of the race-specific genes for stem rust resistance (*Sr* genes) or leaf rust resistance (*Lr* genes) catalogued by McIntosh (1973, 1979) have originated from mutagenic treatments. This observation also applies to the major genes for stem rust resistance (*Pg* genes) and crown rust (*Pc* genes) resistance in oats (Simons *et al.*, 1978).

D. DETECTION AND EVALUATION OF SOURCES OF RESISTANCE

The evaluation of a large and diverse collection of cereal-crop germ plasm may be divided into two phases: initial screening of accessions for resistance to prevalent races or biotypes of the pathogen, and the identification of resistance genes and their relationships to other genes of the same source species and to known resistance genes of the host crop. Effective procedures have been developed for detecting resistance of the race-specific type (Dinoor, 1977; Parlevliet, 1981). Preliminary screening and selection can be done in field nurseries, where large populations can be tested for resistance to natural occurring inoculum, or under an artificially created epidemic to a race or to a composite of biotypes. More reliable information can be obtained by controlled testing of plants to specific pathogen cultures under greenhouse conditions. Some information of genetic variability among accessions of a crop or related species can be gained by noting the differential reactions expressed when tested to a series of critical cultures of the pathogen. This will often permit the classification of the material into groupings based on similarities and differences in reaction (phenotypes) to the rust cultures. These differential reactions can be related to those expressed by host cultivars with known race-specific genes, as, for example, the leaf rust (*Lr*) and stem rust (*Sr*) resistance genes in wheat. This information may indicate whether the resistance being evaluated is different from that already available.

Computerized methods have been devised in which data on host reaction to specific pathogen variants are processed to provide an indication of the host genotype (Dinoor, 1977; Gerechter-Amitai and Loegering, 1977). This information is useful for categorizing new sources of resistance from which selections can then be made for detailed genetic study. The precise identification of genes for race-specific resistance in the host involves crosses and genetic analysis based on the classification of segregating generations into phenotypes as determined from the reaction of plants (F_2 backcross or F_3 families, for example) to specific pathogen cultures. Although the genetic identification of a source of resistance is not essential prior to its use, this information is important for the strategic employment of race-specific genes. In addition, comparative genetic analysis involving known genes for resistance avoids wasteful duplication in the utilization of identical types of resistance.

VI. Use of Specific-Type Resistance

A. Breeding Strategy

The early years of breeding for resistance to the cereal rusts were filled with great hopes; cultivars with single genes for resistance were released, but virulent strains of the rust organism present in low frequency would then increase rapidly and spread over an entire area devoted to the new cultivar. Thus, cultivars with single-gene specific resistance were in most instances short-lived.

Gradually it became apparent that a crop required greater diversity in genetic resistance if more stability to disease resistance was to be achieved. Watson and Singh (1952) were among the first to propose the use of multiple-gene resistance to control stem rust. They suggested the development of cultivars with pairs of genes, each giving resistance to all of the prevalent races. If the origin of new pathogen races is by mutation only, which they indicated is the situation in Australia, new virulent strains can arise only through simultaneous or stepwise mutations at all the corresponding loci in the pathogen. The probability of this occurrence would be much less than that of single-gene mutations.

Multigene resistance cannot be expected to last when the genes are also exposed individually as single-gene resistance in other cultivars, due to the selection for strains virulent on the single-gene cultivars. In addition to the rate of mutation toward virulence, other factors that influence the duration of effectiveness of genetically complex resistance are the size of the rust population, during both the growing and the overwintering seasons, and the degree of selection pressure exerted. Multigene resistance may also be overcome more readily when genetic variability in the rust fungus originates through either sexual or somatic recombination of several mutant loci present in different rust strains. However, wheat cultivars with a more complex resistance appear to have stabilized the stem rust population in several regions including North America and Australia.

Multigenic resistance should not be considered permanent; eventually new rust strains can be expected to appear. Reference is frequently made to Caldwell's observation (1968) that polygenic mature-plant resistance derived from Chinese Spring was gradually overcome by the leaf rust organism over a 5-year period. Polygenic resistance was inferred from an observation of "a continuous array of infection severities from 0 to 100%" in F_3 populations. As few as two interacting adult-plant genes could give such a distribution (P.L. Dyck and D. J. Samborski, unpublished observations). Chinese Spring is reported to have gene *Lr12* for mature-plant resistance on chromosome 4A (McIntosh and Baker, 1966; Dyck and Kerber, 1971) plus one or two modifiers or additional genes for adult-plant resistance (Dyck and Kerber, 1971). Piech and Supryn (1978) found a second gene for adult-plant resistance on chromosome 7D. Failure of resistance due to only two genes is not a good example of breakdown of polygenic resistance.

The breeder can usually combine or pyramid several specific resistance genes into a cultivar. In wheat, most identified genes for resistance have been located on specific chromosomes and assigned gene symbols (McIntosh, 1973). Their typical phenotype and effectiveness is known (Browder, 1980; Roelfs and McVey, 1979). Such information permits planning as to which genes can be combined without the interference of linkage problems. Allelism may limit the number of host genes that can be combined; however, in the parasite there is no clear evidence for allelism of virulence genes, although linkage of virulence genes has been reported (Samborski and Dyck, 1976; Statler, 1979; Lawrence *et al.*, 1981).

If the use of specific genes in a breeding program is to be successful, the rust population must be surveyed regularly so that changes in virulence can be detected. A thorough survey should detect the appearance in the rust population of genes for virulence corresponding to the specific host genes combined in a cultivar. Also, surveys may detect new virulent strains several years before they reach epidemic proportions and so allow time for cultivar replacement. In such a case the cultivar should be replaced by one with a different combination of resistance genes before the pathogen can cause appreciable damage.

Combinations of specific genes permits the exploitation of additive or complementary interallelic interactions that have in some cases been shown to enhance the level of resistance. It also allows the utilization of any existing residual or ghost effect of specific genes. A cultivar with a combination of several effective and ineffective specific genes may result in complex and stable resistance. Leaving the ineffective specific genes in a commercial cultivar is contrary to the suggestion (Person *et al.*, 1976) that they should be removed from exposure to the rust population. Consequently, if stabilizing exists (Vanderplank, 1968), the corresponding virulence gene would disappear from the rust population, and the host gene could then be recycled.

B. BREEDING METHODS

When using the pedigree method of breeding, single crosses usually consist of a well adapted but rust-susceptible cultivar and a rust-resistant but frequently poorly adapted cultivar. The genetic diversity for rust resistance is then limited to that available from the one parent. To obtain greater genetic diversity, a double cross can be made to combine resistance from different sources. However, the likelihood of obtaining well adapted and highly rust resistant selections from a double cross of an adapted and three unadapted rust-resistant parents is poor. To increase the probability of obtaining desirable, highly resistant selections, double crosses can be made between advanced lines derived from other crosses, each with different types of rust resistance. These lines may have compensating weaknesses and strengths in other important characters. The breeding strategy employed at the International Maize and Wheat Improvement Centre (CIMMYT) emphasizes pedigree breeding with multiple or double crosses that lead to a rapid increase in genetic diversity (Dubin and Rajaram, 1981).

In a pedigree breeding program where rust resistance is a major concern, the parents should be carefully chosen. Ideally they have different types of genetic resistance that can be combined under the selection procedures available. The potential parents should be evaluated for resistance under field conditions and, if possible, to specific races in the greenhouse. If the genetics of resistance is not known, parents with different ancestry should be chosen.

Rust resistance is a highly heritable character that is usually selected for in early generations. However, since resistance is frequently dominant or partially dominant, further selections must be made in later generations to ensure that homozygous lines are obtained. If the genes for resistance being used are additive or interacting, selection for resistance should also be made in later generations when more genes are homozygous and interactions are more apparent. In the wheat breeding program at the Winnipeg Research Station (Green and Campbell, 1979), F_2, F_4, and F_6 generations and all lines being yield-tested are grown in the rust nursery. Thus, lines are selected over several years and subjected to repeated artificial rust epidemics under different environments.

The bulk method of breeding differs from the pedigree in that segregating generations, usually F_2 to F_6, are grown in bulk and exposed to a disease epidemic. Mechanical separation of seed according to size eliminates the smaller, shrunken seeds of susceptible plants, aiding in the selection for rust resistance. With this method, many crosses can be handled with a minimum of labor. Selection over several years can result in highly resistant cultivars.

The backcross method is frequently used to improve or correct a defect, such as susceptibility to rust, in an otherwise well adapted and high-yielding cultivar. With this method the choice of the recurrent parent is very important, since an increase in the inherent yielding ability of the cultivar is not expected. The superior adaptibility of some cultivars may be due to genetic heterogeneity, which is not always recovered with backcrossing. Thus, it is important to use a number of plants of the recurrent parent in each backcross, particularly in the final one. To develop a cultivar with multigene resistance, several different known genes must be transferred into the recurrent parent in separate backcross series. When completed, the backcross lines are intercrossed and selections made for lines with various gene combinations.

Mac Key (1959) has described a modified convergent backcrossing scheme in which an adapted cultivar is crossed with four different resistant sources and the F_1 of each combination is backcrossed to the adapted cultivar. Double crosses are then made between resistant selections from the four backcrosses. In subsequent segregating generations, selections for maximum resistance and adaptibility are made. Since many breeding programs must meet some specific requirements—i.e., milling and baking or malting quality or adaptation—additional backcrosses can be made to the adapted cultivar before the double crosses are made. If selection methods are available that ensure the selection of lines with a number of genes for resistance, modified backcrossing may permit selection for increased yield and a complex resistance.

The use of male-sterile facilitated recurrent selection is a way of combining several different additive or complementary genes from a large number of potential sources (Driscoll, 1981; Ramage, 1977; Sharp, 1979). A backcrossing scheme using male sterility can be used to transfer these additive genes into an adapted cultivar.

Regardless of the breeding method used, the segregating populations should be subjected to a timely rust epidemic. Reliance on naturally occurring epidemics is usually not adequate. The breeding material is interspersed with spreader rows of highly rust-susceptible plants that are inoculated with a mixture of rust races. The rust nursery should be planted at a time and place so that the appropriate stages of plant development will coincide with conditions most favorable for rust development. If natural dew formation is not adequate, sprinkler irrigation may facilitate infection and uniform spread of the disease. The inoculum used should be representative of the indigenous rust population. Green and Campbell (1979) used "the most prevalent and widely virulent races available."

The development of a cultivar with complex genetic resistance may be obtained to some extent by combining different types of resistance. For example, specific genes, either the hypersensitive or rate-reducing type, could be combined with genes for adult-plant resistance or with additive, interacting genes. If resistance considered as durable exists, it could perhaps be combined with specific resistance genes. Johnson (1981a) suggests that to be certain that the final selections possess durable resistance, all specific gene sources present in the breeding program should be susceptible to the rust culture used to test for the presence of durable resistance.

Selection for seedling resistance can also be done in the greenhouse, particularly during the off season or winter, or on subsamples of lines grown in field nurseries. Preferably, rust races with known genes for virulence should be used in seedling tests so that lines can be selected for combinations of known specific genes. In Australia, a National Rust Control Program has been established (Watson, 1977) whereby a central laboratory screens wheat lines submitted by participating breeders from throughout the country. On the basis of seedling and adult-plant tests, selections are made and breeders are advised on the types of resistances that are present.

Various methods have been proposed to verify or detect the presence of specific genes when they are combined in advanced lines derived from a pedigree breeding program or from different backcross lines.

1. Differential rust cultures obtained from disease surveys, or through mutagenic and genetic studies can be used; however, care should be taken so that virulent cultures do not escape into commerical fields. The establishment of an international exchange would be desirable whereby testing for combinations of genes in breeder's lines could be done in another country with isolates from a different rust population.
2. Enhanced resistance resulting from gene interactions may make it possible to detect various gene combinations. Since additive genes do not generally show epistasis, several of them could be accumulated in a cultivar by merely selecting for the most resistant lines.
3. Genes conferring different infection types can be combined by selecting plants with the lower infection type from lines that are segregating for low infection type and are homozygous for the higher infection type.
4. If one of the genes is temperature-sensitive, manipulating the temperature may make it possible to detect a gene(s) hidden by a temperature-sensitive gene.
5. Seedling resistance genes can be backcrossed into recurrent parents with adult-plant resistance, thus combining genes for both seedling and adult resistance.
6. A member of a pair of linked specific genes may be used to select for the other; i.e., since *Sr24* and *Lr24* are inherited as a unit, backcrossing *Lr24* into a line already possessing *Sr26* will simultaneously result in combining *Sr24* and *Sr26*.
7. Flor and Comstock (1971) developed three-gene lines in flax by first developing two-gene lines with one gene in common, for example, *LLm3m3N1N1* and *llM3M3N1N1*. From an intercross of these two lines, selection would be made only for genes *L* and *M3*.
8. If none of the above options is available, resistant F_3 lines from intercrosses of several backcross lines or breeding lines, each with

different genes, may have to be test-crossed to lines with the genes being used or to a susceptible tester, as outlined by Johnson and Gilmore (1980).

VII. Conclusions

The development of cultivars with race-specific rust resistance usually involves the application of routine plant-breeding methods. However, because of the continual evolution of new races or biotypes of the pathogen, the strategy of utilizing this type of resistance is not routine or straightforward. Although it is generally agreed that the use of cultivars with single-gene, race-specific resistance should be avoided, the total abandonment of major resistance genes, as proposed by some pathologists and breeders, is unwarranted. The key to the development of cultivars with long-lasting resistance is diversity—genetic diversity in the types of resistance, and diversity in their strategic deployment. This would include a combination of race-specific with non-race-specific resistance. Strategies to increase the durability or longevity of resistance include the pyramiding of genes into a cultivar, multiline cultivars, regional deployment of different resistance genes, and the diversification of types of resistance among cultivars. No single overall strategy can be recommended. The primary objective of the various approaches and strategies should be to reduce the selection pressure for virulence in the pathogen.

Although a sufficient number of race-specific genes appears to be available at present for effective control of the rusts, for the future, greater emphasis will have to be placed on the exploitation of resistance known to exist among the relatives of the cereal crops.

References

Anderson, R. G. (1966). Studies on the inheritance of resistance to leaf rust of wheat. *Hereditas Suppl.* **2,** 144–155.

Andrews, J. E. (1956). Inheritance of reaction to loose smut, *Ustilago nuda,* and to stem rust, *Puccinia graminis tritici,* in barley. *Can. J. Agric. Sci.* **36,** 356–370.

Ayad, G., Toll, J., and Williams, J. T. (1980). "Directory of Germplasm Collections. III. Cereals. 1. Wheat," pp. 1–28. Int. Board Plant Genet. Resour., FAO, Rome.

Baker, E. P. (1966). Isolation of complementary genes conditioning crown rust resistance in the oat variety Bond. *Euphytica* **15,** 313–318.

Biffin, R. H. (1905). Mendels laws of inheritance and wheat breeding. *J. Agric. Sci.* **1,** 4–48.
Borojevic, K. (1979). Induced variability for leaf rust resistance in *Triticum aestivum. Proc. Int. Wheat Genet. Symp., 5th, 1978,* Vol. 1, pp. 559–564.
Browder, L. E. (1980). A compendium of information about named genes for low reaction to *Puccinia recondita* in wheat. *Crop Sci.* **20,** 775–779.
Caldwell, R. M. (1968). Breeding for general and/or specific plant disease resistance. *Proc. Int. Wheat Genet. Symp., 3rd, 1968,* pp. 263–272.
Clifford, B. C. (1975). Stable resistance to cereal disease: Problems and progress. *Rep.—Welsh Plant Breed. Stn. 1974, (Aberystwyth, Wales),* pp. 107–113.
Creech, J. L., and Reitz, L. P. (1971). Plant germ plasm now and for tomorrow. *Adv. Agron.* **23,** 1–49.
Dinoor, A. (1977). Genes for disease resistance from native populations. *In* "Induced Mutations Against Plant Diseases," pp. 15–27. IAEA, Vienna.
Driscoll, C. J. (1981). New approaches to wheat breeding. *In* "Wheat Science—Today and Tomorrow" (L. T. Evans and W. J. Peacock, eds.), pp. 97–106. Cambridge Univ. Press, London and New York.
Dubin, H. J., and Rajaram, S. (1981). The strategy of the International Maize and Wheat Improvement Center (CIMMYT) for breeding disease resistant wheat: an international approach. *In* "Strategies for the Control of Cereals Disease" (J. F. Jenkyn and R. T. Plumb, eds.), pp. 27–35. Blackwell, Oxford.
Dvořak, J. (1977). Transfer of leaf rust resistance from *Aegilops speltoides* to *Triticum aestivum. Can. J. Genet. Cytol.* **19,** 133–141.
Dyck, P. L. (1977). Genetics of leaf rust reaction in three introductions of common wheat. *Can. J. Genet. Cytol.* **19,** 711–716.
Dyck, P. L., and Kerber, E. R. (1970). Inheritance in hexaploid wheat of adult-plant leaf rust resistance derived from *Aegilops squarrosa. Can. J. Genet. Cytol.* **12,** 175–180.
Dyck, P. L., and Kerber, E. R. (1971). Chromosome location of three genes for leaf rust resistance in common wheat. *Can. J. Genet. Cytol.* **13,** 480–483.
Dyck, P. L., and Samborski, D. J. (1970). The genetics of two alleles for leaf rust resistance at the *Lr14* locus in wheat. *Can. J. Genet. Cytol.* **12,** 689–694.
Dyck, P. L., and Samborski, D. J. (1974). Inheritance of virulence in *Puccinia recondita* on alleles at the *Lr2* locus for resistance in wheat. *Can. J. Genet. Cytol.* **16,** 323–332.
Dyck, P. L., and Samborski, D. J. (1982). The inheritance of resistance to *Puccinia recondita* in a group of common wheat cultivars. *Can. J. Genet. Cytol.* **24,** 273–283.
Ellingboe, A. H. (1975). Horizontal resistance: An artifact of experimental procedure? *Aust. Plant Pathol. Soc. Newsl.* **4,** 44–46.
Ellingboe, A. H. (1978). A genetic analysis of host-parasite interactions. *In* "The Powdery Mildews" (D. M. Spencer, ed.), pp. 159–181. Academic Press, New York.
Feldman, M. (1979). Genetic resources of wild wheats and their use in breeding. *Monogr. Genet. Agrar.* **4,** 9–26.
Flor, H. H. (1955). Host-parasite interactions in flax rust - its genetics and other implications. *Phytopathology* **45,** 680–685.
Flor, H. H., and Comstock, V. E. (1971). Flax cultivars with multiple rust - conditioning genes. *Crop Sci.* **11,** 64–66.
Gassner, G., and Straib, W. (1931). Zur frage der konstanz des infektionstypus von *Puccinia triticina* Erikss. *Phytopathol. Z.* **4,** 57–64.
Gerechter-Amitai, Z. K., and Loegering, W. Q. (1977). Genes for low reaction to *Puccinia graminis tritici* in *Aegilops* and *Triticum. Crop Sci.* **17,** 830–832.
Green, G. J., and Campbell, A. B. (1979). Wheat cultivars resistant to *Puccinia graminis*

tritici in western Canada: Their development, performance, and economic value. *Can. J. Plant Pathol.* **1,** 3–11.

Green, G. J., Knott, D. R., Watson, I. A., and Pugsley, A. T. (1960). Seedling reactions to stem rust of lines of Marquis wheat with substituted genes for rust resistance. *Can. J. Plant Sci.* **40,** 524–538.

Hare, R. A., and McIntosh, R. A. (1979). Genetic and cytogenetic studies of durable adult-plant resistances in 'Hope' and related cultivars to wheat rusts. *Z. Pflanzenzuecht.* **83,** 350–367.

Harlan, R. J., and Zohary, D. (1966). Distribution of wild wheats and barley. *Science* **153,** 1074–1080.

Hayes, H. K., Parker, J. H., and Kurtzweil, C. (1920). Genetics of rust resistance in crosses of varieties of *Triticum vulgare* with varieties of *T. durum* and *T. dicoccum. J. Agric. Res. (Washington, D.C.)* **19,** 523–542.

Hooker, A. L. (1977). A plant pathologist's view of germplasm evaluation and utilization. *Crop. Sci.* **17,** 689–694.

Hooker, A. L., and Saxena, K. M. S. (1971). Genetics of disease resistance in plants. *Annu. Rev. Genet.* **5,** 407–424.

International Maize and Wheat Improvement Center (1981). "CIMMYT Review," p. 52. CIMMYT, Mexico, D. F.

Johnson, D. A., and Gilmore, E. C. (1980). Breeding for resistance to pathogens in wheat. *Tex. Agric. Exp. Stn. [Misc. Publ.] MP* **MP-1451,** 263–275.

Johnson, R. (1978). Practical breeding for durable resistance to rust diseases in self-pollinating cereals. *Euphytica* **27,** 529–540.

Johnson, R. (1981a). Durable resistance: Definition of, genetic control, and attainment in plant breeding. *Phytopathology* **71,** 567–568.

Johnson, R. (1981b). Durable disease resistance. *In* "Strategies for the Control of Cereal Disease" (J. F. Jenkyn and R. T. Plumb. eds.), pp. 55–63. Blackwell, Oxford.

Johnson, T. (1961). Rust research in Canada and related plant-disease investigations. *Publ.—Can. Dep. Agric.* **1098,** 1–69.

Johnson, T., and Newton, M. (1946). The occurrence of new strains of *Puccinia triticina* in Canada and their bearing on varietal reaction. *Sci. Agric.* **26,** 468–478.

Jones, D. R., and Deverall, B. J. (1977). The effect of the *Lr20* resistance gene in wheat on the development of leaf rust, *Puccinia recondita. Physiol. Plant Pathol.* **10,** 275–284.

Joshi, B. C., and Singh, D. (1979). Introduction of alien variation into bread wheat. *Proc. Int. Wheat Genet. Symp., 5th, 1978,* Vol. 1, pp. 342–348.

Kerber, E. R., and Dyck, P. L. (1969). Inheritance in hexaploid wheat of leaf rust resistance and other characters derived from *Aegilops squarrosa. Can. J. Genet. Cytol.* **11,** 639–647.

Kerber, E. R., and Dyck, P. L. (1973). Inheritance of stem rust resistance transferred from diploid wheat (*Triticum monococcum*) to tetraploid and hexaploid wheat and chromosome location of the gene involved. *Can. J. Genet. Cytol.* **15,** 397–409.

Kerber, E. R., and Dyck, P. L. (1979). Resistance to stem rust and leaf rust of wheat in *Aegilops squarrosa* and transfer of a gene for stem rust resistance to hexaploid wheat. *Proc. Int. Wheat Genet. Symp., 5th, 1978,* Vol. 1, pp. 358–364.

Kerber, E. R., and Green, G. J. (1980). Suppression of stem rust resistance in the hexaploid wheat cv. Canthatch by chromosome 7DL. *Can. J. Bot.* **58,** 1347–1350.

Knott, D. R. (1957). The inheritance of rust resistance. III. The inheritance of stem rust resistance in nine Kenya varieties of common wheat. *Can. J. Plant Sci.* **37,** 366–384.

Knott, D. R. (1971). The transfer of genes for disease resistance from alien species to wheat by induced translocations. *In* "Mutation Breeding for Disease Resistance," pp. 67–77. IAEA, Vienna.

Knott, D. R. (1979). The transfer of genes for rust resistance to wheat from related species. *Proc. Int. Wheat Genet. Symp., 5th, 1978,* Vol. 1, pp. 354–357.

Knott, D. R., and Anderson, R. G. (1956). The inheritance of rust resistance. I. The inheritance of stem rust resistance in ten varieties of common wheat. *Can. J. Agric. Sci.* **36,** 174–195.

Knott, D. R., and Dvořak, J. (1976). Alien germ plasm as a source of resistance to disease. *Annu. Rev. Phytopathol.* **14,** 211–235.

Knott, D. R., and Dvořak, J. (1981). Agronomic and quality characteristics of wheat lines with leaf rust resistance derived from *Triticum speltoides. Can. J. Genet. Cytol.* **23,** 475–480.

Koo, F. K. S., Moore, M. B., Myers, W. M., and Roberts, B. J. (1955). Inheritance of seedling reaction to races 7 and 8 of *Puccinia graminis avenae* Eriks. and Henn. at high temperature in three oat crosses. *Agron. J.* **47,** 122–124.

Krull, C. F., and Borlaug, N. E. (1970). The utilization of collections in plant breeding and production. *In* "Genetic Resources in Plants—Their Exploitation and Conservation" (O. H. Frankel and E. Bennett, eds.), Int. Biol. Programme Handb. No. 11, pp. 427–439. Blackwell, Oxford.

Lawrence, G. J., Mayo, G. M. E., and Shepherd, K. W. (1981). Interactions between genes controlling pathogenicity in the flax rust fungus. *Phytopathology* **71,** 12–19.

Leppik, E. E. (1970). Gene centers of plants as sources of disease resistance. *Annu. Rev. Phytopathol.* **8,** 323–344.

Lewellen, R. T., and Sharp, E. L. (1968). Inheritance of minor reaction gene combinations in wheat to *Puccinia striiformis* at two temperature profiles. *Can. J. Bot.* **46,** 21–26.

Loegering, W. Q., and Harman, D. L. (1969). Wheat lines near-isogenic for reaction to *Puccinia graminis tritici. Phytopathology* **59,** 456–459.

Luig, N. H., and Rajaram, S. (1972). The effect of temperature and genetic background on host gene expression and interaction to *Puccinia graminis tritici. Phytopathology* **62,** 1171–1174.

Lupton, F. G. H., and Macer, R. C. F. (1962). Inheritance of resistance to yellow rust (*Puccinia glumarum* Erikss. & Henn.) in seven varieties of wheat. *Trans. Br. Mycol. Soc.* **45,** 21–45.

McFadden, E. S. (1930). A successful transfer of emmer characters to *vulgare* wheat. *Agron. J.* **22,** 1020–1034.

McIntosh, R. A. (1973). A catalogue of gene symbols for wheat. *Proc. Int. Wheat Genet. Symp., 4th, 1973,* pp. 893–937.

McIntosh, R. A. (1979). A catalogue of gene symbols for wheat. *Proc. Int. Wheat Genet. Symp., 5th, 1978,* Vol. 2, pp. 1299–1309.

McIntosh, R. A., and Baker, E. P. (1966). Chromosome location of mature plant leaf rust resistance in Chinese Spring wheat. *Aust. J. Biol. Sci.* **19,** 943–944.

McIntosh, R. A., and Dyck, P. L. (1975). Cytogenetical studies in wheat. VII. Gene *Lr23* for reaction to *Puccinia recondita* in Gabo and related cultivars. *Aust. J. Biol. Sci.* **28,** 201–211.

McKenzie, R. I. H., Martens, J. W., Fleishmann, G., and Samborski, D. J. (1968). An association of stem rust and crown rust resistance in Jostrain oats. *Can. J. Genet. Cytol.* **10,** 190–195.

McKenzie, R. I. H., Martens, J. W., and Rajhathy, T. (1970). Inheritance of oat stem rust resistance in a Tunisian strain of *Avena sterilis. Can. J. Genet. Cytol.* **12,** 501–505.

Mac Key, J. (1959). II. Morphology and genetics of oats. *In* "Handbuch der Pflanzenzuechtung," Vol. 2, pp. 467–494. Parey, Berlin.

Martens, J. W., McKenzie, R. I. H., and Harder, D. E. (1980). Resistance to *Puccinia graminis avenae* and *P. coronata avenae* in the wild and cultivated *Avena* populations of Iran, Iraq, and Turkey. *Can. J. Genet. Cytol.* **22,** 641–649.

Martens, J. W., Rothman, P. G., McKenzie, R. I. H., and Brown, P. D. (1981). Evidence for complementary gene action conferring resistance to *Puccinia graminis avenae* in *Avena sativa. Can. J. Genet. Cytol.* **23,** 591–595.

Martin, T. J., and Ellingboe, A. H. (1976). Differences between compatible parasite/host genotypes involving the *Pm4* locus of wheat and the corresponding genes in *Erysiphe graminis* f. sp. *tritici. Phytopathology* **66,** 1435–1438.

Mayama, S., Daly, J. M., Rehfeld, D. W., and Daly, C. R. (1975). Hypersensitive response of near-isogenic wheat carrying the temperature-sensitive *Sr6* allele for resistance to stem rust. *Physiol. Plant Pathol.* **7,** 35–47.

Mayo, G. M. E., and Shepherd, K. W. (1980). Studies of genes controlling specific host-parasite interactions in flax and its rust. I. Fine structure analysis of the M group in the host. *Heredity* **44,** 211–227.

Moseman, J. G., Kilpatrick, R. A., and Porter, W. M. (1979). Evaluation and documentation of pest resistance in wheat germplasm collections. *Proc. Int. Wheat Genet. Symp., 5th, 1978,* pp. 143–148.

Nass, H. A., Pedersen, W. L., MacKenzie, D. R., and Nelson, R. R. (1981). The residual effects of some "defeated" powdery mildew resistance genes in isolines of winter wheat. *Phytopathology* **71,** 1315–1318.

Nelson, R. R. (1978). Genetics of horizontal resistance to plant diseases. *Annu. Rev. Phytopathol.* **16,** 359–378.

Nelson, R. R.(1981). Disease resistance breakthrough. Resistance not black and white, but various shades of gray. *Crops Soils Mag.* **34,** 7–9.

Parlevliet, J. E. (1981). Disease resistance in plants and its consequences for plant breeding. *In* "Plant Breeding II" (K. J. Frey, ed.), pp. 309–364. Iowa State Univ. Press, Ames.

Pasquini, M. (1980). Disease resistance in wheat: II. Behaviour of *Aegilops* species with respect to *Puccinia recondita* f. sp. *tritici, Puccinia graminis* f. sp. *tritici,* and *Erysiphe graminis* f. sp. *tritici. Genet. Agrar.* **34,** 133–148.

Person, C., Groth, J. V., and Mylyk, O. M. (1976). Genetic change in host-parasite populations. *Annu. Rev. Phytopathol.* **14,** 177–188.

Piech, J., and Supryn, S. (1978). Location of adult-plant leaf rust resistance on chromosome 7D in Chinese Spring wheat. *Cereal Res. Commun.* **6,** 367–375.

Rajhathy, T., and Thomas, H. (1974). Cytogenetics of oats (*Avena* L.). *Genet. Soc. Can., Misc. Publ.* **2,** 1–90.

Ramage, R. T. (1977). Varietal improvement of wheat through male sterile facilitated recurrent selection. *Tech. Bull., ASPAC Food Fert. Technol. Cent.* **37,** 1–5.

Riley, R., and Kimber, G. (1966). "The Transfer of Alien Genetic Variation to Wheat," Annu. Rep., 1964–65, pp. 6–36. Plant Breed. Inst., Cambridge, England.

Riley, R., Chapman, V., and Johnson, R. (1968). Introduction of yellow rust resistance of *Aegilops comosa* into wheat by genetically induced homoeologous recombination. *Nature (London),* **217,** 383–384.

Robbelen, G., and Sharp, E. L. (1978). "Mode of Inheritance, Interaction and Application of Genes Conditioning Resistance to Yellow Rust," Adv. Plant Breed., Vol. 9. Parey, Berlin.

Robinson, R. A. (1976). "Plant Pathosystems." Springer-Verlag, Berlin and New York.

Roelfs, A. P. (1978). Estimated losses caused by rust in small grain cereals in the United States - 1918–76. *Misc. Publ.—U.S., Dep. Agric.* **1363.** 1–85.

Roelfs, A. P., and McVey, D. V. (1979). Low infection types produced by *Puccinia graminis* f. sp. *tritici* and wheat lines with designated genes for resistance. *Phytopathology* **69,** 722–730.

Rohringer, R., Kim, W. K., and Samborski, D. J. (1979). A histological study of interactions between avirulent races of stem rust and wheat containing resistance genes *Sr5, Sr6, Sr8,* or *Sr22. Can. J. Bot.* **57,** 324–331.

Samborski, D. J., and Dyck, P. L. (1976). Inheritance of virulence in *Puccinia recondita* on six backcross lines of wheat with single genes for resistance to leaf rust. *Can. J. Bot.* **54,** 1666–1671.

Samborski, D. J., and Dyck, P. L. (1982). Enhancement of resistance to *Puccinia recondita* by interactions of resistance genes in wheat. *Can. J. Plant Pathol.* **4,** 152–156.

Samborski, D. J., and Ostapyk, W. (1959). Expression of leaf rust resistance in Selkirk and Exchange wheats at different stages of plant development. *Can. J. Bot.* **37,** 1153–1155.

Saxena, K. M. S., and Hooker, A. L. (1968). On the structure of a gene for disease resistance in maize. *Proc. Natl. Acad. Sci. U.S.A.* **61,** 1300–1305.

Schafer, J. F., Caldwell, R. M., Patterson, F. L., and Compton, L. E. (1963). Wheat leaf rust resistance combinations. *Phytopathology* **53,** 569–573.

Sears, E. R. (1956). The transfer of leaf-rust resistance from *Aegilops umbellulata* to wheat. *Brookhaven Symp. Biol.* **9,** 1–22.

Sears, E. R. (1972). Chromosome engineering in wheat. *Stadler Genet. Symp.* **4,** 23–38.

Sears, E. R. (1973). *Agropyron*-wheat transfers induced by homoeologous pairing. *Proc. Int. Wheat Genet. Symp., 4th, 1973,* pp. 191–199.

Sears, E. R. (1977). An induced mutant with homoeologous pairing in common wheat. *Can. J. Genet. Cytol.* **19,** 585–593.

Sears, E. R. (1981). Transfer of alien genetic material to wheat. *In* "Wheat Science—Today and Tomorrow" (L. T. Evans and W. J. Peacock, eds.), pp. 75–89. Cambridge Univ. Press, London and New York.

Sharp, E. L. (1973). Wheat. *In* "Breeding Plants for Disease Resistance" (R. R. Nelson, ed.), pp. 110–131. Pennsylvania State Univ. Press, University Park.

Sharp, E. L. (1979). Male sterile facilitated recurrent selection populations for developing broad-based resistance by major and minor effect genes. *Phytopathology* **69,** 1045 (abstr.).

Sharp, E. L., Sally, B. K., and Taylor, G. A. (1976). Incorporation of additive genes for stripe rust resistance in winter wheat. *Phytopathology* **66,** 794–797.

Simons, M. D., and Frey, K. J. (1977). Induced mutations for tolerance of oats to crown rust. *In* "Induced Mutations Against Plant Diseases," pp. 499–512. IAEA, Vienna.

Simons, M. D., Martens, J. W., McKenzie, R. I. H., Nishiyama, I., Sadanaga, K., Sebesta, J., and Thomas, H. (1978). Oats: A standardized system of nomenclature for genes and chromosomes and catalogue of genes governing characters. *U.S., Dep. Agric., Agric. Handb.* **509,** 1–40.

Skorda, E. A. (1977). Stem and stripe rust resistance in wheat induced by gamma rays and thermal neutrons. *In* "Induced Mutations Against Plant Diseases," pp. 385–392. IAEA, Vienna.

Stakman, E. C. (1955). Progress and problems in plant pathology. *Ann. Appl. Biol.* **42,** 22–33.

Stakman, E. C., Stewart, D. M., and Loegering, W. Q. (1962). Identification of physiologic races of *Puccinia graminis* var. *tritici. U.S., Agric. Res. Serv. ARS* **E617,** 1–53.

Stalker, H. T. (1980). Utilization of wild species for crop improvement. *Adv. Agron.* **33,** 111–147.
Statler, G. D. (1979). Inheritance of pathogenicity of culture 70-1, race 1, of *Puccinia recondita tritici. Phytopathology* **69,** 661–663.
The, T. T. (1973). Chromosome location of genes conditioning stem rust resistance transferred from diploid to hexaploid wheat. *Nature (London)* **241,** 256.
The, T. T. (1976). Variability and inheritance studies in *Triticum monococcum* for reaction to *Puccinia graminis* f. sp. *tritici* and *P. recondita. Z. Pflanzenzuecht.* **76,** 287–298.
The, T. T., and Baker, E. P. (1975). Basic studies relating to the transference of genetic characters from *Triticum monococcum* L. to hexaploid wheat. *Aust. J. Biol. Sci.* **28,** 189–199.
Vanderplank, J. E. (1968). "Disease Resistance in Plants." Academic Press, New York.
Voronkova, A. A. (1980). Manifestations of a complex interaction of genes for brown rust resistance. *Genetika* **16,** 935–938; *Sov. Genet. (Engl. Transl.)* **16,** 1466–1472.
Wall, A. M., Riley, R., and Chapman, V. (1971a). Wheat mutants permitting homoeologous meiotic chromosome pairing. *Genet. Res.* **18,** 311–328.
Wall, A. M., Riley, R., and Gale, M. D. (1971b). The position of a locus on chromosome 5B of *Triticum aestivum* affecting homoeologous meiotic pairing. *Genet. Res.* **18,** 329–339.
Watson, I. A. (1970). The utilization of wild species in the breeding of cultivated crops resistant to plant pathogens. *In* "Genetic Resources in Plants—Their Exploitation and Conservation" (O. H. Frankel and E. Bennett, eds.), Int. Biol. Programme Handb. No. 11, pp. 441–457. Blackwell, Oxford.
Watson, I. A. (1977). "The National Wheat Rust Control Programme in Australia." University of Sydney, Sydney, Australia.
Watson, I. A. (1981). Wheat and its rust parasites in Australia. *In* "Wheat Science—Today and Tomorrow" (L. T. Evans and W. J. Peacock, eds.), pp. 129–147. Cambridge Univ. Press, London and New York.
Watson, I. A., and Singh, D. (1952). The future for rust resistant wheat in Australia. *J. Aust. Inst. Agric. Sci.* **18,** 190–197.
Zohary, D. (1970). Centers of diversity and centers of origin. *In* "Genetic Resources in Plants—Their Exploitation and Conservation" (O. H. Frankel and E. Bennett, eds.), Int. Biol. Programme Hand. No. 11, pp. 33–42. Blackwell, Oxford.

16

Resistance of the Non-Race-Specific Type

J. E. Parlevliet
Department of Plant Breeding (I.v.P.), Agricultural University, Wageningen, The Netherlands

I. Introduction

Host populations vary greatly for resistance to disease, while pathogen populations may contain variants (races) capable of inciting disease in resistant host cultivars. These basic observations date back to the early years of this centruy (Day, 1974; Ellingboe, 1981). With the accumulation of scientific information, it gradually became clear that high levels of resistance based on single genes were very vulnerable to the occurrence and spread of new races, while so-called field resistance

THE CEREAL RUSTS,
Volume II

ISBN 0-12-148402-5

did not seem to be troubled by this problem of "being overcome by new races."

Vanderplank (1963) assumed the existence of two kinds of resistance: vertical and horizontal resistance. This heralded a new era in which all resistance was considered to be either race-specific or non-race-specific. In the 1970s publications started to appear reporting small race-specific effects in systems hitherto thought to be typical of a horizontal or non-race-specific nature (Caten, 1974; Clifford and Clothier, 1974; Parlevliet, 1976a). Gradually the concept is emerging that most if not all resistance against specialized pathogens is of a specific nature (Parlevliet and Zadoks, 1977; Parlevliet, 1981a; Ellingboe, 1981).

II. Terminology

The terminology dealing with host-pathogen systems is often far from consistent. An explanation of the meaning of the terms used in this chapter seems useful.

Host resistance is defined as the ability of the host to hinder or arrest the growth and/or development of the pathogen. Complete resistance prevents the multiplication of the pathogen; the spore production is zero. In case of incomplete resistance there is some spore production. Partial resistance, sometimes indicated as nonhypersensitive resistance, is a form of incomplete resistance whereby the individual lesions are characterized by a susceptible infection type. Hypersensitivity is defined as the rapid death of the host cells surrounding the infected site. It is accompanied by a restriction of the growth of the pathogen. Hypersensitivity of cereals to rust is commonly manifested as small necrotic flecks, in the center of which some sporulation may occur. This is often described as a resistant or low infection type, in contrast to the susceptible or high infection type. Adult-plant resistance is resistance that cannot be identified in the seedling stage. Slow rusting describes a reduced rate of epidemic development. Any form of incomplete resistance can result in slow rusting; it is not necessarily evidence of non-race-specific resistance. Partial resistance, nonhypersensitive resistance, and slow rusting are often used to indicate the same type of resistance, the one that is assumed to be polygenic and non-race-specific in nature. Race-specific resistance is characterized by the presence of differential genetic interactions between host and pathogen genotypes. It implies the possibility to identify races. Non-race-specific resistance is characterized by the absence of genetic interactions between host and pathogen genotypes. It is not possible to discern pathogen races.

Because Vanderplank (1978) distinguishes vertical resistance (VR) and horizontal resistance (HR) on the basis of presence or absence of differential interactions, VR and HR do not coincide anymore with race-specific and non-race-specific resistance, respectively, making the terms rather impractical. In this chapter, therefore, the concepts of race-specific and non-race-specific resistance as described in the next section are used.

III. Specificity

Vanderplank (1963, 1968, 1978) identified two types of resistance, and many workers, sometimes for the sake of convenience (Clifford, 1975), joined in this. This inevitably leads to classifying all resistance that has not been proven to be race-specific as non-race-specific. This is rather unsatisfactory, especially because there appear to be more than two levels of specificity. Parlevliet (1981a,b) discerned non-host resistance, broad or general resistance, pathogen-specific resistance, and non-race-specific and race-specific resistance. So called non-host resistance is often absence of pathogenity rather than presence of resistance. Broad resistance is resistance effective against whole groups of parasites, such as terpenes against insects, tannins against a wide range of parasites, and phytoalexins against microorganisms. Some parasites have evolved mechanisms to neutralize, tolerate, or circumvent such broad resistance mechanisms. Cereal rusts are highly specialized pathogens that suppress these broad resistances. The cereals have evolved more specific resistance mechanisms to them. These resistances, whether race-specific or non-race-specific, operate to one pathogen species only; they are pathogen specific. The race-specific *Sr*, *Lr*, and *Yr* genes in wheat, for instance, are effective only to some races of stem, leaf, or stripe rust, respectively. Similarly, the partial resistance of barley to leaf rust does not operate against yellow rust, and the partial resistance of wheat to leaf rust is independent of that to stem rust.

In cereal breeding, the resistances to cereal rusts that are presently employed belong to the category of pathogen-specific resistance, whether of the race-specific or of the non-race-specific type. Other types of usable resistance have not yet been reported.

IV. Theoretical Aspects of Non-Race-Specificity

Both the cereals and the cereal rusts vary with respect to one another. This variation may be specific: with differential cultivars races can be identified, and with differential races one can identify resistance genes in the host cultivars. Or it may be non-specific, whereby the recognition of races through host cultivars or vice versa is not possible. The recognition of races is possible in case of genetic interactions between host and pathogen genotypes based on a gene-for-gene system. Table I shows this. The four pathogen races can be identified unambiguously only in the model B with large gene effects. Each race has its own pattern of disease severity on the four host cultivars. The genetic interaction effects have the same size as the effects of the individual genes. When the gene effects and therefore the interaction effects become smaller, the experimental error, absent in Table I, starts to play a role. As soon as the size of the gene effects falls within the range of the

Table I

Hypothetical Example Demonstrating Disease Severity When the Gene Effects[a] between Host and Pathogen Are Additive (Model A) or When They Interact According to a Gene-for-Gene Relationship (Model B)

			Gene effects large				Gene effects small			
			Pathogen race							
Model[b]	Host cultivar		—	a-	b-	ab	—	a-	b-	ab
A	—	—	40	60[c]	60	80	50	55[c]	55	60
	R	—	20	40	40	60	45	50	50	55
	—	S	20	40	40	60	45	50	50	55
	R	S	0	20	20	40	40	45	45	50
B	—	—	80	80	80	80	50	50	50	50
	R	—	40	80	40[d]	80	45	50	45[d]	50
	—	S	40	40	80	80	45	45	50	50
	R	S	0	40	40	80	40	40	40	50

[a]Only the resistance (R, S) and pathogenicity (a, b) alleles are shown.

[b]In Model A the resistance alleles reduce the disease severity with 20% (large) or 5% (small), and the pathogenicity alleles add 20% or 5% to the disease severity. In Model B the resistance alleles reduce the disease severity with 40% (large) or 5% (small); the pathogenicity alleles a and b cancel the effects of R and S, respectively.

[c]Variation in absence of interactions.

[d]Variation due to differential interaction.

experimental error, one cannot discern interaction effects any more; there are no races to be recognized and one speaks of non-race-specific resistance. Typical race-specific and typical non-race-specific resistance form the extremes of a continuum; the size of the race-specific effects depends on the size of the gene effects. This is nicely supported by the wheat–stripe rust system. The effects of the many resistance genes vary from small to large (Robbelen and Sharp, 1978). Zadoks (1972), studying various wheat cultivar-stripe rust race combinations, observed a continuum between instances of near non-race-specific resistance to instances of clear race-specific resistance.

Non-race-specificity may have another basis. If there is no gene-for-gene relationship as in Table I, model A, it is also impossible to identify races. The races *a*- and -*b*, for instance, cannot be discerned. Independently of the size of the gene effects, one must speak of non-race-specific resistance because there are no genetic interaction effects present. All variation is between cultivars and between races. The size of the gene (allele) effects (R, S, a, b) were taken the same in Table I for the sake of simplicity. Varying the size of the gene effects does not change the fact that model A is characterized by the absence and model B by the presence of genetic interactions.

Apparently non-race-specific resistance can arise from two genetic systems: when the host and the pathogen genes have small effects and operate on a gene-for-gene basis (model B) and when the host and pathogen genes, whether small or large in effect, do not operate on a gene-for-gene basis (model A). Genes with large effects operating in a non-specific way (model A) are not believed to exist in host-specialized pathogen systems (Parlevliet and Zadoks, 1977; Parlevliet, 1981b; Ellingboe, 1981). In cereal–cereal rust systems resistance genes with large effects most likely operate on a gene-for-gene system and are race-specific; resistance genes with small effects, too small to study them individually (polygenes), give a non-race-specific pattern whether they operate on a gene-for-gene system or not.

V. Resistance of the Non-Race-Specific Type

A. EXPRESSION

In 1963 and 1968, Vanderplank concluded that horizontal resistance was characterized by a reduced apparent infection rate, *r*. This view was apparently widely accepted, and resistances that varied in a quantitative way and resulted in slow rusting, were assumed to be non-race-specific. This is too simple an assumption. Simply inherited, race-specific resistance can also be expressed as slow rusting (Johnson and Taylor, 1972, 1980; Parlevliet, 1979a). To differentiate alleged non-race-specific resistances from typical race-specific resistance with greater accuracy, Parlevliet and van Ommeren (1975) also used the infection type. Cultivars with a high or susceptible infection type that rusted slowly were considered to carry the alleged non-race-specific resistance (partial resistance).

Despite the difficulties in recognizing non-race-specific resistance, it is clear that the cereals vary, in partial resistance as they do for the typical race-specific resistances. This partial resistance is assumed or hoped to be by and large non-race-specific, and has been reported in wheat to stem rust (Wilcoxson *et al.*, 1974, 1975; Rees *et al.*, 1979a), leaf rust (Caldwell *et al.*, 1970; Young, 1970; Statler *et al.*, 1977; Rees *et al.*, 1979b; Shaner and Finney, 1980), and stripe rust (Hendriksen and Pope, 1971; Dehne, 1977); in barley to leaf rust (Clifford, 1972; Parlevliet and van Ommeren, 1975; Parlevliet *et al.*, 1980) and stripe rust (Parlevliet, 1980b); in oats to crown rust (Luke *et al.*, 1972); in rye to leaf rust (Parlevliet, 1977a); and in maize to *Puccinia sorghi* (Hooker, 1969; Kim and Brewbaker, 1977).

B. MEASURING

The assessment of partial resistance in the field can be accomplished in various ways. Basically, the proportion of leaves or stems affected by the rust is measured either once at the peak of epidemic development or several times from the beginning to the end of the epidemic. The former is assumed to represent the cumulative result of all resistance factors during the progress of the epidemic (Parlevliet and van Ommeren, 1975); the latter makes it possible to compute the apparent infection rates r (Vanderplank, 1963, 1968) or the area under the disease progress curve (AUDPC) (Wilcoxson *et al.*, 1975). Rees *et al.* (1979a,b) compared various methods and concluded that the r value was the least suitable parameter to describe the slow rusting of the cultivars. J. E. Parlevliet (unpublished) came to a similar conclusion because the r value depended not only on the partial resistance to leaf rust of the barley cultivars, but also on the stage of the development of the epidemic. Shaner and Finney (1980) found that the AUDPC was a better criterion to measure partial resistance than was r.

In order to evaluate partial resistance, cultivars are normally grown on relatively small plots adjacent to one another. The rust population to which the cultivars are exposed varies. Often it is a mixture of races, either introduced or natural. In other experiments, the rust population exists predominantly of one race either because it is possible to start an epidemic well in advance of the natural one or because the natural occuring population consists of mainly one race. Under such conditions assessments may lead to an over- or underestimation of partial resistance or to a wrong evaluation altogether, as shown next.

1. Assuming that all cultivars are assessed on the same day, it is easily seen that the partial resistance of late cultivars is overestimated and that of early cultivars under evaluated.
2. Interplot interferences can reduce the real differences in partial resistance very considerably in case of leaf-borne pathogens like rusts (Vanderplank, 1968; Parlevliet and van Ommeren, 1975; Parlevliet, 1979b). Table II illustrates this. Five barley cultivars were exposed to one race of *P. hordei*. Some plots were isolated from one another by wide strips of winter wheat, and others were adjacent. The former estimates the partial resistance as experienced by the farmer, and the latter as seen experimentally (Parlevliet, 1983a). In the adjacent plots, the partial resistance of Julia and Vada are underestimated considerably compared with the simulated fields of the farmer represented by the isolated plots. Although the differences were much smaller, the ranking order

Table II

Number of *Puccinia hordei* Uredia per Tiller on Five Spring Barley Cultivars Approximately 4 Weeks after Heading at Three Different Field Plot Situations in 1973

	Field plots		
		Adjacent	
Cultivar	Isolated 3 × 4 m	2 × 2 m	$\frac{1}{4}$ × 1 m
L98	1000	500	2300
Sultan	750	250	1700
Volla	110	40	700
Julia	17	12	450
Vada	1	15	130
Range	1000×	42×	18×

Table III

Amount of Disease One Might Observe If Cereal Cultivars Are Affected by a Rust When the Cultivars Carry Different Race-Specific Resistance Genes and the Rust Population Consists of Various Races

Resistance genes of host cultivars	Virulence genes of pathogen races[a]				Accumulated amount of disease[b]
	V1 (30)	*V1V3* (40)	*V1V4* (25)	*V1V2V3V4* (5)	
R1	+[c]	+	+	+	70%
R3		+		+	50%
R1R4			+	+	40%
R1R2R4				+	15%

[a]Percentage in the intial inoculum given in parentheses.
[b]Expressed as leaf area affected by infections of a susceptible type.
[c]A + sign indicates that the host is susceptible for that race.

remained the same enabling efficient selection in small plots (Parlevliet and van Ommeren, 1975; Parlevliet *et al.*, 1980).

3. The presence of major race-specific resistance genes that are partially effective may thoroughly confound the assessment of partial resistance (Parlevliet, 1983b). When cultivars varying in race-specific genes (R genes) are exposed to a mixture of races, some of the R genes may be expressed partially against a part of the population (Table III). Depending on the frequency of the races that are virulent on the various cultivars, one may get large differences in amount of rust (of a susceptible type). These differences, though, do not reflect differences in partial resistance; they merely show which major genes are still most effective to that mixture of races.

C. Components

Factors that affect the rate *r* of epidemic development may operate before or after the penetration and formation of infection hyphae. Of the factors that operate before penetration, little is known. Most authors concluded that differences in resistance between cultivars were only observable after penetration. In a few cases, though, prepenetration differences were reported. Russell (1975) observed that on wheat cultivars with erect leaves, less stripe rust spores were deposited than on wheat plants with less erect leaves. Neervoort and Parlevliet (1978) showed data suggesting that the crop morphology is of some importance. The susceptibility to barley leaf rust increased when the barley cultivars were shorter and had a denser plant stand (the more modern cultivars). The rust spores apparently can reach the higher leaf canopies more easily in shorter cultivars, and get lost less easily in a dense crop. Hart (1929) reported that the stomatal behavior may have an effect on stem rust development in wheat.

The majority of rate-reducing factors, though, operate after penetration and formation of infection hyphae. The rate of epidemic development is reduced when a given initial inoculum produces lesions that (1) are less in number, (2) sporulate later, and/or (3) sporulate less profusely. These components are generally indicated as (1) infection frequency, infection density, infection efficiency, or receptivity, (2) latent period, and (3) spore production.

1. *Receptivity*

Differences between cultivars have been found (Clifford, 1972, 1974a; Luke *et al.*, 1972; Kochman and Brown, 1976; Ohm and Shaner, 1976; Kuhn *et al.*, 1978). The differences in receptivity between barley cultivars to barley leaf rust varied with the development stage (Parlevliet and Kuiper, 1977a). The reduced receptivity of partially resistent barley cultivars results mainly from arresting part of the barley leaf rust colonies in a very early stage, before haustoria are formed (Niks, 1982, 1983).

2. *Latent Period*

Here again, variation for latent period (LP) is the rule (Clifford, 1972; Luke *et al.*, 1972; Parlevliet, 1975, 1977a, 1980b; Ohm and Shaner, 1976; Dehne, 1977; Martin *et al.*, 1979). Differences in LP probably reflect differences in growth rate of the pathogen in the host. This, however, is not necessarily so. Five *Avena sterilis* accessions and the oat cultivar Fulghum were tested with five isolates of oat stem rust. The hosts did not differ for LP, despite differences in colony size (Szteynberg and Wahl, 1976). The plant stage and the age of the leaves can play an important role. Barley cultivars infected with barley leaf rust show the greatest differences in LP at heading. At this stage the LP is longest in the flag leaf and decreases with the leaves down the stem. Younger and older plants have shorter LPs, with the seedling stage showing the shortest LPs and the smallest differences (Parlevliet, 1975). In the rye *P. recondita* f. sp. *recondita,* a similar pattern was observed (Parlevliet, 1977a).

3. *Spore Production*

This can be expressed in various ways, such as spore production (SP) per unit leaf area, SP per uredium, SP per unit area of lesion, or SP per unit area of sporulating surface. The SP per unit area can be measured in units of time or over the entire infectious period. Because SP is strongly affected by the number of uredia per unit area, it is very difficult to study SP without studying receptivity. Despite the problems in estimating SP, it is clear that considerable differences in SP per lesion per unit time exist (Szteynberg and Wahl, 1976; Dehne, 1977; Neervoort and Parlevliet, 1978).

Because SP is difficult to measure, it is often estimated by the size of the uredium, assuming a close association between SP and size. Considerable differences in uredium size have been reported (Ohm and Shaner, 1976; Johnson and Wilcoxson, 1978; Kuhn *et al.*, 1978).

4. *Infectious Period*

The pustules of the cereal rusts sporulate over extended periods, although the bulk of the spores tends to be produced in the early phase of the infectious period (IP) (Szteynberg and Wahl, 1976; Neervoort and Parlevliet, 1978). In mature plants the uredia may stop producing spores because of the formation of telia. The IP, like SP, shows a negative association with receptivity. At high receptivities, the IP is shorter, apparently because the leaves are exhausted sooner. Despite the negative association, in barley clear differences in IP were observed to barley leaf rust (Neervoort and Parlevliet, 1978). The influence of a shorter IP in partial resistance is probably small compared with the effect of other components, because the period over which the epidemic develops is often not much larger than the IP of an individual pustule.

5. *Associated Variation of the Components*

These studies where two or more components have been investigated show that components tend to vary in association. Partial resistance tends to go together with reduced receptivity, longer LP, and reduced SP (Szteynberg and Wahl, 1976; Dehne, 1977; Johnson and Wilcoxson, 1978; Kuhn *et al.*, 1978; Neervoort and Parlevliet, 1978).

6. *Effect of the Components on Partial Resistance*

Because the components tend to vary in association, it is difficult to find out the contribution of each component to partial resistance.

From studies simulating epidemics, it can be concluded that for pathogens with short LPs, variations in LP contribute very strongly to variations in partial resistance (Zadoks, 1971). This may be especially so for rusts with a short LP and many discrete, small infections, like barley leaf rust, and less so for a more systemic one like stripe rust. In barley, variation in partial resistance is largely explained by variation in LP (Parlevliet and van Ommeren, 1975; Neervoort and Parlevliet, 1978).

The associated variation of the components is probably one of the main reasons why the component analyses necessarily done in the greenhouse generally show good agreement with the partial resistance in the field. Component analyses done on adult plants tend to correspond better with partial resistance than those carried out on seedlings (Martin and Miller, 1974; Parlevliet and van Ommeren, 1975; Martin *et al.*, 1979; Parlevliet, 1980b; Parlevliet *et al.*, 1980; Wahl *et al.*, 1980).

D. INHERITANCE

The genetics of slow rusting in cereals were recently reviewed by Wilcoxson (1981). Slow rusting is generally the same type of resistance as partial resistance. In most studies the progenies of crosses between slow and fast rusters were investigated. The segregation patterns were quantitative in nature, and transgression was observed regularly. The number of genes assumed to be involved varied from a few in maize to *P. sorghi* (Kim and Brewbaker, 1977), in wheat to *P. recondita* f. sp. *tritici* (Kuhn *et al.*, 1980), and in oats to *P. coronata* (Luke *et al.*, 1975), to several in wheat to *P. recondita* f. sp. *tritici* (Gavinlertvana and Wilcoxson, 1978) and *P. graminis* f. sp. *tritici* (Skovmand *et al.*, 1978a). In barley, partial resistance to *P. hordei* appeared polygenic (Johnson and Wilcoxson, 1979).

In a few cases one component of partial resistance was studied. Brennan (Wilcoxson, 1981; Brennan, 1977) suggested that receptivity of wheat to *P. graminis* f. sp. *tritici* was governed by several genes; one line, though, showed a simple inheritance. Parlevliet (1976b, 1978a) studied LP, the most important component of partial resistance in barley to *P. hordei,* in detail. The estimated number of polygenes for increased LP was zero in L94 (LP 8.0 days), one in L92 (LP 8.6 days), three in Sultan and Volla (LP 10.5 days), five in Julia (LP 13 days), and six in Minerva and Vada (LP 15.5 days). One of the genes in Minerva and Vada had a somewhat larger effect. The genes present in the cultivars were to a large extent the same, with only seven loci assumed to be involved. An unrelated cultivar, Cebada Capa, appeared to carry three to four minor genes for increased LP different from those in Vada (Parlevliet and Kuiper, 1977b; Parlevliet, 1980a). J. E. Parlevliet (unpublished) collected information suggesting that receptivity and latent period are pleiotropically associated in the barley–*P. hordei* system.

In wheat infected with *P. striiformis,* minor genes have been found that additively reduce the infection type. These minor genes appear to be temperature-sensitive and to act independently of the race-specific major genes (Pope, 1968; Lupton and Johnson, 1970; Sharp and Violin, 1970; Hendriksen and Pope 1971; Sharp *et al.,* 1976; Röbbelen and Sharp, 1978; Krupinsky and Sharp, 1979).

Minor or polygenes apparently are as much a part of the resistance complex of cereals to rusts as the major genes are. They are, however, less easily detected, and their contribution to overall resistance is probably greatly underrated.

E. RELATIONSHIP TO RACE-SPECIFIC TYPE RESISTANCE

Several studies have investigated the relationship between the genes for partial resistance and those for typical race-specific resistance. In wheat, partial resistance to stem rust appeared to be inherited independently of the major *Sr* genes *5, 6, 7a, 7b, 8, 9b, 11, 12, 16,* and *Tc,* while it seemed linked to a few other *Sr* genes such as *6* and *36* (*Tt-1*) (Skovmand *et al.,* 1978b; Ayers *et al.,* 1981; Wilcoxson, 1981). Similar observations were obtained in the barley–*P. hordei* system. The partial resistance genes giving a longer LP segregated independently from the race-specific major gene *Pa7* in Cebada Capa and similar cultivars (Parlevliet and Kuiper, 1977b; and J. E. Parlevliet, unpublished).

Other observations support the view that partial resistance and the race-specific major gene resistance are based on independent genetic systems. Cultivars carrying race-specific major genes that are no longer effective may show levels of partial resistance that vary from very low to very high. Brennan (Wilcoxson, 1981) showed that near-isogenic Marquis wheat lines carrying noneffective *Sr* genes were fast rusting. In barley, the cultivars Sudan and Mamie are extremely susceptible to barley leaf rust, while Armelle and Sundance have a fair level of partial resistance. All four cultivars carry one or more *Pa* genes (J. E. Parlevliet, unpublished).

The expression of the genes for partial resistance also seems to be independent of that of the major race-specific genes. The partial resistance genes that reduce the receptivity and LP in barley to barley leaf rust appear to act even in the presence of effective *Pa* genes (Clifford, 1974a; Parlevliet, 1980a). In the presence of *Pa* genes, the partial resistance genes express themselves by a reduced number of necrotic flecks that appear later. Histological studies showed that partial resistance is caused by the very early abortion of a proportion of the infections (colonies) resulting in a reduced receptivity, and in a slower growth of the colonies not aborted early, giving a longer LP and a reduced rate of sporulation (Niks, 1982; Niks and Kuiper, 1983). *Pa* genes also cause colonies to grow slowly or to stop growth. The difference between partial resistance and Pa gene resistance is that early abortion and reduced colony growth is accompanied by extensive host-cell collapse in the latter but not in the former (Niks and Kuiper, 1983). Early abortion due to partial resistance occurs before haustorium formation, and early abortion due to *Pa* gene resistance after haustoria formation (Niks, 1983). The partial resistance genes and the *Pa* genes come to their normal expression when they occur together, indicating the independency of the two mechanisms. When both are present, the colonies that abort early because of partial resistance still do so without cell collapse; the *Pa* genes do not operate yet. But the colonies that continue to grow do show the effects of both the *Pa* genes and of the partial resistance genes. There is host-cell collapse due to the *Pa* genes, and a reduced growth of the colonies reflecting both the effects of the *Pa* genes and the effects of the partial resistance genes (Niks and Kuiper, 1983).

These observations indicate that partial resistance and race-specific major gene resistance represent different systems, governed by different genes.

F. DURABILITY

Partial resistance is considered to be durable, but pertinent information is scant. The wheat cultivars Thatcher and Lee have been known to rust slowly with stem rust for 55 and 30 years, respectively (Wilcoxson, 1981). The oat cultivar Red Rustproof has been reported to be slow rusting to crown rust for more than 100 years already (Luke *et al.*, 1972). The slow rusting of the wheat cultivar Knox to leaf rust has been effective for some 20 years (Ohm and Shaner, 1976). In the United States, maize has never known severe epidemics of *P. sorghi* due to partial resistance (Hooker, 1969).

Many West European barley cultivars have good levels of partial resistance to barley leaf rust (Parlevliet *et al.*, 1980). Several of the minor genes involved occur at a high frequency in the West European barleys (Parlevliet, 1978a), and these genes must have been present for a considerable time as they have not been consciously introduced or selected for recently. As they are still effective, they must be durable despite being race-specific (Section V,C). Habgood and Clifford (1981), considering the English experience, concluded that partial resistance of barley to barley leaf rust is durable enough to be useful.

Other indications that partial resistance can have considerable durability came from studies where old and new cultivars were compared. In areas of the United States where breeders did not exploit specific resistance in wheat to leaf rust, modern cultivars have less rust than the older cultivars (Young, 1970), indicating that an accumulation of partial resistance has endured. Similar observations were made in Australia with wheat and stem rust (Rees *et al.*, 1979a). Cultivars in use near the beginning of this century appeared, on average, to be faster rusting then modern cultivars without effective major genes. Although wheat breeding for stem rust resistance has been for race-specific major resistance genes for the last 40 years, the level of partial resistance apparently increased at the same time. This is not surprising. Parlevliet (1983a) showed that strong selection for resistance tend to increase both the race-specific and the assumed non-race-specific resistance.

Partial resistance apparently is of considerable durability. No cases of erosion of this resistance have been reported yet.

G. SPECIFICITY

In the typical race-specific major-gene resistances, such as governed by *Sr, Lr, Yr, Pc,* and *Pa* genes, and also in the polygenic resistances, partial resistance is pathogen-specific (Parlevliet, 1981a,b). Pathogen-specific resistance operates against one pathogen species only. The partial resistances of wheat to stem rust and leaf rust, of maize to *P. sorghi* and *P. polysora,* and of barley to barley leaf rust and stripe rust are all independent of one another (Parlevliet, 1981b; Wilcoxson, 1981). The spring wheat cultivar Thatcher, for instance, rusts slowly with stem rust and rapidly with leaf rust, while Lee rusts moderately with both pathogens. From a cross between the two cultivars, lines were obtained that rusted slowly with both pathogens (Gavinlertvana and Wilcoxson, 1978). The barley cultivars Vada and Berac have a high and fair level of partial resistance to barley leaf rust, respectively. Vada is very susceptible and Berac is extremely susceptible to stripe rust.

Armelle, Mazurka, and Midas have the same level of moderate susceptibility to stripe rust; their levels of partial resistance to barley leaf rust are high, moderate, and low, respectively (Parlevliet, 1983b). The minor genes that are effective to one *Puccinia* species are apparently not effective to another *Puccinia* species adapted to the same host. The effects of these minor genes are far from general: they are highly species-specific.

Pathogen-specific resistance may be race-specific as governed by the major, hypersensitive genes, or may be more or less non-race-specific. Partial or nonhypersensitive resistance or slow rusting is often considered to be non race-specific. This assumption, however, is difficult to maintain in its absolute sense, both in terms of theoretical reflections (Section IV) and in terms of reported observations. Resistance genes with small effects tend to give a non-race-specific pattern when cultivars differing in partial resistance are tested against different isolates or races of the pathogen. If, however, the experimental error is small enough, the effect of individual minor genes becomes visible. In such cases one might observe cultivar × isolate interaction when the host and pathogen genes operate on a gene-for-gene basis. Such interactions have been reported for the barley–*P. hordei* system (Clifford and Clothier, 1974; Parlevliet, 1976a, 1977b; Aslam and Schwarzbach, 1980), for the wheat–*P. graminis* f. sp. *tritici* system (Brennan, 1977; Mortensen and Green, 1978), and for the wheat–*P. recondita* f. sp. *tritici* system (Kuhn *et al.*, 1978; Milus and Line, 1980). Kuhn *et al.* (1978), testing four cultivars against 22 *P. recondita* isolates, concluded that the host–pathogen patterns were non-race-specific, but they looked only for race-specific increases in susceptibility. The race-specific interaction present toward higher resistance was not taken into account. Szteynberg and Wahl (1976) observed only non-race-specific effects in the *Avena sterilis*–*P. graminis* f. sp. *avenae* system. Ashagari and Rowell (1980) also observed only non-race-specific effects when studying receptivity for stem rust in wheat.

With stripe rust in wheat, it is very difficult to discern typical hypersensitive or low infection type resistance from partial or high infection type resistance. Any increase in resistance seems to occur with a decrease in infection type (Hendriksen and Pope, 1971; Dehne, 1977). It is therefore not possible to clearly distinguish between the two types of resistance. The minor genes, inherited in an additive way (as discussed in Section V,D), have so far exhibited non-race-specific behavior. British research, though, suggests that small differences in resistance can be of a race-specific nature (Johnson and Taylor, 1972, 1980; Johnson and Law, 1975).

According to the theoretical reasoning (Section IV), non-race-specific resistance in the host can arise from minor genes that do not interact with genes in the pathogen or from minor genes that do interact in a gene-for-gene way with genes in the pathogen. Small interactions (race-specific effects) can be expected with the latter, but not with the former gene action. When the genes governing partial resistance act independently of the pathogen genes, it is difficult to explain why they are effective against only one *Puccinia* species, although the species are very closely related and adapted to the same host. This pathogen-specificity, together with the presence of small race-specific effects, suggests that the partial resistance genes too operate on a gene-for-gene basis with minor genes in the pathogen. The observations of Parlevliet (1978b) in the barley–*P. hordei* system support this. The partially resistant cultivar Julia shows a somewhat increased susceptibility and a concomitant reduction in LP to race 18. This small race-specific effect could be explained very well by assuming that one of the five minor genes for LP was neutralized by race 18. This gene-for-gene action does not automatically mean lack of durability for such resistance. Eenink (1976), Meiners (1981), and Parlevliet (1981b, 1983a) clearly showed that durability and race-specificity can be independent qualities of resistance.

There are indications that specialized pathogenic fungi employ two gene-for-gene systems that are the reverse of each other (Parlevliet, 1981b, 1983b; Ellingboe, 1981). In the well-known one, the recognition (interaction) is for incompatibility and occurs between the products of the resistance and the avirulence alleles (Flor, 1960; Ellingboe, 1981; Bushnell and Rowell, 1981). The race-specific resistance genes in the cereal–rusts systems, normally of a hypersensitive or low infection type, are considered to belong to this gene-for-gene system. Resistance and avirulence are positive functions. Mutations from avirulence to virulence then are easy, because they involve a loss of function; there is no recognition reaction anymore, and the basic pathogenicity is restored. Because this change to virulence can occur so easily, the resistances are highly elusive (Parlevliet, 1981b, 1983b). In the second gene-for-gene system, assumed to govern basic pathogenicity (Ellingboe, 1981; Parlevliet, 1981b), the recognition (interaction) is for compatibility occurring between the products of the susceptibility and pathocinicity alleles. Loss of function leads to loss of compatibility. The minor genes for partial resistance are thought to operate within this second gene-for-gene system interfering with basic pathogenicity (Parlevliet, 1981b, 1983b). In this gene-for-gene system, neutralizing a resistance gene requires a gain mutation. The frequencies of gain mutations are very low, and this could be the reason for the durability of partial resistance.

It is of course possible that both gene actions described in Section IV occur. Apart from minor genes that operate on a gene-for-gene basis, there could be genes with a nonspecific action. Like in any other organism, pathogen genotypes may vary slightly in inherent growth rates; host genotypes may vary somewhat in their suitability to sustain a good growth of the pathogen. These differences are similar to those shown by plant–soil relationships for instance. The genotypes may vary in growth rate, while soils may vary in their effects on these growth rates. The long-lasting and close coexistence between cereal rusts and their hosts makes it likely that these inherent differences in both rusts and cereals are small. There are some indications for the existence of such assumed nonspecific effects. The barley cultivar L94 does not seem to have any genes for an increased LP to *P. hordei* (Parlevliet, 1978a). If all partial resistance and partial pathogenicity followed the second gene-for-gene system, L94 should have an identical LP for all *P. hordei* isolates. This is not the case. With different isolates, very small, but real, differences in LP were observed on L94 (Parlevliet, 1976a).

Some of the differences in the observed partial resistance could be due to nonspecific effects of this nature. Most of the partial resistance, though, cannot be explained with such nonspecific effects, because partial resistance is predominantly pathogen-specific.

VI. Selection for Partial Resistance

A. FREQUENCY AND STABILITY OF PARTIAL RESISTANCE

Partial resistance to cereal rusts seems to occur in all cereals to all rusts (Section V,A), and seems to be very common within each species. Barley cultivars or lines that do not carry any partial resistance to barley leaf rust are rare, while cultivars that have moderate to high levels of partial resistance are common (Parlevliet and van Ommeren, 1975; Niks and Parlevliet, 1978; Parlevliet *et al.*, 1980). Rees *et al.* (1979a,b) reported similar observations in wheat to stem and leaf rust in Australia. Most cultivars tested had moderate to high levels of partial resistance. According to Young (1970), most wheat cultivars grown in the central plains of the United States carry sufficient partial resistance to leaf rust to prevent serious damage.

Another aspect of partial resistance is its stability. Stability is not used here in the sense of durability, but in the sense of sensitivity to variations in the physical environment. In Minnesota, the slow rusting of wheat to stem rust appears a stable characteristic, being quite insensitive to the environment. Nine cultivars have been tested annually for 8 years, and the rank order for slowness of rusting was the same each year. Variations in fertilizers, initial inoculum, and sowing date also did not affect the rank order in slowness of rusting (Wilcoxson, 1981). Very similar results have been obtained with partial resistance of barley to barley leaf rust. Over the years various cultivars have been tested at different sites (Table IV). The epidemics varied greatly with the years, due to year and site differences. The rank order of the cultivars for partial resistance, however remained the same. When barley is sown in the early summer rather then in spring, the barley leaf rust develops much faster and earlier. The rank order for partial resistance, however, is not changed. Cultivars tested in Wageningen showed the same rank order for partial resistance as in England, according to the English lists of recommended cultivars. (Parlevliet, 1979b, also unpublished). Partial resistance in barley to barley leaf rust thus appears a very stable characteristic.

The rank order for partial resistance to leaf rust in wheat has also

Table IV

Number of *Puccinia hordei*, Race 1–2, Uredia per Tiller of Four Barley Cultivars in 7 Years[a]

Year and site[b]	Cultivar				
	L94	Sultan	Berac	Julia	Vada
1973, FE	2800	750	—	17	1.1
1974, FE	750	40	1.9	1.5	0.5
1976, FS	—	—	310	180	60
1977, FE	—	1300	—	65	7
1978, W	2300[c]	400	45	20	7[d]
1979, W	—	—	250	200	90
1980, FE	—	—	20	18	2

[a]Data from isolated plots where interplot interference is small to negligible. One percent leaf area affected corresponds with 100–200 uredia.

[b]FE = East Flevo Polder; FS = South Flevo Polder; W = Wageningen.

[c]Data from a cultivar with a susceptibility equal to L94.

[d]Data from the cultivar Varunda, derived from Vada, and with a similar partial resistance to barley leaf rust.

remained the same over the years (Statler *et al.*, 1977; Mehta and Igarashi, 1979).

Very little information is available about the expression of partial resistance when tested in widely different areas. Wilcoxson (1981) reports that the rank order for slow rusting to stem rust in wheat cultivars in South Africa differed from the one in Minnesota, suggesting some instability of partial resistance. This is not necessarily true, as the wheat cultivars were exposed to different stem rust populations in the different areas. Partial resistance can be quite easily confounded with the effects of major race-specific genes that are still effective to some extent, as shown in Table III.

B. SELECTION

Selection for higher levels of resistance to cereal rusts is not difficult (see also Section V,B). The main problem is that selection for race-specific major-gene resistance is even easier than selection for partial resistance. In the absence of such major genes, the differences in partial resistance are readily recognized, and selection for increased levels of such resistance should be easy. This is the case with barley and barley leaf rust. The great majority of barley cultivars carry no effective *Pa* genes (Clifford, 1974b; Parlevliet, 1975; Niks and Parlevliet, 1978; Rintelen, 1979; Parlevliet *et al.*, 1980). The cultivar differences for partial resistance are highly reproducible—i.e., heritability is high. Selection for it should be successful (Parlevliet and van Ommeren, 1975; Johnson and Wilcoxson, 1978, 1979). This was confirmed in actual selection experiments. It was easy to increase the level of partial resistance by selection in the seedling stage, in the individual adult-plant stage, and in the small-plot stage (Parlevliet, 1976c; Parlevliet *et al.*, 1980).

The results of Luke *et al.* (1975) also suggest that in oats, partial resistance to crown rust will respond readily to selection. The heritability reported by them was 0.87. In wheat selection for a higher level of minor-gene resistance to stripe rust was relative easy (Lupton and Johnson, 1970; Sharp *et al.*, 1976; Krupinsky and Sharp, 1979).

Selection for partial resistance in the presence of effective or partly effective race-specific major genes is more difficult due to the confounding effects as discussed in Section V,B, Table III. Parlevliet (1983a) discussed this comprehensively and concluded that any selection towards higher resistance tends to select partial resistance, even in the presence of major genes. If one wishes to increase partial resistance and to get rid of race-specific major genes, one must expose the plant population to be selected to a widely virulent race if possible and not to a broadly virulent mixture of races. The plants selected should be neither the most susceptible ones nor those that remain rust-free.

VII. Usefulness of Partial Resistance

Partial resistance is often suggested as the alternative for major-gene resistance. This suggests resistance is either one or the other, but that does not represent reality. Partial resistance is probably present and effective in all cereal–rust systems. The disease pressure of some cereal–rust systems, however, can be so severe that the partial resistance present is insufficient. In such cases the major-gene approach is the easiest and fastest breeding solution, but not always the best one. Even easier is to spray with fungicides (wheat stripe rust in northwestern Europe).

There is little doubt that the use of race-specific resistance genes and of fungicides can be restricted if more emphasis is laid on partial resistance. This might in the long run be the cheapest way of controlling the rusts. It is, however, quite possible that the level of partial resistance obtainable with a reasonable breeding effort is insufficient for areas extremely suitable for a certain rust. In such cases one might consider the restricted use of fungicides and/or the use of race-specific major genes within strategies that retard the development of new races.

References

Ashagari, D., and Rowell, J. B. (1980). Postpenetration phenomena in wheat cultivars with low receptivity to infection by *Puccinia graminis* f. sp. *tritici*. *Phytopathology* **70,** 624–627.

Aslam, M., and Schwarzbach, E. (1980). Specificity of quantitative interactions between brown rust isolates and barley varieties. *Proc.—Eur. Mediterr. Cereal Rusts Conf., 5th,* 1980, pp. 123–127.

Ayers, J. E., Southern, J. W., Roelfs, A. P., and Wilcoxson, R. D. (1981). Inheritance of slow rusting and the relationship of *Sr* genes to slow rusting in the wheat line FKN. *Phytopathology* **71,** 835–838.

Brennan, P. S. (1977). Non-hypersensitive resistance to wheat stem rust and its race-specificity. *Proc. Int. Congr. SABRAO, 3rd, 1977,* pp. 4(a)3–5.

Bushnell, W. R., and Rowell, J. B. (1981). Suppressors of defense reactions: A model for roles in specificity. *Phytopathology* **71,** 1012–1014.

Caldwell, R. M., Roberts, J. J., and Eyal, Z. (1970). General resistance ("slow rusting") in

Puccinia recondita f. sp. *tritici* in winter and spring wheats. *Phytopathology* **60,** 1287 (abstr.).

Caten, C. E. (1974). Intra-racial variation in *Phytophthora infestans* and adaptation to field resistance for potato blight. *Ann. Appl. Biol.* **77,** 259–270.

Clifford, B. C. (1972). The histology of race non-specific resistance to *Puccinia hordei* Otth. in barley. *Proc.—Eur. Mediterr. Cereal Rusts Conf., 3rd, 1972,* Vol. I, pp. 75–78.

Clifford, B. C. (1974a). Relation between compatible and incompatible infection sites of *Puccinia hordei* on barley. *Trans. Br. Mycol. Soc.* **63,** 215–220.

Clifford, B. C. (1974b). The choice of barley genotypes to differentiate races of *Puccinia hordei* Otth. *Cereal Rusts Bull.* **2,** 5–6.

Clifford, B. C. (1975). Stable resistance to cereal disease: problems and progress. *Rep.—Welsh Plant Breed. Stn. (Aberystwyth, Wales), 1974,* pp. 107–113.

Clifford, B. C., and Clothier, R. B. (1974). Physiologic specialization of *Puccinia hordei* on barley hosts with non-hypersensitive resistance. *Trans. Br. Mycol. Soc.* **63,** 421–430.

Day, P. R. (1974). "Genetics of Host-parasite Interactions." Freeman, San Francisco, California.

Dehne, D. (1977). Untersuchungen zur Resistenz von Sommerweizen genenüber Gelbrost (*Puccinia striiformis* West.) im Feld und unter kontrolierten Bedingungen während der Ontogenese der Wirtspflanze. Ph.D. Dissertation, University of Göttingen.

Eenink, A. H. (1976). Genetics of host-parasite relationships and uniform and differential resistance. *Neth. J. Plant Pathol.* **82,** 133–145.

Ellingboe, A. H. (1981). Changing concepts in host-pathogen genetics. *Annu. Rev. Phytopathol.* **19,** 125–143.

Flor, H. H. (1960). The inheritance of X-ray-induced mutations to virulence in a urediospore culture of race 1 of *Melampsora lini. Phytopathology* **50,** 603–605.

Gavinlertvatana, S., and Wilcoxson, R. D. (1978). Inheritance of slow rusting of spring wheat by *Puccinia recondita* f. sp. *tritici* and host parasite relationships. *Trans. Br. Mycol. Soc.* **71,** 413–418.

Habgood, R. M., and Clifford, B. C. (1981). Breeding barley for disease resistance: The essence of compromise. *In* "Strategies for the Control of Cereal Disease" (J. F. Jenkyn and R. T. Plumb, eds.), pp. 15–25. Blackwell, Oxford.

Hart, H. (1929). Relation of stomatal behavior to stem-rust resistance in wheat. *J. Agric. Res. (Washington, D.C.)* **39,** 929–948.

Hendriksen, G. B., and Pope, W. K. (1971). Additive resistance to stripe rust in wheat. *Crop Sci.* **11,** 825–827.

Hooker, A. L. (1969). Widely based resistance to rust in corn. *Spec. Rep.—Iowa Agric. Home Econ. Exp. Stn.* **64,** 28–34.

Johnson, D. A., and Wilcoxson, R. D. (1978). Components of slow-rusting in barley infected with *Puccinia hordei. Phytopathology* **68,** 1470–1474.

Johnson, D. A., and Wilcoxson, R. D. (1979). Inheritance of slow rusting of barley infected with *Puccinia hordei* and selection of latent period and number of uredia. *Phytopathology* **69,** 145–151.

Johnson, R., and Law, C. N. (1975). Genetic control of durable resistance to yellow rust (*Puccinia striiformis*) in the wheat cultivar Hybride de Bersée. *Ann. Appl. Biol.* **81,** 385–391.

Johnson, R., and Taylor, A. J. (1972). Isolates of *Puccinia striiformis* collected in England from the wheat varieties Maris Beacon and Joss Cambier. *Nature (London),* **238,** 105–106.

Johnson, R., and Taylor, A. J. (1980). Pathogenic variation in *Puccinia striiformis* in relation to the durability of yellow rust resistance. *Ann. Appl. Biol.* **94,** 283–286.

Kim, S. K., and Brewbaker, J. L. (1977). Inheritance of general resistance in maize to *Puccinia sorghi* Schw. *Crop Sci.* **17,** 456–461.

Kochman, J. K., and Brown, J. F. (1976). Host and environmental effects on the penetration of oats by *Puccinia graminis avenae* and *Puccinia coronata avenae. Ann. Appl. Biol.* **82,** 251–258.

Krupinsky, J. M., and Sharp, E. L. (1979). Reselection for improved resistance of wheat to stripe rust. *Phytopathology* **69,** 400–404.

Kuhn, R. C., Ohm, H. W., and Shaner, G. E. (1978). Slow leaf-rusting resistance in wheat against twenty-two isolates of *Puccinia recondita. Phytopathology* **68,** 651–656.

Kuhn, R. C., Ohm, H. W., and Shaner, G. E. (1980). Inheritance of slow leaf-rusting resistance in Suwon 85 wheat. *Crop Sci.* **20,** 655–659.

Luke, H. H., Chapman, W. H., and Barnett, R. D. (1972). Horizontal resistance of red rustproof oats to crown rust. *Phytopathology* **62,** 414–417.

Luke, H. H., Barnett, R. D., and Pfahler, P. L. (1975). Inheritance of horizontal resistance to crown rust in oats. *Phytopathology* **65,** 631–632.

Lupton, F. G. H., and Johnson, R. (1970). Breeding for mature-plant resistance to yellow rust in wheat. *Ann. Appl. Biol.* **66,** 137–143.

Martin, C. D., and Miller, J. D. (1974). Development of the stem rust fungus in seedling plants of slow rusting wheats. *Proc. Am. Phytopathol. Soc.* **1,** 127 (abstr.).

Martin, C. D., Miller, J. D., Busch, R. H., and Littlefield, L. J. (1979). Quantitation of slow rusting in seedling and adult spring wheat. *Can. J. Bot.* **57,** 1550–1556.

Mehta, Y. R., and Igarashi, S. (1979). Partial resistance in wheat against *Puccinia recondita*—A new view on its detection and measuring. *Summa Phytopathol.* **5,** 90–100.

Meiners, J. P. (1981). Genetics of disease resistance in edible legumes. *Annu. Rev. Phytopathol.* **19,** 189–209.

Milus, E. A., and Line, R. F. (1980). Characterization of resistance to leaf rust in Pacific Northwest Wheats. *Phytopathology* **70,** 167–172.

Mortensen, K., and Green, G. J. (1978). Assessment of receptivity and urediospore production as components of wheat stem rust resistance. *Can. J. Bot.* **56,** 1827–1839.

Neervoort, W. J., and Parlevliet, J. E. (1978). Partial resistance of barley to leaf rust, *Puccinia hordei.* V. Analysis of the components of partial resistance in eight barley cultivars. *Euphytica* **27,** 33–39.

Niks, R. E. (1982). Early abortion of colonies of leaf rust, *Puccinia hordei,* in partially resistant barley seedlings. *Can. J. Bot.* **60,** 714–723.

Niks, R. E. (1983). Haustorium formation by *Puccinia hordei* in leaves of hypersensitive, partially resistant and nonhost plant genotypes. *Phytopathology* **73,** 64–66.

Niks, R. E., and Kuiper, H. J. (1983). Histology of the relation between minor and major genes for resistance of barley to leaf rust. *Phytopathology* **73,** 55–59.

Niks, R. E., and Parlevliet, J. E. (1978). Variation for partial resistance to *Puccinia hordei* in the barley composite XXI. *Cereal Rusts Bull.* **6,** 3–10.

Ohm, H. W., and Shaner, G. E. (1976). Three components of slow leaf-rusting at different growth stages in wheat. *Phytopathology* **66,** 1356–1360.

Parlevliet, J. E. (1975). Partial resistance of barley to leafrust, *Puccinia hordei.* I. Effect of cultivar and development stage on latent period. *Euphytica* **24,** 21–27.

Parlevliet, J. E. (1976a). Evaluation of the concept of horizontal resistance in the barley *Puccinia hordei* host-pathogen relationship. *Phytopathology* **66,** 494–497.

Parlevliet, J. E. (1976b). Partial resistance of barley to leaf rust, *Puccinia hordei.* III. The

inheritance of the host plant effect on latent period in four cultivars. *Euphytica* **25,** 241–248.
Parlevliet, J. E. (1976c). Screening for partial resistance in barley to *Puccinia hordei* Otth. *Proc.—Eur. Mediterr. Cereal Rusts Conf., 4th, 1976,* pp. 153–155.
Parlevliet, J. E. (1977a). Variation for partial resistance in a cultivar of rye, *Secale cereale,* to brown rust, *Puccinia recondita* f. sp. *recondita. Cereal Rusts Bull.* **5,** 13–16.
Parlevliet, J. E. (1977b). Evidence of differential interaction in the polygenic *Hordeum vulgare–Puccinia hordei* relation during epidemic development. *Phytopathology* **67,** 776–778.
Parlevliet, J. E. (1978a). Further evidence of polygenic inheritance of partial resistance in barley to leaf rust, *Puccinia hordei. Euphytica* **27,** 369–379.
Parlevliet, J. E. (1978b). Race-specific aspects of polygenic resistance of barley to leaf rust, *Puccinia hordei. Neth. J. Plant Pathol.* **84,** 121–126.
Parlevliet, J. E. (1979a). Components of resistance that reduce the rate of epidemic development. *Annu. Rev. Phytopathol.* **17,** 203–222.
Parlevliet, J. E. (1979b). Interplot interference between experimental plots and its consequence for the evaluation of partial resistance (in Dutch), *Zaadbelangen* **33,** 329–332.
Parlevliet, J. E. (1980a). Minor genes for partial resistance epistatic to the Pa 7 gene for hypersensitivity in the barley - *Puccinia hordei* relationship. *Proc. — Eur. Mediterr. Cereal Rusts Conf., 5th, 1980,* pp. 53–57.
Parlevliet, J. E. (1980b). Variation for latent period, one of the components of partial resistance, in barley to yellow rust, caused by *Puccinia striiformis. Cereal Rusts Bull.* **8,** 17–22.
Parlevliet, J. E. (1981a). Race-non-specific disease resistance. *In* "Strategies for the Control of Cereal Disease" (J. F. Jenkyn and R. T. Plumb, eds.), pp. 47–54. Blackwell, Oxford.
Parlevliet, J. E. (1981b). Disease resistance in plants and its consequences for plant breeding. *In* "Plant Breeding II" (K. J. Frey, ed.), pp. 309–364. Iowa State Univ. Press, Ames.
Parlevliet, J. E. (1983a). Durable resistance in self-fertilizing annuals. *In* "Durable Resistance in Crops" (F. Lamberti, J. M. Waller, and N. A. Vander Graaff, eds.), pp. 347–362. Plenum, New York.
Parlevliet, J. E. (1983b). Models explaining the specificity and durability of host resistance derived from the observations on the barley - *Puccinia hordei* system. *In* "Durable Resistance in Crops" (F. Lamberti, J. M. Waller, and N. A. Vander Graaf, eds.), pp. 57–80. Plenum, New York.
Parlevliet, J. E., and Kuiper, H. J. (1977a). Partial resistance to leaf rust, *Puccinia hordei.* IV. Effect of cultivar and development stage on infection frequency. *Euphytica* **26,** 249–255.
Parlevliet, J. E., and Kuiper, H. J. (1977b). Resistance of some barley cultivars to leaf rust, *Puccinia hordei;* polygenic, partial resistance hidden by monogenic hypersensitivity. *Neth. J. Plant Pathol.* **83,** 85–89.
Parlevliet, J. E., and van Ommeren, A. (1975). Partial resistance of barley to leaf rust, *Puccinia hordei.* II. Relationship between field trials, micro plot tests and latent period. *Euphytica* **24,** 293–303.
Parlevliet, J. E., and Zadoks, J. C. (1977). The integrated concept of disease resistance; a new view including horizontal and vertical resistance in plants. *Euphytica* **26,** 5–21.
Parlevliet, J. E., Lindhout, W. H., Van Ommeren, A., and Kuiper, H. J. (1980). Level of

partial resistance to leaf rust, *Puccinia hordei,* in West-European barley and how to select for it. *Euphytica* **29,** 1–8.
Pope, W. K. (1968). Interaction of minor genes for resistance to stripe rust in wheat. *Proc. Int. Wheat Genet. Symp., 3rd, 1968,* pp. 251–257.
Rees, R. G., Thompson, J. P., and Mayer, R. J. (1979a). Slow rusting and tolerance to rusts in wheat. I. The progress and effects of epidemics of *Puccinia graminis tritici* in selected wheat cultivars. *Aust. J. Agric. Res.* **30,** 403–419.
Rees, R. G., Thompson, J. P., and Goward, E. A. (1979b). Slow rusting and tolerance to rusts in wheat. II. The progress and effects of epidemics of *Puccinia recondita tritici* in selected wheat cultivars. *Aust. J. Agric. Res.* **30,** 421–432.
Robbelen, G., and Sharp, E. L. (1978). "Mode of Inheritance, Interaction, and Application of Genes Conditioning Resistance to Yellow Rust," Adv. Plant Breed., Suppl. Parey, Berlin.
Russell, G. E. (1975). Deposition of *Puccinia striiformis* uredospores on adult wheat plants in laboratory experiments. *Cereal Rusts Bull.* **3,** 40–43.
Shaner, G., and Finney, R. E. (1980). New sources of slow leaf rusting resistance in wheat. *Phytopathology* **70,** 1183–1186.
Sharp, E. L., and Violin, R. B. (1970). Additive genes in wheat conditioning resistance to stripe rust. *Phytopathology* **60,** 1146–1147.
Sharp, E. L., Sally, B. K., and Taylor, G. A. (1976). Incorporation of additive genes for stripe rust resistance in winter wheat. *Phytopathology* **66,** 794–797.
Skovmand, B., Wilcoxson, R. D., Shearer, B. L., and Stucker, R. E. (1978a). Inheritance of slow rusting to stem rust in wheat. *Euphytica* **27,** 95–107.
Skovmand, B., Roelfs, A. P., and Wilcoxson, R. D. (1978b). The relationship between slow-rusting and some genes specific for stem rust resistance in wheat. *Phytopathology* **68-** 491–499.
Statler, G. D., Watkins, J. E., and Nordgaard, J. (1977). General resistance displayed by three hard red spring wheat (*Triticum aestivum*) cultivars to leaf rust. *Phytopathology* **67,** 759–762.
Szteynberg, A., and Wahl, I. (1976). Mechanisms and stability of slow stem rusting resistance in *Avena sterilis. Phytopathology* **66,** 74–80.
Vanderplank, J. E. (1963). "Plant Diseases: Epidemics and Control." Academic Press, New York.
Vanderplank, J. E. (1968). "Disease Resistance in Plants." Academic Press, New York.
Vanderplank, J. E. (1978). "Genetic and Molecular Basis of Plant Pathogenesis." Springer-Verlag, Berlin and New York.
von Rintelen, J. (1979). Verfügen unsere Gerstensorten über spezifische Resistenzen gegen physiologische Rassen des Zwergrostes. *Bayer. Landwirtsch. Jahrb.* **56,** 391–397.
Wahl, I., Wilcoxson, R. D., and Rowell, J. B. (1980). Slow rusting of wheat with stem rust detected in the glasshouse. *Plant Dis.* **64,** 54–56.
Wilcoxson, R. D. (1981). Genetics of slow rusting in cereals. *Phytopathology* **71,** 989–993.
Wilcoxson, R. D., Atif, A. H., and Skovmand, B. (1974). Slow rusting of wheat varieties in the field correlated with stem rust severity on detached leaves in the greenhouse. *Plant Dis. Rep.* **58,** 1085–1087.
Wilcoxson, R. D., Skovmand, B., and Atif, A. H. (1975). Evaluation of wheat cultivars for ability to retard development of stem rust. *Ann. Appl. Biol.* **80,** 275–281.
Young, H. C. (1970). Variation in virulence and its relation to the use of specific re-

sistance for the control of wheat leaf rust. *Plant Dis. Probl., Proc. Int. Symp., 1st, 1966–1967,* pp. 3–8.

Zadoks, J. C. (1971). System analysis and the dynamics of epidemics. *Phytopathology* **61,** 600–610.

Zadoks, J. C. (1972). Modern concepts of disease resistance in cereals. *In* "The Way Ahead in Plant Breeding" (F. G. H. Lapton, G. Jenkins, and R. Johnson, eds.), pp. 89–98. Adland & Son Ltd., Bartholomew Press, Dorking.

17

Genetic Diversity and Cereal Rust Management

C. C. Mundt

Department of Plant Pathology, North Carolina State University, Raleigh, North Carolina

*J. A. Browning**

Department of Plant Pathology and Microbiology, Texas A&M University, College Station, Texas

I. Rust Development in Agricultural versus Natural Ecosystems—The Call for Diversity

It has been some 80 years since Biffen (1905) demonstrated that resistance of wheat (*Triticum aestivum*) to a cereal rust can follow

*Present address: Department of Plant Pathology, Seed and Weed Sciences, Iowa State University, Ames, Iowa 50011

ISBN 0-12-148402-5

Mendelian inheritance. During that time, more resistance research has been conducted with the cereal rusts than any other group of pathogens, and great strides have been made in resistance theory. Thus, it is ironic that devastating rust epidemics still occur and that agricultural scientists are still debating how to best manage host resistance to the rusts.

Until recently, small-grains pathologists and breeders concentrated on the use of single race-specific genes for resistance to the cereal rusts. Such use of resistance often results in ephemeral disease control, because virulent genotypes are rapidly selected from the pathogen population (Browning and Frey, 1969; Browning *et al.*, 1969; Johnson, 1961; Stakman and Christensen, 1960). Pure-line selection and hybridization worsen this situation with self-pollinated crops such as the small grains, because homogenous cultivars with a narrow base of resistance are often grown extensively (Frey *et al.*, 1973). The results of such mismanagement are large yield losses and a reduced life expectancy of crop cultivars. For example, a recent wheat leaf rust epidemic in northern Mexico caused yield losses of up to 40% (Dubin and Torres, 1981). Kilpatrick (1975) presented data indicating that the average useful life of race-specific genes for resistance to the stem (*Puccinia graminis*), leaf (*P. recondita*), and stripe (*P. striiformis*) rusts of wheat is ~5 years on a global basis.

In the past, agriculturalists have given little consideration to the development of disease in natural ecosystems. Recently, however, some workers have realized that knowledge gained from such ecosystems can contribute to disease management in agroecosystems, and studies of natural ecosystems are now being conducted by several plant pathologists. Very pertinent to this chapter are studies conducted in the Mid-East, a center of coevolution of the cereal rusts and the putative progenitors of cultivated small grains (Anikster and Wahl, 1979; Browning, 1974; Segal *et al.*, 1980). Two major points can be derived from these studies. First, in the indigenous ecosystem, rusts and their hosts are in a state of dynamic equilibrium. Rust epidemics are rare despite the fact that host, parasite, and a favorable environment are all present. Second, diversity abounds in both the host and parasite populations. This diversity is expressed in many traits, including resistance/susceptibility and avirulence/virulence.

Several mechanisms contribute to the protection of cereal populations from rusts in natural ecosystems. These include specific and general resistance, tolerance, and "disease escape" mechanisms such as early maturity and late rusting (Anikster and Wahl, 1979; Segal *et al.*, 1980). It is of particular relevance to this chapter that race-specific resistance often occurs in abundance in the center of coevolution of the small grains and their rusts. In contrast to agroecosystems, however, this resistance occurs in a diverse pattern; different plants within a population possess different race-specific resistance genes. Thus, rust populations (which are diverse for avirulence/virulence) are buffered because no single genotype is virulent on all plants in the population. In transect studies of wild oat (*Avena sterilis*) populations, only 30% of the plants were resistant to the most predominant crown rust (*P. coronata*) race (summarized by Browning, 1974). Thus, Browning (1974) concluded that 30% resistance, present in a diverse pattern and backstopped by general resistance, was sufficient to reduce the development of crown rust and allow *A. sterilis* to thrive. In contrast to its intended role in agroecosystems, race-specific resistance does not function to protect the host by eradicating the pathogen in natural ecosystems. In fact, Parlevliet (1981a) suggested that, in natural ecosystems, the presence of specific resistance in the host population is advantageous to obligate plant parasites because it decreases their aggressiveness as a population and facilitates maintaining their hosts, upon which they depend for survival. Thus, knowledge gained from natural ecosystems indicates that specific resistance has been ephemeral in agroecosystems because genes were used "in an ecologically unnatural and, therefore, unsound way. They did all that could be expected of them under the circumstances; namely, they assured resistance to races they were released to thwart" (Browning, 1974).

One might assume that diversity per se is responsible for the stability of natural ecosystems and that agroecosystems can be adequately stabilized merely by increasing their diversity. This is not entirely true. We accept the view of May (1975) that "The instability of so many man-made agricultural monocultures is likely to stem not from their simplicity, as such, but rather from their lack of any significant history of coevolution with pests and pathogens." Thus, coevolution of the rusts and cereals resulted in a balanced, stable system. One result of this coevolution is genetic diversity, but in the words of Schmidt (1978), it is a functional diversity; haphazard or unplanned diversity will not necessarily provide stability. For example, Schmidt (1978) noted that *Endothia parasitica* (an introduced pathogen) devastated the chestnut (*Castanea dentata*) population in the hardwood forests of the eastern United States despite the extreme diversity of that ecosystem. Because suscept and pathogen had become separated, there was no selection for a functional diversity, i.e., diversity that would protect the chestnut population from *E. parasitica*. The purpose of this chapter is to discuss strategies to provide functional genetic diversity to protect the cereal crops from rust fungi.

Before discussing the use of genetic diversity in agriculture, it is necessary to make an important distinction between natural ecosystems and agroecosystems. In natural ecosystems, coevolution allows for the selection of resistance/susceptibility and avirulence/virulence loci in frequencies that enable host and parasite to coexist (Mode, 1958; Person, 1966). In agroecosystems, evolution of the *host* population is "frozen" as it is replanted each year with no opportunity for disease to influence resistance gene frequencies (Parlevliet, 1979). Thus, to provide functional genetic diversity, agriculturalists must emulate the coevolution of natural ecosystems by managing resistance genes in an ecologically sound manner based on knowledge of epidemiology and population genetics.

There are many methods of deploying resistance genes to increase diversity, but none is necessarily independent or clearly distinct from another. For the purpose of organization, we will discuss three classifications of spatial diversity (intrafield, interfield, and interregional) that are arbitrary, but convenient, points along a continuum.

II. Intrafield Diversity

A. INTRASPECIFIC MIXTURES

An alternative to the culture of pure-line cultivars is to grow mixtures of different genotypes that are phenotypically similar for important agronomic traits. Jensen (1952) called such populations multiline varieties. The use of multiline cultivars to control disease has been reviewed (Browning and Frey, 1969) and recently updated (Frey, 1982). Borlaug (1981) summarized the use of multilines in international wheat-breeding programs.

1. Terminology

There is confusion in the literature concerning the definition of a multiline cultivar. A few authors have considered a multiline to be a population of near-isogenic lines (isolines) possessing different race-specific resistance genes (Barrett and Wolfe, 1980; Browning and Frey, 1981; Wolfe *et al.*, 1981). Barrett and Wolfe (1980) and Wolfe *et al.* (1981) attributed this method of multiline development to Jensen

(1952), who is considered the originator of the multiline concept. However, Jensen did not propose the use of isolines as multiline components and, in fact, criticized the isoline approach for not allowing maximum genetic diversity within the population (Jensen and Kent, 1962). In addition, Jensen (1952) did not limit the multiline concept to pathological traits. Groenewegen and Zadoks (1979) made a distinction between multiline cultivars (mixtures of near-isogenic lines) and "mixtures of related lines" (mixtures of lines developed by restricted backcrossing to a common parent). However, breeders have produced small-grain populations via similar methods but have referred to them as multilines (see Section II,A,4,c). In this chapter we will forego use of the term multiline. Instead, we will use the more general term mixture, and attempt to indicate the composition of specific mixtures as we discuss them.

2. *Mechanisms of Disease Control in Mixed Populations*

The density of a given plant genotype is of importance to the development of epidemics (Burdon, 1978; Burdon and Chilvers, 1982). In mixed populations, the density of any one genotype is decreased. If plants within the population possess different race-specific resistance genes, the "efficiency" of any given pathogen race is reduced because distances among plants of the same genotype are increased. In addition, plants of one genotype may function as "barriers" by physically blocking spore dispersal among plants of a different genotype (Burdon, 1978; Trenbath, 1977). The combined effect of these two mechanisms on epidemic development has been studied in simple, two-component mixtures where one genotype is susceptible and the other is resistant to (or immune from) the pathogen population (Browning, 1957; Browning and Frey, 1969, 1981; Burdon and Whitbread, 1979; Chin and Wolfe, 1984; Clifford, 1968; Cournoyer, 1970; Elliot *et al.*, 1980, Fried *et al.*, 1979; Leonard, 1969a). For example, Browning and Frey (1969) found that the number of stem rust uredia on susceptible oats was reduced by 40% in a 1:1 mixture of susceptible:resistant plants as compared to a pure stand of the susceptible component.

In addition to "reduced density" and "barrier" effects, interactions among different rust races may contribute to the effectiveness of mixtures when more than one race is present. Induced resistance occurs when a resistance mechanism in the host is triggered by inoculation with an avirulent race and protects the plant from infection by races that would normally be virulent. Induced resistance has been demonstrated for rusts and other diseases under controlled conditions and for *P. striiformis* in the field (Johnson, 1978). The effect of induced resistance seemed to be localized around infection sites (Johnson, 1978), and some have felt that induced resistance will be a significant mechanism in mixtures only at higher levels of disease (Burdon and Shattock, 1980; Parlevliet, 1979). More recent results with powdery mildew, however, suggest that the effects of induced resistance may be much less localized than previously thought (Chin and Wolfe, 1984; Hwang and Heitêfuss, 1984). Competitive inhibition among virulent races may also retard epidemic development in mixed populations (Leonard, 1969a).

A few studies have been conducted to determine the relative importance of the mechanisms that contribute to disease reduction in mixed populations. In controlled-environment chambers, reducing the density of barley plants resulted in a much larger reduction in the development of powdery mildew than did the barrier effect of resistant plants (Burdon and Chilvers, 1977). By altering the composition and spatial arrangement of barley cultivar mixtures in the field, Chin and Wolfe (1984) estimated the relative contributions of reduced density, the barrier effect of incompatible plants, and induced resistance to the total reduction of powdery mildew. In the beginning of the season, most of the disease reduction was due to reducing the density of compatible genotypes. Later in the season, the barrier effect and induced resistance increased in importance.

Most recent work with mixtures has involved the use of race-specific resistance. However, research with *Septoria nodorum* on wheat and *Rhynchosporium secalis* on barley suggests that race-specificity is not required to obtain a benefit from mixing different genotypes. When cultivars with different levels of general resistance were mixed, disease severity was reduced as compared to the mean of the cultivars grown separately. With *S. nodorum*, the presence of only 25% of the more resistant cultivar in a mixture reduced the disease severity of the less resistant genotype to near that of the more resistant cultivar grown in pure stand (Jeger *et al.*, 1981b,c). Jeger *et al.* (1981a,b) also developed mathematical models to describe the increase of nonspecialized pathogens in two-component mixtures based on differences in infection frequency (number of lesions/spore) and sporulation rate between the two genotypes. The models predict that mixtures may either decrease or increase epidemic development depending on the ranking and magnitude of differences in infection frequency and sporulation rate between the two genotypes. However, Jeger *et al.* noted that the conditions under which mixtures are predicted to result in an increase in disease development will probably occur very infrequently.

3. *Epidemiological Effects of Mixtures*

Vanderplank (1963) proposed that the increase of "compound interest" diseases (those that increase through successive generations of the pathogen) can be described by the equation

$$X_t = X_0 e^{rt}$$

where X_t = the proportion of disease at time t, X_0 = the initial proportion of disease, r = the apparent infection rate, and t = time from initiation of the epidemic. This equation describes exponential growth and is valid for the early stages of epidemic development when uninfected host tissue is not limiting. When the equation is adjusted to account for previously infected tissue, it describes logistic growth. Based on this equation, Vanderplank proposed three strategies to control plant disease. One may reduce the amount of initial inoculum (thus resulting in a decrease of initial disease); one may reduce the rate of epidemic progression; and one may reduce time (for example, by planting an early-maturing cultivar to "escape" disease). Vanderplank (1963, 1968) also proposed two types of disease resistance and their epidemiological effects. Vertical (race-specific) resistance reduces X_0 by "filtering out" avirulent races of the pathogen; however, Vanderplank said that it generally does not affect the rate of increase of virulent races. The effectiveness of vertical resistance is decreased when it is used extensively because virulent races are selected from the pathogen population. Horizontal resistance reduces the apparent infection rate through its effects on the components of resistance (infection efficiency, latent period, sporulation rate, and infectious period); it was defined as being equally effective against all races, but usually does not provide a high level of resistance or immunity as is often the case with vertical resistance.

The epidemiological effects of host mixtures have been hypothesized. They should decrease X_0 as compared to a pure-line susceptible population because a given race will not be virulent on all plants in the mixture (Browning and Frey, 1969; Vanderplank, 1968). However, if one compares a mixed population with the mean of its components grown separately and exposed to the same pathogen population, there should be no reduction in X_0, as is supported by both theoretical and experimental studies (Barrett, 1980; Barrett and Wolfe, 1980; Wolfe and Barrett, 1980). In contrast to a pure-line cultivar, inoculum produced on one plant in a mixture may not be virulent on other plants in that population. Thus, the infection efficiency of the pathogen is reduced in the secondary cycles, resulting in a decrease of the apparent infection rate (Browning and Frey, 1969; Vanderplank, 1968). In simple mixtures of resistant and susceptible oats, the apparent infection rate of stem rust on the susceptible genotype decreased in proportion to the logarithm of the percentage of resistant plants in the mixtures (Leonard, 1969a). A similar logarithmic relationship has been found in other studies (Burdon and Chilvers, 1977; Chin, 1979; Elliott *et al.*, 1980; Luthra and Rao, 1979a,b).

In studies with powdery mildew (Fried *et al.*, 1979) and leaf rust (Luthra and Rao, 1979a,b) in wheat mixtures, very large reductions of X_0 and relatively small reductions of r were reported. Based on such studies, some researchers have concluded that the primary epidemiological effect of mixtures is to decrease X_0 (Frey, 1982; Fried *et al.*, 1979). The relative importance of reducing X_0 versus r by host mixtures is not merely an academic point. Vanderplank (1963) stated that, for compound interest diseases, reducing X_0 would be very effective only for epidemics in which the quantity rt in his equation is small, i.e., epidemics that increase slowly and/or are of short duration. Thus, if mixtures have little or no effect on the rate of epidemic progression, one would expect their usefulness to be limited.

There are several reasons that may explain why the effect of host mixtures on the rate of epidemic progression may be more important than the evidence cited above suggests. First, r is an exponent in Vanderplank's equation, while X_0 is not. Thus, a small reduction of r will have a much greater effect on epidemic progression than the same percentage decrease of X_0. For example, Luthra and Rao (1979a) reported an average 64% and 7% reduction of X_0 and r, respectively, for six wheat mixtures infected with leaf rust. However, they calculated that reductions of X_0 and r had nearly equivalent effects on delaying the time till 50% disease severity was attained in the mixtures as compared to pure-line populations. If epidemic development is described more accurately by the Gompertz than by the logistic growth equation as Berger (1981) reported for 113 epidemics, then the effectiveness of decreasing X_0 will be even less. Second, three studies (Gregory *et al.*, 1981; Plaut and Berger, 1981; Rouse *et al.*, 1981) involving five different suscept–pathogen systems indicate that X_0 and r may not be independent as is indicated by Vanderplank's equation. In all five pathosystems, decreasing levels of X_0 were associated with large increases of calculated apparent infection rates. It is yet unclear whether this correlation between X_0 and r is a biological phenomenon and/or an artifact of mathematical transformations used to calculate r. Nevertheless, if a mixture decreases X_0 as compared to a pure-line, susceptible population, then one may expect to calculate a relatively larger r for the mixture than if the two epidemics had been initiated with equal amounts of disease. Third, if a significant proportion of inoculum entering a mixture plot is from an independent source (e.g., from an adjacent, infected field or from "spreader plants" placed in or around the plot), the rate-reducing effect of the mixture may be masked, especially if the epidemic is of short duration (Barrett, 1981). Finally, with mixtures of susceptible and resistant genotypes, disease severity will be reduced in direct proportion to the percentage of resistant plants in the population if the mixture does not decrease epidemic development on the susceptible genotype. To determine if the presence of resistant plants in a mixture decreases epidemic development on the susceptible genotype, one must adjust disease severity ratings to account for the percentage of resistant plants in the mixture. Chin (1979) suggested that the failure to make this adjustment may have led Fried (1978) to erroneously conclude that reductions of the apparent infection rate do not contribute significantly to the effectiveness of multilines.

In extensive field studies, three-component barley cultivar mixtures reduced the severity of powdery mildew by 50% or more as compared to the mean of the components grown separately (Wolfe, 1978; Wolfe and Barrett, 1980). Because a mixture does not reduce X_0 as compared to the mean of its components grown separately and exposed to the same pathogen population, we attribute the effectiveness of these mixtures to a reduction in the rate of epidemic progression.

4. *Composition and Development of Intraspecific Mixtures*

a. The "Clean-Crop" versus the "Dirty Crop" Approach. A dichotomy that has been discussed relative to mixtures is the "clean-crop" versus the "dirty-crop" approach (Marshall, 1977). With the clean-crop approach, typified by Borlaug's (1959) program, each component line is resistant to all prevalent races and a line is replaced when a virulent race appears. It is expected that only one or a small number of components will become susceptible at any one time. Because the susceptible lines will be at a reduced density, rust will develop more slowly and less yield loss will occur than if the susceptible lines were grown in a pure stand (Borlaug, 1959). This hypothesis was confirmed with a stem and stripe rust resistant mixture in Colombia which was composited according to the clean-crop approach (Rockefeller Foundation, 1965). The CIMMYT (Centro Internacional para Mejoramiento de Maiz y Trigo) program still follows the clean-crop approach to mixture composition (Rajaram *et al.*, 1979; Dubin and Rajaram, 1982). The clean-crop approach has been criticized as being impractical to maintain (Frey, 1982) and unnecessarily wasteful of resistance genes (Frey, 1982; Frey *et al.*, 1977; Marshall, 1977). With the dirty-crop approach to mixtures, typified by the Iowa program (Frey *et al.*, 1973, 1977), each component line is resistant to some, but not all of the pathogenic races. Although rust may develop on many or all component lines in a given year, the rate of epidemic development is decreased because the density of any single host genotype is reduced (see Section II,A,2). Thus, the multiline provides synthetic horizontal resistance (Browning and Frey, 1969). The dirty-crop approach to mixture composition mimics the natural ecosystem because it allows for the coexistence of host and parasite and it incorporates the philosophy of integrated pest management that a certain level of disease can be tolerated without resulting in significant yield loss.

The clean-crop versus dirty-crop classification signifies a difference in philosophy. In practice, however, most mixtures may not fit perfectly into either category. For example, Borlaug (in a discussion addended to Borlaug, 1959) indicated that he would consider incorporating lines with resistance to leaf rust even though they may be susceptible to stem rust. On the other hand, with the Iowa oat isoline mixtures there may be one one or more lines included that are resistant to all prevalent crown rust races; the concern has been that the genes used be different and functional against the prevalent crown rust races, not that the mixtures be "clean" or "dirty" per se.

b. Percentage Resistance and Number of Components. The percentage of resistance required to protect a mixed population against rust is not well known. Evidence from indigenous ecosystems indicates that when only 30% of the population is resistant to a given crown rust race and when this specific resistance is backstopped by general resistance, the population will be sufficiently protected (Browning, 1974). This 30% figure is supported by studies conducted with oat crown rust in agricultural systems (Browning, 1980). Jensen and Kent (1963) suggested that 40% resistance is adequate to protect a population. A level of resistance that is adequate to protect a population in one environment may be inadequate in an environment more conducive for disease development (Browning and Frey, 1981; Frey,

1982). Thus, agriculturalists may choose to err on the conservative side. For example, in the Iowa program, oat isoline mixtures are composited in an attempt to get at least 60% of the plants resistant to each of the prevalent crown rust races (Frey *et al.*, 1973, 1977).

Less is known about the number of lines that are desirable in a mixture than is known about the percentage of resistance required to protect a host population. Several studies have shown that two-component mixtures can provide a considerable reduction of rust development (Browning, 1957; Browning and Frey, 1969, 1981; Suneson, 1960; Clifford, 1968), and three-component barley cultivar mixtures provide adequate protection from powdery mildew (Wolfe and Barrett, 1980). Vanderplank (1968) warned against expounding the belief that "if a little heterogeneity is good, more must be better" because breeders might be deterred by the prospect of developing a very complex and heterogeneous mixture, and it has been proposed that the gain in disease control attained by adding components to a mixture will follow the law of diminishing returns (MacKenzie, 1979). Nevertheless, there may be advantages to incorporating a large number of components in a mixture. Browning and Frey (1981) reported that an oat isoline mixture composed of several lines provided adequate protection against a severe crown rust epidemic, while a two-component mixture with approximately the same percentage of resistance did not; in a lesser epidemic, the two-way mixture was adequate. Increasing the number of effective resistance genes in a mixture will decrease the probability of any single rust race being virulent on a large percentage of the host population and may slow the evolution of complex races that are virulent on more than one component of the mixture (Groth, 1976; Groth and Person, 1977; Parlevliet, 1979). Borlaug (1959) found that minor agronomic differences among mixture components were masked when 12–16 lines were used as compared to four to eight lines. Practical considerations such as the number of genes available and the resources of the breeding program will probably set the upper limit to the number of components used (Frey, 1982; Rajaram *et al.*, 1979).

The efficacy of a mixture might be increased by utilizing components that completely suppress reproduction of races possessing the corresponding gene(s) for avirulence (Parlevliet, 1979; Wolfe, 1978). However, this practice may severely limit the number of genes available for use in mixtures. In addition, the use of resistance genes that do not completely suppress reproduction of "avirulent" races may contribute to the stabilization of the pathogen population by allowing simple races (those virulent on one or a few components of the mixture) to predominate (Marshall and Burdon, 1981b). Many of the genes used in the Iowa oat isoline mixtures do not completely suppress reproduction of "avirulent" crown rust races (Browning and Frey, 1981; Frey, 1982; Frey *et al.*, 1973, 1977).

c. Development of Mixture Components. Multiline varieties as Jensen originally proposed would consist of mixtures of lines phenotypically similar for important agronomic characteristics but as genetically diverse as possible for all other traits, including disease resistance (Jensen, 1952; Jensen and Kent, 1963). This approach is similar to what Barrett and Wolfe (1980) referred to as line mixtures (mixtures of lines developed from different parents but selected toward a common phenotype). Borlaug (1959) attained agronomic uniformity among mixture components by backcrossing stem rust resistance genes from donor parents to a recurrent parent of desirable agronomic type. By crossing and selecting plants in the field, he found that two or three backcrosses were required to obtain lines of sufficient phenotypic similarity. In the Iowa program, five backcrosses of greenhouse-grown plants were used to develop near-isogenic oat lines because of the high standards of uniformity in the United States, because many of the donor parents were wild oats (*A. sterilus*) or unadapted oat lines (Browning and Frey, 1981), and because these populations were also being used as a tool to study the effectiveness of multilines without the "confounding" effects of differences in genetic background among the components. However, the most recent Iowa oat mixture has been developed with just two backcross generations because adequate crown rust resistance genes are now present in agronomically adapted genotypes. In recent years, CIMMYT has developed mixture components via a double-cross system of the type (A × B) × (A × C), where A represents Siette Cerros (the desired agronomic type) and B and C represent donor parents chosen for their resistance to rusts and Septoria diseases (Rajaram and Dubin, 1977). Mutagenesis can be used to develop isolines differing in their spectrum of race-specific resistance (Pal, 1979). Mixtures of pure-line barley cultivars possessing different race-specific resistance genes have been used to control powdery mildew (White, 1982; Wolfe, 1978; Wolfe and Barrett, 1980; Wolfe *et al.*, 1981).

d. Heterogeneity among Mixture Components. Intraspecific mixtures will, by definition, be more heterogeneous than pure-line cultivars. However, cultivar uniformity has probably been ingrained too deeply in the minds of agriculturalists (Groenewegen and Zadoks, 1979; Harlan, 1972). In practice, a crop population needs to be reasonably uniform for only a few traits such as height and maturity date; it can be diverse for other characteristics (Browning, 1974; Jensen and Kent, 1963). Borlaug (1959) noted that many of his mixtures of back-crossed lines were so phenotypically similar to the recurrent parent that farmers would be unable to distinguish them from conventional cultivars. Wolfe and Barrett (1980) have apparently been able to choose and composite different pure-line barley cultivars that will form agronomically acceptable mixtures, even for malting.

There are advantages to maintaining or developing a larger degree of genetic diversity among the components of a mixed population than is attained with the isoline approach. First, with isoline mixtures, agronomically superior genotypes may become available while the isolines are being developed, thus making the mixture outclassed agronomically (Wolfe, 1978; Wolfe and Barrett, 1980; Groenewegen and Zadoks, 1979). For example, the availability of agronomically superior cultivars may have been an important factor contributing to the decline in popularity of oat isoline mixtures in Iowa (Browning and Frey, 1981) and may outclass a wheat isoline mixture developed in the Netherlands before it is released (Groenewegen and Zadoks, 1979). When less time and effort are invested in obtaining genetically uniform lines, agronomically improved genotypes can be incorporated into mixtures more rapidly (Groenewegen and Zadoks, 1979; Wolfe, 1978; Wolfe and Barrett, 1980). Second, there are yield synergisms associated with the mixing of diverse genotypes in addition to those attributable to disease control (Borlaug, 1959; Browning and Frey, 1969; Jensen, 1952, 1965; Simmonds, 1962; Wolfe and Barrett, 1980). For example, the average yield of three-component barley cultivar mixtures was 3% greater than the mean of the components grown in pure stand even when the "target disease" was of little consequence (Wolfe and Barrett, 1980). Yield synergisms associated with mixing diverse genotypes can be increased by choosing genotypes for their specific mixing ability (Jensen, 1965). Third, genetically diverse mixtures may cause disruptive selection within the population of a race virulent on more than one component of the mixture, thereby reducing selection for complex races (Leonard, 1969b; Wolfe and Barrett, 1980) (see Section VI,B). Fourth, differences in genetic background among mixture components may provide buffering against the "target" disease in addition to that provided by the identified race-specific resistance genes (Wolfe, 1978; Wolfe *et al.*, 1981). For example, Wolfe *et al.* (1981) compared barley cultivar mixtures in which the components contained either the same or different identified, race-specific genes for resistance to powdery mildew. Using this information, they estimated that differences in genetic background among the components accounted for about 25% of the total mixture effect. They hypothesized that the disease-reducing effectiveness of genetic background" may have been due to differences in resistance of the components to the same pathogen race, or to each cultivar selecting its 'own' sub-race from the pathogen population." Fifth, differences in genetic background among the components of a mixture may provide protection from "nontarget" diseases and abiotic stresses (Borlaug, 1959; Browning and Frey, 1969; Groenewegen and Zadoks, 1979; Wolfe and Barrett, 1980). For example, barley cultivar mixtures composited to control powdery mildew often reduce the severity of stripe rust, leaf rust (incited by *P. hordei*), and *Rhynchosporium*-induced disease (Wolfe and Barrett, 1980). Chin and Wolfe (1984) reported that, although reaction to stripe rust was not considered in the selection of components for barley cultivar mixtures, one of the components was (fortuitously) resistant to a prevalent stripe rust race and provided protection to the other components in the mixtures. On the other hand, rust infection was increased slightly (but not significantly at $p = 0.05$) by mixing two wheat cultivars that possess different levels of partial resistance to stripe rust (Groenewegen and Zadoks, 1979).

Cultivar mixtures have been proposed as a method to avoid disadvantages associated with the isoline approach to mixture composition (Wolfe, 1978; Wolfe and Barrett, 1980). As a compromise between cultivar and isoline mixtures, Groenewegen and Zadoks (1979) proposed the culture of mixtures of related lines that would be composed of progeny from the first and second backcross generations of diverse donor parents to a single recurrent parent. The use of double or complex crosses to develop mixture components will provide similar advantages (Gill *et al.*, 1980). The approach that Groenewegen and Zadoks (1979) suggested is similar to that used in India (Gautam *et al.*, 1979; Gill *et al.*, 1980) and in the CIMMYT program (Rajaram and Dubin, 1977). It may be possible to reap the benefits of transgressional gains by developing related lines (Gautam *et al.*, 1979; Groenewegen and Zadoks, 1979). For example, the components of multiline KSML 3 (a mixture of related lines) were found to possess improved tillering ability and larger kernel size as compared to the recurrent parent (Gill *et al.*, 1980).

e. Number of Target Pathogens. Multiline cultivars often are visualized as being effective against only one pathogen. However, several mixtures have been developed that possess resistance genes effective against more than one. Miramar 63, a wheat mixture released by the

Rockefeller Foundation in Colombia, incorporates resistance to both stem and stripe rust (Rockefeller Foundation, 1963). In the CIMMYT mixture program, donor parents are chosen for their resistance to stem, stripe, and leaf rusts as well as Septoria diseases (Rajaram and Dubin, 1977). Suneson (1960) described the development of a barley population that incorporates resistance to the four major diseases of barley in California.

f. Mixtures in Practice. In 1950, the Rockefeller Foundation Agricultural Program in Mexico (later to be incorporated into CIMMYT) began a wheat mixture project under the leadership of Dr. Borlaug. Two mixtures of backcrossed lines were ready for release in the early 1960's but were withheld (Borlaug, 1981). Failure to release these mixtures was not due to inadequacy of the mixtures or to a lack of interest in them by the Rockefeller Foundation. Rather, the development of high-yielding, photoperiod-insensitive, semidwarf wheats made the mixtures agronomically obsolete (Borlaug, 1981; Rajaram and Dubin, 1977). In the meantime, the first multilines to be used commercially were developed in the Rockefeller Foundation Agricultural Program in Colombia. Miramar 63 is a 10-component mixture developed by restricted backcrossing and incorporates resistance to both stripe and stem rusts (Rockefeller Foundation, 1963). Released in 1963, this mixture was reported to have been widely accepted by Colombian farmers and to have more than doubled yields in some areas (Rockefeller Foundation, 1964). In keeping with the clean-crop philosophy, Miramar 63 was recomposited to form Miramar 65 when a stem rust race virulent on two of the components appeared (Rockefeller Foundation, 1965).

The high-yielding semidwarf wheats were rapidly accepted by farmers in some Third World countries. At one point, semidwarf cultivars derived from CIMMYT cross 8156 accounted for ~75% of the 15 million hectares planted to CIMMYT wheat cultivars on the Asian subcontinent (Rajaram and Dubin, 1977). In some cases, however, virulent rust races have been selected within several years after the release of these cultivars (Borlaug, 1981; Gautam *et al.*, 1979; Gill *et al.*, 1979). These agronomic phenotypes can continue to provide high yields if their rust resistance can be maintained, however (Gill *et al.*, 1979). Thus, in 1970, CIMMYT began to develop mixture components based on the 8156 genotype. These components are made available to wheat breeding programs around the world (Borlaug, 1981; Rajaram and Dubin, 1977). In yield tests, mixtures composed of these lines have maintained the yield potential of 8156 while incorporating genetic diversity for disease resistance (Rajaram and Dubin, 1977; Rajaram *et al.*, 1979). Mixtures are being developed and researched at three plant breeding stations in India (Borlaug, 1981). KSML 3 (a wheat mixture based on the cultivar Kalyonsona and resistant to both leaf and stripe rust) was released by the University of Punjab, India, in 1978 (Gill *et al.*, 1979; Borlaug, 1981). Mixtures from two other Indian plant breeding stations were scheduled to be released in 1979 (Borlaug, 1981).

A stripe-rust-resistant isoline mixture has been developed in the Netherlands (Groenewegen, 1977; Groenewegen and Zadoks, 1979). Crew, a semidwarf winter wheat mixture, was released by the Idaho and Washington Experiment Stations in 1982. Crew is composed of 10 closely related lines that incorporate at least nine different race-specific genes for resistance to stripe rust and also provide some diversity for reaction to leaf rust and powdery mildew (Allan *et al.*, 1983a,b). In comparisons of Bayleton-treated versus nontreated plots, the maximum yield loss for Crew was 7% as compared to 9–27% for its components, and in extensive yield trials, the mixture has exceeded or equalled the yield of all but one commercially grown club wheat cultivar (Allan *et al.*, 1983).

The mixture program of longest continuous duration is that of the Iowa Agricultural Experiment Station. This program has developed two series of isolines with resistance to crown rust, one based on an early-maturing recurrent parent and the other based on a midseason-maturing parent (Browning and Frey, 1969; Frey *et al.*, 1973, 1977). From these components, eight early and five midseason mixtures have been composited and released since 1968. In the first several years after their release, the mixtures were grown on 0.4 million hectares, comprising greater than 50% of the entire Iowa oat area. No crown rust damage was ever reported or observed in the mixtures, despite occasional yield losses of up to 30% for susceptible cultivars in severe epidemic years. By 1979, however, the mixtures were grown on only 10% of the total Iowa oat area. Iowa workers postulated that the decline in popularity of the mixtures may be due to a decline in crown rust pressure and also to the release of agronomically superior oat cultivars that possess increased levels of general resistance and tolerance to crown rust and resistance to barley yellow dwarf virus (Browning and Frey, 1981). A new Iowa oat mixture that was developed with a high-yielding cultivar, resistant to barley yellow dwarf, as the recurrent parent was released in the winter of 1983/84 (Frey *et al.*, 1985). In 1982, the new mixture yielded 25% higher than its recurrent parent at locations where crown rust was severe (K. J. Frey, personal communication, Iowa State University, Ames, Iowa).

B. COMPOSITE CROSSES

A second method of obtaining intraspecific genetic diversity is the use of composite-cross populations. Composite crosses are developed by crossing a large number of diverse genotypes, bulking the F_1 progeny, and subjecting the bulked progeny to prolonged natural selection in environments in which the crop is to be grown. Natural selection should increase the frequency of alleles for traits such as high yield and resistance to naturally occurring diseases (Suneson, 1956). For example, barley composite-cross populations (especially the later generations) are less affected by *Rhynchosporium secalis* (incitant of the barley scald disease) than are pure-line cultivars. This protection may be due to selection for general resistance within the populations, in addition to diversity of race-specific genes (Jackson *et al.*, 1978). Suneson (1956) presented data indicating that a yellow dwarf epidemic resulted in an increased frequency of yellow dwarf resistance genes in subsequent generations of a barley composite population. We are unaware of studies concerning the effects of composite crosses on rust pathogens, although we expect that such populations would be effective in controlling rusts.

C. INTERSPECIFIC MIXTURES

Mixtures of different crop species should reduce disease development by mechanisms very similar to those operating in intraspecific mixtures. Mixtures of different small grain species have been shown to decrease the severity of spot blotch (incited by *Cochliobolus sativus*) (Clark, 1980), powdery mildew (Burdon and Whitbread, 1979), and rust (Atkinson, 1900). Interspecific mixtures can contribute significantly to rust control were mixed crops are commonly grown. For example, in 1977, 330,000 hectares of mixed grains were sown in the Province of Ontario, Canada. This represents about twice the area sown to either oats or barley alone (Ontario Ministry of Agriculture and Food, 1977). In addition, the use of interspecific mixtures to control the cereal rusts has special appeal due to renewed interest in multiple cropping systems.

III. Interfield Diversity

Interfield diversity is attained by deploying resistance genes in different fields on a farm or larger area. This approach will be more successful if genes are deployed in a planned and controlled manner and if the pathogen's virulence pattern is monitored (Parlevliet, 1981a). Interfield diversity is presently being conducted, in a planned manner, in the United Kingdom (Doodson, 1982; Priestly, 1981). Wheat cultivars that possess similar genes for race-specific resistance to stripe rust are placed within the same diversity group. The grouping of cultivars is updated annually based on virulence surveys. To achieve effective diversity, farmers deploy cultivars so that crops in neighboring fields come from different diversity groups. In 1979, 56% of farmers surveyed indicated that they were using the stripe rust diversification scheme (Priestly, 1981). The same procedure is conducted with barley cultivars based on their race-specific resistances to powdery mildew.

Intuitively, one might expect that interfield diversity would be less effective than intrafield diversity because genotypes are more aggregated and there is less opportunity for inoculum exchange among host genotypes. The effect of host aggregation on the disease-reducing effectiveness of mixed populations was studied by using oat crown rust as a model (Mundt and Browning, 1985). The proportion of genotypes was the same in all mixture treatments, but the planting arrangement was altered to obtain different-sized genotype units (areas occupied by the same genotype). Increasing the size of genotype units from 0.003 to 0.84 m^2 did not reduce the efficacy of the mixtures as compared to a pure-line susceptible population. Subsequent studies with oat crown rust, however, suggest a strong effect of host aggregation on the efficacy of mixtures if initial inoculum is distributed uniformly rather than in a single focus, as was the case in Mundt and Browning's (1985) study (C. C. Mundt and K. J. Leonard, unpublished data). Zadoks and Kampmeijer (1977) used a computer simulator (Kampmeijer and Zadoks, 1977) to evaluate the effectiveness of small versus large fields for the control of airborne plant pathogens, which was discussed earlier by Vanderplank (1949) and Waggoner (1962). They concluded that the relative effectiveness of small versus large fields would depend on several factors, including (1) the frequency, size, and arrangement of susceptible fields; (2) the number, position, and size of inoculum sources; (3) the steepness of the spore dispersal gradient; and (4) wind characteristics. The effects of interfield diversity on the increase and spread of the cereal rusts may depend on similar factors.

It is important to note that, due to the continental movement of cereal rust uredospores (see Section IV,A), the effectiveness of interfield diversity should also be considered in the context of rust increase and spread over large areas and not just from the standpoint of one or a small number of farms.

IV. Regional Deployment of Resistance Genes

A. THE CONTINENTAL NATURE OF THE CEREAL RUSTS

"A basic dichotomy in the epidemiology of plant diseases is whether the pathogen is *residual* or *continental*" (Browning *et al.*, 1969). A residual pathogen is one that perpetuates itself within a relatively small area. A continental pathogen, on the other hand, originates outside the area and is disseminated over very large hectarages. The small grain rusts can be either residual or continental (Browning *et al.*, 1969). However, in several geographic regions of the world, e.g., North America (Browning *et al.*, 1969; Stakman and Harrar, 1957), India (Mehta, 1933), and Europe (Ogilvie and Thorpe, 1961; Zadoks, 1961), the continental nature of the rusts is well known. In these areas, the uredial stage of rust pathogens overseasons in a limited geographical region and spreads by prevailing winds along a known path. In some cases, the occurrence of rust epidemics has become increasingly dependent on continental spore movements due to eradication of alternate hosts (Browning *et al.*, 1969; Stakman and Harrar, 1957; Ogilvie and Thorpe, 1961).

B. REGIONAL DEPLOYMENT SCHEMES

The continental nature of the cereal rusts has led several plant scientists to propose regional deployment of resistance genes to control them. It has been hypothesized that race-specific resistance could adequately protect spring wheat in the Great Plains of the United States and Canada if the same gene was not used in the South (Vanderplank, 1968). Browning *et al.* (1969) suggested breaking the unity of the "*Puccinia* path" for oat crown rust by deploying different race-specific resistance genes in three different geographical zones of North America. These three zones correspond to differences in climate and crop ontogeny that allow for the uninterrupted movement of rust when a homogeneous crop is grown. Similar schemes have been proposed for wheat stem rust in North America (Knott, 1972) and wheat leaf rust in India (Reddy and Rao, 1979). Earlier suggestions for the use of interregional diversity were reviewed by Browning *et al.* (1969). The basis for such schemes is that inoculum produced in one geographical zone would be avirulent on the crop in a different zone. The success of regional deployment depends on the amount of inoculum that is disseminated among regions, the importance of this inoculum to the development of epidemics, and the decline in frequency of virulence genes in regions where the corresponding gene for resistance is not used.

Regional gene deployment implies that the use of resistance genes will be controlled by mutual consent among plant scientists or by some governing body. The only epidemiological system in which a regional deployment scheme has been attempted is for oat crown rust in North America. Two experiment stations in the United States (Iowa and Texas) and one in Canada (Manitoba) entered into an agreement to deploy crown rust resistance genes. To initiate the program, 10 resistance genes were assigned among the southern, central, and northern regions such that no known crown rust race would possess corresponding virulence in more than one of the regions (Frey *et al.*, 1973, 1977). Because of a decrease in crown rust "pressure" (possibly caused by a precipitous decline in oat area during the 1960s and 1970s), the oat workers decided against investing the administrative effort required to actually carry out the deployment program. This illustrates the necessity of considering the entire cropping system when designing a rust management program.

There are cases where regional gene deployment has been "tested" unintentionally. For example, in the 1940s, oat cultivars possessing the gene *Pg-4* for resistance to stem rust (*Puccinia graminis* f. sp. *avenae*) were grown extensively in both the United States and Canada. Because of the susceptibility of these genotypes to *Helminthosporium victoriae*, breeders in the United States switched to the use of Bond oat derivatives that are resistant to *H. victoriae* and possess gene *Pg-1* for resistance to oat stem rust. Inoculum produced on these new cultivars in the United States was avirulent on the Canadian cultivars. Thus, the Canadian oat crop was protected from stem rust (Browning *et al.*, 1969; Johnson, 1958). Browning *et al.* (1969) discussed other examples of unintentional testing of regional gene deployment.

V. Temporal Diversity

Genetic diversity can be attained by deploying different resistance genes at different times. Such diversity will impose disruptive selection on the pathogen population by forcing it to cycle between crops with different resistance genes (Wolfe, 1973). Cultivars with different resistance genes can be deployed in different years. With crops such as small grains that are planted in both fall and spring, different resistance genes can be used in winter and spring cultivars (Vanderplank, 1968; Wolfe, 1973). Temporal diversity can also be achieved by altering or rotating the components of a mixture over time.

VI. Effects of Host Diversity on the Population Genetics of the Cereal Rusts

The successful use of host diversity to manage the cereal rusts depends on the maintenance of diversity within the pathogen population. There is concern among plant scientists that the use of host diversity will result in selection of complex races of the pathogen that will be virulent on many or all of the components of the diverse population. Thus, it is feared that rust control will be eroded and that many resistance genes will become ineffective, possibly simultaneously, for use in agriculture.

A. STABILIZING SELECTION AND HOST MIXTURES

Vanderplank (1963, 1968) proposed that a cost in pathogen fitness is associated with virulence genes. He suggested that stabilizing selection would decrease the frequency of unnecessary virulence genes (virulence genes not needed to incite disease on a given host genotype) in the pathogen population. Vanderplank claimed that the strength of this selection would depend on the magnitude of the cost associated with the unnecessary virulence gene(s). Vanderplank's theory of stabilizing selection has had a large impact on plant pathology and has been debated frequently in the literature (Crill, 1977; Leonard, 1977; Leonard and Czochor, 1980; Nelson, 1972, 1973; Parlevliet, 1981b; Wolfe, 1973). We will not attempt to evaluate this controversy other than to mention that there is not yet a clear consensus regarding the validity of Vanderplank's hypothesis. It has been claimed that Vanderplank used the term "stabilizing selection" differently than have population geneticists (Browning and Frey, 1981; Crill, 1977; Leonard and Czochor, 1980). Stabilizing selection in this chapter refers to stabilizing selection sensu Vanderplank (1963, 1968). In addition, we recognize the distinction between Leonard's (1977) cost of virulence [the mechanism which Vanderplank (1963, 1968) hypothesized to be responsible for reduced fitness of virulent genotypes] and stabilizing selection (the selective force hypothesized to cause a decline of virulent genotypes in absence of corresponding resistance).

Frey *et al.* (1973, 1977) proposed that intraspecific mixtures should stabilize the virulence pattern of the pathogen population. Because components of a mixture usually differ by only one effective resistance gene, they hypothesized that stabilizing selection would allow both simple and complex races to be maintained. Many mathematical models that address this hypothesis have been published. In general, these models quantify the relationship between two opposing processes. First, there is an increase in fitness associated with increasing virulence because a race with several virulence genes is able to reproduce on several components of the mixture. Second, since the models assume a cost of virulence, there will be a countering reduction in fitness associated with the accumulation of virulence. Thus, a complex race can reproduce on a greater percentage of the host population than a simple race but its fitness is less than that of a simple race on any one mixture component. The predictions of these mathematical models vary; however, some indicate that it will be difficult to prevent the dominance of a complex race if mixtures are used. Space limitations preclude an adequate description of these models. We refer interested readers to a recent review by Leonard and Czochor (1980).

When studying mathematical models, one must make a clear distinction between the models and reality. Because of the complexity of biological systems, models describing them must often begin simply and increase in complexity as our knowledge of the system and of modeling increases. Simplified models allow scientists to identify components of the system that need to be studied more closely and to identify new variables that need to be included in the model. However, one should not assume that a model will predict reality very closely, especially in the early stages of development. For example, results of a very simple model indicate that mixtures developed via the dirty-crop approach "will provide stable disease control only in limited and relatively rare circumstances" (Marshall and Pryor, 1978). The authors assumed that the cost of virulence would be the only factor contributing to a decreased fitness of complex races, that the magnitude of the cost of virulence would be constant with respect to time and environment, that the composition of the multiline would be static, and that the mixture would be grown over a large area, thus being "the major factor influencing virulence in the local pathogen population." As Marshall and Pryor (1978) recognized, these assumptions are unlikely to be fulfilled in the practical use of mixtures. Thus, the model might be viewed as a "worst case" model and possibly a good one from which to add other variables, as Marshall and co-workers have done in subsequent publications (Marshall and Pryor, 1979; Marshall and Burdon, 1981a,b).

B. OTHER FACTORS INFLUENCING RUST GENETICS IN DIVERSE HOST POPULATIONS

Mechanisms other than the cost of virulence may contribute to the maintenance of genetically diverse rust populations. Heterozygote superiority is considered the dominant mechanism by which balanced polymorphisms occur in natural populations (Wilson and Bossert, 1971). Isolates of *P. graminis* from North America have been shown to be very heterozygous for virulence (Johnson and Newton, 1940; Newton *et al.*, 1930), and there is evidence that *P. graminis* genotypes that are heterozygous for virulence may be more fit than genotypes homozygous for avirulence (Leonard, 1977). Genetic polymorphisms may also be maintained if different alleles are selected in different environments in which an organism survives (Falconer, 1960). With cereal rusts, a complex race may be selected in a diverse population, but not in other fields within an epidemiological unit or during the survival or sexual stages. Selection for complex virulence may be less strong if the pathogen population becomes adapted to the host genotype on which it reproduces. Under such conditions, complex races may be less fit than simple races on any one mixture component because it can attack several different host genotypes on which it is virulent (Leonard, 1969b; Wolfe and Barrett, 1980). A significant degree of adaptation to different cultivars has been demonstrated for stem rust of oats (Leonard, 1969b), and for leaf rust (Clifford and Clothier, 1974) and powdery mildew (Chin, 1979) of barley. Barrett and Wolfe (1980) and Wolfe and Barrett (1980) discussed evidence indicating that host adaptation may prevent the dominance of complex powdery mildew races in barley cultivar mixtures. However, a complex race may eventually acquire the necessary genes to be fit on all components of a mixture if the same host population is grown for extended periods of time (Wolfe and Barrett, 1980).

Leonard and Czochor (1980) concluded that, if selection against unnecessary virulence is "soft" (i.e., the intensity of selection depends on competitive interactions among pathogen genotypes), the use of host mixtures may result in stable equilibria among simple and complex races. They noted that, in a study conducted with mixtures of wheat stem rust races (Ogle *et al.*, 1973), "soft" selection seems to have played a role in the selection of rust genotypes. Nearly all mathematical models of host mixtures assume that selection is "hard" (i.e., the intensity of selection is constant and independent of competitive interactions) (Leonard and Czochor, 1980).

The effect of gene flow (migration) on rust populations has been given little consideration, but is likely to be very important (Watson, 1979), especially where the continental nature of the rusts is well expressed. For example, the effect of gene flow makes the present mathematical models irrelevant to the Iowa isoline mixtures. Iowa is part of a vast, unified epidemiological unit extending from Northern Mexico to Canada (Browning *et al.*, 1969). The contribution of inoculum produced in the Iowa mixtures to the entire North American crown rust population is very small (Browning and Frey, 1981), especially since little rust develops in the mixtures. Of course, the effect of gene flow will become increasingly less important as a given mixture is grown more extensively.

The development of an aggressive and virulent pathogenic race depends on the accumulation of the proper fitness genes in addition to the necessary genes for virulence (Groth, Vol. 1; Parlevliet, 1981b). Because of the polygenic nature of fitness, the accumulation of fitness genes may slow the evolution of a complex rust race significantly. Frey (1982) hypothesized that the time required for a pathogen to accumulate the necessary virulence and aggressiveness genes to become a dominant superrace is likely to be longer than any one mixture would be grown.

Host resistance can be managed to prevent or slow the evolution of complex rust races. Because genetic shifts toward complexity in the pathogen population are likely to be gradual, there will be ample time to change the components of mixed populations to counter these shifts (Parlevliet, 1979). In practice, this task is easy to accomplish. For example, the Iowa E-Series isoline mixture has been recomposited eight times since 1968 (Browning and Frey, 1981). Other methods for avoiding selection of pathogen complexity include using gene combinations rather than single resistance genes in mixture components (Marshall and Pryor, 1979) and ensuring that no single host mixture is grown extensively (Wolfe and Barrett, 1980). Although a cost may not always be associated with virulence, particular virulence genes and combinations of virulence genes are known to be associated with reduced fitness (Wolfe, 1973; Wolfe and Barrett, 1977). It may be possible to increase the probability of selection against complex races, at least temporarily, by utilizing the corresponding resistance genes in diversification programs (Wolfe and Barrett, 1977). Reddy and Rao (1979) suggested increasing the stability of a proposed regional deployment scheme for wheat leaf rust by deploying resistance genes in multigene cultivars and by rotating the genes among three geographical regions.

C. Empirical Evidence

In natural ecosystems, there are mechanisms, yet unidentified, that result in genetic homeostasis and maintain diversity in both the host and parasite populations (Browning, 1974). For example, in the indigenous *A. sterilis* populations of Israel race-group 276-264 of *P. coronata* possesses virulence effective against most sources of resistance in hexaploid oats, yet it neither causes epidemics nor dominates the crown rust population. In fact, in plots of *A. sterilis* exposed to the indigenous crown rust population, 37 different races were identified; only 25% of these belonged to race-group 276-264 (Segal *et al.*, 1982). Genetic changes in natural ecosystems seem to progress more slowly relative to what has been experienced in pure-line populations in agroecosystems. In summarizing research conducted with *P. coronata* in the natural ecosystems of Israel, Anikster and Wahl (1979) concluded that "despite the continuous production of new races, the composition of race populations has remained rather stable over two decades of race surveys." Will similar self-regulatory mechanisms also operate in diverse populations used in agricultural systems? This question will be answered only with extensive use of diverse populations in agroecosystems. However, it is significant that, in 4 years of experiments at the same location, a powdery mildew race virulent on all components of three-way barley mixtures has neither dominated the population nor eroded the effectiveness of the mixture (Wolfe and Barrett, 1980). Because *E. graminis* is haploid, mutants will be expressed immediately, and recombinations via the sexual cycle are concommitant with each disease season. In spite of this, it has been observed that racial changes of powdery mildew seemed slower in barley mixtures as compared to pure-line populations and that selection seemed to favor races virulent on more than one, *but not all* components of the mixtures (Wolfe, 1978). In a more artificial test, Munk (1983) found evidence suggesting a strong selection for complex powdery mildew races in a seven-component barley cultivar mixture. It is unclear, however, what effect this selection would have on the efficacy of the mixture. Cournoyer (1967) found that inoculation of an oat multiline with a crown rust race virulent on all six components did not decrease the efficacy of the multiline as compared to a pureline susceptible population.

Clearly, plant pathologists have underestimated the complexity of genetic interactions in diverse populations. We have suggested some additional factors that should be considered. However, we, and most other plant pathologists, lack the genetic training necessary to treat this subject adequately. Some of the factors we discussed may be unimportant; there certainly are others that we have not considered.

VII. Concluding Remarks

In our introduction, we mentioned Parlevliet's (1981a) emphasis that, in natural ecosystems, race-specific resistance protects the host by rendering the pathogen *population* less aggressive. Epidemiologically, "there is no difference between the effects of increased horizontal resistance and those of reduced aggressiveness" (Vanderplank, 1968). Browning (1980) considered this point more fully and concluded that "nature uses primarily a single type of epidemiological resistance—dilatory resistance—to protect populations, but it uses many different genetic systems and population structures to achieve it." Similarly, there are many strategies available to plant breeders and pathologists that will provide equivalent epidemiological effects. These strategies include pure-line general or horizontal resistance and also race-specific resistance *when deployed as part of a diverse population.* The "best" strategy will depend on many interacting factors that may be unique to each epidemiological unit and breeding program. Nevertheless, plant breeders and pathologists should recognize the advantages of diversification in addition to that of rust control. These include yield synergisms and buffering against secondary pests and abiotic stresses (see Section II,A,4,d), yield stabilization (Pfahler and Linskens, 1979; Shorter and Frey, 1979; Wolfe and Barrett, 1980), and protection from genetic vulnerability to future risks. Although general resistance is certainly desirable, resistance that appears to be general has sometimes been found to be race-specific (Browning *et al.*, 1977; Johnson, 1979). Therefore, it may be wise to derive general resistance from several diverse sources rather than to depend solely on a single genotype.

Cereal populations with adequate levels of durable resistance (Johnson, 1981) are most likely to be developed by combining different tactics, as is the case in natural ecosystems. The epidemiological advantage of combining practices that reduce the apparent infection rate has been well demonstrated for potato late blight (incited by *Phytophthora infestans*) (Fry, 1975) and stripe rust of wheat (Stubbs and de Bruin, 1970, cited in Zadoks and Schein, 1979) by applying fungicides to cultivars with general resistance. A similar effect can be obtained by combining general resistance with diversity. For example, the Iowa E-series isoline mixtures were developed from a recurrent parent possessing a moderate level of general resistance to oat crown rust (Frey *et al.*, 1973). These mixtures also benefit from early maturity which allows for rust "escape" in the short, but intense, disease season of Iowa (Frey *et al.*, 1973, 1977). Studies conducted with powdery mildew of wheat suggest that some genotypes may complement mixtures better than others. Thus, genotypes with the highest levels of general resistance will not necessarily be the best genotypes to incorporate into a diverse population (Elliott *et al.*, 1981). Intrafield diversity and regional gene deployment can be combined by using different resistance genes in diverse populations grown in different geographical regions. Such integration will increase the epidemiological effectiveness of the tactics and should reduce the degree of selection for complex races. Rather than asking if any single tactic will provide adequate control of the cereal rusts, agriculturalists should inquire how a given tactic can contribute to an entire disease-management system.

One principle of integrated pest management is that pest populations are monitored so sound management decisions can be made. Thus, when possible, virulence surveys should be an integral part of resistance-gene management. For example, in the Iowa program, the virulence pattern of the crown rust population is one of three major factors used to determine the composition of oat mixtures, and the composition of the mixtures is altered in response to changes in the rust population over time (Frey *et al.*, 1973, 1977).

Finally, we should emphasize that race-specific resistance will continue to provide disappointments until plant scientists recognize the function of these genes in natural ecosystems and use this knowledge to design gene-management systems that will provide functional diversity.

References

Allan, R. E., Line, R. F., Peterson, C. J., Jr., Rubenthaler, G. L. Morrison, K. J., and Rohde, C. R. (1983a). Crew, a multiline wheat cultivar. *Crop Sci.* **23**, 1015–1016.

Allan, R. E., Line, R. F., Rubenthaler, G. L., and Pritchett, J. A. (1983b). Registration of eight club wheat germplasm lines resistant to stripe rust. *Crop Sci.* **23,** 603–604.

Anikster, J., and Wahl, I. (1979). Coevolution of the rust fungi on Graminae and Liliaceae and their hosts. *Annu. Rev. Phytopathol.* **17,** 367–403.

Atkinson, J. (1900). Field experiments. *Iowa Agric. Exp. Stn. Bull.* **45,** 216–229.

Barrett, J. A. (1980). Pathogen evolution in multilines and variety mixtures. *Z. Pflanzenkr. Pflanzenschutz* **87,** 383–396.

Barrett, J. A. (1981). Disease progress curves and dispersal gradients in multilines. *Phytopathol. Z.* **100,** 361–365.

Barrett, J. A., and Wolfe, M. S. (1980). Pathogen response to host resistance and its implication in breeding programmes. *Bull. OEPP* **10,** 341–347.

Berger, R. D. (1981). Comparison of the Gompertz and logistic equations to describe plant disease progress. *Phytopathology* **71,** 716–719.

Biffen, R. H. (1905). Mendel's laws of inheritance in wheat breeding. *J. Agric. Sci.* **1,** 4–48.

Borlaug, N. E. (1959). The use of multilineal or composite varieties to control airborne epidemic diseases of self-pollinated crop plants. *Proc. Int. Wheat Genet. Symp., 1st, 1958,* pp. 12–27.

Borlaug, N. E. (1981). Increasing and stabilizing food production. *In* "Plant Breeding II" (K. J. Frey, ed.), pp. 467–492. Iowa State Univ. Press, Ames.

Browning, J. A. (1957). Studies on the effect of field blends of oat varieties on stem rust losses. *Phytopathology* **47,** 4–5 (abstr.).

Browning, J. A. (1974). Relevance of knowledge about natural ecosystems to development of pest management programs for agroecosystems. *Proc. Am. Phytopathol. Soc.* **1,** 191–199.

Browning, J. A. (1980). Genetic protective mechanisms of plant-pathogen populations: Their coevolution and use in breeding for resistance. *Tex. Agric. Exp. St. [Misc. Publ.] MP* **MP-1451,** 52–75.

Browning, J. A., and Frey, K. J. (1969). Multiline cultivars as a means of disease control. *Annu. Rev. Phytopathol.* **7,** 355–382.

Browning, J. A., and Frey, K. J. (1981). The multiline concept in theory and practice. *In* "Strategies for the Control of Cereal Disease" (J. F. Jenkyn and R. T. Plumb, eds.), pp. 37–46. Blackwell, Oxford.

Browning, J. A., Simons, M. D., Frey, K. J., and Murphy, H. C. (1969). Regional deployment for conservation of oat crown rust resistance genes. *Spec. Rep.—Iowa Agric. Home Econ. Exp. Stn.* **64,** 49–56.

Browning, J. A., Simons, M. D., and Torres, E. (1977). Managing host genes: Epidemiologic and genetic concepts. *In* "Plant Disease: An Advanced Treatise" (J. G. Horsfall and E.B. Cowling, eds.), Vol. 1, pp. 191–212. Academic Press, New York.

Burdon, J. J. (1978). Mechanisms of disease control in heterogeneous plant populations—an ecologist's view. *In* "Plant Disease Epidemiology" (P. R. Scott and A. Bainbridge, eds.), pp. 193–200. Blackwell, Oxford.

Burdon, J. J., and Chilvers, G. A. (1977). Controlled environment experiments on epidemic rates of barley mildew in different mixtures of barley and wheat. *Oecologia* **28,** 141–146.

Burdon, J. J., and Chilvers, G. A. (1982). Host density as a factor in plant disease ecology. *Annu. Rev. Phytopathol.* **20,** 143–166.

Burdon, J. J., and Shattock, R. C. (1980). Disease in plant communities. *Appl. Biol.* **5,** 145–219.

Burdon, J. J., and Whitbread, R. (1979). Rates of increase of barley mildew in mixed stands of barley and wheat. *J. Appl. Ecol.* **16,** 253–258.

Chin, K. M. (1979). Aspects of the epidemiology and genetics of the foliar pathogen, *Erysiphe graminis* f. sp. *hordei,* in relation to infection of homogeneous and heterogeneous populations of the barley host (*Hordeum vulgare*). Ph.D. Thesis, University of Cambridge.

Chin, K. M., and Wolfe, M. S. (1984). The spread of *Erysiphe graminis* f. sp. *hordei* in mixtures of barley varieties. *Plant Pathol.* **33,** 89–100.

Chin, K. M., Wolfe, M. S., and Minchin, P. N. (1984). Host-mediated interactions between pathogen genotypes. *Plant Pathol.* **33,** 161–171.

Clark, R. V. (1980). Comparison of spot blotch severity in barley grown in pure stands and in mixtures with oats. *Can. J. Plant Pathol.* **2,** 37–38.

Clifford, B. C. (1968). Relations of disease resistance mechanisms to pathogen dynamics in oat crown rust epidemiology. Ph.D. Thesis, Purdue University, Lafayette, Indiana.

Clifford, B. C., and Clothier, R. B. (1974). Physiologic specialization of *Puccinia hordei* on barley hosts with non-hypersensitive resistance. *Trans. Br. Mycol. Soc.* **63,** 421–430.

Cournoyer, B. M. (1967). Crown rust intensification within and dissemination from pure line and multiline varieties of oats. M.S. Thesis, Iowa State University, Ames.

Cournoyer, B. M. (1970). Crown rust epiphytology with emphasis on the quantity and periodicity of spore dispersal from heterogeneous oat cultivar-rust race populations. Ph.D. Thesis, Iowa State University, Ames.

Crill, P. (1977). An assessment of stabilizing selection in crop variety improvement. *Annu. Rev. Phytopathol.* **15,** 185–202.

Doodson, J. K. (1982). Disease testing at NIAB, Britain's unique agricultural institute. *Plant Dis.* **66,** 875–879.

Dubin, H. J., and Rajaram, S. (1982). The CIMMYT's international approach to breeding disease-resistant wheat. *Plant Dis.* **66,** 967–971.

Dubin, H. J., and Torres, E. (1981). Causes and consequences of the 1976–1977 wheat leaf rust epidemic in northwest Mexico. *Annu. Rev. Phytopathol.* **19,** 41–49.

Elliot, V. J., MacKenzie, D. R., and Nelson, R. R. (1980). Effect of the number of component lines on powdery mildew epidemics in a wheat multiline. *Phytopathology* **70,** 461–462 (abstr.).

Elliot, V. J., MacKenzie, D. R., and Nelson, R. R. (1981). Selection of cultivars for conversion to multilines. *Phytopathology* **71,** 872 (abstr.).

Falconer, D. S. (1960). "Introduction to Quantitative Genetics." Ronald Press, New York.

Frey, K. J. (1982). Breeding multilines. *In* "Plant Improvement and Somatic Cell Genetics" (I. K. Vasil, W. R. Scowcroft, and K. J. Frey, eds.), pp. 43–72. Academic Press, New York.

Frey, K. J., Browning, J. A., and Simons, M. D. (1973). Management of host resistance genes to control diseases. *Z. Pflanzenkr. Pflanzenschutz* **80,** 160–180.

Frey, K. J., Browning, J. A., and Simons, M. D. (1977). Management systems for host genes to control disease loss. *Ann. N.Y. Acad. Sci.* **287,** 255–274.

Frey, K. J., Browning, J. A., Simons, M. D., Murphy, J. P., and Michel, L. J. (1985). Registration of Webster oats. *Crop Sci.* **25,** (in press).

Fried, P. M. (1978). Epidemiology of *Erysiphe graminis.* f. sp. *tritici* in Chancellor wheat and multilines. Ph.D. Thesis, Pennsylvania State University, University Park (Diss. Abst. **38,** 5674–B).

Fried, P. M., MacKenzie, D. R., and Nelson, R. R. (1979). Disease progress curves of *Erysiphe graminis* f. sp. *tritici* on Chancellor wheat and four multilines. *Phytopathol. Z.* **95,** 151–166.

Fry, W. E. (1975). Integrated effects of polygenic resistance and a protective fungicide on development of potato late blight. *Phytopathology* **65,** 908–911.

Gautam, P. L., Malik, S. K., Pal, S., and Singh, T. B. (1979). Synthesis of wheat multilines. *Indian J. Genet. Plant Breed.* **39,** 72–77.

Gill, K. S., Nanda, G. S., Singh, G., and Aujula, S. S. (1979). Multilines in wheat—a review. *Indian J. Genet. Plant Breed.* **39,** 30–37.

Gill, K. S., Nanda, G. S., Singh, G., and Aujula, S. S. (1980). Studies on multilines in wheat (*Triticum aestivum* L.). 12. Breeding a multiline variety by convergence of breeding lines. *Euphytica* **29,** 125–128.

Gregory, L. V., Ayers, J. E., and Nelson, R. R. (1981). Reliability of apparent infection rates in epidemiological research. *Phytopathol. Z.* **100,** 135–142.

Groenewegen, L. J. M. (1977). Multilines as a tool in breeding for reliable yields. *Cereal Res. Commun.* **5,** 125–132.

Groenewegen, L. J. M., and Zadoks, J. C. (1979). Exploiting within-field diversity as a defense against cereal diseases: A plea for "poly-genotype" varieties. *Indian J. Genet. Plant Breed.* **39,** 81–94.

Groth, J. V. (1976). Multilines and "superraces": A simple model. *Phytopathology* **66,** 937–939.

Groth, J. V., and Person, C. O. (1977). Genetic interdependence of host and parasite in epidemics. *Ann. N.Y. Acad. Sci.* **287,** 97–106.

Harlan, J. R. (1972). Genetics of disaster. *J. Environ. Qual.* **1,** 212–215.

Hwang, B. K., and Heitefuss, R. (1982). Induced resistance of spring barley to *Erysiphe graminis* f. sp. *hordei. Phytopathol. Z.* **103,** 41–47.

Jackson, L. F., Kahler, A. L., Webster, R. K., and Allard, R. W. (1978). Conservation of scald resistance in barley composite cross populations. *Phytopathology* **68,** 645–650.

Jeger, M. J., Griffiths, E., and Jones, D. G. (1981a). Disease progress of non-specialised fungal pathogens in intraspecific mixed stands of cereal cultivars. I. Models. *Ann. Appl. Biol.* **98,** 187–198.

Jeger, M. J., Griffiths, E., and Jones, D. G. (1981b). Effects of cereal cultivar mixtures on disease epidemics caused by splash-dispersed pathogens. *In* "Strategies for the Control of Cereal Disease" (J. F. Jenkyn and R. T. Plumb, eds.), pp. 81–88. Blackwell, Oxford.

Jeger, M. J., Jones, D. G., and Griffiths, E. (1981c). Disease progress of non-specialised pathogens in intraspecific mixed stands of cereal cultivars. II. Field experiments. *Ann. Appl. Biol.* **98,** 199–210.

Jensen, N. F. (1952). Intra-varietal diversification in oat breeding. *Agron. J.* **44,** 30–34.

Jensen, N. F. (1965). Multiline superiority in cereals. *Crop Sci.* **5,** 566–568.

Jensen, N. F., and Kent, G. C. (1962). Line forming techniques for multiline populations. *Agron. Abstr.,* p. 69.

Jensen, N. F., and Kent, G. C. (1963). New approach to an old problem in oat production. *Farm Res.* **29,** 4–5.

Johnson, R. (1978). Induced resistance to fungal diseases with special reference to yellow rust of wheat. *Ann. Appl. Biol.* **89,** 107–110.

Johnson, R. (1979). The concept of durable resistance. *Phytopathology* **69,** 198–199.

Johnson, R. (1981). Durable resistance: Definition of, genetic control, and attainment in plant breeding. *Phytopathology* **71,** 567–568.

Johnson, T. (1958). Regional distribution of genes for rust resistance. *Robigo* **6,** 16–17.

Johnson, T. (1961). Man-guided evolution in plant rusts. *Science* **133,** 357–362.

Johnson, T., and Newton, M. (1940). Mendelian inheritance of certain pathogenic characters of *Puccinia graminis tritici. Can. J. Res. Sect. C* **18,** 599–611.

Kampmeijer, P., and Zadoks, J. C. (1977). "EPIMUL, a Simulator of Foci and Epidemics in Mixtures of Resistant and Susceptible Plants, Mosaics and Multilines." Centre for Agricultural Publishing and Documentation, Wageningen.
Kilpatrick, R. A. (1975). New wheat cultivars and longevity of rust resistance, 1971–75. *U.S., Agric. Res. Serv., Northeast. Reg. [Rep.] ARS-NE* **NE-64.**
Knott, D. R. (1972). Using race-specific resistance to manage the evolution of plant pathogens. *J. Environ. Qual.* **1,** 227–231.
Leonard, K. J. (1969a). Factors affecting rates of stem rust increase in mixed plantings of susceptible and resistant oat varieties. *Phytopathology* **59,** 1845–1850.
Leonard, K. J. (1969b). Selection in heterogeneous populations of *Puccinia graminis* f. sp. *aveneae. Phytopathology* **59,** 1851–1857.
Leonard, K. J. (1977). Selection pressures and plant pathogens. *Ann, N. Y. Acad. Sci.* **287,** 207–222.
Leonard, K. J., and Czochor, R. J. (1980). Theory of genetic interactions among populations of plants and their parasites. *Annu. Rev. Phytopathol.* **18,** 237–258.
Luthra, J. K., and Rao, M. V. (1979a). Escape mechanism operating in multilines and its significance in relation to leaf rust epidemics. *Indian J. Genet. Plant Breed.* **39,** 38–49.
Luthra, J. K., and Rao, M. V. (1979b). Multiline cultivars—How their resistance influence leaf rust diseases in wheat. *Euphytica* **28,** 137–144.
MacKenzie, D. R. (1979). The multiline approach in controlling some cereal diseases. *In* "Rice Blast Workshop," pp. 199–216. Int. Rice Res. Inst., Los Banos, Philippines.
Marshall, D. R. (1977). The advantages and hazards of genetic homogeneity. *Ann. N.Y. Acad. Sci.* **287,** 1–20.
Marshall, D. R., and Burdon, J. J. (1981a). Multiline varieties and disease control. III. Combined use of overlapping and disjoint gene sets. *Aust. J. Biol. Sci.* **34,** 81–95.
Marshall, D. R., and Burdon, J. J. (1981b). Multiline varieties and disease control. IV. Effects of reproduction of pathogen biotypes on resistant hosts on the evolution of virulence. *SABRAO J.* **13,** 116–126.
Marshall, D. R., and Pryor, A. J. (1978). Multiline varieties and disease control. I. The "dirty-crop" approach with each component carrying a single unique resistance gene. *Theor. Appl. Genet.* **51,** 177–184.
Marshall, D. R., and Pryor, A. J. (1979). Multiline varieties and disease control. II. The "dirty crop" approach with components carrying two or more genes for resistance. *Euphytica* **28,** 145–159.
May, R. M. (1975). Stability in ecosystems: Some comments. *In* "Unifying Concepts in Ecology" (W. H. van Dobben and R. H. Lowe-McConnell, eds.), pp. 161–168. Junk, The Hague/PUDOC, Wageningen.
Mehta, K. C. (1933). Rusts of wheat and barley in India. *Indian J. Agric. Res.* **3,** 939–962.
Mode, C. J. (1958). A mathematical model for the co-evolution of obligate parasites and their hosts. *Evolution* **12,** 158–165.
Mundt, C. C., and Browning, J. A. (1985). Development of crown rust epidemics in genetically diverse oat populations: Effect of genotype unit area. *Phytopathology* **75,** (in press).
Munk, L. (1983). Response of a powdery mildew population to a barley variety mixture. *In* "Durable Resistance in Crops" (F. Lamberti, J. M. Waller, and N. A. Van der Graaff, eds.), pp. 105–107. Plenum Press, New York.
Nelson, R. R. (1972). Stabilizing racial populations of plant pathogens by use of resistance genes. *J. Environ. Qual.* **1,** 220–227.
Nelson, R. R. (1973). The use of resistance genes to curb population shifts in plant

pathogens. *In* "Breeding Plants for Disease Resistance" (R. R. Nelson, ed.), pp. 49–66. Pennsylvania State Univ. Press, University Park.

Newton, M., Johnson, T., and Brown, A. M. (1930). A preliminary study on the hybridization of physiologic forms of *Puccinia graminis tritici*. *Sci. Agric.* **10,** 721–731.

Ogilvie, L., and Thorpe, I. G. (1961). New light on epidemics of black stem rust of wheat. *Sci. Prog. (Oxford)* **49,** 209–227.

Ogle, H. J., Taylor, N. W., and Brown, J. F. (1973). A mathematical approach to the prediction of differences in the relative ability of races of *Puccinia graminis tritici* to survive when mixed. *Aust. J. Biol. Sci.* **26,** 1137–1143.

Ontario Ministry of Agriculture and Food (1977). "Agricultural Statistics for Ontario, 1977," Publ. No. 20. Ontario Ministry of Agriculture and Food.

Pal, B. P. (1979). Opening address (Symposium on use of multilines for reducing rust epidemics). *Indian J. Genet. Plant Breed.* **39,** 1–2.

Parlevliet, J. E. (1979). The multiline approach in cereals to rusts. Aspects, problems and possibilities. *Indian J. Genet. Plant Breed.* **39,** 22–29.

Parlevliet, J. E. (1981a). Disease resistance in plants and its consequences for plant breeding. *In* "Plant Breeding II" (K. J. Frey, ed.), pp. 309–364. Iowa State Univ. Press, Ames.

Parlevliet, J. E. (1981b). Stabilizing selection in crop patho-systems: An empty concept or a reality? *Euphytica* **30,** 259–269.

Person, C. (1966). Genetic polymorphism in parasitic systems. *Nature (London)* **212,** 266–267.

Pfahler, P. L., and Linskens, H. F. (1979). Yield stability and population diversity in oats. *Theor. Appl. Genet.* **54,** 1–5.

Plaut, J. L., and Berger, R. D. (1981). Infection rates in three pathosystem epidemics initiated with reduced disease severities. *Phytopathology* **71,** 917–921.

Priestly, R. H. (1981). Choice and deployment of resistant cultivars for cereal rust control. *In* "Strategies for the Control of Cereal Disease" (J. F. Jenkyn and R. T. Plumb, eds.), pp. 65–72. Blackwell, Oxford.

Rajaram, S., and Dubin, H. J. (1977). Avoiding genetic vulnerability in semidwarf wheats. *Ann. N.Y. Acad. Sci.* **287,** 243–254.

Rajaram, S., Skovmand, B., Dubin, H. J., Torres, E., Anderson, R. G., Roelfs, A. P., Samborski, D. J., and Watson, I. A. (1979). Diversity of rust resistance of the CIMMYT multiline composite, its yield potential, and utilization. *Indian J. Genet. Plant Breed.* **39,** 60–71.

Reddy, M. S. S., and Rao, M. V. (1979). Resistance genes and their deployment for control of leaf rust of wheat. *Indian J. Genet. Plant Breed.* **39,** 359–365.

Rockefeller Foundation (1963). "Program in Agricultural Science, Annual Report 1962–63." Rockefeller Found., New York.

Rockefeller Foundation (1964). "Program in Agricultural Science, Annual Report 1963–64." Rockefeller Found., New York.

Rockefeller Foundation (1965). "Program in Agricultural Science, Annual Report 1964–65." Rockefeller Found., New York.

Rouse, D. I., MacKenzie, D. R., and Nelson, R. R. (1981). A relationship between initial inoculum and apparent infection rate in a set of disease progress data for powdery mildew on wheat. *Phytopathol. Z.* **100,** 143–149.

Schmidt, R. A. (1978). Diseases in forest ecosystems: The importance of functional diversity. *In* "Plant Disease: An Advanced Treatise" (J. G. Horsfall and E. B. Cowling, eds.), Vol. 2, pp. 287–315. Academic Press, New York.

Segal, A., Manisterski, J., Fishbeck, G., and Wahl, I. (1980). How plant populations defend themselves in natural ecosystems. *In* "Plant Disease: An Advanced Treatise" (J. G. Horsfall and E. B. Cowling, eds.), Vol. 5, pp. 75–102. Academic Press, New York.

Segal, A., Manisterski, J., Browning, J. A., Fishbeck, G., and Wahl, I. (1982). Balance in indigenous plant populations. *In* "Resistance to Diseases and Pests in Forest Trees" (H. M. Heybroek, B. R. Stephan, and K. von Weissenberg, eds.), pp. 361–370. Centre for Agricultural Publishing and Documentation, Wageningen.

Shorter, R., and Frey, K. J. (1979). Relative yields of mixtures and monocultures of oat genotypes. *Crop Sci.* **19,** 548–553.

Simmonds, N. W. (1962). Variability in crop plants, its use and conservation. *Biol. Rev. Cambridge Philos. Soc.* **37,** 422–465.

Stakman, E. C., and Christensen, J. J. (1960). The problem of breeding resistant varieties. *In* "Plant Pathology: An Advanced Treatise" (J. G. Horsfall and A. E. Dimond, eds.), Vol. 3, pp. 567–624. Academic Press, New York.

Stakman, E. C., and Harrar, J. G. (1957). "Principles of Plant Pathology." Ronald Press, New York.

Stubbs, R. W., and de Bruin, T. (1970). Bestrijding van gele roest in wintertarw met het systemische fungicide oxycarboxin "Plantvax." *Gewasbescherming* **1,** 99–103.

Suneson, C. A. (1956). An evolutionary plant breeding method. *Agron. J.* **48,** 188–191.

Suneson, C. A. (1960). Genetic diversity—a protection against plant diseases and insects. *Agron. J.* **52,** 319–321.

Trenbath, B. R. (1977). Interactions between diverse hosts and diverse parasites. *Ann. N.Y. Acad. Sci.* **287,** 124–150.

Vanderplank, J. E. (1949). The relation between the size of fields and the spread of plant-diseases into them. Part II. Diseases caused by fungi with air-borne spores; with a note on horizons of infection. *Empire J. Exper. Agric.* **17,** 18–22.

Vanderplank, J. E. (1963). "Plant Diseases: Epidemics and Control." Academic Press, New York.

Vanderplank, J. E. (1968). "Disease Resistance in Plants." Academic Press, New York.

Waggoner, P. E. (1962). Weather, time, and chance of infection. *Phytopathology* **52,** 1100–1108.

Watson, I. A. (1979). The recognition and use in multilines of genes for specific resistance to rust. *Indian J. Genet. Plant Breed.* **39,** 50–59.

White, E. M. (1982). The effect of mixing barley cultivars on the incidence of powdery mildew (*Erysiphe graminis*) and on yield in Northern Ireland. *Ann. Appl. Biol.* **101,** 539–545.

Wilson, E. O., and Bossert, W. H. (1971). "A Primer of Population Biology." Sinauer Associates, Stamford, Connecticut.

Wolfe, M. S. (1973). Changes and diversity in populations of fungal pathogens. *Ann. Appl. Biol.* **75,** 132–136.

Wolfe, M. S. (1978). Some practical implications of the use of cereal variety mixtures. *In* "Plant Disease Epidemiology" (P. R. Scott and A. Bainbridge, eds.), pp. 201–207. Blackwell, Oxford.

Wolfe, M. S., and Barrett, J. A. (1977). Population genetics of powdery mildew epidemics. *Ann. N.Y. Acad. Sci.* **287,** 151–163.

Wolfe, M. S., and Barrett, J. A. (1980). Can we lead the pathogen astray? *Plant Dis.* **64,** 148–155.

Wolfe, M. S., Barrett, J. A., and Jenkins, J. E. E. (1981). The use of cultivar mixtures for disease control. *In* "Strategies for the Control of Cereal Disease" (J. F. Jenkyn and R. T. Plumb, eds.), pp. 73–80. Blackwell, Oxford.

Zadoks, J. C. (1961). Yellow rust on wheat. Studies in epidemiology and physiologic specialization. *Tijdschr. Plantenziekten* **67,** 69–256.

Zadoks, J. C., and Kampmeijer, P. (1977). The role of crop populations and their deployment, illustrated by means of a simulator, EPIMUL76. *Ann. N.Y. Acad. Sci.* **287,** 164–190.

Zadoks, J. C., and Schein, R. D. (1979). "Epidemiology and Plant Disease Management." Oxford Univ. Press, London and New York.

18

Evaluation of Chemicals for Rust Control

J. B. Rowell

Cereal Rust Laboratory, Agricultural Research Service, U.S. Department of Agriculture, Department of Plant Pathology, University of Minnesota, St. Paul, Minnesota

I. Introduction

The epidemics of wheat stem rust due to race 15B of *Puccinia graminis* Pers. f. sp. *tritici* Eriks. and E. Henn. during the 1950s in the central United States stimulated a renewed interest in the search for an effective and economically feasible chemical control for wheat rusts. A report of control of leaf rust on wheat in field tests with calcium sulfamate applied at low volume (47 liter/hectare) by aircraft (Livingston, 1953) raised expectations that powerful chemotherapeutants with the capacity to eradicate rust and halt epidemics would be found. An informal conference on control of cereal rusts by chemotherapeutants at the University of Minnesota in 1954 was attended by 112 public and private scientists. Cooperative field testing of promising compounds, which had shown potential for field control of wheat rusts in preliminary testing, was organized by the U. S. Department of

THE CEREAL RUSTS,
Volume II

ISBN 0-12-148402-5

Agriculture in 1955 (Loegering, 1955). These tests were conducted between 1956 and 1960 in up to 12 locations, which included 10 states in the United States and one location each in Canada and Mexico. Data were recorded on the time of first appearance and final severity of leaf and stem rusts in the test plots, average severity of leaf and stem rusts, grain yield, and test weight. A mimeographed summary of test results was prepared annually and distributed to interested parties. A second conference on chemical control of cereal rusts held in 1959 at the University of Minnesota was cosponsored by the U. S. Department of Agriculture and supported by 20 chemical companies. By then the threat of catastrophic stem rust epidemics was diminished because Selkirk spring wheat, which had high resistance to leaf and stem rust, predominated throughout the highly vulnerable spring wheat region of the central United States and Canada.

A virulent race of *P. recondita* Rob. ex Desm. f. sp. *tritici* Eriks. caused a severe epidemic of leaf rust on Selkirk wheat in 1965, and this cultivar was supplanted in the United States by other hard red spring wheat cultivars with effective resistance to stem and leaf rust. Selkirk wheat is still effectively resistant to the North American population of *P. graminis* f. sp. *tritici.*

The early tests with systemic chemicals demonstrated that toxicologic and economic evaluations are as important as pathologic evaluation in the development of a successful chemical control. The promising effectiveness against wheat rusts of calcium sulfamate, cycloheximide, and inorganic salts of nickel were nullified, respectively, by adverse effects on grain quality, a narrow range between the effectiveness and phytotoxic doses, and the presence of appreciable residues of a heavy metal element in a basic food (Rowell, 1968a). Presently, chemical control of foliar diseases has only limited use in cereal production in the United States (Bissonette *et al.*, 1969). In Great Britain and Western Europe, where powdery mildew and various other foliar diseases frequently cause losses in wheat and barley, systemic fungicides have a role in cereal production (Cook *et al.*, 1981; Jenkins and Lescar, 1980; Rathnell and Skidmore, 1982). The following presentation deals only with the pathological evaluation of compounds for control of the stem and leaf rusts of wheat, which historically were the major limiting diseases in wheat production in the United States. Pathological evaluation is only a first stage in the development of a practical chemical control, since a chemical for use on the cereal rusts must also pass toxicological, economical, ecological, social, and political evaluations.

II. *In Vitro* Tests

Rust fungi have not been used routinely in testing compounds *in vitro* for fungicidal activity. The nonpolar urediospore and germ-tube walls (Rowell, Vol. 1) resist wetting by water, giving erratic growth responses by germinating spores to toxicants in aqueous systems. The special requirements and slow mycelial growth of rust fungi in axenic culture (Vol. 1, Williams) make such cultures unsuitable for tests of the toxicity of large numbers of compounds on aqueous media. Thus, conventional methods have had little use in testing for fungicidal activity against rust fungi.

A. SPORE GERMINATION TESTS

The standard slide-germination method of the American Phytopathological Society (Anonymous, 1943) for evaluating fungicides is unsatisfactory for determining the dosage response of germinating urediospores of the cereal rusts. When rust urediospores are deposited or mixed with aqueous solutions, the spores float, so only a small portion of the wall surface is in direct contact with the liquid surface and many of the germ tubes grow aerially. Thus, a uniform challenge of spore protoplasts by test compounds is not achieved. Furthermore, spores wetted and submerged by adding surfactants to the test solutions germinate poorly. These difficulties can be circumvented by germinating urediospores in the interface between the test solutions and a floating section of thin polyethylene film.

Pure polyethylene film is highly inert chemically and unlikely to react with test substances or affect the metabolism of germinating spores. Such film at 1 mil (0.001 inch or 0.025 mm) thickness is permeable to dry oxygen and carbon dioxide gases (against air) at the rate of 0.74 and 1.3 ml/cm^2 in 24 hr, respectively, at 25°C and 760 mm Hg. Commercial films used as packaging material often contain additives to improve transparency and surface texture that may affect fungal development.

My test procedure for germination tests with rust urediospores was as follows: (1) Freshly collected urediospores were deposited in a settling tower at a density of about 25 spores/mm^2 on 15 × 15 mm squares of 0.5 mil pure polyethylene film. (2) Two milliliters of the test solution were dispensed into Pyrex dishes 2.5 cm diameter × 0.6 cm deep. (3) The film was placed on the surface of the test solution with the spores in contact with the aqueous surface. (4) The dish was incubated in the dark at 18°C for 24 hr. (5) About 0.2 ml of a stain mixture (to be described) was added to the dish. (6) The number of germinated and ungerminated spores were counted in five fields per dish at 100× magnification. Double-glass-distilled water was used to prepare all solutions. Each compound was tested in a dilution series of five doses in comparison to a distilled water control. The six dishes for a test series were placed in a covered petri dish and three replicates were used per compound.

In this method, rust urediospores adhered to the film and the germ tubes grew in the interface between the test solution and the polyethylene film, which results in a uniform exposure of spores and germ tubes to the test solutions. The stain mixture, which stains walls blue and protoplasts red, was as follows:

20% Acetic acid	40 ml
1% Cotton blue in 95% ethanol	1 ml
1% Acid fuchsin in 95% ethanol	2 ml
Water	57 ml

Spores were counted as germinated if germ-tube growth was sufficient for the protoplast to migrate out of the spore, so that the spore then stained blue and appeared empty. The percent reduction in spore germination was calculated and the dosage response was plotted on a log-probit graph (Horsfall, 1956).

Spore germination tests have limited utility for evaluating the potential usefulness of systemic compounds for controlling cereal rusts. Table I lists the effective dose for 50% inhibition (ED_{50}) of urediospore germination for some compounds that control wheat stem rust. Cupric ions, cycloheximide, and zineb are strong inhibitors of spore germination and are effective protectants in the field. Sulfur, although an effective protectant fungicide against rust in the field, is a weak inhibitor of spore germination in this assay system. The higher toxicity of sulfamic and sulfanilic acids than that of the salts of these acids to germinating urediospores presumably was due to the injury from acidity rather than to the toxicity of the compound per se. Even though the salts of sulfamate, sulfanilate, and sulfadiazine were weak inhibitors of urediospore germination, these compounds were effective eradicants of wheat stem rust infections. Apparently the critical system in parasitic mycelia that is affected by these compounds is not essential for spore germination

Table I

Effective Dose of Various Fungicides to Inhibit Germination in 50% (ED_{50}) of *Puccinia graminis* f. sp. *tritici* Urediospores in the Interface between Pure Polyethylene Film and the Test Solution

Fungicides	ED_{50} (ppm)
Cu^{2+} (from $CuSO_7$)	12.70
Hg^{2+} (from $HgCl_2$)	3.69
Ni^{2+} (from $NiCl_2$)	0.05
S (as Sulfuron)	500.00
Zineb	1.50
Sulfamic acid	20.00
Sodium sulfamate	>500.00
Calcium sulfamate	>500.00
Sulfanilic acid	15.00
Sodium sulfanilate	>500.00
Sodium sulfadiazine	>500.00
Oxycarboxin	16.00
Triarimol	15.30
Benomyl	8.60
Cycloheximide	0.38
Triazbutil	>300.00

and germ-tube growth. Other systemic fungicides such as oxycarboxin, triarimol, and benomyl were moderately toxic to germinating urediospores of *P. graminis* but differ widely in effectiveness in the field. Triazbutil, which is specifically effective against wheat leaf rust, not only gave negligible inhibition of urediospore germination for *P. graminis* but also of *P. recondita* f. sp. *tritici.* Thus, spore germination tests may or may not indicate the toxicity of a compound to the rust fungus, presumably because of the differences in the metabolism of the germinating spore and the parasitic mycelia.

B. MYCELIAL GROWTH TESTS

Harvey and Grasham (1979) devised a test to evaluate compounds as inhibitors of mycelial growth in axenic cultures of *Cronartium ribicola* J. C. Fisch. Appropriate amounts of test concentrations of compounds were added to sterile basal medium at 50°C, and about 10 ml was poured into sterile petri dishes. Test and control plates were inoculated with 5-mm-diameter agar discs with actively growing mycelia of the rust culture. After 30 days of incubation under standard conditions in a growth chamber, the mycelial mat was removed from the surface of the medium and weighed. If no growth was evident, the original disc of inoculum was transferred to fungicide-free medium and incubated for 30 days as a test for fungistatic or fungicidal activity.

Nineteen potential antifungal compounds were tested in this assay. Triarimol was the most fungitoxic of the test compounds and completely inhibited mycelial growth at 0.1 mg/liter; thiophanate-methyl and benomyl were completely inhibitory at 1 mg/liter; and thiobendazole and oxycarboxin were completely inhibitory at 5 mg/ml. Although cycloheximide prevented mycelial growth on media containing 1 mg/liter, the rust fungus continued to grow on the surface of the inoculum disc and vigorously grew out onto the agar surface after transfer to fungicide-free media. Thus, cycloheximide appears to be fungistatic to *C. ribicola* at the tested concentrations. Jones (1973) also has developed an assay for fungicidal activity using an axenic culture of *Uromyces dianthi* (Pers.) Niessel.

Many questions exist about how *in vitro* exposure of rust mycelia to fungicides resembles exposure to the same toxicants in the host. The external concentration of a fungicide required for a lethal dose is not always a direct measure of the level of the actual toxicity of a chemical because the fungicidal action of a compound is a function of both its ability to be taken up and its innate toxicity (McCallen, 1956). Thus, the true measure of fungicidal activity is the internal concentration required for cell death. Haustoria appear to be the principal means of nutrient absorption in parasitic rust thalli, but mycelia in axenic cultures lack these structures and apparently absorb nutrients through the mycelial walls. In the control of a cereal rust, the dose of a systemic fungicide delivered to the pathogen is determined by the factors that affect the uptake, transport, and deactivation of the compound in the host (Crowdy and Jones, 1956, 1958; Crowdy *et al.*, 1958). Systemic fungicides predominantly move through the apparent free space (the apoplast) of the host in the transpirational stream and therefore would be in direct contact with rust mycelia. Active accumulation of a systemic fungicide by host cells, however, may result in efficient and rapid uptake by rust haustoria. Thus, the dosage response of axenic rust mycelia to a toxicant may differ markedly from that of the parasitic thallus. The principal value of axenic rust cultures for studying fungicides would be in investigations of the mode and mechanisms of fungicidal activity against rust fungi.

III. *In Vivo* Tests

The ultimate determination of the efficacy of fungicides for control of cereal rusts requires comparative tests in commercial production fields. Evaluation of candidate fungicides by a variety of seedling tests, however, can define the characteristics for activity as a protectant and an eradicant, uptake by foliage and roots, activity after application to soil, and phytotoxicity to the host. Prior knowledge of such properties can suggest the manner, timing, and dosage of applications for optimal performance in the field.

A. SEEDLING ASSAYS

1. *General Procedures*

The following seedling tests are essentially *in vivo* assays I have used to determine the relative effectiveness of systemic fungicides against *P. graminis* and *P. recondita.* Essentially the same basic experimental design, disease measurement, and data analyses were used in the various types of tests. Two seeds were planted at each of five equally spaced points around the periphery of 7.5-cm-diameter clay pots, with the exceptions detailed later for seed and soil treatments. Pots were numbered sequentially for randomization of treatments. The planting medium was a 3:3:3:1 (v/v) mixture of field soil, sand, peat, and manure. Two wheat cultivars, Little Club and Thatcher, which produce erect first foliar leaves needed for manipulation and uniform inoculation, were used for tests with stem and leaf rusts, respectively. Uniform test seedlings were produced under controlled conditions as given earlier (Rowell, Vol. 1). Seven days after planting, the five most uniform seedlings in each pot were selected for testing, the remainder were removed, and the schedule of test procedures given with the descriptions of the various assays was initiated. Plants were inoculated by controlled procedures and exposed to 16 hr of dew under the programmed environment for stem rust described previously (Rowell, Vol. 1). Plants inoculated with *P. recondita* were exposed to 16 hr of dew in the dark at 18°C. On removal from the dew chamber, all plants were held in a growth chamber under a 14-hr light period at 26°–27°C and 10-hr dark period at 20°–21°C until flecks appeared. Plants then were removed to a glasshouse to prevent dissemination of spores and reinfection in the growth chamber.

At 10 days after inoculation, the number of uredia per leaf was counted, visible effects on infection development were noted, and any visible symptoms of phytotoxicity were recorded. Four days later the plants were reexamined for changes in the relative effects of the treatments on disease development, and the test was terminated.

Each test compared the effects of five dosages of a test compound to a control. Four replicate pots randomized by serial number were used per treatment. An initial trial of a compound tested dosages ranging over five magnitudes (10-fold intervals) in concentration. This trial generally indicated a narrow concentration range in which the pathogen responded differentially to dosage. The five test dosages used in subsequent trials were selected as those most likely to fall within the range between 10 and 90% control. The logit of the percentage of reduction in number of uredia for each test dosage in comparison to the untreated control was plotted against log (dosage) to obtain by interpolation the effective dosage for 50% disease control (ED_{50}) (Horsfall, 1956) and the slope (Rich and Horsfall, 1952). The variation in ED_{50} values obtained in repeated trials with a given compound could vary as much as ± 50% of the mean; therefore, each assay method was repeated three times with those systemic fungicides that appear to have promising effectiveness against the rust pathogens.

2. *Seed Treatment*

In view of the small profit margin for small-grain cereal production in North America, control of rust diseases by treating seed with an effective systemic fungicide is highly attractive because of economies in the use of material, labor, and energy. The feasibility of this approach for control of some foliar pathogens has been established by the control of powdery mildew on barley (Brooks, 1972) and wheat (Johnston, 1972) with ethirimol and the control of wheat leaf rust by triazbutil (Rowell, 1976).

Hansing (1978) gives directions for treating small quantities of seed with different types of fungicides formulations for field testing. I have used modifications of these methods for treating the small amounts of seed used to determine the dosage response of systemic fungicides on seedling plants. In my experience, the maximum amount of a dry treatment retained by wheat is about 400 g material/100 kg seed (0.4 lb/cwt). Therefore, a dosage series of the initial dry formulation is prepared by dilution with diatomaceous earth to maintain this fixed ratio of treatment material to seed. Ten grams of seed are placed in a 40 mm diameter × 65 mm clear plastic, snap-cap vial with 40 mg of dry treatment for each dosage. Controls are treated with the same amount of diatomaceous earth.

The vials are shaken horizontally on a reciprocal shaker for 6 min. Some wastage of fungicide occurs as the material coats both the seed and the inner surfaces of the vial; therefore, the first lot of treated seed is discarded. The vials are reloaded with 10 g fresh seed and 40 mg dry treatment material and shaken for 6 min to achieve approximate retention of the intended dosage. This procedure can be used for liquid formulations by pipetting aqueous dilutions of the material onto the side of the vial and spreading it evenly by rotating the vial before adding the seed and shaking. For all treatments, however, chemical analysis of the seed is necessary to determine accurately the amount of fungicide adhering to the seed.

Treatment with quantities of dry-formulated material exceeding 400 g/100 kg seed is accomplished by pelleting the seed with a solution of methyl cellulose (Hansing, 1978; Rowell, 1973a). Preparation of a completely dispersed solution of methyl cellulose requires special procedures (Hansing, 1978). Seed is coated with methyl cellulose first by adding 0.55 g of a 2% solution to the walls of the vial, shaking 10 g seed in the vial for 10 min, and then adding the required quantity of dry fungicide formulation to the vial and shaking the seed for another 6 min. The treated seed is dried overnight before planting. After treatment, six seeds are planted per 10-cm square plastic pot. Watering is controlled to prevent leaching of the chemical from the soil. Plants are thinned to three uniform test seedlings at emergence and inoculated 7 days after planting, and the assay is completed as described under general procedures.

This method provides a relative assessment of the effectiveness of a systemic fungicide applied as a seed treatment. Potentially useful compounds will have an extremely low dosage response against the pathogen and negligible phytotoxicity at high rates of treatment. For example, in this assay the ED_{50} (g/100 kg) for control of leaf rust on Thatcher wheat seedlings was 0.84 g triazbutil, 12.3 g triarimol, 108 g oxcarboxin, and >200 g benomyl (Rowell, 1976). Triazbutil effectively controlled leaf rust on Thatcher wheat in the field at 125 g/100 kg (Rowell, 1976). Triadimefon also appears promising as a seed treatment for rust. At 100–200 g/100 kg, it controlled stripe rust on wheat through the heading stage of development in tests in the Pacific Northwest (Line and Rakotondradona, 1980).

3. *Soil Treatment*

Investigators have used various test procedures to determine systemic activity of chemicals against cereal rusts by root uptake. Gassner and Hassebrauk (1936) sought fungicides with systemic activity against cereal rusts by adding solutions of chemicals to the soil at inoculation. Sempio (1936) demonstrated systemic protective activity of nickel against wheat leaf rust by placing the roots in a nutrient solution containing $Ni(NO_3)_2$ for 8–10 days before inoculation. Von Meyer *et al.* (1970) mixed triazbutil into soil prior to planting to demonstrate systemic protective activity against leaf rust.

Although soil is a potential reservoir site for maintaining an effective dose of systemic fungicide in the host plant, it is an inhospitable environment for the fungicide to retain effective activity. Mineral, organic, and biologic components of the soil may irreversibly adsorb, absorb, or react with the fungicide to render it either unavailable for uptake or inactive against the pathogen. Fluctuations in soil moisture may either suppress fungicide availability under dry conditions or leach the fungicide out of the root zone. Systemic insecticides are maximally effective when soil moisture is at maximum field capacity (Chapman and Naoum, 1966), and systemic fungicides would probably have the same relationship to moisture. Under field conditions, spray deposits of systemic fungicides on cereal foliage are washed by rain into the soil. Therefore, systemic fungicides such as triazbutil that remain active in the soil environment and are absorbed in effective concentrations by roots (Rowell, 1973b) have an extra advantage for controlling rust.

The procedure I have used to evaluate the effectiveness against cereal rusts of fungicides in soil was to apply the test chemical in 50 ml distilled water to the surface of a fixed volume of the soil mixture in a 10-cm square plastic pot. Leaching of the fungicide from the pot is avoided by adding increments of distilled water in amounts sufficient to wet the soil without drainage from the pot. Protective activity was assayed by applying the test chemical at planting followed by inoculation of the seedling seven days later. Eradicant activity was assayed by applying the test solution to the seedlings three days after inoculation. A dose of 660 μg/pot is equivalent to 1 kg/hectare based on the area of the soil surface. The ED_{50} values for various systemic fungicides, applied as soil drenches at planting, against *P. graminis* f. sp. *tritici* on Little Club wheat are listed in Table II. Oxycarboxin has shown limited activity against wheat stem rust epidemics in the field when applied to the soil at planting (Rowell, 1967, 1968b). Triazbutil effectively controlled epidemics of leaf rust on Thatcher wheat in the field when sprays were applied to plants in the three-leaf growth stage about 3 weeks after planting and most of the spray was deposited on the soil surface (Rowell, 1973b, 1976).

Generally a systemic fungicide is more effective in these assays as a

Table II

Effective Dose of Various Fungicides for 50% Control of Infection (ED_{50}) in Soil-Drench Assays of Protectant Activity with *Puccinia graminis* f. sp. *tritici* and Little Club Wheat Seedlings

Fungicides	ED_{50} (μ/pot)[a]	Source
Oxycarboxin	750	Rowell, 1972
Triazbutil	>2000	Rowell, 1972
Triarimol	208	Rowell, 1972
Benomyl	3100	Rowell, 1972
Triadimefon	85	Rowell[b]
Fenapanil	2700	Rowell[b]

[a]Based on 50 ml test solutions applied at planting to a fixed volume of a 3:3:3:1 (v/v) mixture of field soil, sand, peat, and manure in a 10-cm^2 plastic pot.
[b]Unpublished.

protectant than as an eradicant (Rowell, 1976), presumably due to the greater vulnerability of the rust fungus before than after parasitism is established. The ED_{50} values for triazbutil, which is a highly effective protectant fungicide specifically active against wheat leaf rust, were 3.6 and >2000 μg/pot in the protectant and eradicant assays, respectively. This wide disparity in the results from the two assays was due to the greater protective than eradicant activity or triazbutil against wheat leaf rust. Conversely, the ED_{50} values for oxycarboxin against wheat leaf rust were 420 and 380 μg/pot, respectively, in the protectant and eradicant assays (Rowell, 1976). Since results from other assay procedures indicated that oxycarboxin has greater activity as a protectant than as an eradicant, the results from the soil drench assays suggest that gradual inactivation of this fungicide in the soil reduced effectiveness in the assay for protectant activity.

4. Foliar Treatment

Foliar spraying is a popular test method for evaluating fungicides against cereal rusts in seedling tests (Livingston, 1953; Keil *et al.*, 1958; Davis *et al.*, 1960). Dosage generally is reported as the concentration of the applied spray mixture, and fungicides are compared by the lowest concentration required for effective disease control. Unfortunately, the amount of spray deposited on plant surfaces can vary widely unless special procedures are used. As much as 80% of the spray accumulated on plant surfaces just prior to run-off can be lost after run-off occurs. Furthermore, the rate of drying of the liquid deposits markedly affects the uptake of systemic chemicals by plant tissues. Ideally, a spray method to evaluate test fungicides should duplicate the spray mixtures and deposits that occur under field conditions. The concentration and amounts of spray mixtures for 1 kg fungicide/hectare ranges from 2.1% in 46.8 liters (5 gal/acre) for aerial application to 0.36% in 281 liters (30 gal/acre) for ground sprayers. The spray deposits in field applications consist of scattered drops that vary over a wide range in size. Thus, special equipment and procedures are needed to standardize fungicide deposition for reasonably accurate evaluations of systemic fungicides.

A spray chamber used for controlled inoculation of wheat seedlings with rust urediospores (Rowell and Olien, 1957) also was used for assays of fungicides applied as foliar sprays. Spray deposition was calibrated by colorimetric analysis at 585 nm of the deposit produced by 1 ml of a 0.1% aqueous solution of crystal violet atomized on first leaves of seedlings in a pot rotated on a turntable at 38 cm from the spray nozzle. A mean deposit of 1.67 $\mu g/cm^2$ was produced by this amount of crystal violet, which was estimated to be equivalent to the complete coverage produced by 1.3 kg fungicide evenly distributed on the plant surface present in 1 hectare of mature wheat. Under the test conditions in this spray chamber, 1 ml of spray solution covered the seedling leaves uniformly with a deposit of fine droplets that dried within 5 min. Fungicides were evaluated as eradicants by spray applications to infected wheat seedlings 3 days after inoculation. Thus, this method determined the response of the established pathogen to spray mixtures approximating the concentrations used in field application.

The ED_{50} values (Table III) for some systemic fungicides evaluated in this assay indicated the potential of the compound in an aqueous deposit on leaf surfaces to penetrate leaf tissue rapidly and eradicate established infections with minimal movement of the compound inside the plant. Cycloheximide was the most effective eradicant in this assay, but the effective dose was close to the threshold of phytotoxicity. The other two highly effective eradicants, triarimol and nickel ions, had similar high eradicant activity in the field at rates of 0.14 and 0.28 kg/hectare, respectively. The lack of eradicant activity of triazbutil against *P. graminis* f. sp. *tritici* was not surprising because it also lacks protectant activity against the stem rust pathogen. With *P. recondita* f. sp. *tritici,* however, triazbutil has weak eradicant activity in foliar spray assays in contrast to strong activity as a protectant.

Table III

Effective Dose of Various Fungicides for 50% Control of Infection (ED_{50}) in Foliar Spray Assays of Eradicant Activity with *Puccinia graminis* f. sp. *tritici* and Little Club Wheat Seedlings

Fungicides	ED_{50} ($\mu g/cm^2$)[a]
Ni^{2+} (as chloride)	0.045
Cycloheximide	0.014
Sulfamic acid	11.500
Oxycarboxin	50.000
Triazbutyl	>500.000
Triarimol	0.025
Benomyl	35.000

[a]Based on the deposition obtained on foliar surfaces sprayed with 1 ml test solution 3 days after inoculation.

Spray applications prior to inoculation are undesirable for evaluating test materials as systemic protectants because this test cannot distinguish between the external inhibition of urediospore germination by fungicidal residues on leaf surfaces and the internal disruption of the infection process in the host by systemically absorbed fungicide. Furthermore, surface residues of systemic fungicides are reservoirs for additional uptake of the compound during the dew period for rust infection and thereby give erroneous data on the persistance of systemic activity.

The eradicant and protectant activity of internal doses of fungicides with minimal interference by surface residues was evaluated by immersing wheat leaves in treatment solutions (Andersen and Rowell, 1962). Fully elongated first foliar leaves of five wheat seedlings in 7.5-cm-diameter clay pots were immersed in 250 ml test solution plus 100 ppm Tween 20 in a glass container of the same diameter. The leaves were dipped several times in the test solution to remove air bubbles clinging to the leaf and thoroughly wet leaf surfaces. Then the rim of the inverted pot rested on the rim of the glass container with the leaves completely immersed for 4 hr. After the exposure period was completed, the plants were removed from the test solution and shaken to remove excess liquid and washed with agitation in distilled water for 10 sec to remove surface residues. Controls were similarly treated with

250 ml distilled water containing 100 ppm Tween 20. Protective and eradicant activities were evaluated by this method with treatments applied 2 days before and after inoculation, respectively (Table IV). The ED_{50} for nickel ions, sodium sulfadiazine, and potassium sulfanilate in these tests was lower for eradicant than for protectant activity, which indicated that these fungicides lost activity rapidly inside the host. Other fungicides such as cycloheximide and oxycarboxin, however, had lower ED_{50} values as protectants than as eradicants. The most effective protectant found in my tests of various systemic fungicides was triazbutil with an ED_{50} of 0.014 μg/ml against *P. recondita* f. sp. *tritici* (Rowell, 1976).

The duration of effectiveness of systemic chemicals inside the host was evaluated by Andersen and Rowell (1962) by inoculating plants treated by immersion in a single dosage of the fungicide at various times after treatment. Separate sets of plants were treated with the dosage for 90% control of infection on treatment 24 hr before inoculation on each of 4 successive days or other suitable intervals and were inoculated on the day following the last treatment. Controls were treated on each day with 100 ppm Tween 20. The percentage of rust control observed at each time of treatment was plotted against time to determine the rate of loss in protective activity of the test dosage. The

Table IV

Effective Dose of Various Fungicides for 50% Control of Infection (ED_{50}) and the Duration of Protectant Activity in Leaf Immersion Assays with *Puccinia graminis* f. sp. *tritici* in Little Club Wheat Seedlings

	ED_{50} (ppm)[a]		DI[b]		
Fungicides	Eradicant	Protectant	hr	ppm	Source[c]
Ni^{2+} (as chloride)	1.05	3.0	31	6.18	(A)
Cycloheximide	5.5	2.6	278	10.0	(A)
Sodium sulfadiazine	100.0	160.0	80	500.0	(A)
Potassium sulfanilate	160.0	280.0	92	300.0	(A)
Calcium sulfamate	—[d]	—	49	1500.0	(A)
Oxycarboxin	390.0	193.0	57	850.0	(A)
Benomyl	—	2.6	175	40.0	(B)
Triarimol	—	0.82	76	1.4	(B)

[a]Treatment applied 2 days after inoculating in test as eradicant; 2 days before inoculating as protectant.

[b]Duration index, the number of hours required for the ED_{90} treatment at 24 hr before inoculation to lose 50% of the protectant activity. All treatments, including controls, had 100 ppm Tween 20.

[c]Sources: (A) Andersen and Rowell, 1962; (B) Rowell, unpublished.

[d]Indicates tested but no mean calculated.

duration index (DI)—i.e., the number of hours required for a treatment to lose 50% of the protective activity observed for treatment at 24 hr before inoculation—was used to compare the longevity of effectiveness of various test compounds (Table IV).

The method provides an evaluation of the rapidity with which an internal dosage of a systemic fungicide loses effectiveness against rust. Most systemic fungicides evaluated in this test lose effectiveness in several days, and long DI values such as 278 hr for cycloheximide and 175 hr for benomyl are exceptional.

The drop test is another foliar method for evaluating the effectiveness of systemic fungicides against cereal rusts (Rowell, 1972, 1976). In this method a 10-μl drop of the test compound in a solution of Tween 20 at 500 μg/ml is placed on the abaxial surface at 5 cm from the tip of fully elongated first foliar leaves of seedlings. Prior to treatment, the site of drop treatment is rubbed firmly between thumb and index finger to alter the wettability of the cuticle and improve drop retention by decreasing the contact angle between drop and cuticle. The point of treatment on each leaf is marked with India ink on the adaxial surface, and the pots are placed on their sides to orient the leaves horizontally to prevent drop run-off. Treatments were made in a walk-in growth chamber with 50% relative humidity, where the applied drops dried slowly over a period of about 2 hr. Protectant and eradicant activities were evaluated by applying treatments 24 hr before and 48 hr after inoculation, respectively. Uredia developing between the ink mark and the leaf tip and base were counted separately 10 days after inoculation. All systemically active compounds that I have tested controlled rust distally (toward the leaf tip) at much lower concentrations than proximally (toward the base) from the point of application. In general, proximal effects of systemic fungicides were limited to a small zone adjacent to the site of treatment with high doses, which suggests that the compound moved by diffusion against the transpirational stream. Thus, the method is most useful for measuring the ability of a compound to enter the plant and move distally in the transpirational stream in dosages effective against the pathogen.

The ED_{50} value per leaf for distal movement of a compound in the eradicant drop assay (Table V) is generally, but not always, lower than that for the eradicant spray assay. (Multiply values in Table III by 10 for an approximate comparison.) In the drop assay, the longer period for uptake, younger infection, and smaller leaf volume than in the spray test would tend to reduce the external dosage required for the same level of disease control. Triadimefon, triarimol, and fenapanil were the best systemic eradicants in this assay.

Table V

Effective Dose of Various Fungicides for 50% Control of Infection (ED_{50}) in Drop Assays of Eradicant Activity against *Puccinia graminis* f. sp. *tritici* Infections in the First Foliar Leaf of Little Club Wheat Seedlings

Fungicide	ED_{50}[a] (μg/leaf)	Source
Sulfamic acid	10.5	Rowell[b]
Calcium sulfamate	16.0	Rowell[b]
Sodium sulfanilate	14.0	Rowell[b]
Sodium sulfadiazine	11.8	Rowell[b]
Ni^{2+} (as chloride)	7.0	Rowell[b]
Cycloheximide	1.4	Rowell[b]
Oxycarboxin	1.25	Rowell, 1972
Triazbutil	>100.0	Rowell, 1972
Triarimol	0.012	Rowell, 1972
Benomyl	8.9	Rowell, 1972
Triadimefon	0.0036	Rowell, 1981
Fenapanil	0.016	Rowell, 1981

[a]Treatments applied 2 days after inoculation.
[b]Unpublished.

The protective drop test is less precise than the eradicant drop test because additional uptake of chemical occurs from residues at the site of application during the dew period required for rust infection. Thus, the ED_{50} value for many compounds will be lower in the protective than in eradicant drop tests (Rowell, 1976). Those systemic fungicides that have distinctively better eradicant than protectant activity, such as triadimefon and fenapanil, have lower ED_{50} values in the eradicant assay (Rowell, 1981). Triazbutil had an exceptionally large difference in protectant and eradicant activity with ED_{50} value of 0.0012 and 190 μg/leaf, respectively, in assays with wheat leaf rust (Rowell, 1976). This large difference indicates that triazbutil is an unusual systemic fungicide. Histologic studies (Watkins *et al.*, 1977; J. B. Rowell, unpublished) of the development of penetrants of *P. recondita* f. sp. *tritici* in wheat seedlings treated with triazbutil revealed that the protectant effect was achieved by inhibiting the growth of infection hyphae and suppressing the formation of haustoria. That most systemic fungicides with activity against plant rusts are effective eradicants probably is due to the use of eradicant screening tests in the search by industrial scientists for potentially effective compounds.

B. Field Evaluation

The ultimate evaluation of the efficacy of a chemical for control of a cereal rust must be made in commercial production fields. Wheat, however, is produced over a wide geographic range, using many different cultivars under diverse climatic and cultural conditions. Thus, extensive small plot testing is required in many locations to determine the feasibility of large-scale trials. Small-plot tests also are useful for evaluating new fungicides and for defining parameters for dosage, timing, and formulation.

1. Application Equipment

The ideal spray equipment for small plot testing would duplicate the spray deposition obtained with the spray equipment used in commercial fields. Most field crop farms in North America lack appropriate ground equipment for spraying fungicides on large fields of adult wheat. Most fungicides have low water solubility and are formulated as wettable powders that form suspensions in the spray tank. Spraying such mixtures with ground equipment requires spray nozzles with large orifices that apply about 280 liters/hectare (30 gal/acre) of water to achieve adequate coverage and to prevent suspended solids from settling in spray lines and clogging nozzles. A 100-hectare (247-acre) field, which is not an unusually large field for wheat in the Great Plains, requires 28,000 liters (7400 gal) of water at this spray rate, and such a large amount of water requires frequent trips to a water source to refill spray tanks. Furthermore, up to 4% crop loss is caused by wheel damage with ground equipment after flag-leaf emergence (Cook *et al.*, 1981), which increases the economic threshold at which fungicidal use becomes feasible for control of cereal rust. Thus, spraying by aircraft, which achieves uniform coverage with reduced spray volumes by high-speed application, is the most feasible means of spraying the large production fields in the Great Plains. Sprays of the protectant fungicides zineb and mancozeb currently in limited use for controlling foliar diseases of cereals are applied by aircraft at 47 liters/hectare (Bissonnette *et al.*, 1969).

Ground spraying, however, is the only feasible means for small plot tests of fungicides for control of cereal rusts. The use of more water than would be used in aircraft application theoretically increases the potential for uptake of systemic fungicides. The greater dilution in the large volume of water dissolves more of the fungicide, and deposition of a greater liquid volume at a slow speed gives more complete coverage of plant surface than that obtained by aircraft spraying.

A small 12-hp lawn and garden tractor was adapted as a mechanized sprayer for my fungicide trials on small field plots. A corrosion resistant Hypro 6 roller pump with a capacity of 21 liters/min was run by the power take-off of the tractor. An adjustable spray boom, 3 m long, was attached to the tractor chassis on the right side just forward of the front axle. Nine nozzles were mounted by piping clamps on the boom so nozzle spacing and direction could be adjusted for different spacings of plant rows. The nozzles were equipped with Spraying Systems D-2 discs and number 13 cores that gave a cone spray pattern. Experience showed that the best spray pattern was obtained with alternating nozzles pointed down at 45° forward and backward from the vertical plane of the boom. The spray nozzles were connected by pressure hose to a control valve with a bypass that directed all spray liquid back to the spray tank in the "shut-off" position and in the "open" position directed the liquid to the spray line with the excess bypassing through the valve. The bypassing liquid agitated the spray mixture sufficiently to prevent settling of suspended solids in the spray tank. Spray pressure was regulated at 21 kg/cm^2 by a diaphram pressure relief valve and a pressure gauge connected to the control valve. Racks for two removable cylindrical spray tanks were mounted on the rear of the tractor. A series of spare tanks was used to premix treatments.

To apply a spray with this equipment, the operator aligned the tractor in the upwind alley along the appropriate plot to avoid exposure to drifting spray. The boom was aligned in the cross alley facing the plot to be sprayed. Then the power take-off was engaged, the throttle and gear positions were set for the required speed, the spray valve was opened, spray pressure was adjusted if needed, the spray pattern was checked for nozzle function, the clutch was engaged, and the sprayer was moved at a constant speed while applying spray to the plot, and the spray was shut off as the boom emerged into the alley at the opposite end of the plot.

2. *Experimental Plan and Plot Design*

The mechanics for developing experimental plans and plot design concerned with fungicidal control of plant diseases are excellently presented by Nelson (1978). Two particular concerns in conducting field trials on the cereal rusts are the choice of a suitable host cultivar and the influence of inoculum interference between plots.

Most fungicidal experimentation in plant pathology is concerned with plant diseases for which chemicals are the sole or preferred method of control. The prevailing host cultivars for these diseases are susceptible to the pathogen and are not readily replaceable by resistant cultivars. Thus, the diseases are highly likely in most years to cause significant losses, i.e., exceed the economic threshold at which the economic benefits are greater than the costs of spraying. The cereal rusts, however, are highly erratic in occurrence and severity, partly due to the extensive development and use of resistant cultivars and otherwise due to the variables affecting the appearance, overwintering, long distance dissemination, and development of epidemics of virulent races of the pathogens. It is highly desirable to use cultivars in current use to test the feasibility of chemical control of cereal rusts. Presently, stem and leaf rusts are not problems on the prevalent cultivars of spring wheat in production throughout the north central states in the United States where the last important losses were caused by stem rust on durum wheats in 1963 and by leaf rust in 1965 (Roelfs, 1978). Likewise, stem rust has caused negligible damage on the prevalent cultivars of the hard red winter wheats planted in the northern Great Plains since 1965. Thus, older cultivars or special wheat lines susceptible to the prevailing races of stem and leaf rusts are used as hosts in field tests of fungicides. These hosts generally have the disadvantages of low yield capacity, susceptibility to both stem and leaf rusts, or high receptivity to rust that gives an unrepresentative evaluation of the efficacy of chemical control. In the Palouse area of the Pacific Northwest, most of the wheat cultivars in use are susceptible to the prevalent races of leaf rust and range from partially resistant to completely susceptible to the prevalent races of stripe rust, which permits direct evaluation of the efficacy of fungicidal controls in production fields (Line, 1976).

Field testing of fungicides for control of cereal rusts must contend with the capacity of these pathogens to produce copious amounts of airborne inoculum. As much as 10,000 urediospores/cm^2 have been trapped during a 24-hr period just above the canopy of a susceptible wheat cultivar at the peak of epidemics of leaf and stem rusts (Knutson, 1965). This number of spores weighs about 20 μg, a biomass that exceeds fungicide deposits.

Vanderplank (1963) has discussed the importance of interplot interference between adjacent plots with severely and lightly rusted plants in assessing the effectiveness of fungicidal treatments. When plot size is too small, epidemic severity in untreated or ineffectively treated plots is reduced by the significant loss of inoculum from the plot. In effectively treated plots, however, epidemic severity is increased by the influx of inoculum from adjacent severely rusted plots. Similarly, the use of borders of susceptible plants as a means of increasing the uniformity of inoculum across the experimental area will atypically increase epidemic severity in plots with effective treatments.

Roelfs (1972) studied the gradients of horizontally dispersed urediospores of leaf and stem rusts in and around a circular plot of a susceptible wheat cultivar. The average number of spores/cm^2 trapped during the season at the edge of the plot for stem and leaf rust were 63 and 72%, respectively, of the number trapped 9 m inside the plot. His data suggest that the concentration of rust inoculum at the center of a plot of 20 m diameter would be representative of the cereal rust epidemic. At 36 m outside of the rusted wheat plot, the mean number of urediospores/cm^2 trapped during the season for stem and leaf rust was 10 and 16%, respectively, of the number trapped at the plot center. These data indicate that considerable separation of plots is needed to appreciably reduce inoculum interference. Knutson (1965) investigated the dispersal of urediospores of stem and leaf rusts of wheat throughout the experimental area of a field test of fungicides in which circular plots, 5.8 m in diameter, were separated by 30.5-m borders planted to a nonrusted cover crop. In rust-free control plots, the seasonal totals of stem and leaf rust urediospores trapped were 6.5 and 9.3%, respectively, of the untreated controls. This burden of exogenous inoculum was 72 and 85%, respectively, of the stem and leaf rust spores trapped in the center of plots with the most effective fungicidal treatments. Thus, wide separation of treatment plots can appreciably reduce inoculum interference but has the disadvangage of large space requirements, which limits the number of treatments evaluated and may introduce significant variation due to other factors.

I have found that the conventional field plan of replicated, randomized small plots arranged in adjacent rows adequately screens candidate fungicides for potentially useful activity against wheat rusts. This experimental layout, however, will not reliably rank treatments that are approximately similar in effectiveness, due to variation in rust severity induced by inoculum interference. In evaluating the effects of dosage, formulation, and timing of applications, the use of small plots separated by wide rust-free borders reduces inoculum interference and improves the discrimination of small differences in effectiveness between treatments.

3. *Timing Applications*

Fungicidal performance is dependent on the timing of applications to the stage of rust and host development. The original expectation that powerful eradicants could eliminate rust, halt the epidemic, and reverse the crop damage after plants were severely rusted was overly optimistic. Latent infection of wheat stem rust is approaching 100% severity when conditions favor a severe rust epidemic and visible infection exceeds 10% severity. Even though a powerful eradicant can eliminate many of these infections, an appreciable number of infections survive because the fungicide is not thoroughly and uniformly deposited on all infected host tissues. Although applying an effective eradicant fungicide at rust severities of 10% or greater will briefly slow an epidemic, plant productivity has been so badly impaired that little improvement in crop yield and quality is obtained. Most fungicides that have shown promise for use in wheat production have both eradicant and persistent protectant activity. Even the best fungicides must be applied early in the epidemic or they will not delay disease development enough to minimize crop loss.

Generally, epidemics of leaf and stem rusts of wheat peak and cause crop loss during the last month of crop growth in the north and central United States. Thus, the prevalent strategy for chemical control is to prevent severe rusting of the flag leaf, which contributes a major portion of the photosynthate required for kernel development. Fungicidal control is potentially most effective on flag leaves because deposition by either air or ground application is heaviest on this leaf. Furthermore, fungicides applied prior to head emergence leave less residue in and on the harvested grain. However, in the Pacific Northwest, fungicide applications at growth stages after heading often controlled stripe rust better than earlier applications (Line, 1976), even though stripe rust is favored there by the cool temperatures of spring.

The degree of control of rust on the flag leaf is affected to a great extent by the timing of fungicide application in relation to the stage of rust development. The severity of leaf and stem rust epidemics in the spring wheat region of the Great Plains in the United States varies with the time and abundance of primary infection, the pathogenicity of the rust, the reaction of the host, and the prevailing weather. A record of the occurrence of losses caused by rust on small grains in the United States since 1918 (Roelfs, 1978) shows that devastatingly severe epidemics of wheat stem rust occurred about every 20 years. These epidemics were characterized by abundant primary infection early in the growing season by a virulent race on a highly susceptible cultivar that predominated throughout a wide region and by prevailing weather favorable for rust development. In intervening years, the extent of rust development varied widely from negligible amounts to moderate epidemics as one or more of the essential factors were less propitious for rust development. Predictors for the development of rust epidemics and associated crop losses have been developed for stem (Buchenau, 1970; Eversmeyer *et al.*, 1973), leaf (Eversmeyer and Burleigh, 1970), and stripe rust (Coakley and Line, 1981) of wheat under cropping conditions in the United States. These systems will be useful to public and private pest management services in forecasting the need for and timing of chemical control measures.

The optimum timing of applications of a systemic fungicide is governed not only by the stage of the rust epidemic, but also by the properties of the fungicide. Highly effective protectants but poor eradicants such as triazbutil control leaf rust on spring wheat best when applied before rust is present on the wheat (Rowell, 1976). Conversely, highly effective eradicants but poor protectants such as nickel ions control rust best when applied at about 10% rust severity (Rowell, 1964). Timing is less critical for fungicides such as triadimefon, which is effective both as protectant and eradicant (Rowell, 1981). Two sprays of this fungicide gave good control of wheat stem rust when applied in various schedules at 7-day to 20-day intervals during the period from primary infection to early logarithmic increase of rust. A single spray applied when rust was about 1% severity also controlled rust effectively.

The nature of the rust attack also affects the optimum timing of systemic fungicides. Severe leaf rust generally is restricted to leaf blades of wheat, whereas stem rust can severely attack blades, sheaths, peduncles, and heads. Consequently leaf rust is easier to control than stem rust because the deposits of fungicides are greatest on the uppermost leaf blades (Livingston, 1953). Thus, effective control of leaf rust was achieved when the initial application of a mixture of nickel sulfate with maneb was made when disease incidence was under 100 uredia per tiller, whereas control of stem rust was achieved only when the initial application was made before rust exceeded 10 uredia per tiller (Knutson, 1965).

My evaluations of systemic compounds for control of stem and leaf rusts on wheat included a timing series of applications schedules according to the stage of rust development. Three times of application were selected about 5–7 days apart and roughly based on the initial appearance of rust, a rust incidence of about 1–10 uredia per tiller, and a rust incidence of about 10–100 uredia per tiller, respectively. Six treatments were used to evaluate a compound: three single-spray treatments each applied at one of the three times, and three double-spray treatments applied at the three possible pairs of the single-spray times. Inclement weather occasionally disrupted this schedule. This procedure evaluated the protectant and eradicant activities of the test compound and indicated the degree of latitude permissible in effective use.

4. *Disease Observations*

The rapid multiplication of rust infections in an epidemic of cereal rust destroys host productivity through the cumulative effect of the numerous parasitic colonies. Each infection site is a metabolic sink that accumulates the products of photosynthesis and nitrogen metabolism and thereby competes with the growth needs of the host (Durbin, Vol. 1). Growth and differentiation of each succeeding leaf of the cereal plant is supported principally by photosynthate from the preceding leaf. The adventitious root system requires a continuous flow of photosynthate for vigorous development and function (Bushnell and Rowell, 1968). Thus, the effect of rust on crop development is affected by the timing of the epidemic in relation to the stage of host development. If leaf rust becomes epidemic on wheat seedlings, as occasionally occurs in the southern hard red winter wheat region of the United States, the plants may be killed or so severely stunted and weakened that few survive the winter. When rust becomes severe during the mid-growth stages, tillering is reduced, subsequent development is stunted, and seed set is reduced. Severe rust development after heading affects seed filling, which results in small seed from leaf and stripe rusts or shriveled seed from stem rust.

Generally, chemical control only alters the course of the rust epidemic, and rarely are treated plants completely free of rust. Therefore, repeated observations of rust incidence are required to evaluate the efficacy of the fungicides. Counting the number of rust infections on treated and untreated inoculated plants gives a direct measure of fungicide effectiveness in glasshouse tests. This procedure is too time consuming for use in field experiments after rust incidence exceeds 100 uredia per culm (Rowell, Vol. 1). Generally the modified Cobb scale (Peterson *et al.*, 1948) or a similar system is used to estimate rust severity in field tests. Frequent observations at intervals of 5–10 days are required to determine effectiveness of each treatment for suppressing rust development. These data can be used to compare disease progress curves or the area under the disease progress curves (Vanderplank, 1963) of different treatments.

Estimating rust severity is complicated by the many factors that cause tissue necrosis of cereals under field conditions. Accurate determination of rust incidence and severity on dead plant tissue is exceedingly difficult. The wheat leaf has a normal life span of about 40 days under optimal conditions in the north central states of the United States. This lifespan, however, is often shortened by stress due to drought, frost, or foliar pathogens, singly or in combinations. Under these conditions, observations on leaf rust severity are often made only on the flag leaf or the flag and penultimate leaves. Furthermore, severe incidence of leaf rust causes premature leaf death, and severe stem or stripe rust infection causes early plant death, which is hastened by other stress factors. Untreated control plants in field tests with cereal rusts necrose earlier and at faster rates than plants treated with effective fungicides. Thus, in the advanced stages of the rust epidemic, a control treatment in which rust is almost completely controlled by frequent treatments with an effective fungicide such as mancozeb is valuable as a reference for the potential development of rust-free plants under the prevailing conditions. Based on this reference, the percentage of necrosis induced by rust can be estimated in the various treatments.

The present system for estimating rust severity gives an approximate measure of rust damage to crop production (Calpouzos *et al.*, 1976). A precise system for estimating damage from rust on a given cultivar probably requires a model of the productivity of individual leaf blades, leaf sheaths, exposed stem tissue, glumes, palea, and awns at various stages of plant development. Observations on rust severity on each of these tissues at progressive stages of host development would permit calculation of the cumulative effect of the progressive increase of rust on host productivity.

5. *Yield and Quality Measurements*

The ultimate evaluation of the effectiveness of a chemical treatment rests on the quantitative comparisons of the yield and quality of the harvested grain to those of the untreated controls and of the potential crop produced by nearly rust-free plants in the frequently treated control. Thus, the relative effectiveness of chemical control of cereal rusts is determined by how well the treatment prevented rust damage to the crop. In general, the degree of rust control is correlated with a proportionate increase in yield and kernel weight. The chemical treatment itself, however, may be injurious to the crop. For example, phytotoxicity of cycloheximide could reduce crop yields (Wallen, 1958), and adverse effects of calcium sulfamate have reduced germination and milling properties of wheat grain (Acosta C. and Livingston, 1955; Mattern and Livingston, 1955).

One difficulty in evaluating chemical treatments by yield is the relatively large standard error in the amount of grain harvested from small plots. In my experience with the statistical analysis of data from field tests of chemical treatments, the least significant difference in yield between treatments ranges from about 201 to 336 kg/hectare. When frequently treated controls indicate a yield potential of 2700 kg/hectare, yields between treatments must differ by 7–12% to be significant. Consequently, it is difficult to separate good treatments that have perceivable differences in degree of rust control in light epidemics. For example, in developing a model for the effect of stem rust severity on yield, Calpouzos *et al.* (1976) omitted data from plots with 5% or less rust because of high variability in yield loss at low rust severities. Numerous factors such as unevenness in plant stand, head damage cause by birds and rodents, plant lodging caused by disease or weather, and seed shattering during harvest apparently contribute to the variation in the amounts of harvested grain between plots. Rust severity can also vary between replicate plots of moderately effective treatments, presumably due to interplot interference. Theoretically, the variation in yield between plots of the same treatment could be reduced by increasing plot size and/or the number of replications, a procedure that either increases other sources of variation due to enlargement of the experimental area or requires a reduction in the number of possible treatments.

IV. Concluding Statement

The technical evaluation of chemicals for control of cereal rusts has revealed a number of effective and potentially useful compounds, such as calcium sulfamate, sodium sulfanilate, inorganic salts of nickel, oxycarboxin, triarimol, triazbutil, fenapanil, and triadimefon. The number of applications and dosage required for effective control of wheat rusts has decreased from two to three sprays of 3.4 kg/hectare for calcium sulfamate (Livingston, 1953) and formulations of inorganic salts of nickel with zineb or maneb (Rowell, 1964) to single applications of 0.14–0.28 kg/hectare for triazbutil (Rowell, 1976) and triadimefon (Line, 1976; Rowell, 1981).

Unfortunately, none of these compounds are registered at the present time for use on cereals in the United States, due in part to the difficulties in toxicologic and economic evaluations. Concern about the consequences of chemical residues in and on cereal grains has been a deterrent in the registration process. The economics of small grain production in the United States severely limits the market potential for fungicides. In Western Europe the intensive cropping systems, high yield, and high level of grain prices permit economic use of chemical control of rusts at crop losses as low as 2% of yield for winter wheat and barley (Rathnell and Skidmore, 1982). The extensive cropping systems, relatively low yields, and often depressed level of small-grain prices in the United States limit economic use of chemical control of cereal rusts to situations in which crop damage by rust exceeds losses of 10–20%.

In recent years, epidemics of wheat rusts in the central United States rarely attained the economic threshold. Use of resistant cultivars in the spring wheat region has controlled stem and leaf rusts since 1965. In the hard red winter wheat region, the use of early-maturing and resistant cultivars has controlled stem rust and reduced damage by leaf rust. In the Pacific Northwest, however, stripe rust is often epidemic on susceptible cultivars. Furthermore, most cultivars in use in that region lack effective resistance against leaf rust, which occasionally becomes epidemic late in the growing season. Chemical control is feasible in the high-yielding areas of the Pacific Northwest, where yields range from 5 to 7 metric tons/hectare. Epidemics were predicted for both rusts in 1981, and an emergency registration for use of triadimefon on wheat in the Pacific Northwest was obtained (Line, 1982). Production was increased by 27,000 metric tons on the 40,000 hectares of wheat that was sprayed. This is the first extensive commercial use of chemical control of rust on small gains in the United States.

References

Acosta, C. A., and Livingston, J. E. (1955). Effects of calcium sulfamate and sodium sulfanilate on small grains and on stem rust development. *Phytopathology* **45,** 503–506.

Andersen, A. S., and Rowell, J. B. (1962). Duration of protective activity in wheat seedlings of various compounds against stem rust. *Phytopathology* **52,** 909–913.

Anonymous (1943). The slide-germination method of evaluating protectant fungicides. *Phytopathology* **33,** 627.

Bissonnette, H. L., Rowell, J. B., and Calpouzos, L. (1969). Chemical control of foliar diseases of cereals. *Minn. Sci.* **25,** 8–10.

Brooks, D. H. (1972). Results in practice—I. Cereals. *In* "Systemic Fungicides" (R. W. Marsh, ed.), pp. 186–205. Wiley, New York.

Buchenau, G. W. (1970). Forecasting profits from spraying for wheat rusts. *S.D. Farm Home Res.* **21,** 31–34.

Bushnell, W. R., and Rowell, J. B. (1968). Premature death of adult rusted wheat plants in relation to carbon dioxide evolution by root systems. *Phytopathology* **58,** 651–658.

Calpouzos, L., Roelfs, A. P., Madson, M. E., Martin, F. B., Welsh, J. R., and Wilcoxson, R. D. (1976). A new model to measure yield losses caused by stem rust in spring wheat. *Univ. Minn., Agric. Expt. Stn. Tech. Bull.* **307,** 1–23.

Chapman, R. K., and Naoum, A. (1966). The influence of soil moisture on the uptake of systemic insecticides. *Proc. North Cent. Branch Entomol. Soc. Am.* **21,** 122–123.

Coakley, S. M., and Line, R. F. (1981). Qualitative relationships between climatic variables and stripe rust epidemics on winter wheat. *Phytopathology* **71,** 461–467.

Cook, R. J., Jenkins, J. E. E., and King, J. E. (1981). The deployment of fungicides in cereals. *In* "Strategies for the Control of Cereal Disease" (J. F. Jenkyn and R. T. Plumb, eds.), pp. 91–99. Blackwell, Oxford.

Crowdy, S. H., and Jones, D. R. (1956). The translocation of sulphonamides in higher plants. I. Uptake and translocation in broad beans. *J. Exp. Bot.* **7,** 335–346.

Crowdy, S. H., and Jones, D. R. (1958). The translocation of sulphonamides in higher plants. III. Acetylation and deactylation of sulphanilamide in broad beans and wheat. *J. Exp. Bot.* **9,** 220–228.

Crowdy, S. H., Jones, D. R., and Witt, A. V. (1958). The translocation of sulphonamides in higher plants. II. Entry into the leaves of wheat. *J. Exp. Bot.* **9,** 206–219.

Davis, D., Chaiet, L., Rothrock, J. W., Deak, J., Halmos, S., and Garber, J. D. (1960). Chemotherapy of cereal rusts with a new antibiotic. *Phytopathology* **50,** 841–843.

Eversmeyer, M. G., and Burleigh, J. R. (1970). A method of predicting epidemic development of wheat leaf rust. *Phytopathology* **60,** 805–811.

Eversmeyer, M. G., Burleigh, J. R., and Roelfs, A. P. (1973). Equations for predicting wheat stem rust development. *Phytopathology* **63,** 348–351.

Gassner, G., and Hassebrauk, K. (1936). Untersuchungen zur Frage der Getreiderost bekampfung mit chemischen mitteln. *Phytopathol. Z.* **9,** 427–454.

Hansing, E. D. (1978). Techniques for evaluating seed-treatment fungicides. *In* "Methods for Evaluating Plant Fungicides, Nematicides, and Bactericides," pp. 88–92. Am. Phytopatol. Soc., St. Paul, Minnesota.

Harvey, A. E., and Grasham, J. L. (1979). The effects of selected systemic fungicides on the growth of *Cronartium ribicola in vitro. Plant Dis. Rep.* **63,** 354–358.

Horsfall, J. G. (1956). "Principles of Fungicidal Action." Chronica Botanica, Waltham, Massachusetts.

Jenkins, J. E. E., and Lescar, L. (1980). Use of foliar fungicides on cereals in Western Europe. *Plant Dis.* **64,** 987–994.

Johnston, H. W. (1972). Control of powdery mildew of wheat by systemic seed treatments. *Can. Plant Dis. Surv.* **52,** 82–84.

Jones, D. R. (1973). A bioassay of systemic fungicides using axenic cultures of the carnation rust fungus, *Uromyces dianthi. Can. J. Bot.* **51,** 2004–2006.

Keil, H. L., Frohlich, H. P., and Van Hook, J. O. (1958). Chemical control of cereal rusts. I. Protective and eradicative control of rye leaf rust in the greenhouse with various chemical compounds. *Phytopathology* **48,** 652–655.

Knutson, D. M. (1965). Air-borne inoculum and the rate of disease development in epidemics of leaf and stem rusts of wheat. Master's Thesis, University of Minnesota, Minneapolis.

Line, R. F. (1976). Chemical control of *Puccinia striiformis* and *Puccinia recondita* on wheat in Northwestern United States. Proc.—*Eur. Mediterr. Cereal Rusts Conf. 4th,* 1976, pp. 105–108.

Line, R. F. (1982). Chemical control of stripe rust and leaf rust of wheat in the Pacific Northwest. *Phytopathology* **72,** 972 (abstr.).
Line, R. F., and Rakotondradona, R. (1980). Chemical control of *Puccinia striiformis* and *Puccinia recondita. Proc. Eur. Mediterr. Cereal Rusts Conf. 5th, 1980,* pp. 239–241.
Livingston, J. E. (1953). The control of leaf and stem rust of wheat with chemotherapeutants. *Phytopathology* **43,** 496–499.
Loegering, W. Q. (1955). The chemical cereal rust control project of the United States Department of Agriculture. *Plant Dis. Rep., Suppl.* **234,** 1–126.
McCallan, S. E. A. (1956). Mechanisms of toxicity with special reference to fungicides. *Proc. Int. Plant Prot. Conf., 2nd, 1956,* pp. 77–95.
Mattern, P. G., and Livingston, J. E. (1955). The effect of three leaf and stem rust chemotherapeutants on the baking behavior of wheat. *Cereal Chem.* **32,** 208–211.
Nelson, L. A. (1978). Use of statistics in planning, data analysis, and interpretation of fungicide and nematicide tests. *In* "Methods for Evaluating Plant Fungicides, Nematicides, and Bactericides," pp. 2–14. Am. Phytopathol. Soc., St. Paul, Minnesota.
Peterson, R. F., Campbell, A. B., and Hannah, A. E. (1948). A diagrammatic scale for estimating rust intensity of leaves and stems of cereals. *Can. J. Res., Sect. C* **26,** 496–500.
Rathmell, W. G., and Skidmore, A. M. (1982). Recent advances in the chemical control of cereal rust diseases. *Outlook Agric.* **11,** 37–43.
Rich, S., and Horsfall, J. G. (1952). The relation between fungitoxicity, permeation, and lipid solubility. *Phytopathology* **42,** 457–460.
Roelfs, A. P. (1972). Gradients in horizontal dispersal of cereal rust uredospores. *Phytopathology* **62,** 70–76.
Roelfs, A. P. (1978). Estimated losses caused by rust in small grain cereals in the United States. 1918–76. *Misc. Publ.—U.S., Dep. Agric.* **1363,** 1–85.
Rowell, J. B. (1964). Factors affecting field performance of nickel salt plus dithiocarbamate fungicide mixtures for the control of wheat rusts. *Phytopathology* **54,** 999–1008.
Rowell, J. B. (1967). Control of leaf and stem rust of wheat by an 1,4-oxathiin derivative. *Plant Dis. Rep.* **51,** 336–339.
Rowell, J. B. (1968a). Chemical control of the cereal rusts. *Annu. Rev. Phytopathol.* **6,** 243–262.
Rowell, J. B. (1968b). Control of leaf and stem rusts of wheat by combinations of soil application of an 1,4-oxathiin derivative with foliage sprays. *Plant Dis. Rep.* **52,** 856–858.
Rowell, J. B. (1972). Fungicidal management of pathogen populations. *J. Environ. Qual.* **1,** 216–220.
Rowell, J. B. (1973a). Control of leaf and stem rusts of wheat by seed treatment with oxycarboxin. *Plant Dis. Rep.* **57,** 567–571.
Rowell, J. B. (1973b). Effect of dose, time of application, and variety of wheat on control of leaf rust by 4-*n*-butyl-1,2,4-triazole. *Plant Dis. Rep.* **57,** 653–657.
Rowell, J. B. (1976). Control of leaf rust on spring wheat by seed treatment, with 4-*n*-butyl-1,2,4-triazole. *Phytopathology* **66,** 1129–1134.
Rowell, J. B. (1981). Control of stem rust on spring wheat by triadimefon and fenapanil. *Plant Dis.* **65,** 235–236.
Rowell, J. B., and Olien, C. R. (1957). Controlled inoculation of wheat seedlings with urediospores of *Puccinia graminis* var. *tritici. Phytopathology* **47,** 650–655.

Sempio, C. (1936). Influenza di varie sostanze sul parassitamento: Ruggine e Fagiolo, ruggine e mal bianco del Frumento. *Riv. Patol. Veg.* **26,** 201–278.

Vanderplank, J. E. (1963). "Plant Diseases: Epidemics and Control." Academic Press, New York.

von Meyer, W. C., Greenfield, S. A., and Seidel, M. C. (1970). Wheat leaf rust: Control by 4-*n*-butyl-1,2,4-triazole, a systemic fungicide. *Science* **169,** 997–998.

Wallen, V. R. (1958). Control of stem rust of wheat with antibiotics. II. Systemic activity and effectiveness of derivatives of cycloheximide. *Plant Dis. Rep.* **42,** 363–366.

Watkins, J. E., Littlefield, L. J., and Statler, G. D. (1977). Effect of the systemic fungicide 4-*n*-butyl-1,2,4-triazole on the development of *Puccinia recondita* f. sp. *tritici* in wheat. *Phytopathology* **67,** 985–989.

Index

U

V

W